MATHEMATICAL RESEARCH
MATHEMATISCHE FORSCHUNG

Systems Analysis and Simulation 1985 II

edited by
A. Sydow
M. Thoma
R. Vichnevetsky

Band 28

Akademie-Verlag Berlin

In this series original contributions of mathematical research in all fields are contained, such as

— research monographs
— collections of papers to a single topic
— reports on congresses of exceptional interest for mathematical research.

This series is aimed at promoting quick information and communication between mathematicians of the various special branches.

In diese Reihe werden Originalbeiträge zu allen Gebieten der mathematischen Forschung aufgenommen wie

— Forschungsmonographien
— Sammlungen von Arbeiten zu einem speziellen Thema
— Berichte von Tagungen, die für die mathematische Forschung besonders aktuell sind.

Die Reihe soll die schnelle Information und gute Kommunikation zwischen den Mathematikern der verschiedenen Fachgebiete fördern.

Manuscripts
in English and German comprising at least 100 pages and not more than 500 pages can be admitted to this series. With respect to a quick publication the manuscripts are reproduced photomechanically. Authors who are interested in this series please turn directly to the 'Akademie-Verlag'. Here you will get more detailed information about the form of the manuscripts and the modalities of publication.

Manuskripte
in englischer und deutscher Sprache, die mindestens 100 Seiten und nicht mehr als 500 Seiten umfassen, können in diese Reihe aufgenommen werden. Im Interesse einer schnellen Publikation werden die Manuskripte auf fotomechanischem Weg reproduziert. Autoren, die an der Veröffentlichung entsprechender Arbeiten in dieser Reihe interessiert sind, wenden sich bitte direkt an den Akademie-Verlag. Sie erhalten dort genauere Informationen über die Gestaltung der Manuskripte und die Modalitäten der Veröffentlichung.

Systems Analysis and Simulation 1985

II

Mathematical Research **Mathematische Forschung**

Wissenschaftliche Beiträge
herausgegeben von der
Akademie der Wissenschaften der DDR
Institut für Mathematik

Band 28

Systems Analysis and Simulation 1985

II

Systems Analysis and Simulation 1985

II: Applications

Proceedings of the International Symposium held in Berlin (GDR), August 26–31, 1985

edited by

Achim Sydow
Manfred Thoma
Robert Vichnevetsky

Akademie-Verlag Berlin 1985

Herausgeber:
Prof. Dr. Achim Sydow, Zentralinstitut für Kybernetik und Informationsprozesse der Akademie der Wissenschaften der DDR, Berlin
Prof. Dr. Manfred Thoma, Technische Universität Hannover, Institut für Regelungstechnik, Hannover
Prof. Dr. Robert Vichnevetsky, Dept. of Computer Science Rutgers, The State University of N. I., New Brunswick

---

Die Titel dieser Schriftenreihe werden vom Originalmanuskript der Autoren reproduziert.

ISSN 0138 - 3019

Erschienen im Akademie-Verlag Berlin, DDR-1086 Berlin, Leipziger Str.3-4

Lizenznummer: 202 · 100/510/85
Printed in the German Democratic Republic
Gesamtherstellung: VEB Kongreß- und Werbedruck, 9273 Oberlungwitz
LSV 1095
Bestellnummer: 763 475 6 (2182/28)
06200

P R E F A C E

The present volume contains the papers which were accepted for presentation at the 2nd International Symposium on Systems Analysis and Simulation, held in Berlin (GDR), August 26-31, 1985.
The first International Symposium was also held in Berlin (GDR) 1980. It reflected already the great international activities and interests of the broad international scientific community working in the field of systems analysis, modelling and simulation. Since that time we can realize a big progress in methodology, computational tools and application.
Following the first, the second symposium is devoted to the art and techniques and applications of modelling and simulation in systems analysis.
Systems Analysis as an interdisciplinary activity connects mathematicians, physicists, biologists, economists, system scientists, computer specialists, control engineers and many others. The aim of these activities is the creation of a useful mathematical or simulation model for control and decision making.
Especially for large-scale and complex systems and systems with nonlinearities simulation models play an important role as bridge between theoretical modelling and practical use. The knowledge and the use of such system properties as largeness complexity and nonlinearities bear fully new knowledge on systems behaviour and control as well as decision strategies.
Such fields for systems analytical research have been emerged in environmental protection, macroeconomics, agricultural production, regional planning, biosciences, traffic control, large engineering systems, resources distribution, management systems - just to mention a few. These pracitcal demands give impacts to develop system sciences, mathematical methods, simulation systems and other computational tools.
In the very past a lot of activities could be realized in the field of developing powerful tools for computer-aided modelling and decision making. Computer sciences and computer techniques have built the foundation.
At present the practical and social demands and expectations to systems analysis, modelling and simulation join the developed systems sciences and applied mathematics and reapidly developing powerful computer techniques.
This volume - one of two - contains **papers concerning the special fields** of simulation techniques and the applications of systems analysis and simulation to industrial and technological systems, environmental, ecological, medical and biotechnological systems and the related invited papers.

The 2nd International Symposium is organized by the Central Institute of Cybernetics and Information Processes of the Academy of Sciences of the GDR (ZKI)
with cosponsorship of the

- International Association for Mathematics and Computers in Simulation (IMACS)
- International Federation of Automatic Control (IFAC)
- International Institute for Applied Systems Analysis (IIASA)
- International Research Institute of Management Sciences (IRIMS)
- Scientific Society of Measurement and Automation (WGMA) in the Chamber of Technology (KdT) of the GDR
- Mathematical Society (MG) of the GDR.

The papers included in these Proceedings were not formally refereed. The authors are fully responsible.
The international Program Committee consisted of:
W. Ameling (FRG), F. Conrad (USA), A.A. Dorodnicyn (USSR), S.V. Emelyanov (USSR), K.H. Fasol (FRG), W. Findeisen (Poland), O.I. Franksen (Denmark), J. Halin (Switzerland), V. Hamata (Czechoslovakia), A. Javor (Hungary), K. Kabes (Czechoslovakia), V.V. Kalashnikov (USSR), V. Kempe (GDR), E.J.H. Kerckhoffs (The Netherlands), R. Klötzler (GDR), G. Koch (Italy), P. Kopacek (Austria), R. Kulikowski (Poland), Kurzhanski (USSR), N. Levan (USA), T.I. Oren (Canada), M. Peschel (GDR), Phan Dinh Dieu (Vietnam), F. Pichler (Austria), V. S. Pugachev (USSR), K. Reinisch (GDR), F. Stanciulescu (Romania), M. Thoma (FRG), I. Troch (Austria), S.G. Tzafestas (Greece), G.C. Vansteenkiste (Belgium), R. Vichnevetsky (USA), A. Sydow (Chairman).

I am pleased to thank the members of this committee for the excellent cooperation. Special thanks are devoted to Prof. Dr. V. Kempe again, Director of the ZKI for his great support in preparing and performing this symposium.

Furthermore, great gratitude is to express to

Prof. Dr. R. Vichnevetsky (USA), IMACS-President
Prof. Dr. M. Thoma (FRG), IFAC-President
Prof. Dr. V. Kaftanov (USSR), IIASA-Vice-Director
Prof. Dr. S.V. Emelyanov (USSR), IRIMS-Director
Prof. Dr. Richter (GDR), Chairman of WGMA

for their help and proposals.

In preparing the symposium I had many discussions with authorities of the Academy of Sciences. Thanks should be also given to Prof. Dr. M. Peschel, Chairman of the research area for mathematics and informatics.
Most of the preparation work was done by the department for systems analysis and simulation of the Central Institute of Cybernetics and Information Processes. The editor says many thanks to all collegues who were very much engaged in the preparation. I would like to name Dr. U. Grote and Mr. A. Wittmüß who helped me to prepare the proceedings. In this connection should be named also Mrs. U. Hartung and Mrs. I. Essegern for speedy service in preparing the manuscript.

Last not least I would give my thanks to the publishers, especially Mrs. G. Reiher, for their assistance and cooperation.

Finally, I would like to express my hope also on behalf of the co-editors Prof. Dr. M. Thoma and Prof. Dr. R. Vichnevetsky that also the second symposium will be a contribution to the further development in systems analysis, modelling and simulation as well as a place for cooperation and communication.

March 1985

Achim Sydow
On behalf of the editors

C o n t e n t s

ADVANCED SIMULATION : ADVANCED DATA/KNOWLEDGE-PROCESSING

E.J.H. Kerckhoffs[1] and G.C. Vansteenkiste[2]

ABSTRACT

Large-scale and complex simulation studies require the use of advanced information processing (AIP) techniques, such as number crunching facilities and database-management capabilities. Recently there is an increasing interest to incorporate another AIP-technique in simulation : artificial intelligence (in particular its applied branch "knowledge engineering"). Without having the pretention to be complete, in this paper we try to survey AIP in simulation.

Keywords : parallel processing, parallel algorithms, parallel simulation, database-management, artificial intelligence, knowledge engineering.

1. INTRODUCTION

In many references the requirements have been stated for advanced modelling and simulation [1,2]. In ref. [3] the needs in advanced simulation are summarized for one of the world's most difficult simulation problems : medical and biotechnological engineering. Besides more advanced methodologies and new mathematical concepts, advanced information processing (AIP) techniques are felt to be of major importance to the full exploitation of simulation as a practical design and analysis tool. Such techniques will allow research workers to more and more concentrate on the disciplinary aspects of the implemented problem, rather than on computer implementation and programming aspects. AIP encloses among others nonconventional data-processing (such as multiprocessor and pipelined techniques), the automation of the data acquisition (informationbase management systems), as well as the automation of knowledge acquisition and of the reasoning process (knowledge engineering, knowledge based systems). Note in this respect that in the information processing domain recently a fundamental transition started from data processing to knowledge processing, forming the key for the next generation computers [4].

In the sections 2, 3 and 4 respectively nonconventional data-processing, database-management and artificial intelligence, and their relationships with simulation are discussed.

2. NONCONVENTIONAL DATA-PROCESSING IN SIMULATION

2.1. Multiprocessing and pipelining

Signal processing and large-scale simulation, that ask for number crunching facilities, have stimulated to a large extent the development and utilisation of computer systems with architectures substantially different from those of the conventional von Neumann computers. Especially worthwhile in this respect is the introduction of the multiprocessing and pipelining concepts in some new computer architectures [1,5].

In multiprocessing an important approach is the attainment of parallelism through the replication of one or more of the major functional units (ALU, memory unit, control unit) in some computers, in order to increase the computational speed and power. In SIMD (single instruction stream, multiple data stream) computers a number of processing elements, each consisting of an arithmetic/logic unit and a memory unit, are interconnected to form an array of processors under control of a single control unit [1,5,6]. SIMD computing systems were designed in the 1960's and early 1970's to achieve high computing speed, particularly in the treatment of partial differential equations by finite difference methods, without the need to replicate the relatively expensive control unit. The classic example of such SIMD computers was the (now dismantled) ILLIAC IV [7,8].

With the advent of more advanced hardware technology and decreasing cost of digital hardware, replication of the control unit became feasible ; as a result, a variety of computer systems have been constructed containing an array of processing elements, where each element now includes an arithmetic/logic unit as well as a control unit. Unlike the SIMD systems, the different processing elements can carry out different logic and arithmetic operations. For this reason, systems of this type are referred to as MIMD (multiple instruction stream, multiple data stream) computers [6]. Except for the commercially available HEP manufactured by Denelcor Inc. [9], most MIMD computers are developed, built and operational within a university or other research environment [10,11,12].

---

1) Delft University of Technology, Delft, The Netherlands.

2) University of Ghent, Ghent, Belgium

Pipeling techniques are particularly effective in enhancing execution speeds in the case of processing loops. Pipelining is applied in most so-called supercomputers (such as CRAY-I and CYBER 205), developed for very effective vector processing [5,13]. A basic aim in constructing algorithms for vector processing is to perform the bulk of the computational tasks on one-dimensional arrays, i.e. vectors. Notwithstanding the value of vector operations in simulating continuous fields, many important large-scale problems cannot be organized into vector form efficiently (e.g. problems that call for much searching and sorting, or that are dominated by conditional branching). Pipelining techniques are also employed in the so-called peripheral array processors [14]. Although nearly all array processors are signal processing oriented (such as e.g. the AP120B manufactured by Floating Point Systems [1,15]), they are also extensively used for simulation purposes [15]. However, from the simulation point of view array processors have the disadvantage that they are especially designed to perform relatively compact operations on large blocks of data, while in simulation normally during each frame time lengthy and elaborate computations must be executed on small sequences of numbers ; moreover, no facilities are available for direct access to the processor from external communication lines. An array processor, especially developed for real time simulation, is the AD10 manufactured by Applied Dynamics International [1,15].

In this section the advanced simulation computer SIMSTAR, manufactured by EAI, is still mentioned, which is a simulation-oriented multiprocessor that includes both advanced analog and digital techniques [16].

2.2. Parallel algorithms in simulation

The employment of parallel and pipeline computers implies the use of special methods and algorithms in order to exploit the offered possibilities of parallelism as much as possible. A number of good textbooks and articles are available, that deal with parallel algorithms for MIMD-computers, vector processing on supercomputers and algorithms for array-processors [17,18,19,20]. In the context of continuous systems simulation parallel algorithms for ODE's and PDE's are of special interest. In this paper some global notes on the solution of ODE's on MIMD-structured parallel computers are considered ; for the treatment of PDE's the reader is referred to for instance references [21,22]. In this respect it is remarked that semidiscretization methods, where all independent variables - but one - are discretized, convert a given PDE into a system of ODE's, which can be solved on a parallel computer accordingly ; these so-called methods of lines are used in some software packages, such as LEANS III, PDELAN, DSS/2, PDECOL, MOLID, DISPEL [22].

Employing a MIMD-type parallel computer, a straightforward approach to solving the system of ODE's

$$\frac{dY}{dt} = F(Y,t) \; ; \; Y(t_0) = Y_0 \; ; \; Y,F \in R^n$$

is to partition the set of equations and then to allocate a certain part of the n equations to each of the available processing elements ; each processing element is responsible for performing the function evaluations and integrations associated with its assigned equations. The operations performed in the various processing elements can occur in parallel, with the y-values necessary to do succeeding function evaluations being communicated periodically between the processing elements. This method can be referred to as "equations segmentation" (ES) method [23,24]. As an illustrating example of the ES-method, in ref. [25] a problem-flexible implementation of r-th order Runge-Kutta integration algorithms on the DPP81 (see section 2.3) is considered for the special case that each differential equation runs in a separate processing element ; under the same condition, also the DPP81-implementation is outlined of a nonself-starting integration method, such as the 4-th order PC (Adams-Moulton pair) integration algorithm, where fruitful use can be made of the "reconfigurable parallel structuring" of the DPP81.

The ES-method can be augmented with a partitioning procedure (on a large or small task level) of the necessary function evaluations. An example of partitioning on a large task level is to distribute the function evaluations (or parts of them) over the available processing elements in such a way, that the workload is to some extent equally spread (additional information exchanges between the processing elements are required!) ; in this way the frame-time, determined by that processing element evaluating the most time-consuming functions, can be considerably decreased. With respect to the partitioning of arithmetic expressions on a small task level some topics of interest are : reduction by the application of associative/commutative/distributive laws and/or of reordering [17], time-complexities [17] and automatic task partitioning and task allocation [26].
Remark that some integration methods are well-suited to parallel treatment in the sense that to evaluate the right hand members of the differential equations many operations can be done simultaneously, e.g. recursive Taylor series approaches [10].

The approaches considered so far assume the employment of well-explored serially-oriented integration methods ; partitioning is across the system of equations and the function evaluations. Another approach is to make use of methods that are parallel in nature, i.e. partitioning is also across the algorithms concerned. In some parallel methods dependent -- hence inherently serial -- operations (e.g. predictor/corrector calculations, argument/function calculations) are approximated by independent operations which can be performed in parallel. Parallel PC-methods such as the parallel 4-th order Adams-Moulton pair algorithm :

$$Y^p_{n+1} = Y^c_{n-1} + \frac{h}{3}(8F^p_n - 5F^c_{n-1} + 4F^c_{n-2} - F^c_{n-3})$$

$$Y^c_n = Y^c_{n-1} + \frac{h}{24}(9F^p_n + 19F^c_{n-1} - 5F^c_{n-2} + F^c_{n-3})$$

have been reported in the literature [23,27]. In the above algorithm the predictor and corrector are determined in different points and the predictor and corrector equations are solved in parallel. Error and stability properties have been dealt with in respectively the references [27] and [28]. A DPP81-implementation can be found in reference [25] Another example of removing inherent serialism is to approximate the function evaluations on the basis of information obtained in the recent past rather than the current argument values [29]. Consider an integration method :

$$Y_{n+1} = G(Y_n,\ldots,Y_{n-p}, F_n,\ldots,F_{n-q}, h)$$

Instead of F_n we use an approximation $\tilde{F}_n$, e.g. :

$$\tilde{F}_n = F_{n-1} + (F_{n-1} - F_{n-2}) \qquad (1)$$

Consequently :

$$Y_{n+1} = G(Y_n,\ldots,Y_{n-p}, \tilde{F}_n, F_{n-1},\ldots,F_{n-q}, h) \qquad (2)$$

Now the argument Y_{n+1} and corresponding function $\tilde{F}_{n+1}$ can be determined in parallel according to respectively the equations (2) and (1) , resulting in a shorter computing time per integration step. The increase of the local truncation error and decrease of the stability region do not need to be dramatic [29]. A DPP-implementation of the (in this way parallized) 2-nd order Adams-Bashforth algorithm has shown a mean speed advantage of roughly 10%.

There is another class of parallel methods, where a considerable computational overhead has been introduced in order to create parallelisms. These methods make sense if the speed advantage of the parallel treatment exceeds the disadvantage of the included overhead. Examples are the domain decomposition method proposed in ref. [30] and the so-called block-implicit methods [31,32,33]. In the latter the solutions are determined in a block of r points simultaneously on the basis of the solutions in the previous block ; for each block the calculation is in two steps : 1) approximative solutions in r points and 2) implicit procedures to improve the approximations.

2.3. Some parallel simulation research activities in the Benelux countries

a) The DPP-project

At Delft University of Technology (The Netherlands) the MIMD-structured Delft Parallel Processor DPP81 has been built, which is operational now since 1981. The machine's properties can be shortly summarized as follows [34] :

1) Number of processing elements (PE's) : 8
2) At any time each PE can perform an arithmetic operation from an extensive set of operations B = {+,-,*,/,...}. Different operations require different lengths of time.
3) At any time different PE's can perform different operations (from the set B).
4) Full interconnectability between PE's. Each PE can communicate with the host computer.
5) In the so-called synchronous mode of operation, a "processing cycle" is characterized by a common start of all PE's, and is ended as soon as all (!) PE's have finished their program execution. Information exchange between the PE's is always effected at the very end of each processing cycle, but can also be effected a number of times during a processing cycle. In continuous systems simulation one or more processing cycles are used to proceed one step in the integration process.
6) Each PE has a single-valued output and can sample at its input 16 variables.
7) "Reconfigurable parallel structuring" allows loading of four subprograms in each PE and easy switching from one subprogram to another one during run-time.

The DPP81 has been extensively used to among ohters simulation applications (mainly continuous systems simulation but also discrete event simulation [25,35,36,37,38]). Meanwhile research efforts are going on with respect to next DPP-versions [39,40] .
For example, a DPP84/1/n will consist of one PM (processing module) with a number of n (maximally 16) fully interconnected PE's ; the PE's will be equipped with IPP's (inner product processors), which are dedicated processors to perform mappings such as (extended) inner products, transversal and recursive filtering, polynomial mappings, convolutions, auto- and crosscorrelations, FFT, [40] . Much attention is also paid to parallel programming tools ; the research focusses around two main subjects :

1) the construction of a compiler and scheduler for a high level language [26] , where the compiler is meant to transform a program into a task graph, that indicates the dependencies among the tasks and hence the potential parallelisms, and
2) the development of a hierarchical programming system [39] . We remark that, unlike the DPP81, the next versions of the DPP will be of the distributed MIMD (DMIMD) type, where the data-processing power is distributed over a number of levels [39,40] .

In a later stage also artificial intelligence techniques are planned to be incorporated in the Delft Parallel Processor project.

b) The AD10 simulation-oriented peripheral array processor

The AD10 is an example of a multiprocessor system, designed for the simulation of continuous systems. Features as parallel computing and pipelining are incorporated in it to reach the required high computational speed. Each element is dedicated to perform a part of the total simulation task. This results in an architecture with five dissimilar processing units :

1) a fast adder-multiplier for irregularly structured computations (right hand sides of differential equations).
2) an integration processor implements different integration schemes.
3) a special address generator.
4) a decision processor to contribute, in connection with the address-generator, to fast function generation.
5) a control processor to control program execution and I/O channels.

The AD10 further contains a large MOS data memory, with an interleaved organization, to match the speed of the ECL processors and an analog-digital input-output system for real-time simulation. All processors operate on a synchronous base and communicate with each other through a synchronous bus. Programs can run concurrently in the different processors, but must take into account timing for information exchange. The AD10 is normally programmed by means of a high level language, called MPS10. It combines a rather efficient use of the AD10 capabilities with a small programming effort. This is made possible by the limited scope of the application domain. The AD10 can be used in any area where differential equations need to be solved, and is e.g. extensively used for flight simulations. At the University of Ghent the AD10 is generating solutions from large classes of highly nonlinear mathematical model structures describing phenomena encountered in biotechnological processes [41] . For time critical simulation of ill-defined systems, concepts of parallelism in the computation allows to expose the user to the inherent uncertainty included in the models and hence improves simulation realism. Recently the AD10 is performing the arithmetic operations in 64-bit floating-point [42] .

3. SIMULATION ENVIRONMENTS

Advanced simulation environments are of utmost importance in large-scale modelling and simulation. We mention in this respect as illustrating examples : computer generated modelling documentation (e.g. documentation about the method of modelling, the development of equations, calculations of constants, references and assumptions [43]), database management capacities [44,45,46,47,48] and automated generation of input subprograms [49] . In the following we discuss shortly the database management needs in simulation and the development of a first prototype simulation-oriented database management system at the University of Ghent (Belgium).

3.1. Data-management needs in simulation

One of the main activities, peripheral to modelling and model implementation, is the management of data involved in simulation studies. Although a number of database management systems are widely available, they rarely provide many of the specific capabilities needed in simulation. An exception is for instance the database management system called Simulation Data Language [48] . Research into future simulation methodologies has also indicated the importance of utilizing informationbase systems [50] . In ref. [2] a next-generation simulation system is proposed to consist of several functional parts, one of which is an intelligent informationbase system.

In the following the main procedures are shortly considered for building, analyzing and utilizing a simulation model with the aid of an informationbase system [48]. First of all the analyst must study the real world system to specify the relevant à priori knowledge and to gather à posteriori materials. The resulting descriptive and quantitative data can be loaded in the informationbase. Since the relative validity of à priori facts and the precision of the measurement data and thus their quality are crucial in the success of further conclusions and new interpretations, the informationbase should be built to incorporate a validity dimension. A next step in the simulation study is to construct a model, to estimate the model parameters and to validate the model. Model description data and the model parameter values as well as descriptive and quantitative information about model validation can be loaded in the informationbase. In this respect attention is to be paid to the impact of different model formalisms. Considerable efforts have been done in systems theory and model building theory to develop concepts and procedures, that are to a certain extent broadly applicable. The aim is thus to exploit this generality. Finally, the descriptive data are stored with respect to the various simulation experiments performed, as well as the resulting model outputs and/or processed versions of them. When a simulation study involves the assessment of many alternatives (or candidate models), the analyst must repeat the above steps for each alternative. The storage of information allows the analyst to compare easily alternatives with computer assistance. On the basis of this comparison and/or the validation results he may decide to delete the data of irrelevant alternatives.

The informationbase management system could provide a lot of additional capabilities. The informationbase system allows the analyst to separate the analysis of results from the performance of simulation experiments ; additional analyses can be performed without rerunning the model [48]. In an advanced stage of the simulation study the activity "modelling" may simply imply the search of a model with certain specified characteristics, provided this model is already available in the informationbase ; if not, a model with features as close as possible to those wanted can be fetched. In this respect model retrieval differs from normal search of data in a database. Facilities could be developed and incorporated to facilitate the construction of new models from previously stored models. In the design of informationbase systems for simulation the above modelling and simulation principles and kind of queries should be taken into account.

3.2. The PROCCIS-project : towards a data-management system for productivity control and simulation

"Productivity Control Information System - PROCCIS" is based on the above developments [51]. The software allows the data management and control of arbitrary multiple input/multiple output processes and their simulation models. PROCCIS is unique in the sense that it allows productivity improvement of complex bio-industrial processes. It deals with problems encountered when living systems are part of the process, like irreducibility, many measuring methods and their results, many models and their results, etc..
The basic concepts of PROCCIS are :

- PROCCIS integrates both real world system information and computer model information.
- System information and model information are both methodologically separated from information about experiments on the system itself and on its models. Consequently, a systembase, a modelbase, a measurementbase and a simulationbase are considered [46,47].
- The *systembase* contains à priori knowledge about the system, such as the number of inputs/outputs and their names, names and values of the system parameters, textual clarifications of the research goals as well as questions.
- The *modelbase* stores information about the mathematical models, that are used to describe the system involved : input-, state- and output variables ; state- and output operators ; model parameters ; model validation.
- The *measurementbase* is meant to keep information about both the measuring conditions or circumstances and the measurement results. Appropriate methodological concepts are utilized to describe these measuring conditions/circumstances [46,47].
- The *simulationbase* contains both the simulation results (i.e. the results from experimenting on a computer model) and information about the circumstances under which these results have been obtained.
- A careful definition of PROCCIS is done on a conceptual level.
- The (graphically represented) conceptual scheme of PROCCIS (provided with an extensive list of comments) should be the proper basis for user's understanding and evaluation, as well as modification and expansion according to own wishes.
- The information analysis is supported by the information analysis method NIAM to enhance the quality of the final product [46,47,52,53,54]. The NIAM basic concepts are used to describe the objects, activities and constraints in general process control and simulation studies, of which the contents of the informationbases should be an excellent reflection.

- Application programs allow the user to query the informationbases and to enter the information from his specific study. Ample use is made of menu selection and graphical facilities.

As a first step in the realisation of PROCCIS a prototype simulation informationbase management system is currently being developed at the University of Ghent.
The conceptual grammar of the planned prototype simulation information system has been implemented on the QINT DMBS-system, running on a IBM PC/XT, with the aid of the support tools provided by the Information Analysis Services of QINT Database Systems Corporation. Currently, application programs are implemented.

4. SIMULATION AND ARTIFICIAL INTELLIGENCE (AI)

AI can be defined as the subfield of computer science concerned with the use of computers in tasks, which are normally considered to require knowledge, perception, reasoning, learning, understanding and similar cognitive abilities [55]. The matter to consider here is : how can AI and especially its applied branch "Knowledge Engineering" help in simulation ? In the following we shall successively deal with descriptive-oriented applications (representing and using knowledge in a simulation model) and prescriptive-oriented research (AI-supported model synthesis). Finally some additional remarks are made on the relationships between AI and simulation.

4.1. Knowledge-based simulation

The existing modelling theories and methodologies can be classified as extended programming languages (such as Simscript II, Simula -- where the usual programming constructs are augmented with simulation-oriented language constructs (queues, events, entities)-- and GASP, DESPL/1 -- where preprocessors or subroutine packages are added to implement simulation features) as well as special purpose packages, that provide model building facilities without extensive programming knowledge (BORIS, INS). The latter have their own limitations in that they are based upon a particular approach (e.g. Petrinets, queueing networks).

To overcome some drawbacks in the "classical" modelling and simulation (lack of flexibility in expressing model structures, extensive programming effort, batch-oriented modelling) an alternative approach could be using an AI knowledge-representation system to express knowledge in a simulation model (knowledge-based simulation). Research topics of interest in this respect are among others qualitative modelling [56,57,58], using different knowledge representation schemes (Logical Representation Schemes, Procedural Representation Schemes, Network-oriented Representation Schemes) dependent upon the kind of simulation [59], and simulation in LISP and PROLOG [60,61]. A sophisticated way of representing knowledge in a simulation model is offered by the so-called object-oriented modelling languages [62,63,64]. Here the knowledge-bases contain :
- various facts about each entity in the system concerned
- knowledge about the entity's relationship to other entities
- knowledge about the relationship between entities and consisting specifications.

Moreover, knowledge is processed about the effect of actions in the system. Some goal features are : interactivity in model creation and alteration (knowledge representations are both flexible and extendible), little programming effort in the creation of models, automatic examination of consistency and completeness, a.s.o..

4.2. AI-aided model synthesis

One aspect of AI support in the synthesis of simulation models is consulting knowledge-based expert systems, that contain a priori (disciplinary-dependent) knowledge about the system to be modelled. Knowledge-based expert systems are information processing systems, utilizing a combination of symbolic reasoning and data-processing. They require, for expert performance on problems, heuristic knowledge to be combined with the facts of the discipline. Knowledge engineering transfers such human expert knowledge to computer programs through "expertise modelling". The most important problem in this respect is centered around "knowledge acquisition research".

In the context of simulation such expert systems can serve several objectives [59]. First an expert system organizes and conserves knowledge. It allows to access knowledge that previously was dispersed over diverse experts who are not always available for discussion and interaction. Moreover, human experts also tend to forget things they don't use frequently. Secondly, by experimenting with the system the user will be capable of checking his results with the system's results and validate his own reasoning processes. This can provide him with new insights and methods with respect to measurements to perform and modelling. Thirdly, it may allow to avoid the enormous complexity in solving the problem with traditional methods of mathematical modelling by qualitative reasoning about the process. This will build a causal chain of events that can eventually function as a framework for constructing a complete model by composing component models.

Several sophisticated tools exist to develop an expert system (e.g. empty expert systems [65], the open-ended knowledge representation system of the University of Brussels [66]). The latter is used to create an expert system containing à priori knowledge in the field of biotechnological growth processes [59]. Although most existing expert systems are applied to contain a priori knowledge of the domain concerned, to be employed to their full potential in the context of modelling and simulation they should be augmented with modelbases, experimental framebases and databases with experimental results.

The central question in employing knowledge-based expert systems to create simulation models of course is whether the synthesis process (or part of it) can adequately be formalized and captured in rules or that it shall be a pure human skill. Anyway, advanced methodological modelling concepts are a necessary basis and as such for instance the well-known methodologies developed by Zeigler [50,67,68] provide excellent prospectives of applying AI.

As an example of methodology-based AI supported modelling, we shall assume that the modelbase consists of candidate models and that the problem is to find an appropriate model in a given experimental frame [59,69,70]. In the case of non-engineering systems, the model building activity relies almost fully on inductive methods (results from experiments). No general methods exist and the model building activity relies heavily on trial and error. Model construction could however be made easier by an intensive dialogue between the modeller and an expert system. In the first stage a tentative model formalism is chosen. The experimental frame further restricts the set of possible candidate descriptions. The next step will be the choice of a structure. This is done in a decision network, in which each decision step restricts the set of remaining candidate models. The procedure stops when a parametrized model is obtained. Two kinds of knowledge are required : mathematical properties of the model set that results from the decision made (realization theory) and how to choose the correct decision steps, and how they are resolved [1]. Realization theory can be formalized into a number of rules and principles. The decision process depends much on the field studied and the problem considered. It is based on a lot of experience. A much harder task is the formalization of this knowledge from the modelling expert in a set of rules. Heuristics will also be necessary to guide the development of the decision network in case more than one rule is applicable, or in case a dead-point is reached where no further conclusion can be made. In this case backtracking will be required.

One should be aware that the advanced utilization of expert systems in the synthesis of simulation models is in the proposal rather than the implementation phase. A very simple example of a realized implementation can be found in ref. [71], where an expert system is employed to establish one of 12 different mathematical oscillator models, present as hypothyses in the expert system.

4.3. Linking knowledge-based systems to conventional simulation models

Computerized simulation of complicated systems represents a fertile application area for AI. Especially are mentioned here simulations with multiple decision making agents, each performing tasks for which AI is well-suited (such as planning, scheduling, hypothesis formulation, performed in an environment full of uncertainty and incomplete and distorted information). For such systems those subprocesses, which are clearly heuristic and symbolic (e.g. decision making) could be modeled by expert systems, while the subprocesses which are deterministic and continuous (e.g. physical processes) are conventionally modelled as dynamic systems. The fusion of AI and simulation here encloses the proper integration of AI planning and expert system techniques into a more classical simulation methodology. As such its goal is to provide a felicitous environment in which a user may create, analyse and evaluate potential tactics. Requirements for knowledge-based expert systems in such simulations are the abilities to operate in a rapidly time-varying environment and to do so speedily [72], which stimulates the research in parallel knowledge-based systems [73].

4.4. Simulation embedded in symbolic reasoning

Many of the systems, to which humans are confronted with, are of large or very large complexity (crop-pest systems, medical diagnosis and therapy, management and control of complex industrial processes). Such systems cannot be adequately captured by only the "system analysis approach", using an engineering type of models, because of its basic procedural way of thinking. Experts way of working is dominated by symbolic processing (or reasoning) and not by mathematical simulation. To overcome the weakness of the system analysis approach, an extension of the concept of knowledge is needed to take into account also heuristic knowledge, and consequently the system organization should draw on concepts as knowledge bases and inference engines. Data processing and modelling are still required, but now the use of data and models is embedded in a symbolic reasoning process, which dominates and controls the entire system. AI is felt to gain the position of central discipline in the treatment of these very complex systems [74].

4.5. Simulation as an aid to test knowledge-based systems

One of the most important applications of knowledge-based systems is perhaps their use in a process and productivity control environment for purpose of control and analysis. When the control system detects a failure in the process it shall try to determine the cause of the failure. Moreover, an attempt could be made -- based on the knowledge available -- to supply the process with an alternative route or, if no such alternative can be found or even exists, the process should be stopped. Knowledge-based systems can be important for this type of applications.
For the purpose of testing such intelligent control systems, they could be linked to the real process. However, this might be impossible or undesired in many cases. Therefore testing these systems should be done by linking them to a simulation model of the process, in which failures can be artificially created. It provides us with tools to test the system thoroughly and to make an estimate of its capabilities and performance to expect.

Another example of applying simulation to test knowledge-based systems is in the design of particular architectures for knowledge-based systems. For instance, parallel knowledge based system concepts are being studied, where a number of rule processors is simultaneously operating on a common rule base and (multi-accessible) database. Some tentative ideas have been implemented in (concurrent) Modula-2 to run on a PDP 11/45 computer and, in addition, are being implemented to run on the Delft Parallel Processor DPP 81 [73]. The aim of these simulations are to investigate the efficiency of the system in relation to some parameters (such as the number of rule processors, the number of ports of the database memory, the strategy in the Rule Agenda Manager, the strategy in the Question Agenda Manager, the length of the question queue, a.s.o.).

REFERENCES

[1] Spriet, J.A. and G.C. Vansteenkiste : Computer-Aided Modelling and Simulation. International Lecture Notes in Computer Science, Academic Press, London, 1982.

[2] Dekker, L., E.J.H. Kerckhoffs, G.C. Vansteenkiste, J.C. Zuidervaart : Outline of a future parallel simulator. In : L. Dekker, G. Savastano, G.C. Vansteenkiste (eds.) Proceedings of the IMACS World Congress on Simulation of Systems, Sorrento, Italy, 1979. North Holland Publ. Co., pp. 837-864.

[3] Deland, E.C. : Conceptual models in physiology, where are we ? In : G.C. Vansteenkiste and P.C. Young (Eds.) : Proceedings of the IFIP WG 7.1 Working Conference on Modelling and Data Analysis in Biotechnology and Medical Engineering, Brussels, 1982, North Holland Publ. Co., Amsterdam, 1983, pp. XI-XVII.

[4] Kowalski, R. : Software Engineering and Knowledge-Based Systems in New Generation Computing. Future Generations Computer Systems, vol.1, nr.1, july 1984, North-Holland Publ. Co.

[5] Hockney, R.W. and C.R. Jesshope : Parallel Computers. Adam Hilger Ltd., Bristol 1981.

[6] Flynn, M.J. : Some computer organizations and their effectiveness. IEEE Trans. Comp., vol. C-21, nr.9, Sept. 1972, pp. 948-960.

[7] Barnes, J.H. et al. : The Illiac IV Computer. IEEE Trans. Comp., vol. C-17, nr.8, Aug. 1968, pp. 746-757.

[8] Kuck, D.J. : Illiac IV software and application programming. IEEE Trans. Comp., vol. C-17, nr.8, Aug. 1968, pp. 758-770.

[9] Brinton, B. : System grows from 10 to 160 MIPS. Electronics (New Products), Febr. 1982.

[10] Halin, H.J., R. Bührer, W. Hälg, H. Benz, B. Bron, H.-J. Brundiers, A. Isacson and M. Tadian : The ETH multiprocessor project : parallel simulation of continuous systems. Simulation, vol.35, nr.4, Oct. 1980, pp. 109-123.

[11] Halin, H.J. : Das ETH-Multiprocessorsystem EMPRESS. In : F. Breitenecker and W. Kleinert (eds.) : Proceedings of 2nd ASIM Symposium on Simulation Techniques. Springer-Verlag 1984, pp. 309-312.

[12] Dekker, L. : Parallel processors in engineering sciences. In : Proceedings of IMACS Conf. on Simulation in Engineering Sciences, Nantes 1983, North-Holland Publ. Co.

[13] Johnson, P.M. : An introduction to vector processing. Computer Design, vol.17, Febr. 1978, pp. 89-97.

[14] Louie, T. : Array processor : a selected bibliography. Computer, sept. 1981, pp. 53-57.

[15] Karplus, W.J. : Peripheral processors for high-speed simulation. Simulation, vol.29, nr.5, Nov. 1977, pp. 143-153.

[16] Embley, R.W. : The technology behind SIMSTAR^{TM}, an All-new simulation multiprocessor. In : F. Breitenecker and W. Kleinert (Eds.) : Proceedings of the 2nd ASIM-Symposium on Simulation Techniques. Springer-Verlag 1984, pp. 317-327.

[17] Hossfeld, F. : Parallele Algorithmen. Informatik-Fachberichte, Springer-Verlag, Berlin 1983 (in German).

[18] Schendel, U. : Einführung in die parallele Numerik. Publ. Oldenbourg, München 1981 (in German).

[19] Jordan, J. : A guide to parallel computation and some CRAY-1 experiences. In : G. Rodrigue (ed.) : Parallel Computations. Academic Press, New York 1983.

[20] Cohen, D. : A methodology for programming a pipeline array processor. Proceedings of MICRO-11, 1978. IEEE Catalog No.78 CH1411-8, pp. 82-89.

[21] Miranker, W.L. : A survey of parallelism in numerical analysis (section 4 : Differential equations). SIAM Review, vol.13, nr.4, Oct. 1971, pp. 524-547.

[22] Vemuri, V. and W.J. Karplus : Digital computer treatment of partial differential equations. Prentice-Hall, Inc., Englewood Cliffs, 1981.

[23] Franklin, M.A. : Parallel solution of ordinary differential equations. Transactions on Computers, vol. C-27, nr.5, May 1978, pp. 413-420.

[24] Yura, E., R. Yoshikawa, Y. Nara, T. Kimura, H. Aiso : An approach to parallel processing for continuous dynamic system simulation with microprocessors. Proceedings of the 2nd USA-Japan Computer Conference 1975, pp. 8/1/1-8/1/6.

[25] Kerckhoffs, E.J.H. and S.W. Brok : The Delft Parallel Processor DPP81 : properties and utilization in simulation and related fields. Accepted for publication in : Systems Analysis, Modelling and Simulation (Journal of Mathematical Modelling and Simulation in Systems Analysis). Akademie-Verlag, Berlin (GDR), 1985.

[26] Sips, H. : Design aspects of a distributed MIMD processor. Ph-D Thesis. Delft University of Technology (Dept. of Applied Physics). Publ. Dutch Efficiency Bureau, Pijnacker (The Netherlands), 1984.

[27] Miranker, W.L. and W.M. Liniger : Parallel methods for the numerical integration of ordinary differential equations. Mathematics of Computation, vol. 21, 1967, pp. 303-320.

[28] Katz, I.N., M.A. Franklin and A. Sen : Optimally stable parallel predictors for Adams-Moulton correctors. Comput. and Math. with Appl., vol. 3, 1977, pp. 217-233.

[29] Yen, K. and G. Cook : Digital simulation algorithms using parallel processing. IEEE Transactions on Industrial Electronics, vol. IE-29, nr.3, Aug. 1982, pp.217-219.

[30] Nievergelt, J. : Parallel methods for integrating ordinary differential equations. Communications of the ACM, vol.7, nr.12, Dec. 1964, pp. 731-733.

[31] Worland, P.B. : Parallel methods for the numerical solution of ordinary differential equations. IEEE Transactions on Computers, oct. 1976, pp. 1045-1048.

[32] Barkley Rosser, J. : A Runge-Kutta for all seasons. SIAM Review, vol.9, nr.3, July 1967, pp. 417-452.

[33] Shampine, L.F. and H.A. Watts : Block-implicit one-step methods. Mathematics and Computation, vol. 23, 1969, pp. 731-740.

[34] User's Guide for the Delft Parallel Processor. Internal Report. Delft University of Technology (Dept. of Applied Physics), Delft 1983.

[35] Delft Parallel Processor : General Information and Applications. Internal Report. Delft University of Technology (Dept. of Applied Physics), Delft 1983.

[36] Kerckhoffs, E.J.H. : The application of an experimental parallel processor for the simulation of systems in biotechnology and medical engineering. In : Vansteenkiste, G.C., Young P.C. (eds.) : Modelling and Data-Analysis in Biotechnology and Medical Engineering. North-Holland Publ. Co., Amsterdam 1983, pp. 221-231.

[37] Kerckhoffs, E.J.H., J. Potůček and M. Šnorek : Simulation of myocardial fibre action potential generation units on the Delft Parallel Processor. In : Ameling, W. (ed.): Proc. of First European Simulation Congress ESC 83. Springer-Verlag, Berlin 1983, pp. 538-548.

[38] Brok, S.W. and E.J.H. Kerckhoffs : The Delft Parallel Processor in a simulation environment. In : Murray-Smith, D.J. (ed.) : Proc. of the 1984 UKSC Conference on Computer Simulation. Butterworths, London 1984, pp. 133-150;

[39] Brok, S.W., L. Dekker, E.J.H. Kerckhoffs, A.B. Ruighaver, H.J. Sips : Architecture and programmature of the MIMD-structured Delft Parallel Processor. In : W. Ameling (ed.) : Proceedings of the First European Simulation Congress ESC 83. Springer-Verlag, Berlin 1983, pp. 125-139.

[40] Andriessen, J.H.M., S.W. Brok, L. Dekker, A.B. Ruighaver, H.J. Sips : The Delft Parallel Processor / A distributed MIMD processor. In : W.D. Wade (Ed.) : Proc. of the 1984 Summer Computer Simulation Conference. Publ. The Society for Computer Simulation, La Jolla (U.S.A.), 1984, pp. 1241-1246.

[41] Spriet, J.A. and G.C. Vansteenkiste : Structure characterization for ill-defined systems. In : T.I. Oren, B.P. Zeigler and M.S. Elzas (eds.) : Simulation and Model-Based Methodologies - An Integrative view. Ottawa, Canada, 1982, pp. 383-408.

[42] SYSTEM-10 : Applied Dynamics International, Stone School Road, Ann Arbor, Michigan, U.S.A.

[43] Laforest, R. and C. Capparelli : Computer Generated Modelling Documentation. In : W.D. Wade (ed.) : Proceedings of the 1984 Summer Computer Conference, Boston 1984, Publ. : The Society for Computer Simulation, pp. 83-87.

[44] Standridge, C.R., A.A.B. Pritsker and J.J. O'Reilly : TESS : A Simulation System. In : W.D. Wade (ed.) : Proceedings of the 1984 Summer Computer Conference, Boston 1984, Publ. : The Society for Computer Simulation, pp. 68-72.

[45] Smart, P.J. and N.J.C. Baker : SYSMOD - An Environment for Modular Simulation. In : W.D. Wade (ed.) : Proceedings of the 1984 Summer Computer Conference, Boston 1984, Publ. : The Society for Computer Simulation, pp. 77-82.

[46] Calu, J., E.J.H. Kerckhoffs and G.C. Vansteenkiste : On the way to an information-base to manage the measurement, modelling and simulation data obtained from fermentation studies. In : G.C. Vansteenkiste and P.C. Young (eds.) : Proceedings of the IFIP WG 7.1 Working Conference on Modelling and Data Analysis in Biotechnology and Medical Engineering. North-Holland Publ. Co., Amsterdam 1983, pp. 59-70.

[47] Vansteenkiste, G.C. and E.J.H. Kerckhoffs : Informationbase support in simulation of biological systems. In : D.J. Murray-Smith (ed.) : Proceedings of the 1984 UKSC Conference on Computer Simulation, Bath 1984. Publ. : Butterworths, pp. 198-218.

[48] Standridge, C.R. : Using the Simulation Data Language (SDL). Simulation, vol.37, nr.3, 1981, pp. 73-81.

[49] Schechter, B.S. : GENINP - Generator of Input Subprograms. In : W.D. Wade (ed.) : Proceedings of the 1984 Summer Computer Conference, Boston 1984, Publ. : The Society for Computer Simulation, pp. 73-76.

[50] Oren, T.I. and B.P. Zeigler : Concepts for advanced simulation methodologies. Simulation, vol. 32, nr.3, 1979, pp. 69-82.

[51] Kerckhoffs, E.J.H., J.A. Spriet and G.C. Vansteenkiste : PROCCIS : productivity control information system. Internal Report of the University of Ghent (Belgium) 1984. Submitted to QINT Database Systems Corporation.

[52] CDC-publication : Information Analysis support tools reference manual. CDC-publication 60484610, 1982.

[53] CDC-publication : Information management facility / version 2 reference manual. CDC-publication 60484600, 1982.

[54] Falkenberg, E. et al. : Feature analysis of ACM/PCM, CIAM, ISAC and NIAM. In : T.W. Olle, H.G. Sol, C.J. Tully (eds.) : Information Systems Design Methodologies : A Feature Analysis. North Holland Publ. Co., 1983.

[55] Duda, R.O., N.J. Nilsson and B. Raphael : State of technology in artificial intelligence. In : P. Wegner (ed.) : Research Directions in Software Technology. MIT Press Cambridge, Mass., 1980, pp. 729-749.

[56] Forbus, K.D. : Qualitative process theory. Artificial Intelligence Laboratory, AIM-664, Cambridge (U.S.A.), M.I.T., 1982.

[57] Forbus, K.D. : Qualitative reasoning about physical processes. Proceedings of the 7th International Joint Conference on Artificial Intelligence, 1981, pp. 326-330.

[58] Rajagopalan, R. : Qualitative modelling in the turbojet engine domain. M.S. Thesis, CSL Techn. Rept. T-139. Dept. of Electrical Engineering, University of Illinois, Urbana, IL, March 1984.

[59] Calu, J., L. De Wael, E. Esmeijer, E.J.H. Kerckhoffs and G.C. Vansteenkiste : Knowledge-base aspects in advance modelling and simulation. In : W.D. Wade (ed.) : Proceedings of the 1984 Summer Computer Conference, Boston 1984, Publ. : The Society for Computer Simulation, pp. 1247-1253.

[60] Cleary, J., K.-S. Goh and B. Unger : Discrete event simulation in PROLOG. Research Report No. 84/171/29, The University of Galgary (Dept. of Computer Science). Appeared in : Proceedings of Conference on AI, Graphics and Simulation, San Diego, January 1985.

[61] Futo, I. and T. Gergely : TS-PROLOG / A language of logic simulation. To be published in : Simulation.

[62] Klahr, P., D. McArthur and S. Narain : SWIRL : An object-oriented air battle simulator. Proc. of the Second Annual National Conference on Artificial Intelligence, Pittsburgh,1982, pp. 331-334.

[63] Klahr, P. : Artificial intelligence approaches to simulation. In : D.J. Murray-Smith (ed.) : Proceedings of the 1984 UKSC Conference on Computer Simulation. Publ. : Butterworths, London 1984, pp. 87-92.

[64] Reddy, Y.V., M.S. Fox, K. Doyle and J. Arnold : INET : A knowledge based simulation model of a corporate distribution system. Proc. of IEEE Conference on Trends and Applications 83, National Bureau of Standards, Gaithersburg, Maryland, May 1983, pp. 109-118.

[65] De Swaan Arons, H., A. Groen, A.G. Hofland and J.C. Stoop : Building and Consulting of expert systems with DELFI. Delft Progress Report, Vol. 9, nr.1, March 1984, pp. 63-85.

[66] Steels, L. : An open-ended framework for structuring knowledge. Internal report of the Free University of Brussels (Belgium), 1983.

[67] Zeigler, B.P. : Multifaceted modelling methodology : Grappling with the irreducible complexity of systems. In : Behavioral Science, Vol. 29, 1984, pp. 169-178.

[68] Zeigler, B.P. : System-theoretic representation of simulation models. IIE Transactions, Vol. 16, nr.1, March 1984, pp. 19-34.

[69] Vansteenkiste, G.C. : Simulation methodology for improved process interaction. In : W. Ameling (ed.) : Proc. of the First European Simulation Congress ESC 83. Springer-Verlag, Berlin 1983, pp. 32-43.

[70] Vansteenkiste, G.C., L. De Wael, E.J.H. Kerckhoffs : Artificial intelligence in large scale system modelling. In : Intelligence Artificielle et Productique (session : Simulation), Journal of the Association Française d'Intelligence Artificielle et des Systèmes de Simulation (AFIAS), Nov. 1983, pp. 30-32.

[71] De Swaan Arons, H. : Expert systems in the simulation domain. In : Mathematics and Computers in Simulation (Transact. of IMACS), Vol. 25, nr. 1, Febr. 1983, pp. 10-16.

[72] Erickson, S.A., D.J. Balaban, D.O. Nelson and K.D. Olwell : The fusion of AI and Simulation. Report Lawrence Livermore National Laboratory. Livermore, California, U.S.A.

[73] Groen, A., A.G. Hofland, J.C. Stoop, H.J. Van Den Herik, N. Kasabov and E.J.H. E.J.H. Kerckhoffs : Towards parallel knowledge-based systems : A first orientation. To be published in : Delft Progress Report, Delft University Press, The Netherlands.

[74] Argentesi, F., R. Cavalloro and G. Di Cola : A new theoretical approach to the control and management of fruit-fly systems. Proc. of the CEC/IOBC Intern. Symp. on Fruit flies of economic importance. Athens, Nov. 1982, (R. Cavalloro - ed.). Publ. for the Commission of the European Communities by A.A. Balkema, Rotterdam, 1983.

MODELLING AND / FOR / BY SIMULATION

INGE TROCH*)

Various aspects of the close interconnection of the two tasks 'Modelling a System' and 'Simulating a System', shall be discussed for continuous systems. Some typical examples from different areas of application will be given in order to provide some insight into this matter. Then the importance of implicit programming will be discussed shortly which allows for a 'one-to-one-mapping' of the system block structure on the simulation program. Some remarks on available simulation tools will complete the discussions.

Keywords: Modelling - Simulation - Dynamic Systems - Simulation tools

1. INTRODUCTION

It is quite well-known that there is a strong connection between modelling and simulation and that it is not always evident what comes first. At a first glance, modelling the system to be investigated is expected to come before the simulation. But then one realizes in many cases very quickly that the model has to be established in a special way to allow not only for an efficient simulation with the hard- and software tools available but also to get the insight in the system behaviour for which the simulation had been undertaken. Further, often simulation shows that an updating not only of model parameters but of the model itself is a necessity. Besides, there are many problems - especially in the wide field of the so-called soft systems - where simulation is used as a tool to develop a model.

The following discussions shall be based on examples taken from different areas of application. They are chosen in a way not only to illustrate the remarks just made but also to give the occasion to discuss shortly some important aspects of modelling and simulation in a rather general way. Although these examples are chosen from the field of continuous-time systems - with which the author is more familiar - the following statements hold also for discrete-time systems. Many of the following explanations sound rather trivial, nevertheless it seems not superfluous to collect some aspects of the complex task of simulating a given real system.

2. MODELLING AND SIMULATION

Speaking with Bratley et al., [1], '*a model is a description of some system intended to predict what happens if certain actions are taken. Virtually any useful model simplifies and idealizes ... Most forces that impinge on the system must be neglected on a priori grounds to keep the model tractable*'. The result of these simplifications and idealizations is a more or less complex model which can be investigated further analytically or numerically or by simulation. The effort of an analytic investigation usually is substantially higher but results in statements about stability or stability domains, about

*) Professor at the Technical University Vienna, Karlsplatz 13, A-1040 Wien, Austria

system inputs being optimal with respect to a certain well-defined goal etc. Unfortunately, very often these conclusions can be made only at the cost of model simplification which may reduce its credibility. On the other hand simulation allows the use of rather complex models which lead to large computation times and often also to problems of numerical accuracy.

Consequently, a model should be so detailed and complex as necessary to reflect adequately the features of the real system that are important to the application under consideration. But at the same time the model should be as simple as possible not only to reduce the computational burden but also to make it possible to obtain all necessary data and to estimate the required parameters. Especially for soft systems it is of importance to start with a simple model and to learn from experiences with it. If it is suitably modularized, particular modules of it can be specified in more detail later on. An example for this procedure is given e.g. in [6] where the glucagon-stimulated hepatic glucose production of an isolated rat liver was modelled with the help of simulation. A short description will be given in section 4. The modularization leads to a program structure that reflects the systems structure. It may require the use of implicit simulation techniques, which are rather easy to perform on an analog computer but require additional work and - in case the structure is maintained during the simulation - also additional computation time on a digital one.

On the other hand in physics and engineering it is often possible to derive quite accurate and complex models using basic laws. But then it turns out that it is not only costly to obtain the required data but also that the model is to complicated to be used in the application in mind. Then one tries to simplify the model either by simulation or analytically. An example is given in [5,7] and will be discussed shortly in section 3. In the latter case simulation is often used to validate the simplified model. It shall be remarked that any validation of a model by simulation requires a relatively large number of runs which have to be compared to reality. It is always a problem to obtain sufficient data about the input/output response of the real system. If there is not enough such data available then one has to be careful with conclusions especially with extrapolation about the systems future behaviour. In such a case one should speak better about a scenario investigation and be extremely careful not to risk by imposing constraints and assumptions that the simulation output supports preconceived opinions. Here, a sensitivity analysis as e.g. the one reported in example 1 may be of value. As a 'conclusion' these remarks about complexity can be summarized by a statement from [1]:*Parsimony in assumptions and parameters is good'.*

It is always recommandable to test and validate the model and simulation program by using input data different from that used during the development of the model. But, this is not possible in any case especially for soft systems such as the one reported in example 1. On the other hand example 3 gives a good illustration for this recommandation. It shows at the same time that the modelling effort sometimes has - unintentionally - the benefit of improving the understanding of what is really happening in the system being modelled. As illustration of some of the remarks above take the following

Example 1: World models have become of common interest since Forresters report in 1971, [1]. His model has been improved later, [3,4], but these improvements lead more or less to the same conclusions. The problem is not only difficult because 'world dynamics' is a soft system with a large number of variables, parameters and table functions but also because it is not possible to make a second experiment with changed parameters. Unfortunately, one has to expect further that the parameters used in the simulation study (it might be better to speak about a computation because there was not a large number of runs simulating different situations) may differ from their nominal values within a range of ±30%. In [8] a computational sensitivity study had been undertaken with the goal to investigate these parameter uncertanties. It turned out that even very small changes in economic parameters only (which arè relatively reliable compared to others) can change

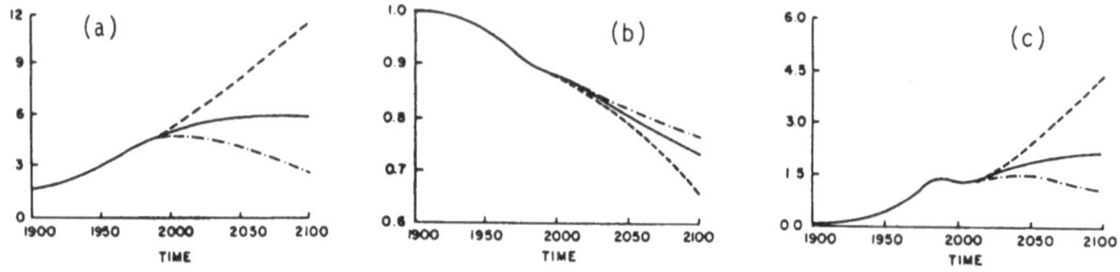

Fig.1: 5% changes to reference run in 1990: (a) World population, (b)Non-renewable resources fraction remeining, (c) Index of persistent pollution (acc.[8])

the result considerably and may consequently lead to conclusions different from the ones in [2-4], Fig.1. This must not necessarily mean that the model is false. If the real system 'world' is very sensitive to some parameters and/or table functions, then a good model must reflect this sensitivity. A definit answer is not possible as one must not experiment with the system, but there are enough indications for such a sensitivity. Nevertheless, one should be very careful with any conclusion drawn from the model, which can give a scenario only.

3. MODELLING FOR SIMULATION

When setting up a model for simulation purposes one should be aware of the hard- and software to be used right from the beginning. Otherwise more or less additional work has to be put in the task of analyzing and transforming the model until the model equations have the required form. Let us illustrate this by the simulation model for the dynamic behaviour of a robot.

Example 2: An industrial robot consists of the arm itself, the driving motors and the gears inbetween. If p denotes the vector of joint coordinates and k_{Dr} the vector of the driving forces and moments, the dynamic behaviour of the arm can be described in the form, [5,7],

$$k_{Dr} = M(p)\ddot{p} + h(p,\dot{p}) \qquad (1)$$

The drives can be modelled in the form

$$\dot{z} = f(z, u_S, k_L) \qquad (2)$$

where z describes the state of the drives, u_S is the control input and k_L the vector of the moments of load. The gears usually are modelled statically by the relations

$$\dot{p} = Tz \tag{3a}$$

$$k_L = S\, k_{Dr} \tag{3b}$$

The system structure is given in Fig.2. Analog (or hybrid) simulation allows a direct use of equs. (1)-(3). This has the advantage that not only $p, \dot{p}, \ddot{p}$ and z are outputs of the simulation but also the forces and moments k_{Dr}. This is impotant in view of the fact that they have to obey certain constraints originating from mechanical aspects.

Contrary to this, direct use of equs. (1)-(3) is not possible in a digital simulation. The programmer will realize this rather quickly in case he writes the program in a 'normal' language as e.g. FORTRAN, ALGOL, PASCAL etc. Straightforward use of a simulation language, e.g. ACSL, will result in the error message 'algebraic loop'. Its existence can be seen

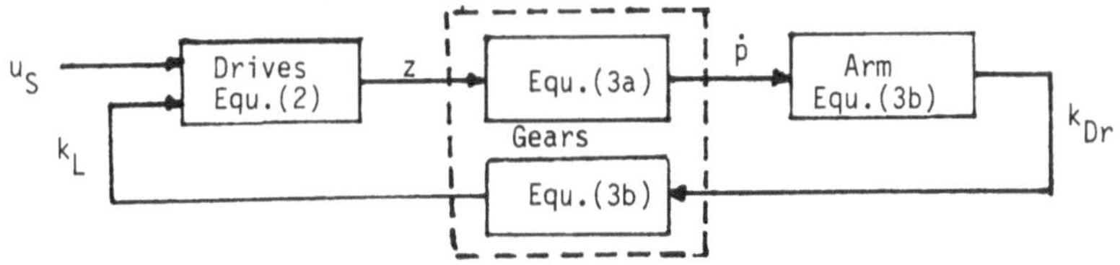

Fig.2: Structure of the system 'robot+drives+gears'

quite easily if the equations are transformed to

$$\begin{aligned} \dot{p} &= Tz \\ \dot{z} &= f(z, u_S, S[M(p)T\dot{z}+h(p,Tz)]) \end{aligned} \tag{4}$$

where $\dot{z}$ appears on both sides whereas the digital integration of differential equations requires their explicit form. In case of an equ. (2) which is linear in k_L, (4) can be transformed directly to it. If this analytic transformation into explicit form is not possible then an algorithm to compute $\dot{p}$ and $\dot{z}$ from equ. (4) has to be incorporated in the simulation program.

This example shows that the simulation hardware can decide whether or not implicit models can be used, which often provide a 'one-to-one-mapping' of the model structure -and consequently of the system structure - on the computer program structure. But also the time needed for a single run, the form of stopping and/or switching conditions depend on this hardware. What can be done easily with a comparator and switches on an analog computer may cause numerical troubles on a digital as is well-known to all who had to handle events. As example think of a bang-bang control where the switching times are determined by a function depending on the state. Concerning computation time (for a single run within the simulation) it shall be remembered only that in a digital simulation, time increases nonlinearly with complexitiy and order of the dynamic equations. This is contrary to analog or hybrid simulations where the available hardware limits the model complexity and order, although in hybrid simulation this can be overcome by an appropriate distribution of tasks between the analog and the digital part. Often, it is claimed, that analog or hybrid simulation requires more preparation time due to scaling etc. This is not true with modern computers which have autopatch capability and where modern simulation languages like HYBSYS can be used, which do the scaling automatically.

4. MODELLING BY SIMULATION

This topic includes establishing a model as well as simplification of a given model. The latter has been discussed for an industrial robot in [5,7]. Consequently, only the first aspect shall be outlined in

Example 3: Infusion of glucagon with stepwise increasing concentration causes in an isolated rat liver the hepatic glucose production to rise rapidly, Fig.3. The type of reaction leads to the idea of modelling the system behaviour as second order lag using thus

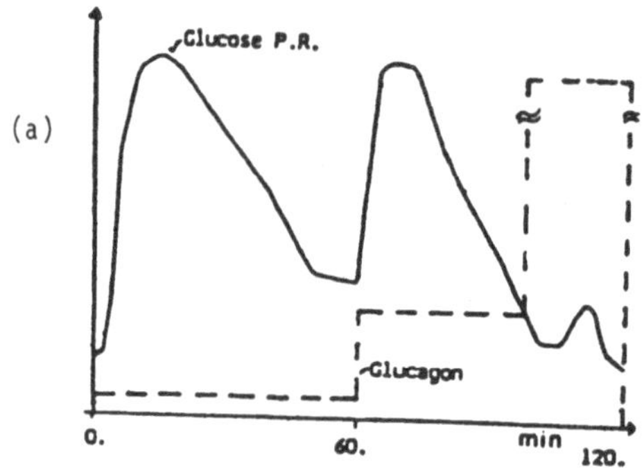

Fig.3: Comparison of measurement (ful line) and simulation (dotted line in (b,c)), acc. [6].

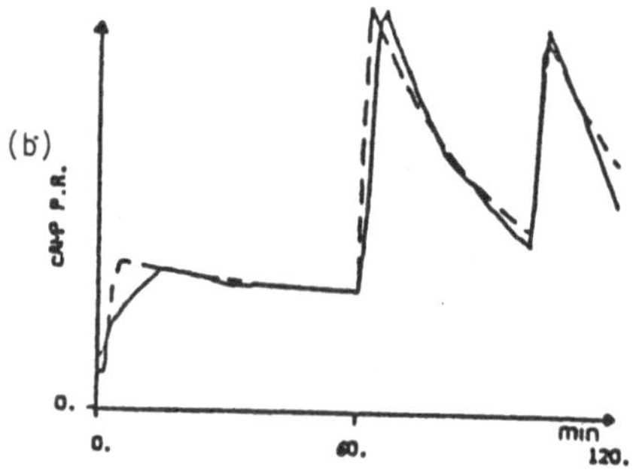

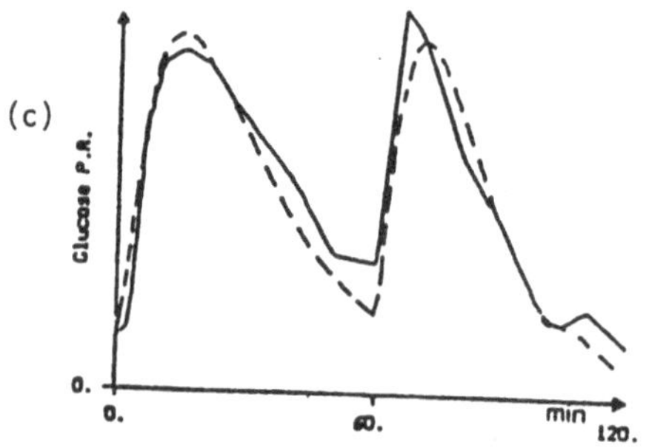

concept fitting as modelling method. During the simulation it turned out that - at least - two compartements have to be modelled: one giving the relation between the infused glucagon dose and the cyclic AMP, the second the relation between the cAMP and the glucose production rate. The resulting model, described in detail in [6] has some essential properties:

- It explains all observed phenomena qualitatively and quantitatively with an accuracy consistent with the one of the measurements
- It is as simple as possible i.e. it does not contain more parameters than necessary
- It was able to predict the existenc of an intermediate signal (later identified as cAMP) and the reaction of the system to a stepwise decrease of the glucagon concentration (later verified by experiment)

As the resulting equations contain limiters and some nonlinearities due to the dependence of some parameters on the actual state of the system, the development of the model as well as the parameter adaption process could be carried out only by simulation. It shall be pointed out, that again the model was developed step by step, starting with one modul. Once knowing that there are to be two compartements there respective models have been developed in more detail again in a step-by-step procedure. By this, example 3 illustrates also quite well that starting with a simple model which is fixed in more details at a later stage of investigations is a recommandable modelling procedures which is - especially for soft systems - often easier to apply than staring with a complicated model (with unknown parameters) and simplifying it later on.

5. REFERENCES

[1] P.Bradley, B.L. Fox, L.E.Schrage, A Guide to Simulation. Springer, New York, 1983

[2] J.W. Forrester, World Dynamics. Wright-Allen Press, Cambridge, Mass., 1971

[3] D.H. Meadows et al., The Limits to Growth. Potomac Ass.-Univ.Books, New York 1972

[4] D.H. Meadows et al., The Dynamics of Growth in a Finite World. Wright-Allen Press, Cambridge, Mass., 1974

[5] I.Troch, P.Kopacek, Erstellung von Modellen für die Dynamik von Industrierobotern mit Hilfe hybrider Simulation. Simulationstechnik (Eds.: F.Breitenecker, W.Kleinert), Springer, Berlin, 1984, 600 - 605

[6] I. Troch et al., Modelling by Simulation of Hepatic Glucose Production in Vitro. Proc. Int.Conf.'IMACS European Simulation Meeting on Simulation in Research and Development' 1984, Eger, Hungary, North-Holland, to appear.

[7] I.Troch et al., Models for Robots - A Simulation Approach. These Proceedings.

[8] P.J. Vermeulen, D.C.DeJongh, Growth in a Finite World - A Comprehensive Sensitivity Analysis. Automatica 13, 1977, 77 - 84

A MODEL FOR EVALUATION OF ACID DEPOSITION IN EUROPE

Leen Hordijk
International Institute for Applied Systems Analysis (IIASA)
A 2361 Laxenburg, Austria.

Abstract

This paper briefly describes a set of models which can be used to assist decision makers in their evaluation of scenarios to reduce the effects of acidification in Europe. The models can be used in an interactive way: the user selects an energy pathway for the 27 European countries and prescribes control strategies like low sulfur fuels and scrubbers in power plants. The model's output include deposition maps for the whole of Europe and maps indicating forest soils where the pH is below a certain specified level. A model analyzing lake acidification is in its final stage of development.

The model's acronym is RAINS which stands for Regional Acidification Information and Simulation.

Contents

Acknowledgements

The author acknowledges the contribution of IIASA's Acid Rain Project team. The research reported here was carried out by E. Runca (Italy, former project leader), J. Alcamo (USA), P. Kauppi (Finland), M. Posch (Austria), J. Kämäri (Finland), A. Mäkelä (Finland) and L. Hordijk (The Netherlands, project leader). The project benefited much from collaboration with the University of Göttingen, FRG; the Norwegian Institute of Meteorology, Oslo; the Institute for Meteorology and Water Management, Warsaw; the University of Helsinki; the Royal Netherlands Meteorological Institute, De Bilt.

The research team is indebted to the Geneva offices of the World Meteorological Organization (WMO) and the UN Economic Commission for Europe (ECE) for providing important data for the project.

This paper draws heavily on an extensive Working Paper by Alcamo et al. (1984).

1. INTRODUCTION

1.1 The Problem of Acidification

Since many decades society has been confronted with increasing negative side effects of human activities. Already during the Industrial Revolution effects of air pollution were known. The character of the pollution problem, however, changed drastically during the past World War II period. Before, the pollution problems were mainly of a local and transitory character. Nowadays we are faced with a transboundary problem of a continuous character. Pollutants may remain airborne for several days and can travel over a distance of thousand kilometers.

Acidic compounds may have a direct and an indirect effect. Direct effects are those resulting from deposition of the compounds on the surfaces. These effects include corrosion of materials, deterioration of monuments and damage to human health and tree foliage. Indirect effects may occur after deposition. Increased acidity of soil can restrict plant growth, acidification of groundwater may increase the solubility of heavy metals which in turn can affect human and animal health. Through a combination of direct and indirect effects forest growth can be reduced. Both direct deposition and soil acidity are viewed as causes of recent forest dieback (ERL, 1983).

It is needless to say that these adverse effects also cause extensive costs for society. In a recent overview of the acid rain issues within the European Community some tentative damage estimates have been presented (ERL, 1983, Ch.11). To give an impression of the magnitude of these damages some of the damage estimates are shown in Table 1. These figures are far from exact estimates and should not be misinterpreted.

Damage category	Hypothesis, year	Costs (10^6US $ p.a.)
Forests	20% of forests affected, 10% drop p.a.	100
Crops	Winter wheat, barley potatoes 1981	900 - 1,100
Fish	Scottish lakes	0.5
Buildings	due to sulfur compounds	540 - 1,100

Table 1. Damage due to acid deposition in EEC countries
Source: ERL (1983, Ch.11).

1.2 The European Context

Wide attention to the transboundary nature of acidification was raised by a Swedish report on this subject presented at the 1972 UN Conference for Human Environment in Stockholm. In 1973 OECD began modeling and monitoring the long range transport of air pollutants in Europe. This project led after being completed in 1977 to a model which estimated the sulfur import-export balance of the European OECD countries. An analysis of costs and benefits of sulfur control was published by OECD in 1980 and 1981.

After 1977 the monitoring and modeling of long-range transport continued within the Cooperative Programme for Monitoring and Evaluation of Long Range Transmission of Pollutants in Europe (EMEP). The United Nations organizations ECE, UNEP and WMO oversee this effort which includes both Eastern and Western European countries.

In 1977 Norway proposed the adoption of an International Convention on Transboundary Pollution. This Convention was signed by 33 countries in 1979 and ratified by the required forum of 24 countries in January 1983. At present all but one country ratified the Convention. The Convention's basic statement is that the countries "shall endeavour to limit and as far as possible, gradually reduce and prevent air pollution, including long-range transboundary air pollution". Within the Convention four programmes are under development: (1) Air quality management; (2) Research and Development; (3) Exchange of Information; (4) EMEP.

1.3 Scope of IIASA's Project

At IIASA it was concluded that within the context of the Geneva Convention there is need for a framework for the analysis of acid rain control scenarios.

Both IIASA's systems approach and East-West background were viewed to be appropriate for this work. The principal objective of the project is to assist decision makers in their evaluation of policies for controlling the impacts of acidification in Europe. To accomplish this we are developing a model and a set of procedures for using it. We hope that the model will serve as a common technical ground in the negotiations on acid rain. Another result of the integrating modeling effort may be that gaps in understanding the acid rain problem can be identified at the same time stimulating research to overcome the gaps.

2. STRATEGY

The design of any model depends very much on the dimension of the problem the model is used for and the users of the model. Regarding the problem of acid rain any modeling will be confronted with the following characteristics of that problem:

1. The acid rain problem is transboundary.
2. Many different scientific disciplines are needed.
3. There exists a wide range of uncertainty in cause-effects relationships.
4. Different time scales are relevant (hours of transport, snowmelt in weeks, acidification of soils in years).
5. Due to the fact that acid rain is an important political issue, research is stimulated which creates a constant stream of new information.
6. Acid rain is expensive to control.
7. Within and between countries the levels of responsibility vary from local authorities to international bodies.

The users of the model will be on the decision making level. Therefore analysis of scenarios, interactive use of the system and quick and readable output form main characteristics of this model.

The above notions led to a model that should

(a) be co-designed by analysts and potential users;
(b) contain submodels that are as simple as possible, but based on more detailed data and complex models;
(c) have interactive input and graphical output;
(d) consist of linked modules which can easily be replaced by other modules that represent another set of hypotheses.
(e) be dynamic in nature.

3. CURRENT SUBMODELS

3.1 Overview

Interactions of human activities and environmental impacts can be linked in a schematic way as is done in Figure 1.

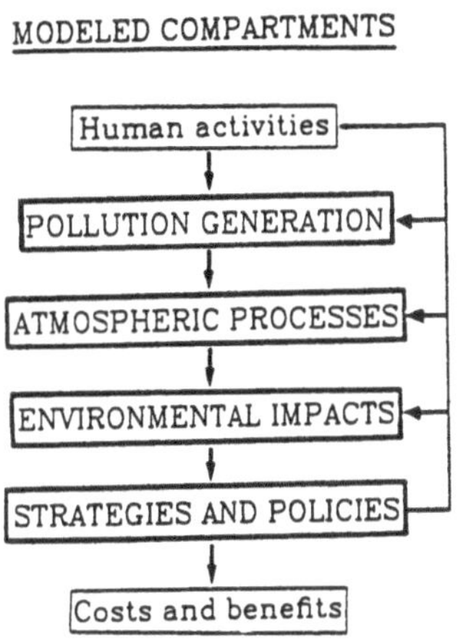

Figure 1. Interaction between human activities and environmental impacts

The present status of the model comprises the four compartments shown with the heavy lines in Figure 1.

For each of these compartments a submodel has been developed. The first compartment Pollution Generation, has been modeled for sulfur emissions based on a selected energy pathway for 27 European countries. The second compartment, Atmospheric Processes, has been modeled with a source-receptor matrix. This submodel computes sulfur deposition in Europe on a grid square which is 150 x 150 km. The sulfur deposition is input to a third submodel, the Forest soil-pH submodel. This model is a first part of a set of submodels that form the third compartment of Figure 1, Environmental Impact.

The model's time horizon is 2030, which permits examination of possible long term impacts. The models time resolution is one year and the spatial resolution is a grid system of approximately 100 x 100 km. The spatial coverage is all of Europe, being 27 countries.

The model is sulfur based since it is generally accepted that sulfur is presently the principal contributor to acidification in Europe. However, other pollutants will be included in the future. The model use is illustrated in Figure 2.

The model user can start with selecting an energy pathway for each country followed by a pollution control program. The model then calculates the sulfur emissions of each country, the atmospheric transport and deposition and the resulting soil-pH. Based on this output the model user may select another energy pathway and/or control program to evaluate with the model. The model allows for comparison of two policies chosen by the user.

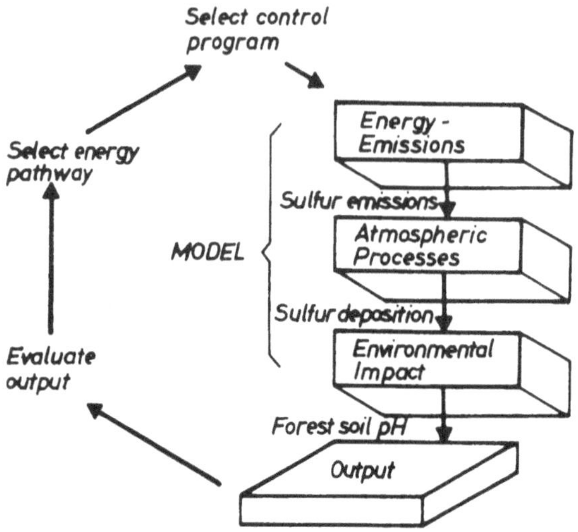

Figure 2. The use of IIASA's Acid Rain Model

3.2 Sulfur emissions

The sulfur emissions are computed on a mass balance basis by summing the sulfur emitted in nine energy sectors in each country together with the contribution from non-combustion sources:

$$S_i^t = \sum_{k=1}^{9} S_{k,i}^t + S_{p,i}^t \tag{1}$$

with $S_{k,i}^t$ = sulfur emissions in country i, sector k, time t (kt/yr)
S_i^t = sulfur emission in country i, at time t (kt/yr)
$S_{p,i}^t$ = sulfur emission from non-combustion processes in country i at time t (kt/yr)

The energy sectors are five for oil (conversion, power plants, domestic use, industrial use, transportation) and four for coal (conversion, power plants, domestic use and industrial use).

Sulfur emissions in each energy sector i are computed as

$$S_k^t = E_k^t \, S_k^e \, (1 - c_k)(1 - a_k)(1 - r_k) \tag{2}$$

with E_k^t = energy used in sector k (PJ/yr)
S_k^e = sulfur content of fuel per energy unit
c_k = fraction of fuel cleaned
a_k = fraction of flue gas desulfurization
r_k = fraction of sulfur retained in the sector

The energy scenarios presently used are taken from ECE (1983).

3.3 Atmospheric processes

The IIASA model uses results from the EMEP model, the Cooperative Program for Monitoring and Evaluation of Long-Range Transmission of Air Pollutants in Europe (see Eliassen and Saltbones, 1983). The EMEP model has been condensed into a source-receptor matrix (SRM), that has as elements deposition in grid i caused by a unit emission in grid j. The grids are 150 x 150 km and cover the 27 largest countries in Europe, including the European part of the USSR.

The SRM currently used in the model RAINS are based on an EMEP model run of 1978-1979 data. New SRM's which take into account inter-annual meteorologic variability and more recent data, are being implemented in the model.

3.4 Forest soil acidification

The EMEP sulfur transport model is linked to the Forest Soil Acidity submodel which estimates the level of soil acidification in natural forest areas due to acidification. Details of this submodel, which was built in cooperation with the University of Göttingen, is reported in Kauppi et al (1984). The Forest Soil Acidification model uses three basic concepts: acid stress, buffer rate and buffer capacity. Acid stress is defined as the input of hydrogen ions to the top layer of the soil (50 cm). Buffer rate is the maximum potential rate of reaction between compounds in the soil and hydrogen ions, and buffer capacity is the total reservoir of buffering compounds.

The chemical state of the soils is classified into four buffer ranges as represented by Ulrich (1983): carbonate, silicate, cation exchange and aluminum ranges. Each range has a specific buffer rate and capacity.

In general, if acid stress exceeds the buffering rate or if the buffer capacity is depleted, the model shifts to the next buffer range. The pH of forest soil is estimated from the computed buffer range.

The model is used by assigning values of buffering variables for each of 88 different soil types in Europe. Each grid element on this submodel contains 1 to 7 soil types and is 1° longitude x 0.5° latitude.

3.5 Lake acidification

The Lake Acidity submodel computes lake acidity levels as a function of catchment characteristics and local acid deposition. Each watershed is conceptually divided into four spatial sectors: snowpack (if consistent with local climate), upper soil layer A, lower soil layer B, and lake volume. The submodel is also subdivided computationally into four modules which compute the hydrology and H^+ flux between the spatial sectors.

The Meteorologic Module transforms computed monthly sulfur deposition into acid stress to various sectors of the catchment. The monthly mean precipitation is broken down into rain and snow according to local mean monthly temperature by assuming threshold temperatures. Snow is stored in snowpack, and melts at a rate proportional to air temperature above a threshold value. Other equations in this module account for storage of wet and dry deposition in snowpack, release with meltwater, and direct acid stress to soil and lake volume.

The Hydrologic Module routes precipitation into quickflow, baseflow, and flow between soil layers. The computation of these flow components is based on rates of precipitation and evapotranspiration, and catchment characteristics such as soil depth, surface slope, hydraulic conductivity and volumetric water content of soil.

The Soil Chemistry Module uses the same analytical approach as the Forest Soil Acidity submodel to estimate H^+ concentration in the A and B soil layers. The loads of ions which contribute to acidity and alkalinity (H^+ and HCO_3^-) of the lake, are then computed for both soil layers. These loads are input to the Lake Response Module which estimates the H^+ concentration of the lake. These loads are assumed to be mixed within a volume defined according to location and season. Finally, the change in lake acidity is calculated according to equilibrium reactions of inorganic carbon species.

4. EXAMPLES OF MODEL USE

As emphasized in section 2 the model has been designed for easy handling by non-technical users. Therefore in each step of model use the user is provided with a clear question on a computer screen together with a number of options he can choose. Another terminal presents the information provided by the model, given the user's particular choice. To give the reader an impression of model use we present an example of model use.

We first assume that all European nations follow the ECE scenario 'Conservation' which levels off after the year 2000, indicated with Energy Pathway 3 in the output shown hereafter. The user then specifies a control strategy and compares it with the case of 'no action'.

The pollution control strategy includes:

(1) 30% removal of sulfur in the domestic coal sector through coal cleaning and 60% removal of sulfur in the domestic oil sector by oil desulfurization.

(2) Phasing in of flue gas control devices in the power plant and industry sectors for coal and oil in the following way:

Year	Fraction of sulfur removed
1990	0.4
2000	0.6
2010	0.8
2020	0.8
2030	0.8

This phasing assumes that 50% of all power plants and industrial boilers in 1990 will have flue gas control devices which have an 80% sulfur removal efficiency (0.5 x 0.8 = 0.4). These devices will be applied to 75% of all plants and boilers in the year 2000 (0.75 x 0.8 = 0.6) and all plants after the year 2000 (1.0 x 0.8 = 0.8).

The figures below summarize the differences in the scenarios. Figure 3 represents a comparison of sulfur deposition in both scenarios for the year 2030. The heavy lines refer to the energy pathway with no pollution controls, the thinner lines represents the sulfur isolines emerging from the scenario with major pollution controls. The user has chosen the 2.0 $g/m^2/yr$ isoline as an indicator. In Figure 4 finally a comparison of areas with low soil pH is shown. The user's choice has been to see the areas with soil-pH below 3.5, being the average of the aluminum buffer range. The darker shaded area represents the scenario with major pollution controls.

TOTAL SULFUR DEPOSITION (G/M**2/YR)

SCENARIO : ENERGY PATHWAY 3 - NO ACTION
COMPARED WITH: ENERGY PATHWAY 3 - MAJOR POLLUTION CONTROLS

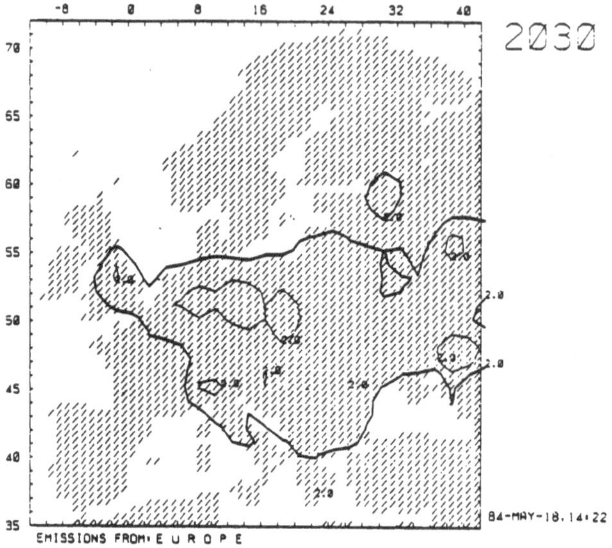

Figure 3. Comparison of computed sulfur deposition

2030: AREAS WITH SOIL-PH < 3.5

SCENARIO: ENERGY PATHWAY 3 - NO ACTION
COMPARED WITH: ENERGY PATHWAY 3 - MAJOR POLLUTION CONTROLS

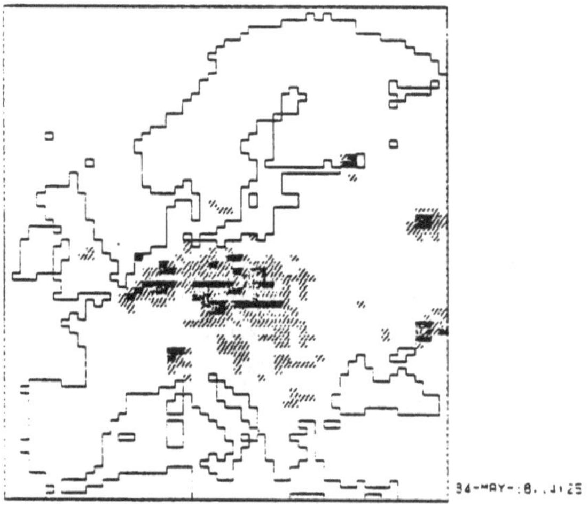

Figure 4. Comparison of computed forest soil pH

5. FUTURE RESEARCH

Present and future research effort focuses on

(1) expanding the model to include cost analysis and additional submodels: NO_x emissions, NO_x transport and deposition, direct forest impacts.

(2) model testing and uncertainty analysis.

(3) development of other operational modes for RAINS, for example, implementing search techniques and optimization algorithms.

(4) applying the model to policy analysis.

(5) distribution of RAINS to international and national institutions for their use in policy analysis.

REFERENCES

Alcamo, J., P. Kauppi, M. Posch and E. Runca. Acid Rain in Europe: a framework to assist decision making. IIASA Working Paper, WP-84-32, 1984.

Economic Commission for Europe (ECE), An Efficient Energy Future. Butterworths, London, 1983.

Eliassen, A. The OECD study of long range transport of air pollutants. Atmospheric Environment 12 (1978), 479-487.

Eliassen, A., and J. Saltbones. 1983. Modelling of long range transport of sulfur over Europe: A two-year model run and some model experiments. Atmospheric Environment. 17(8): 1457-73.

Environmental Resources Limited (ERL). Acid Rain. A review of the phenomenon in the EEC and Europe. Graham and Trotman Ltd., London, 1983.

Kauppi, P., M. Posch, E. Matzner, L. Kauppi, and J. Kämäri. Acidification of forest soils: model development and application for analyzing impacts of acidic deposition in Europe. IIASA Collaborative Paper CP-84-16, 1984.

Ulrich, B. Soil acidity and its relation to acid deposition. In: B. Ulrich and J. Pankrath (eds.), Effects of Accumulation of Air Pollutants in Forest Ecosystems. D. Reidel Publ. Co. Dordrecht, Netherlands, 1983.

MODELING THE TRANSIENT AND EQUILIBRIUM CLIMATE

RESPONSE TO GREENHOUSE GASES

Wilfrid Bach*

Summary

The purpose of climate modeling is to reproduce as realistically as possible the component parts of the climate system. If successful, such models can be used to assess the response of the climate system to e.g. an increase of CO_2 or other trace substances. A hierarchy of models is available, such as radiative-convective models (RCMs) and energy balance models (EBMs) (i.e. 1-D or 2-D thermodynamic models), or general circulation models (GCMs)(i.e. 3-D hydrodynamic models). With the first type transient response studies (e.g. the climate response to a time-dependent CO_2 increase), and with the latter type equilibrium response experiments (e.g. the climate response to a time-independent CO_2 doubling) are conducted. The various methods are applied and a selection of results is shown.

Simulation of climate response, greenhouse gases.

1. Introduction

In the past the Earth's climatic history was governed by natural influences. In recent years an additional factor capable of perturbing the delicate balance of forces which determines the Earth's climate has entered the scene. Through his activities Man is now in a position to influence climate not only locally, but also globally in quite drastic and possibly irreversible ways. One major cause for the man-made climatic change is the enhancement of the natural greenhouse warming in the troposphere through the increase of the carbon dioxide (CO_2) content in the atmosphere due to the burning of fossil fuels (coal, oil, gas), deforestation, forest dieback, and soil degradation through bad cultivation practices. In addition to CO_2 and water vapor (H_2O) other trace gases, such as nitrous oxide (N_2O), methane (CH_4), chlorofluorocarbons (CFCs) and ozone (O_3), plus a number of short-lived gases (e.g. SO_2, CO etc.), as well as aerosols, changes in land use, and waste heat also influence climate. The current state-of-knowledge on the main greenhouse gases is summarized in Table 1 in terms of the rate of change and the climate sensitivity. Many of these trace substances are emitted into the atmosphere more quickly than they can be absorbed by natural sinks. The chain reaction of this 'geophysical experiment' of global extent leads on from increasing greenhouse gases to an increased absorption of longwave radiation (the greenhouse effect) and to an enhanced warming. This, in turn, produces a great variety of regional and seasonal variations in climatic events, which, given the Earth's increasing population, is bound to aggravate the shortages in food, water, and energy supplies.[1] The potential effects can only be simulated with the help of climate modeling.

2. Simulation of the transient climate response

The assessment of the time-dependent or transient response of CO_2 to climate requires energy modeling, carbon cycle modeling, and climate modeling. The historic development of CO_2 emissions can be derived from UN statistics.[5] The future CO_2 emissions are calculated from energy data for a range of plausible energy scenarios as shown in Table 2.

* Center for Applied Climatology and Environmental Studies, Department of Geography, University of Münster, D-4400 Münster, Robert-Koch-Str. 26.

TABLE 1 Overview of the main trace gases influencing climate

Factor	Principal anthropogenic sources	Atmospheric residence time	Principal sinks and removal process	Background concentration in atmosphere (current)	Future development	Potential impact on climate and surface air temp. change	Magnitude and importance of impact
Carbon Dioxide (CO_2)	Fossil fuels, deforestation, soil destruction	10-15 yr (T.) 2 yr (S.)	Oceans Biosphere	342 ppmv	250-290 ppmv (ca. 1850) 342 ppmv(1983) Increase: 0.4 %/yr	Warming in T. Cooling in S. $2xCO_2 \rightarrow 3 \pm 1.5$ K global, 4 to 5 times greater at poles	Global, possibly one of the most important impacts
Nitrous Oxide (N_2O)	Fertilizer, fossil fuels, aircraft	150-175 yr	Little absorption in T., photo-dissipation in S., dry deposition, oxidation to nitrate	310 ppbv	310 ppbv (1980) Increase: 0.2-0.3 %/yr	Greenhouse Effect in T. Impact on O_3-budget in S. $2xN_2O \rightarrow 0.3$-0.4 K	Global, possibly very important
Methane (CH_4)	Fossil fuels, cattle raising, irrigation	7-10 yr	Oxidation with OH-radicals	1.65 ppmv	Increase through growth in fossil fuel and agricultural use Increase: 1-2 %/yr	Impact directly on O_3 of T. and indirectly on Greenhouse Effect; $2xCH_4 \rightarrow 0.25$ K	Global, increasing in importance
Chlorfluorocarbons ($CFCl_3$) (CF_2Cl_2)	Propellants, coolants and insulating materials	≈ 60 yr ≈ 100 yr	No known sink in T., sink in S. through photolysis	0.17 ppbv 0.28 ppbv	Increase: 3-5 %/yr	Increase from 0-1 ppmv: $CFCl_3 \rightarrow 0.15$ K $CFCl_2 \rightarrow 0.13$ K	Global, impact on temperature probably decreasing
Water vapor (H_2O)	Combustion processes, high-flying aircraft	10 dy (T.) 2 yr (S.)	Precipitation	$10\text{-}3x10^4$ ppmv (T.) 3 ppmv (S.)	Increase due to growth in industry and air traffic	Greenhouse Effect and cloud formation/ precipitation in T., influence on trace substances in S. $2xH_2O$ in S. $\rightarrow 0.6$ K	Local and regional in T., global in S.
Ozone (O_3)	Indirectly produced through photochemical reactions with other substances	30-90 dy (T.) 2 yr (S.)	Catalytic reactions with other trace substances, such as NO_x, Cl_x, HO_x in S., dry deposition in T.	0.02-0.3 ppmv (T.) 5-10 ppmv (S. at 30 km)	Decrease of O_3 in upper S. by 0.3-0.4 %/yr in 1970s. Increase of 7% in last decade in layer 2-8 km over NH.	O_3-destruction in S. leads to cooling: $0.75xO_3 \rightarrow -0.4$ K; O_3-production in T. leads to warming: $2xO_3 \rightarrow 0.9$ K	Global, increasing in importance

T. = Troposphere   ppmv = parts per million   K = degree Kelvin
S. = Stratosphere   ppbv = parts per billion   2x = doubling of concentration

After: Bach (1984)[1)] and supplemented.

TABLE 2 Global primary energy supply (TWyr/yr) for different energy sources and a variety of energy scenarios, 1980-2030

Source	Reference Year		IIASA - Scenarios (1981) High Scenario				IIASA - Scenarios (1981) Low Scenario				Colombo/ Bernardini (1979) 16 TW (2030) Scenario[1]				Lovins et al. (1981) Efficiency Scenario			
	1980 TW	%	2000 TW	%	2030 TW	%	2000 TW	%	2030 TW	%	2000 TW	%	2030 TW	%	2000 TW	%	2030 TW	%
Oil	4.20	42	5.89	35	6.83	19	4.75	35	5.02	22	4.26	35	3.58	22	1.77	25	0.24	5
Gas	1.60	16	3.11	19	5.97	17	2.53	18	3.47	16	2.27	18	2.48	16	1.51	21	0.34	6
Coal	2.60	26	4.94	29	11.98	33	3.92	29	6.45	29	3.51	29	4.60	29	1.77	25	0.38	7
Subtotal	8.40	84	13.94	83	24.78	69	11.20	82	14.94	67	10.04	82	10.66	67	5.05	71	0.96	18
Nuclear (LWR)	0.30	3	1.70	10	3.21	9	1.27	10	1.89	8	1.13	10	1.36	8	-		-	
Nuclear (Breeder)	0	0	0.04	0	4.88	14	0.02	0	3.28	15	0.02	0	2.35	15	-		-	
Subtotal	0.30	3	1.74	10	8.09	23	1.29	10	5.17	23	1.15	10	3.71	23	-	0	-	0
Hydro			0.83	5	1.46	4	0.83	5	1.46	6	0.74	6	1.04	6				
Solar[2]	1.30	13	0.10	1	0.49	1	0.09	2	0.30	1	0.08	1	0.20	1	2.02	29	4.27	82
Other[3]			0.22	1	0.81	3	0.17	1	0.52	3	0.16	1	0.36	3				
Subtotal	1.30	13	1.15	7	2.76	8	1.09	8	2.28	10	0.98	8	1.60	10	2.02	29	4.27	82
Total	10.00		16.83		35.63		13.58		22.39		12.17		15.97		7.07		5.23	
Index	100		168		356		136		224		122		160		71		52	

[1]Using the fuel shares of the low IIASA scenario; the 16 TW scenario is also known as the "zero growth scenario" because it assumes that the present per capita energy consumption will be the same in 2030 but that a doubling in world population from 4 to 8 billion will lead to a doubling in total energy consumption.

[2]Mostly soft solar (e.g. rooftop collectors) but also small amounts of centralized solar electricity.

[3]Includes biogas, geothermal, commercial and non-commercial wood use.

Sources: Colombo and Bernardini (1979),[2)] Häfele et al. (1981),[3)] Lovins et al. (1981),[4)] and Rogner (n.y.).[7)]

The IIASA scenarios reflect the past growth philosophy,[3] the Colombo/Bernardini scenario is based on a zero growth rate,[2] and the efficiency scenario follows a least-cost strategy implying that all those efficiency improvements will be carried out which are feasible using available cost-effective technologies.[4] The future emissions of the other greenhouse gases are projected on the basis of the rates given in Table 1, i.e. 0.3 %/yr for N_2O, 1.5 %/yr for CH_4, and 4 %/yr for the CFCs. O_3 is not included in the present calculations because information on its complicated interactions with other substances is not yet adequate.

The atmospheric CO_2 concentrations are calculated with a carbon cycle model consisting of the four reservoirs atmosphere, biosphere, a surface mixed layer of the ocean, and the deep sea, and they are fitted to the observed Mauna Loa data (1958-1980). The observed surface air temperature is that of the Northern Hemisphere, and the reference period is from 1946-1960. The projected temperature changes due to the combined effects of CO_2 and other trace gases are calculated with a parameterized form of a 1-D time-dependent RCM coupled to an atmosphere-ocean EBM.

The results of the simulations show that, compared to base year 1980, the IIASA high scenario results in a 3-4fold increase in CO_2 emission (Fig.1), and a 55 % increase in CO_2 concentration (Fig. 2), leading to a 4fold increase in global mean air temperature from 0.4 to 1.8°C for the CO_2 effect only. For the combined effect, i.e. CO_2 plus other trace gases, an almost 8fold increase from 0.4 to ca. 3.1° C is obtained (Fig. 3). In contrast, over the same time period, less wasteful energy use, the basis of the efficiency scenario, results in a 9fold reduction in CO_2 emission (Fig. 1) and only a 14 % increase in CO_2 concentration (Fig. 2), leading to a mere 2fold temperature increase for the CO_2 effect only, but a 4fold increase for the combined effect (Fig. 3). Notably, high growth energy scenarios could lead to a warming that has not been experienced since the middle Pliocene (ca. 5-3 million years before present).

3. Simulation of the equilibrium climate response

Beside studying the transient response of climate it is also important to investigate the regional and seasonal distribution of perturbed climate parameters. This can only be done with the 3-D GCMs, which compute the equilibrium model climate for a time-independent CO_2 doubling (usually from 300 to 600 ppm). A GCM used in climate studies usually consists of three main components, namely a GCM of the atmosphere, a heat- and water-balance model of the land surface, and a mixed-layer ocean model. The GCM consists of a prognostic system of equations incorporating the physical and dynamical processes which control the climate. These processes are described by a set of nonlinear, partial differential equations, such as the horizontal momentum equation, the continuity equations for mass and water vapor, and the thermodynamic energy equation, which, together with the hydrostatic approximation, the equation of state, and appropriate boundary conditions form a closed system. This permits to determine the rates of change of the prognostic variables, such as surface temperature, pressure, horizontal velocity, water vapor, and soil moisture. GCMs have also many diagnostic variables, such as clouds, surface albedo, and vertical velocity. The governing equations of the GCMs are solved by numerical methods. The variables at the grid points are integrated forward in time

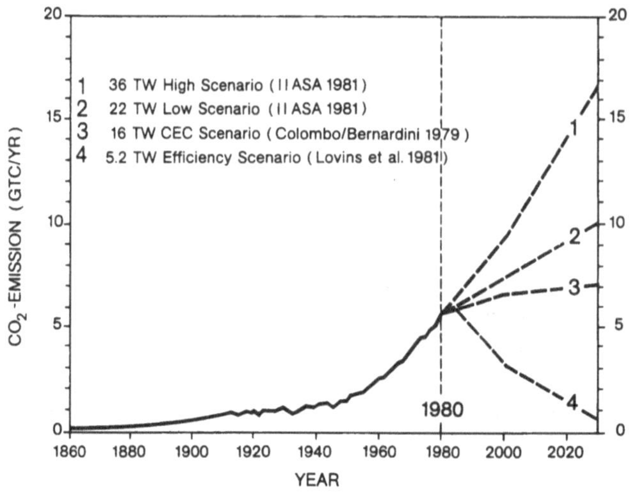

Fig. 1:

Simulation of CO_2 emission from UN fossil fuel statistics (1860 - 1980). Projection to 2030 based on energy scenarios

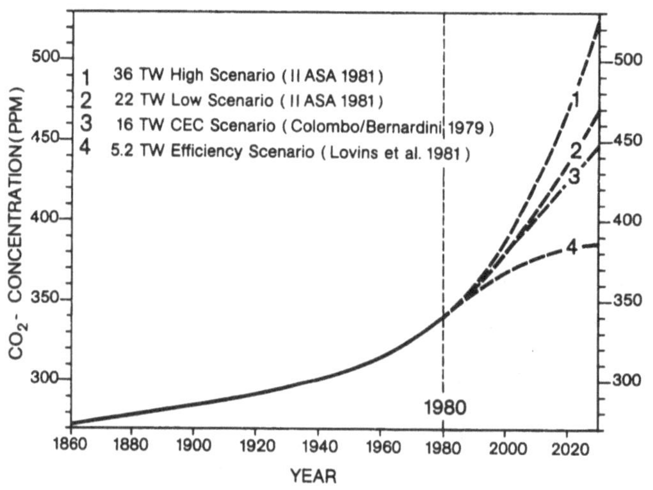

Fig. 2:

Simulation of CO_2 concentration based on CO_2 emission rates shown in Fig. 1

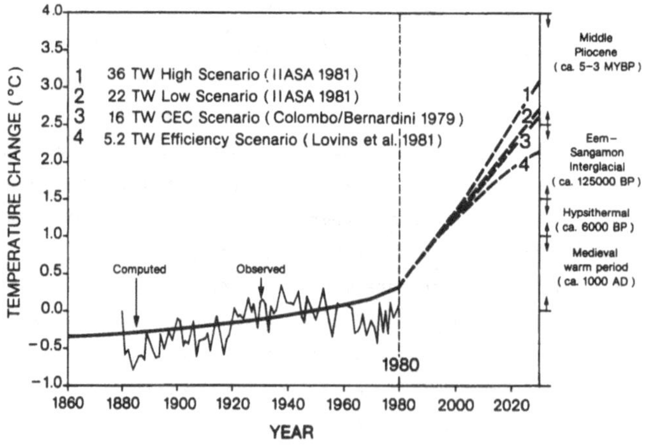

Fig. 3:

Simulation of temperature response to CO_2 effect only (1860 - 1980). Projection to 2030 based on the combined effects of CO_2 and other trace gases until the model approaches a statistical equilibrium. The time required varies from about 300 days (without ocean heat transport) to about 10 years (a GCM coupled to an ocean-mixed-layer model). The simulation of one day requires from one-half minute to a few minutes on the most advanced computers (e.g. the Cray 1 or the Cyber 205). The spatial resolution of GCMs is constrained by the speed and memory capacity of computers. Present GCMs have only a limited horizontal resolution of about 300-400 km, and the number of vertical layers ranges from 2 to 9. Therefore, several of the subgrid-scale physical processes important to climate are not resolved. They must be parameterized, i.e. they are either statistically or empirically related to the scale of those variables which are resolved.[8)]

To compare either model-generated data with observed data, or different GCM results with one another, it is necessary to transfer all data to the same grid system. Spurious small-scale fluctuations and time variations are removed by spatial and temporal filtering. Before a model can be used it must be validated by comparing the computed with the observed climate parameters. Fig. 4a shows for the GISS-GCM that in the eastern half of the study area the model underestimates the temperature distribution (by -2°C), whereas in the western half there is very good agreement (°C). The agreement between observed and modeled values is not that good for precipitation, which, in Fig. 4b, shows practically over the entire area a considerable overestimation of about 1 mm/day.

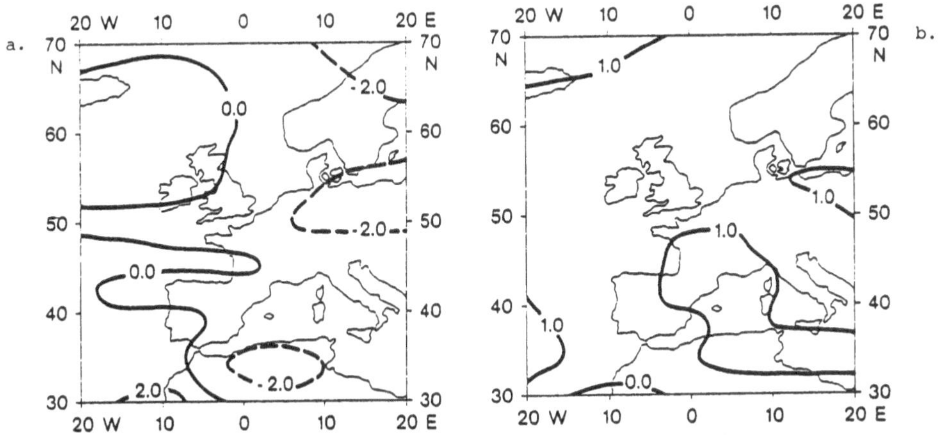

Fig. 4 Model validation in terms of the differences between the model-generated climate by the GISS-GCM and the observed climate; a. temperature (K) (Schutz/Gates data) b. precipitation (mm/day) (Jaeger data)

All of these results must be interpreted with the necessary caution due to the many uncertainties related to the still insufficient representation of the many complicated feedback mechanisms within the climate system. CO_2 simulations based on such validations may, however, still be made because they can give at least a general idea of the changes to be expected for a CO_2-warmed earth, even though details of changes at any given location are presently beyond the capability of the GISS or any other available GCM. With these caveats in mind we can now look at the results of some regional and seasonal temperature and precipitation change scenarios due to a CO_2 doubling.

Fig. 5 shows for the annual mean and the four seasons a temperature increase in the

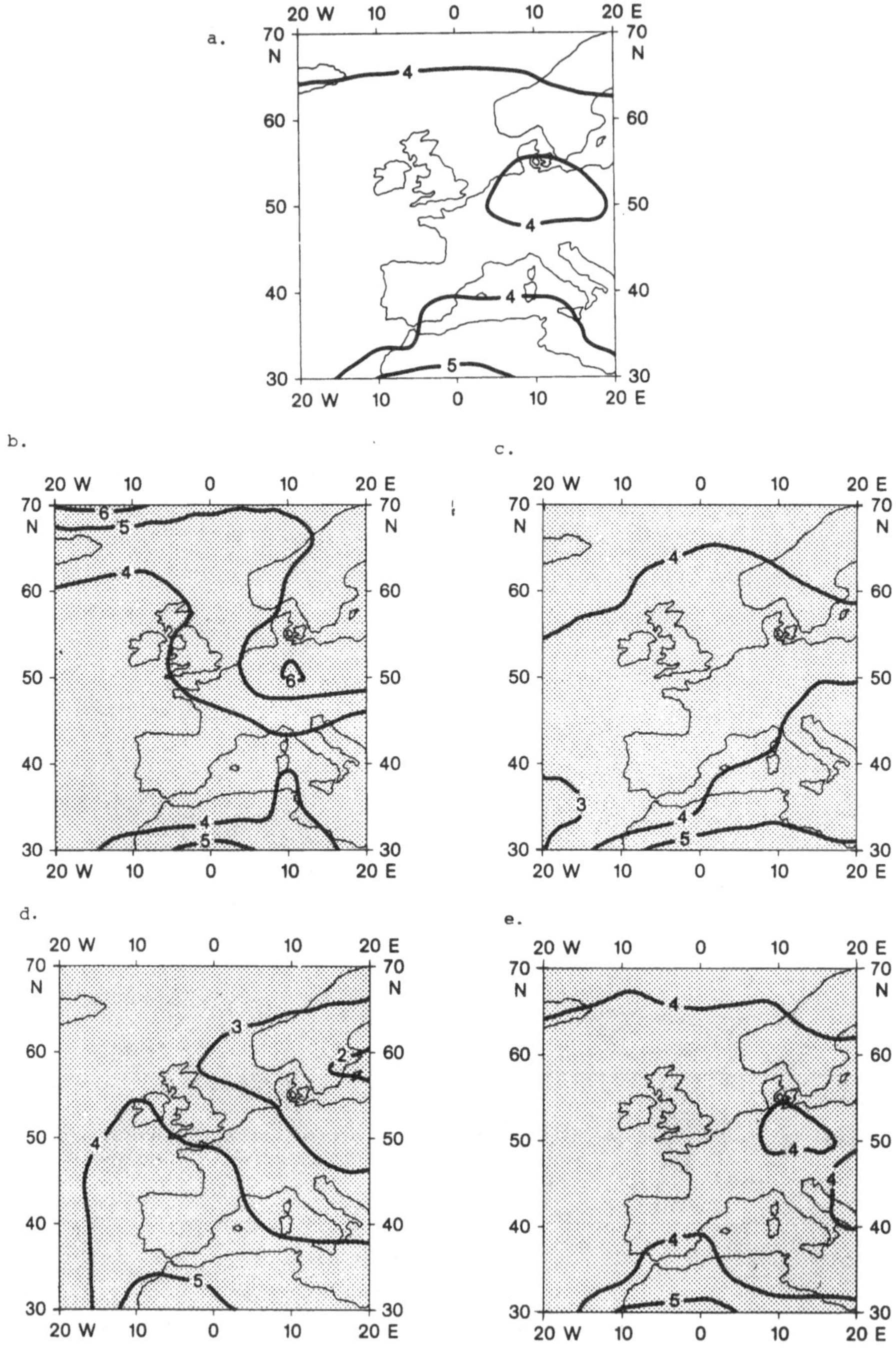

Fig. 5 Regional distribution of temperature change (°C) for the GISS $2xCO_2$ experiment.
a: annual mean; b: winter; c: spring; d: summer; e: autumn

study area which ranges from 3 to 6°C. Temperatures in excess of 5°C are found over northern and north-eastern Europe in winter, and over North Africa in all other instances. The large temperature increase over northern Europe may be related to the reduction in the snow and ice cover from autumn to spring and to an enhanced zonal air flow. The strong temperature rise over North Africa may be due to the poleward shift of the subtropical high pressure belt, which reduces the precipitation rate through enhanced subsidence. Thus both soil moisture and evaporation decrease, making more energy available for heating the surface air.

Fig. 6 shows the simulation results for the precipitation rate changes for a CO_2 doubling. Decreases are represented by dashed isohyets. A precipitation reduction is indicated in south-western parts and a precipitation increase in northern regions in all but the autumn seasons. The change patterns are in fair agreement with the circulation changes, i.e. with increased advection of moist air into northern Europe and an enhanced desiccation through subsidence in south-western parts related to the northward shift of the subtropical high. Of interest are the precipitation rate decreases in spring and in summer over the Mediterranean and south-western Europe, where it is already relatively dry, and the strong precipitation increases over north-eastern Europe, where it is already wet enough. If such changes should come about, they would not be without significance to agriculture. Except for the annual means, statistical tests were done for all the other seasons. The stipled areas on Figs. 5 and 6 indicate significant changes at the 5 % level. In contrast to temperature, precipitation rate changes with their greater inherent variability show smaller areas of statistically significant changes. A more complete discussion of these CO_2 simulation studies is given by Meinl and Bach et al.[6)]

4. Concluding Remarks

To analyse changes in the climate system there is currently available a hierarchy of models ranging from the one- and two-dimensional RCMs and EBMs to the three-dimensional GCMs. Presently transient studies can only be made with the simpler RCMs and EBMs which, because of their greater transparency, will always be valuable tools in sensitivity tests. The reproduction of the complex climate system requires, however, the use of GCMs, which, although our most sophisticated tools, can, at present, only simulate time-independent equilibrium climates. From the standpoint of a decision-maker it is more important to get information on climatic change patterns as they are evolving in time and not at some point in the future as obtained by the present method of arbitrary forcing, such as through a CO_2-doubling or a CO_2-quadrupling. Therefore, the current equilibrium response experiments must be supplemented by addressing the transient nature of the increasing concentrations of these substances in the atmosphere. This work is already under way. To be meaningful, such transient model runs would have to be conducted allowing not only for a variety of energy scenarios, and the major greenhouse gases and aerosols, but also for the intricately changing interactions between the atmosphere, the ocean, and the biosphere, a major undertaking which so far has been constrained by computer capacity. Moreover, a concerted effort at a number of research centers is being made to reduce further the inherent uncertainties and to develop more realistic climatic change scenarios.

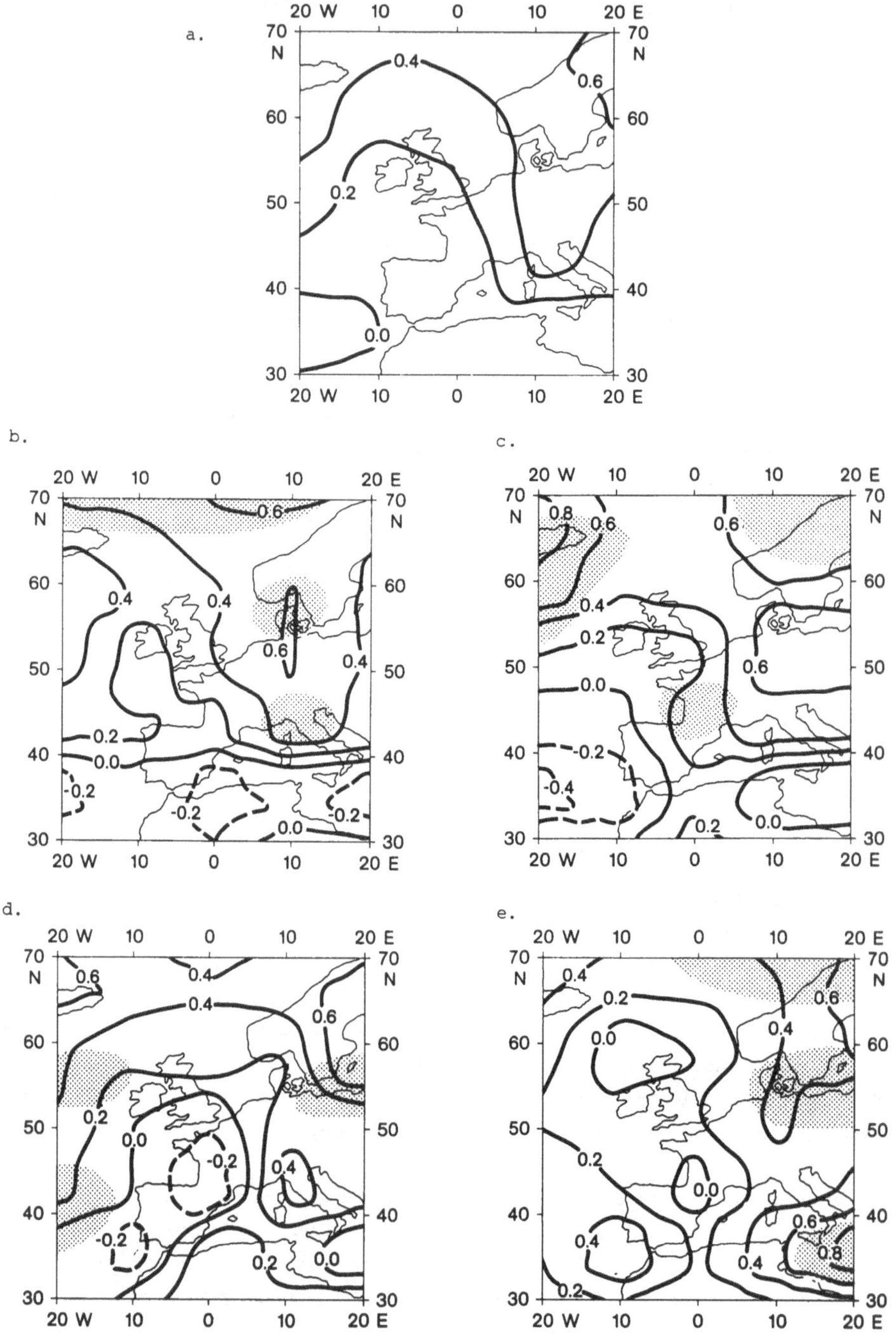

Fig. 6 Regional distribution of the precipitation rate change (mm/day) for the GISS $2xCO_2$ experiment.
a: annual mean; b: winter; c: spring; d: summer; e: autumn

Acknowledgments

I wish to thank Lawrence Gates, Oregon State University, USA, for making available the observed temperature and precipitation data, and James Hansen, Goddard Institute for Space Studies, New York, USA, for supplying the data generated with the GISS-model. The financial support of the Commission of European Communities and of the Deutsche Forschungs- und Versuchsanstalt für Luft- und Raumfahrt is gratefully acknowledged. Some of this material is part of a major study on the "Socio-economic impacts of climatic changes due to a doubling of the atmospheric CO_2 content", carried out in cooperation with Dornier System, Friedrichshafen.

5. Literature

(1) Bach, W.: Our threatened climate. Ways of averting the CO_2 problem through rational energy use. Reidel Publ. Co., Dordrecht 1984.
(2) Colombo, U. and O. Bernardini: A low energy growth 2030 scenario and the perspectives for Western Europe. Rpt. Comm. of the Europ. Communities, Brussels 1979.
(3) Häfele, W. et al.: Energy in a finite world. Ballinger, Cambridge, USA 1981.
(4) Lovins, A.B. et al.: Least-cost energy: Solving the CO_2 problem. Brick House, Andover, USA 1981.
(5) Marland, G. and R.M. Rotty: Carbon dioxide emissions from fossil fuels: A procedure for estimation and results for 1950-1981. Tellus 36 (B) (1984), 232-261.
(6) Meinl, H. and W. Bach et al.: Socio-economic impacts of climatic changes due to a doubling of atmospheric CO_2 content. Res. Report to CEC/DFVLR, Dornier System, Friedrichshafen 1984.
(7) Rogner, J.-H.: IIASA '83 scenario of energy development, summary. IIASA, Laxenburg (o.J.).
(8) Schlesinger, M.E.: Simulating CO_2-induced climatic change with mathematical climate models: Capabilities, limitations and prospects. US DOE 021 (1983) III.3-III.139.

THE ANALYSIS OF LARGE-SCALE SOCIAL SYSTEMS - SOME PROBLEMS AND SOME PROPOSALS

Norbert Müller *

Introduction

In recent years, anthropological approaches are of growing importance in the field of social sciences (e.g., see [2]). In these approaches the description of the functioning of social units is of special interest. This paper focusses on large scale social units; let us speak of a large scale social unit (LSU), if it contains not less than 10^4 members. For instance, a community of 43.000 inhabitants as that one of the project HSDMEL [3] is a LSU. Three features of LSUs render their analysis difficult, complexity, structural change and self-referentiality. In the following, it shall be shown that these features interact resulting in a high probability of structural ruptures. From this, some serious model-theoretical and epistemological problems arise.

Model Construction as usual

In traditional dynamic system modelling, a dynamic variable $y(t)$ is regarded as a stochastic process, for instance given by

$$y(t) = \mu(t) + \varepsilon(t) \qquad (1),$$

with $\mu(t)$ trend function, and $\varepsilon(t)$ random term.
Often $\varepsilon(t) \sim N(o,\sigma^2)$ is assumed, where $\gamma(\tau) = COV[\varepsilon(t), \varepsilon(t+\tau)]$ exists, i.e. γ is independent of t. A process having these properties is called <u>weak stationary</u> [1, chap. 7.2]. Model-theoretically, $Y(t)$ is a set of possible realizations $\{y_\omega(t) / \omega \in \Omega, t \in T\}$, where Ω contains the possible processes, and T is the time-index set.
For estimation and testing of $\mu(t)$ and $y(t)$, traditionally the associated time series $\{y_t / t = o,\dots,n\}$ is taken as data. Being only <u>one</u> realization of the process, using $\{y_t\}$ in this way leads to the following representativity-problems;

- is y_t representative for $\{y_\omega(t)\}_{\omega\in\Omega}$ at a special $t \in T$ (cross-sectional)?
- is $\{y_t\}$ representative for $\{y_\omega(t)\}_{t\in T}$ for the process $\omega\in\Omega$ (longitudinal)?

Traditionally, these two problems are solved by assuming $y(t)$ as weak stationary. Then, for the correlation function $\rho(\tau) = \gamma(\tau)/\gamma(0)$ it holds $\lim_{\tau\to\infty} \rho(\tau) = o$ and a K can be found so that

$$\exists_{\varepsilon>0} \forall_{\tau\geq K} |\rho(\tau)| < \varepsilon \qquad (2)$$

*Albrechtstraße 28, D-4500 Osnabrück,
Systems Research Working Group, University of Osnabrück, FRG

holds. Let $n/K \in \mathbb{N}$; for practical purposes n/K intervals are specified, where in $t^1, \ldots, t^{(n/K)}$, y_t can be regarded as realizations of the stochastic process.

The problem of strucural ruptures

The crucial assumption behind this procedure is that y (t) is unbroken, i.e., the parameters of μ (t) are constant, at least in time-intervals of length K. There is another kind of structural rupture, namely that of the state space itself. Let k be the number of system variables, called the order of the state space. With K from (2), let a system model be defined as free of structural ruptures, if for all $t \in T_K = [t_a, t_a + K]$, where t_a is some starting-point of observation, k and all parameters are constant. Let this be called K-stationarity of the system. Note that the constancy of k is only a necessary condition.

LSU system model and structural ruptures

Generally, LSU system models are complex, i.e. k and the number of parameters are large. Moreover, a LSU is self-referential, i.e. its behavior is to a large extent dependent on the behavior of decision centres (DCs). At least some of the outputs are DC-outputs, and some of the parameters are decision parameters of DCs. Neglecting the irrelevant case, that K-nonstationarity of different DCs compensate each other, K-stationarity implies K-stationarity of DC-models. Trivially, with growing complexity and growing number of DCs, the probability of K-stationarity will rapidly fall. In the region of the HSDMEL project every year at least one serious structural rupture could be observed.

Consequences and proposals

I am aware of four strategies to cope with the before mentioned problems. The first one is that of aggregation by the specification of macro-models hoping that DCs can be ignored. However, DCs are of growing importance in all civilizations. Thus, this strategy often will lead to artificial results. The second one is that of limitation of scope by focussing on mass-phenomena only as in voting or consumer behavior analysis. In its ideal form, due to the existence of pure mass-phenomena, DCs can be neglected effectively. However, the absence of DCs is not clear a priori, this has to be tested in advance, which often is omitted. Moreover, by reduction of scope, many phenomena being perhaps scientifically meaningful, are excluded systematically. The third strategy seems to me one of the leading paradigms of 'modern' science. It is the strategy of constructivism intending to design selected structures of social units in accordance with functioning requirements. Typically, these requirements are formulated by DCs. Thus, research of this type is, explicit or implicit, by order of DCs; it is technology oriented, not description oriented. There are two problematic assumptions behind this strategy. The first one is a rationality assumption, assuming social units to function rationally at least with respect to those aspects selected as relevant by DCs. The second one assumes that effects of self-referentiality can be ignored taking for granted that the functioning of social units can be controlled neglecting the functioning of DCs. This second assumption leads to special problems in the case of LSUs, which are generally characterized by a high DC-influence.

The fourth strategy is that of direct tackling the problem in focussing on the basic research question, whether LSUs are describable at all. From the fact that, in general, LSU-system models are not K-stationary by virtue of DC-influences, the conclusion is drawn that DCs have to be modelled explicitly in descriptive models leading to a decomposition of variables in a DC-fraction and a mass-phenomena-fraction as follows,

$$y(t) = y(t)_{DC} + y(t)_{Mass} \quad (3)$$

There are many other aspects of this anthropologically oriented description methodology called real-structure-modelling (RSM), which cannot be elaborated at this place, see [4],[5],[6],[7],[8].

References

[1] ANDERSON, T. W. (1971): The Statistical Analysis of Time Series, New York: Wiley

[2] BOULDING, K. E. (1978): Ecodynamics - A New Theory of Societal Evolution, Beverly Hills: Sage

[3] MÜLLER/Schön/Thober: Projekt HSDMEL - Abschlußbericht Osnabrück, Frühjahr 1982

[4] MÜLLER,N.: Hierarchic-Sequential Decomposition - A Comprehensive Approach for Real-Structure Modelling of Social Systems, in: Cellier, F.E. (ed.): Progress in Modelling and Simulation, New York 1982

[5] MÜLLER,N.: Real Structure Modelling: Towards a Valid Approach for Social Systems Analysis, in: Horst Wedde (ed.): Adequate Modelling of Systems, Berlin: Springer 1983

[6] MÜLLER,N.: Modelling Standard Actions of Individuals and Institutions for Controlling NO_3-concentration in Drinking Water in a Region of Intensive Agriculture (South Oldenburg FRG) - Evidence from a Pilotmodel, in: Systems Analysis, Modelling, Simulation 2 (1985) in press

[7] MÜLLER, N. (in press): Real-structure-modelling: A Methodology for the Description of Large Scale Social Units, in: Social Science Info., 1985

[8] THOBER, B. (1985): Beschreibung großer sozialer Einheiten durch modularisierte Modellsysteme, Diss., Universität Osnabrück

STOCHASTIC MODELS OF INNOVATION PROCESSES

W. Ebeling [1)]

1. Model assumptions

Technological change is considered here in accordance with an earlier paper (Jimenez Montano and Ebeling, 1980) as a stationary Markov process which is formulated in an occupation number space.
Following Glushkov and Pshenichnyi (1977) which introduced a deterministic mathematical model for evolving economies we assume that the possible and actual technologies form a countable set $i = 1, 2, 3, \ldots$ and further that a given technology i at time t is characterized by its capacity $N_i(t)$, i.e. the number of units of the given technology working at the given time. We may assume e.g. that $N_i(t)$ is the number of autonomous production units or plants which use the technology i at the time t. Since we include in the set of technologies not only the actual technologies but all thinkable technologies, which will be invented only in future, most of the occupation numbers $N_i(t)$ will be zero. An industry state is defined by a set of integers

$$N_1(t),\ N_2(t),\ \ldots\ldots,\ N_i(t),\ \ldots\ldots$$

Let us denote the probability distribution for a given industry state at time t by

$$P(N_1, N_2, \ldots, N_i, \ldots; t)$$

Following the Markov assumption the time evolution is described by a master equation

$$\partial_t P = WP \tag{1}$$

where the evolution operator W is determined by the transition probabilities. We assume 3 basically different elementary processes (Ebeling and Feistel, 1982):

1) Sektion Physik der Humboldt-Universität
DDR-1040 Berlin, Invalidenstr. 42

1) Selfreproduction of a technology, i.e. the number of plants using a given technology is increased: $N_i \rightarrow N_i + 1$. The transition probability is assumed to be

$$W(\ldots N_i + 1 \ldots N_j \ldots N_k \ldots | \ldots N_i \ldots N_k \ldots)$$

$$= A_i^* N_i + A_{ij}^* N_i N_j + B_{ij}^* N_i N_j + C_{ijk}^* N_i N_j N_k \qquad (2)$$

2) Failure of a technology, i.e. the number of plants using a given technology is decreased $N_i \longrightarrow N_i - 1$ with the assumed transition probability

$$W(\ldots N_i - 1 \ldots N_j \ldots N_k \ldots | \ldots N_i \ldots N_j \ldots N_k \ldots)$$

$$= A_i^+ N_i + B_{ij}^+ N_i N_j + B_{ii}^+ N_i (N_i - 1) + B_{ik}^+ N_i N_k \qquad (3)$$

3) Change of a technology, i.e. a plant changes from the technology j to the technology i with the assumed probability

$$W(\ldots N_i + 1 \ldots N_j - 1 \ldots N_k \ldots | \ldots N_i \ldots N_j \ldots N_k \ldots)$$

$$= A_{ij} N_j + B_{ij} N_i N_j + C_{ijk} N_i N_j N_k \qquad (4)$$

The model described above is sufficiently general to describe rather general technological evolution processes up to 3rd order nonlinear couplings; at least from the formal point of view. By appropriate choice of the coefficients one may describe e.g. technological changes due to research and developement, imitation of more successful technologies, confirmation or failure of technologies, autocatalytic enhancement and selfinhibition etc. The model is formally in close analogy to recent stochastic models of the evolution of biomolecules. The most important difference is that in technological change imitation plays a fundamental role (formally this is described by the coefficients B_{ij}). As well known, many decisions to introduce a new technology into a plant are based mainly on the knwoledge that a large number of other plants use already that technology. In other words, the transition probability is proportional to the number of plants using already the target technology.

2. Discussion of innovations

An innovation is a new technology denoted by the number n which is initially present in one exemplar only $N_n(0) = 1$. Let us study the competition with with the main (master) technology $N_m(0) = N$. There are two essentially different situations which may be studied by solutions of eqs. (1 - 4):

(i) Linear growth $A_n^* \neq 0$, $A_m^* \neq 0$, all other coefficients are zero. Then the probability of survival of a new technology is (Allen and Ebeling, 1983)

$$\sigma = \delta \left[(1 + \delta) - (1 + \delta)^{1-N} \right]^{-1} \quad (5)$$

$$\delta = (A_n^* - A_m^*)/A_m^* \quad (6)$$

(ii) Hyperbolic growth $B_{nn}^* \neq 0$, $B_{mm}^* \neq 0$, all other coefficients are zero. Then the probability of survival is (Ebeling et al., 1981)

$$\sigma = (2 + \delta)^{1-N}$$

$$\delta = (B_{nn}^* - B_{mm}^*)/B_{mm}^* \quad (7)$$

As one sees, both cases behave in a fundamentally different way. In the case of linear growth, a new technology with certain advantage $\delta > 0$ has always a good chance to survive and to win the competition with the old master technology; the number of plants N plays a minor role. However in the case of hyperbolic growth, which is typical for processes with selfenhancement, the chances of the new technology depends not only on the advantage δ but also strongly on the number of plants using the old technology N. If N is very large the new technology (even if it is much better) has practically no chance to overcome the high initial difficulties except in conditions, where competition is excluded (at least for certain time e.g. by financial help from governments). The examples present above may show that the stochastic model given by eqs. (1 - 4) is a useful tool for the study of stochastic influences on the behaviour of new technologies. Except a few limiting cases most situations may be studied only by computer simulations.

References

1 Allen P., W. Ebeling: Evolution and the stochastic description of simple ecosystems. Biosystems 16 (1983) 113.

2 Ebeling, W., R. Feistel: Physik der Selbstorganisation und Evolution. Akademie-Verlag Berlin 1982.

3 Ebeling, W., I. Sonntag, L. Schimansky-Geier: On the evolution of biological macromolecules. Catalytic networks. Studia biophysica 84 (1981) 87

4 Jimenez Montano, M.A., W. Ebeling: A stochastic evolutionary model of technological change. Collective Phenomena 3 (1980) 107

BASIC PHYSICAL AND CYBERNETIC PRINCIPLES CONTRIBUTING TO SYSTEMS ANALYSIS IN HYDROLOGY

P. Mauersberger [1])

Instead of analysing water systems exclusively or predominantly on the basis of observational data, it is recommended to deduce the correct type of model equations from basic physical, chemical and biological constraints, using in situ observations and measurements in laboratories mainly for the determination of the values of macroscopic coefficients. The basis principles are derived from generalized thermodynamics of irreversible processes. The entropy principle controls the further development of the water system. It supports the unified treatment of physical, chemical and biological processes in water bodies and allows for the combination of the deterministic, the stochastic and the cybernetic approaches to water quality modelling. The synthesis of these methods is required by systems analysis in hydrology.

Keywords: ADAPTATION, DEDUCTIVE AND INDUCTIVE APPROACH, ENTROPY, OPEN SYSTEM? OPTIMIZATION, RATES OF BIOLOGICAL PROCESSES, SYNTHESIS OF DETERMINISTIC, STOCHASTIC AND CYBERNETIC APPROACHES, THERMODYNAMICS.

INTRODUCTION

It is an essential part of water resources management to represent by mathematical models the response of the aquatic environment to actions of man, since water systems are man-made systems or natural systems deeply influenced by anthropogeneous impact. Involving parts of the biosphere, at least microorganisms, they are capable of adaptation and self-organization processes. Therefore, complex internal control mechanisms and a high degree of structural variability charaterize not only aquatic ecosystems, but also man-made water systems. They belong to the so-called ill-defined systems.

The formulation of a set of equations suited for the mathematical description of the system's behaviour cannot be based only on deductive reasoning and theoretical arguments, the more since a general theory of aquatic ecosystems, based on thermodynamic principles (cf. [5], [6], [8]) and compatible with engineering and economic approaches to water systems modelling and management, is not yet completely developed. Computerized data bases and highly sophisticated interactive information management schemes are indispensible.

On the other hand, analysis and simulation of water systems also cannot be limited to the handling of experimental and field observations. Fur-

[1]) Academy of Sciences of the G.D.R., Institute for Geography and Geoecology, Department of Hydrology. DDR-1162 Berlin, Müggelseedamm 260

ther success in model identification, parameter estimation ana model validation demands for a still better understanding of behaviour and strategies of water systems and of control mechanisms. It is only when we understand causes do we have scientific knowledge. This better understanding cannot be infered only from field observations, since all the measurements in water systems inform us about incomplete sets of variables and about selected processes occuring within a limited time intervall during which the system may not show the full manifold of possible phenomena. Incomplete informationsabout the present structure, state and development of water systems are insufficient for predicting the properties of these systems under deeply changed conditions, the more since self-organization may change the situation drastically.

To a certain extend, knowledge about water systems and, therefore, the basis of decision making in water management will remain incomplete for ever. But in order to minimize the risk, the inductive reasoning approach prefering measurement knowledge must be combined with the deductive reasoning approach using 'a priori knowledge' about possible structures and functioning of water systems, i.e. using basic principles aggregating a great amount of experience.

The contribution of 'a priori knowledge' to analysis and simulation of water systems presents a special challenge to the systems analyst, because it demands for many disciplines. It depends upon hydrodynamics, meteorology, chemistry, biochemistry and hydrobiology. It required knowledge of influences of landuse on water resources, especially about effects resulting from urbanization, agriculture and industry in the catchment area. It calls upon hydraulic engineering and economics. In the following, we restrict ourselves to an outline of basic thermodynamic and cybernetic principles contributing to the analysis of water resources systems. Other aspects of analysis and modelling of aquatic ecosystems and water systems will be discussed by other papers during the Section on these systems of this Symposium.

THE ROLE OF ENTROPY IN WATER SYSTEMS ANALYSIS

Water systems are open systems exchanging energy and matter with their surroundings. In many cases the flows of energy and matter through the system are large compared with the accumulation of matter and/or energy inside the system. These irreversible flows producing entropy give rise to deviations from thermostatic equilibrium. The water system can maintain its state only by exporting the produced entropy. Thus, the structure, state and further development of water systems are regulated by the mutual effects of entropy-producing and entropy-reducing processes inside the system and across its boundaries. Therefore, modelling water sytems it is inevitably necessary to take into account the entropy balance equation and the 2nd law of thermodynamics besides the balances of mass, momentum and energy. For the purpose, thermodynamics of non-linear irreversible processes has been generalized introducing biological variables and processes (cf. [6] ,[8]). Since entropy and entropy production depend upon all important hydrophysical, hydrochemical and hydrobiological variables and processes, the entropy principle reduces a large variety of phenomena to e few equations which can serve as a starting point for further investigations, e.g. for conclusions concerning the existence of stable stationary states or bifurcations, but also the dependence of the rates of biological processes upon physical, chemical and biological state variables.

STOCHASIC PROCESSES

If there exists more than one solution of the nonlinear differential equations used for modelling water systems, the macroscopic equations do not justify preference to one of these solutions. Therefore, we need a finer description involving fluctuations (cf. [12],[4]). The very existence of many degrees of freedom in water systems automatically implies the appearance of fluctuations generated inside the system besides those introduced from outside. The expectation values of the stochastic treatment of system dynamics correspond to the macroscopic description. But the stochastic theory gives additional information especially about variances and other statistical characteristics reflecting the effects fluctuations in nonlinear systems. When the environmental conditions or some internal variables of the water system have undergone such variations that (near a bifurcation) the stability of the state has been lost, the statistical fluctuations can be amplified and finally drive the average values to a new macroscopic state. On account of the significance of fluctuations for the evolution of water systems, theoretical probabilistic formulations must be incorporated into the theory of these systems. Methods of synergetics are also very useful [2] .

There is still another reason for the combination of deterministic and stochastic methods in modelling and management of water systems. For very heterogeneous multiphase systems like aquifers macroscopic balance equations for mass, momenta, energy and entropy are derived from microscopic balances by averaging procedures.

RATES OF BIOLOGICAL PROCESSES

Biological processes play a fundamental role in water systems influencing deeply water quality. In water quality modelling, more or less different formulae are used to describe in the macroscopic level the dependence of one and the same biological process upon state variables (e.g. temperature, light intensity, pH, external nutrient concentrations or internal storage of energy and nutients etc.). The situation is discussed for grazing for instance by Hallan [3] . Examined with regard to their application for the simulation of observed data, these formulae are to a certain extend equally useful. But in view of causal analysis of water resources systems a variety of formulae for one and the same process infered from incomplete field measurements is of very limited value. It is, therefore, inevitably necessary to determine at least the types of mathematical functions which describe the transformation of matter and energy in aquatic ecosystems from fundamentals of macroscopic physics, chemistry and biology [6] .

Biological phenomena, though not completely determined by the fundamental laws of physics and chemistry, are not in contradiction to them. Provided nonlinear irreversible processes are included, physics and chemistry succeed in making out the scope within which biological processes may occur. Therefore, thermodynamics of irreversible physical and chemical processes is generalized introducing biological variables (e.g. biomasses of the species) and biological processes (e.g. uptake, primary production, respiration, grazing) into the balances of mass, momentum, energy and entropy. Furthermore, additional equations typical for biological phenomena are necessary to bring the system of ecological model equations up to the full complement.

For instance, from the entropy balance equation we can infer the 'driving forces' of the biological processes as functions of state variables, but not the types of the nonlinear functional relationships between the rate and the driving forces ('affinities') of these processes. These functions follow from an optimization principle (cf. [7],[10]). It postulates that the deviation of the bioprocess from a stable stationary state ("Fließgleichgewicht") tends to a minimum. The deviation is measured by the time integral over the generalized excess entropy production E of the biocoenosis. During the finite intervall of integration (e.g. between two episodes of stochastic evolution) the biocoenosis is controlled by the affinities in such a way that the time integral over E becomes a minimum value subject to the initial biomasses and to the mass balances of the species. The optimization principle is to be understood in the sense of dynamic programming [1] . By this means follow the rates of biological processes as function of driving forces and, finally, as functions of the above mentioned physical, chemical and biological state variables (cf.[8],[10]). Coefficients which are not determined by the phenomenological theory can be eveluated in accordance with field and laboratory data. Arbitrariness in the choise of mathematical functions for modelling biological processes has been overcome.

CYBERNETICS AND THERMODYNAMICS

Aquatic ecosystems are rather complicated structures. The use of the homology principle for modelling these systems leads to complex systems of equations with solution possibilities limited both by computer time and capacity as well as by insufficient knowledge. For management purposes ecological models must be coupled to economical models and it is often sufficient to analyse and to predict 'integral features' of the water system, e.g. the amount of seston or the total biomass of phytoplankton and its main constituents. Especially the cybernetic approach of ecosytem modelling tends to move from very extensive models to simpler ones. The theory of hierarchical control and multi-objective optimization offers a possibility to 'simplify' ecological modelling by introducing optimality principles from which 'integral features' of the ecosystem can be derived. A goal function is assumed which the system is supposed to be seeking. Optimization procedures are used for determining the model parameters or the species most suited to the given conditions (cf. [15], [16], [17]).

By one of the cybernetic optimization principles the biomass of the predominant species in a trophic or functional group (e.g. phytoplankton) is maximized. In this case the dependence of the rates of the different biomass producing and biomass destructing processes upon the external driving variables and upon internal control variables must be given. It is only in a first approximation that these functions may be obtained from field measurements [14] .

Optimization is also a basis of the thermodynamic approach to ecosystem modelling, c.f. the thermodynamic optimality principle mentioned above. The direct goal tackled is different from the cybernetic goal functions, but not in contradiction to them. On the contrary, the thermodynamic optimization procedure offers a supplement to the cybernetic method, deriving theoretically the relations between process rates, external driving forces and internal control variables needed in the cybernetic optimization principle [11].

The combination of the cybernetic and the thermodynamic approaches to the modelling of water systems seems worth while in order to develop further the methodology of these modelling efforts.

CONCLUSION

Basic thermodynamic principles contribute to the analysis and theory of water systems by supporting the unified treatment of physical, chemical and biological processes in water bodies and by allowing for the combination of the deterministic, the stochastic and the cybernetic approaches to water quality modelling and systems analysis in hydrology.

REFERENCES

[1] Bellman, R.: Dynamic Programming. Princetown UP, Princetown 1957.

[2] Haken, H.: Advanced Synergetics. Springer-Verlag, Berlin/Heidelberg/New York/Tokyo 1983.

[3] Hallam, T.G.: Structural sensitivity of grazing formulations in nutrient controlled plankton models. J. Math. Biol. 5 (1978) 269 - 280.

[4] Horsthemke, W., D.K. Kondpudi (Eds.): Fluctuations and Sensitivity in Nonequilibrium Systems. Proceed. Internat. Conf., Austin/Texas, March 12-16, 1984. Springer-Verlag, Berlin/Heidelberg/New York/Tokyo 1984.

[5] Johnson, L.: The thermodynamic origin of ecosystems. Can. J. Fish. Aquat. Sci. 38 (1981) 571 - 590.

[6] Mauersberger, P.: On the theoretical basis of modelling the quality of surface and subsurface waters. IIASA - IAHS - Sympos., Baden/Austria, Sept. 1978. IAHS - AISH - Publ. 125 (1978) 14- 23.

[7] - Zur Bestimmung der nichtlinearen Beziehungen zwischen Raten und Affinitäten bei Produktions- und Abbauprozessen im aquatischen Ökosystem. Acta hydrophys. 27 (1982) 125 - 130.

[8] - General Principles in Deterministic Water Quality Modeling. In: [13] , Chapter 3.

[9] - Thermodynamic theory of the control of processes in aquatic ecosystems by temperature and light intensity. Gerlands Beitr. Geophys. 93 (1984) 314 - 322.

[10] - Optimal control of biological processes in aquatic ecosystems. Gerlands Beitr. Geophys. 94 (1985), 2, in press.

[11] Mauersberger, P., M. Straskrăba: Two approaches to ecosystem modelling: thermodynamic and cybernetic. Internat. Symp. SISY '84, Praha/CSSR, 12-15 Nov.1984.

[12] Nicolis, G., I.Prigogine: Self-Organization in Nonequilibrium Systems. Wiley-Intersc.Publ., New York/London/Sydney/Toronto 1977.

[13] Orlob, G.T. (Ed.): Mathematical Modeling of Water Quality: Streams, Lakes, and Reservoirs. Wiley-Intersc.Publ.,Chichester/New York /Brisbrane/Toronto/Singapore 1983.

[14] Radtke, E., M. Straškraba: Selfoptimization in a phytoplankton model. Ecol. Modelling 9 (1980) 247 - 268.

[15] Silvert, W., T. Platt: Energy flux in the pelagic ecosystem. Limnol. Oceanogr. 23 (1978) 813 - 816.

[16] Straškraba, M.: Natural control mechanisms in models of aquatic ecosystems. Ecol. Modelling 6 (1979) 305 - 321.

[17] Straškraba, M., A. Gnauck: Aquatische Ökosysteme - Modellierung und Simulation. VEB Gustav-Fischer-Verlag, Jena 1983.

MODELLING AND SIMULATION OF AGROECOSYSTEMS - THE WINTER WHEAT AGROECOSYSTEM "AGROSIM-W"

W. Ebert [1)]

1. Peculiarities of agroecosystems (AES)

AESs are managed ecosystems, i. e. they are controlled anthropogenously, they are maintained in men intended artificial structures, differing strongly from the outochtonic (site specific) ecosystems belonging to the given climatic and edaphic conditions; by this, AES are artificial instable (they are harvested) ecosystems with cutted nutrient cycles.

The AES control inputs fertilization and irrigation (controlling assimilation rate, growth, and development of crop), agrotechnique, pesticide and herbicide spraying (controlling the mortality rate of pests and their crop interaction) in pest management force the system into a state, more profitable for man. These control inputs serve to manage the AES in such a way, that selected state variables of the system (e.g. yield) at given instants are driven als well as to goal values (off-line calculated standard yield, which is obtained in the case of expected weather and exact performance of all necessery treatments).

The structure of an AES is given by the

- set of its components (biotic compartments, environmental variables, pools)
- interactions between (i) the components and (ii) components and environmental variables (including driving forces, control variables).

Often modelling of AES at first serve the understanding of the complex dynamic behaviour of the biotic and abiotic components, of matter cycles and energy flows, in order to derive control policies, later. This control task is called yield programming and two demands arise:

(i) for monitoring of the real AES processes including interactions

(ii) for elaboration of /advices or/recommendations rsp. for AES management and treatment based on its present state and forecasted dynamics.

---

1) Inst. Plant. Protection Res., Acad. Agric. Sci. G.D.R. Kleinmachnow

(ii) leads to model aided decision making in yield programming, esp. in pest management, which we are considering in the following.

fig. 1 PEST MANAGEMENT DECISION MAKING AIDED BY UTILIZATION OF SIMULATION MODEL AGROSIM-WEATH

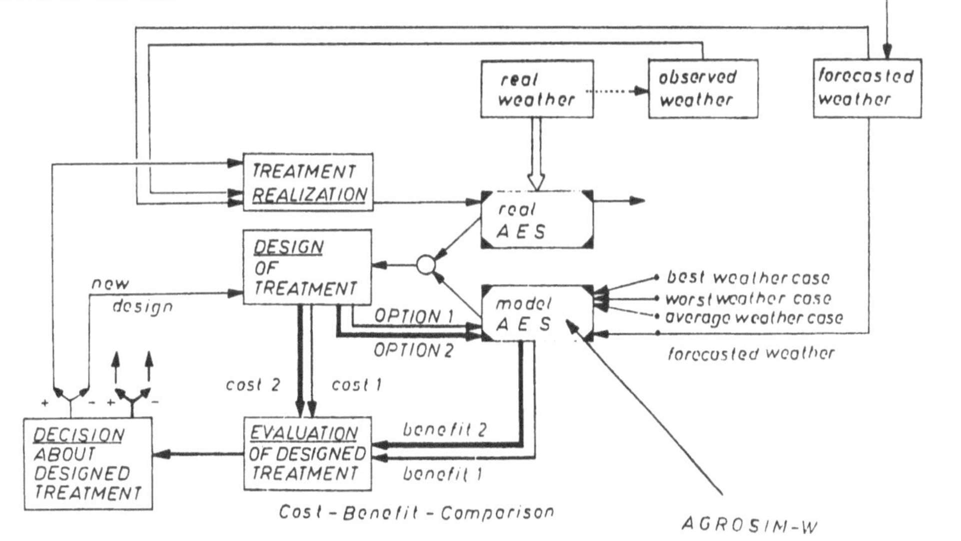

In fig. 1 an example is scatched how to use an AES simulation model for the procedure of selecting treatment variants running the model as long as the most suitable one is found (based on cost benefit analysis) for the real treatment of the field under control.

In fig. 2 the general structure of an AES simulation model is shown (however without considering processes of microorganisms in the soil).

The international trend in pest management is characterized by

solving pest management problems by model based dialogue systems, implemented on minicomputers (e.g. PDP series), utilizing appropriate pest population and crop growth models of sufficient accuracy forecasting.

In order to implement such AES simulation models and to solve problems effectively the following prerequisites are to be ful-filled

(1) existence of an interdisciplinary team (for system analysis, modelling, crop science, phytopathology)

(2) collection and systematization of a priori knowledge corresponding the designed AES frame

(3) flexible and extensive experimental capacities

(4) availability of convenient subject oriented dialogue simulation software

(5) availability of interactive high speed computers with sufficient storage capacity

fig.2

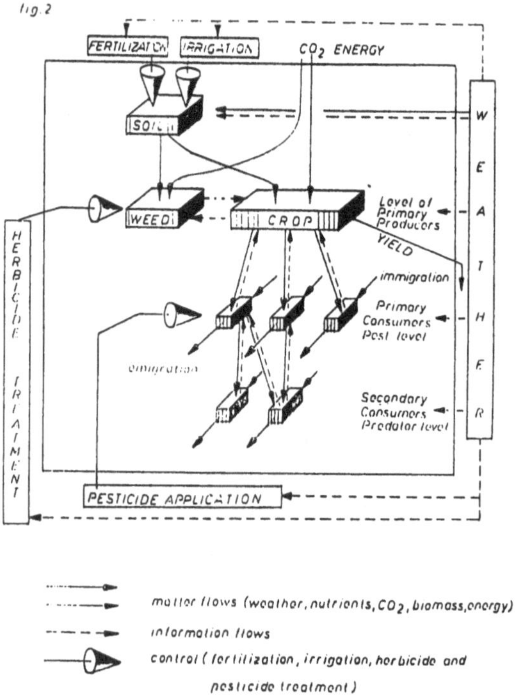

General Structure of an Agroecosystem with regard to Yield Programming

2. Agroecosystem winter wheat simulation model AGROSIM-W

2.1. General characteristics of the component models

AGROSIM-W is a SONCHES implemented software package for simulation of growth and yield dynamics of winter wheat under the influence of a disease (mildew population model), two insect pests (grain aphid and leaf beetle population models), weather, irrigation, and fertilization.

With regard to performance of "first step simulations" (studies of complex dynamics, and elaboration of strategies all model are sufficiently validated, the crop model by gas exchange data for net photosynthesis and dark respiration, sequential harvesting field experiments and the pest and disease models by abundance time series per unit area.

Further work is concentrated to adapt the component models to different soil and climatic conditions (edaphic and climatic site conditions).

AGROSIM-W is the result of an extensive work in experimentation, knowledge classification, system analysis, modelling, computer implementation (with parallel development of the simulation system SONCHES for design, validation and use of ecological systems as a tool aiding the modeller involved in yield programming problems; see contribution Wenzel et al., 1985, in this volume)

AGROSIM-W consists of linked component models for

- crop: winter wheat (Triticum aestivum): TRITSIM
- insect pest: leaf beetle (Oulema spec.): PESTSIM-OUL as a green biomass grazing insect
- insect pest: grain aphid (Macrosiphum avenae): PESTSIM-MAC as a phloem sap sucking insect
- fungal disease: mildew (Erysiphe graminis): PESTSIM-ERY /fig. 3 a/

Each component model is controllable by man's activities and its dynamics depends on weather, too.

Control variables for wheat are irrigation and fertilization (in the AGROSIM-W/85 version nitrogenous fertilization in a restricted manner), for disease and insect pests fungicide and insecticide application, resp. biological control is introduced for grain aphids: predation by lady bird (Coccinellidae).
In addition to pesticide and biological control the grain aphid's population dynamics depends on parasitism and mycosis /fig. 3 b/.

Within the component models the following main processes and state variables are considered

TRITSIM: Simulation model for standing crop with mapped ontogenesis, photogenesis and assimilate storage, respiration, growth of green biomass, grains, roots, yield components (nbr. of ears, nbr. of grains)

PESTSIM-OUL: leaf beetle simulation model with mapped ontogenetic stages: females, eggs, larvae, pupae, mortality (natural and due to control), migration, damage: green biomass grazing. Larvae stages are split into age classes, taking into account infestation waves (see fig. 4).

PESTSIM-MAC: grain aphid population model with mapped ontogenetic stages, immigrantes, larvae, alatae, apterae, nymphae, mortality (natural and due to control), migration parasitism, mycosis, predation by Coccinellidae, damage: phloem sap sucking, crop gas exchange inhibition by honey dew production. Ontogenetic stages are split into age classes.

PESTSIM-ERY: mildew population model with mapped ontogenetic stages infections stage, sporulation stage, mortality (due to control), migration

All processes depend on weather (§ 2.2.). Detailed information about the component models and the mathematical formulations of the interactions are outlined more detailed in the special contributions about component models. (see Tag.-Ber. Akad. Landw.wiss. DDR 1985, in press)

2.2. Interactions in AGROSIM-W

In fig. 3 a AGROSIM is schematically demonstrated with the interactions between the component models dependent on weather and man made control.

fig. 3a

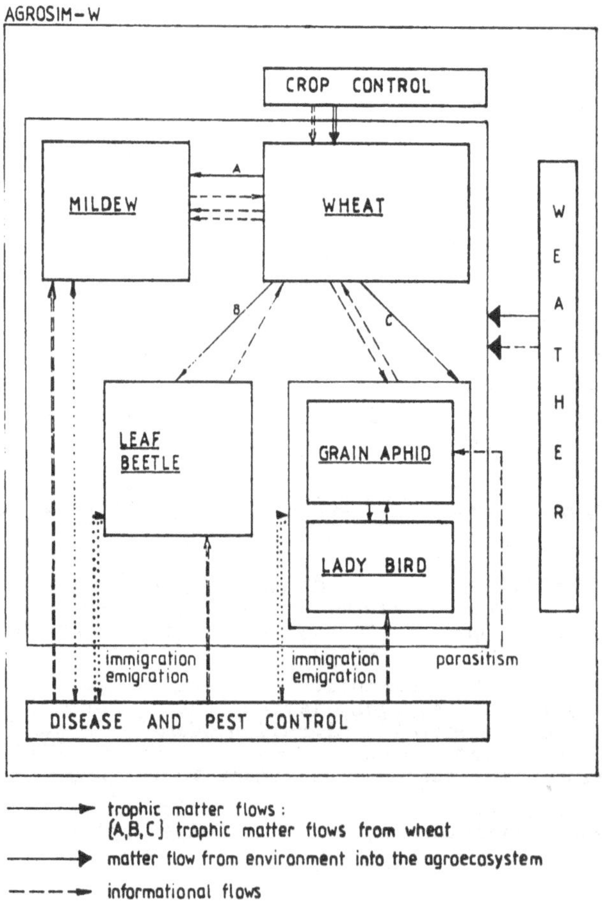

fig. 3b AGROSIM with regard to the interactions between component models and their weather and man made control

AGROSIM-W

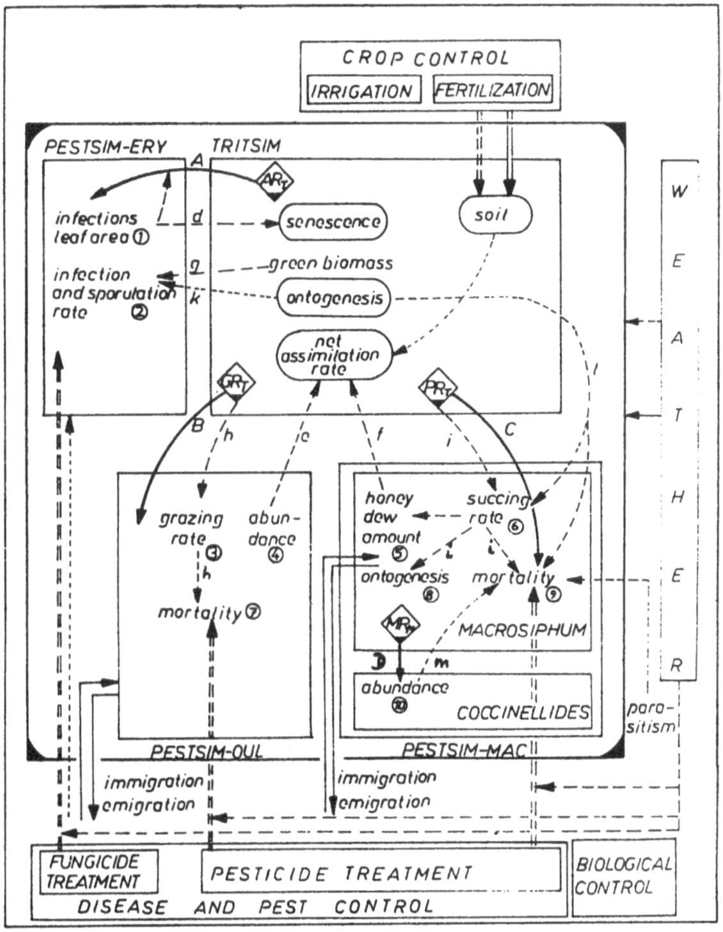

fig.4

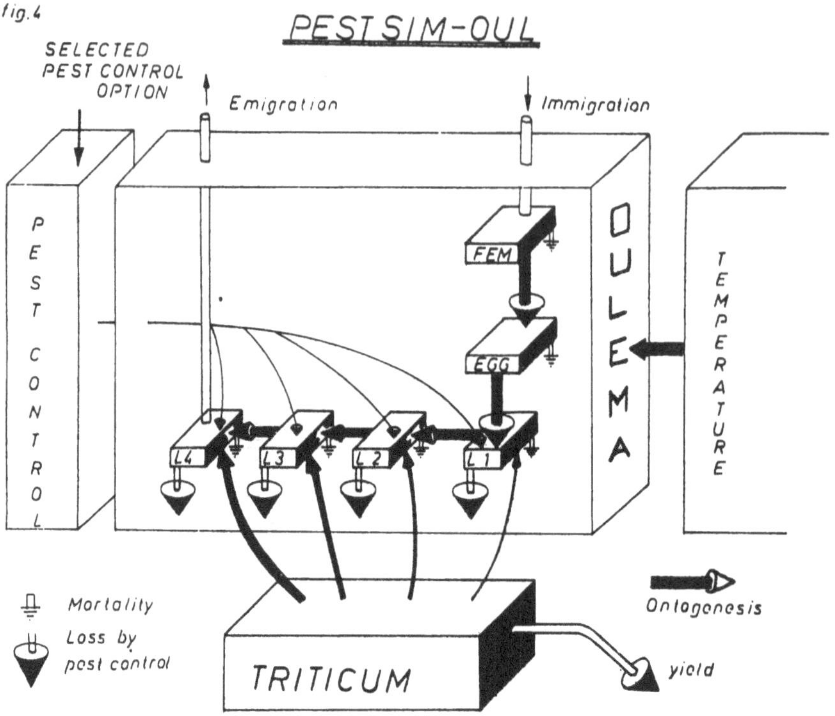

FEM-females, EGG-eggs, L1,L2,L3,L4 - larvae
PESTSIM-OUL is a six compartment submodel

AGROSIM is characterized by three trophic levels:

primary production (crop level): wheat

↓ flow to

primary consumers: mildew, leaf beetles, grain aphids

↓ flow to

secondary consumers (predator): lady bird predation of grain aphids

In fig. 3 b and the following these interactions are described.

- Variables and processes which are involved in the interaction of the component models are

  (1) 3 resources supplied to mildew, leaf beetle, grain aphid in order to describe their effects of some important disease and pest processes

  - AR_T Triticum assimilate resource as supply for assimilate removal by Erysiphe. AR_T is the subset of assimilate pool corresponding to green biomass fraction of the upper 3 leafes.
  - GR_T Triticum green biomass resource supplied for Oulema grazing
  - PR_T Triticum phloem sap resource supplied for Macrosiphum sucking as a competition process parallel to grain growth

  MR_T Macrosiphum abundance supplied for trophic uptake by Coccinelidae.

  (2)-two processes being influenced by diseases and pests: senescence and net assimilation rate;
  -one process influencing disease and pest : wheat ontogenesis.

- Pest-to-crop effects are

  - {A, B, C}, D :
  primary crop damages by trophic <u>matter flows</u> from crop to pest (food uptake of pests)

  A: assimilate removal by mildew from the resource AR_T of wheat, controlled by (1)

  B: leaf biomass grazing by leaf beetle for the resource GR_T controlled by (3)

  C: phloem sap uptake of grain aphids from the resource PR_T controlled by (6)

  - D: matter flow from grain aphid resource to lady birds (10)

  - {d, e, f} :
  secondary crop damages by pests and diseases

  d: acceleration of green biomass senescence (mortality) of wheat controlled by (1) e: increase of net assimilation rate by grazing induced crop thinning leading to a brightening of the standing crop depending on leaf beetle abundance (4)

f: decrease of net assimilation rate of wheat depending on honey dew amount (5) on the green biomass (inhibition of CO_2-exchange and transpiration) controlled by the aphid's sucking rate (6)

- Crop-to-pest effects

  - {g, h, i} :

    wheat resource limitations for disease and pest processes:

    g: decrease of the rate of newly infected leaf area (infection rate) and increase of sporulating leaf area (sporulation rate) (2), due to limited green biomass for mildew growth

    h: increase of leaf beetle's mortality (7) , due to insufficient grazing rate caused by limited supply GR_T for the beetle's grazing

    i: increase of grain aphid's mortality (9), and inhibition of their ontogenesis velocity (reduction of ontogenetic advance rates (8) , due to insufficient sucking rate caused by limited supply PR_T for the aphids' sucking.

  - {k, l} :

    effects of wheat developmental stage to disease and pest processes:

    k: effect of the ontogenetic stage (O_T) of wheat to the relative amount of biomass within the layers of the standing crop, needed for calculation of the mildew infection and sporulation rate (2) , which are different within the standing crop layers

    l: effect ot the ontogenetic stage (O_T) of wheat to aphid's sucking rate (6) , and natural mortality (9).

- predator-to-pest effect

  m: increase of mortality of grain aphids (9) by abundance of lady birds (10) .

the input processes (matter flow and informational flows) mapped in AGROSIM are

- weather: uncontrollable input for each component model
- crop control: irrigation, fertilization - input for the submodel SOIL of the wheat model, controlling mainly netassimilation rate
- disease and pest control: spraying fungizide and insectizide repressing the infectional sporulation rate of mildew and increasing the mortality of the susceptible pest stages, resp.
- biological control: man controlled immigration of lady birds

AGROSIM-W is separable with regard to its component models; each component can be used as single simulation model, and any single pest model can be linked to the crop model.
AGROSIM-W is ready for sale from 1985. It is implementable on PDP like minicomputers (FORTRAN 77).

2.4. Some simulation results in complex disease and pest attacks

The typical damage pattern of the considered pests and disease is summerized in fig. 5, restricted to the grain as the most interesting variable for the practical agriculture. For a normal infestation of mildew and leaf beetle the model shows only changes in the saturation level of the grain filling, but for an aphid infestation the total grain filling period differs from the uneffected crop. This pattern can be easily understood considering the mapped matter flows. The grain filling per kernel can be mapped as a nearly autonomous process, only limited by the available assimilates or translocations from other organs. Therefore all the interactions of pests influencing assimilate or assimilate production lead to a shortening of the grain filling period because of earlier exhausted assimilates, seen as lower yield level and accelerated senescense. Only in the case of strong infestation with an assimilate limitation in earlier stages (shooting) an influence to ear and grain number can be expected leading to an overall lowered grain filling rate per square meter.-The quite different and more direct interaction pattern of aphids on grain is seen in the qualitative jump from the two pest case (mildew, beetle) to the three pest case (mildew, beetle, aphid) underlining the steady removal of grain producing assimilates by "sucking". The interaction pattern for the variables green biomass and assimilates is opposite. For the aphids no primary interaction (matter flow) with these vegetative variables is mapped, therefore the time course of these variables is not influenced by aphid infestation and neglecting photosynthesis inhibition by honeydew. But a primary interaction with these variables is mapped for the mildew and the beetle, this interaction can be phenomenologically characterized by an earlier exhaustion of the green biomass resource or an earlier date of total yellowing compared to the uneffected crop.

For the pest-management advisory system the dependence of yield loss on infestation strength is of central importance. In non-model-aided systems a fictive mostly linear dependence of crop damage on pest population density has to be assumed. Agroecosystem model investigation show a quasi-linear dependence of yield loss on single pests only for weather conditions causing weak infestation. Vegetative biomass loss is always nonlinear with increasing pest population abundance (see fig. 6).

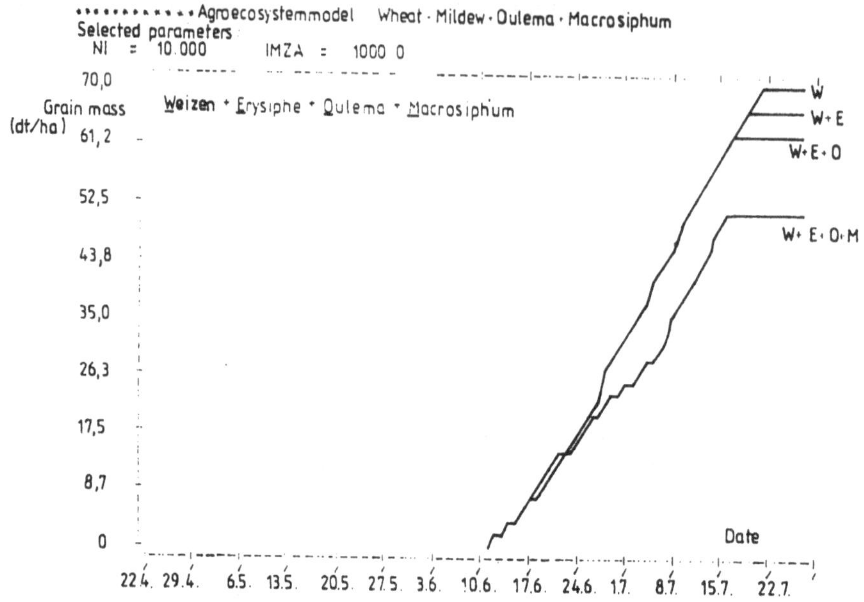

Fig. 5 Influence of complex infestation on grain yield Weather data for 1982

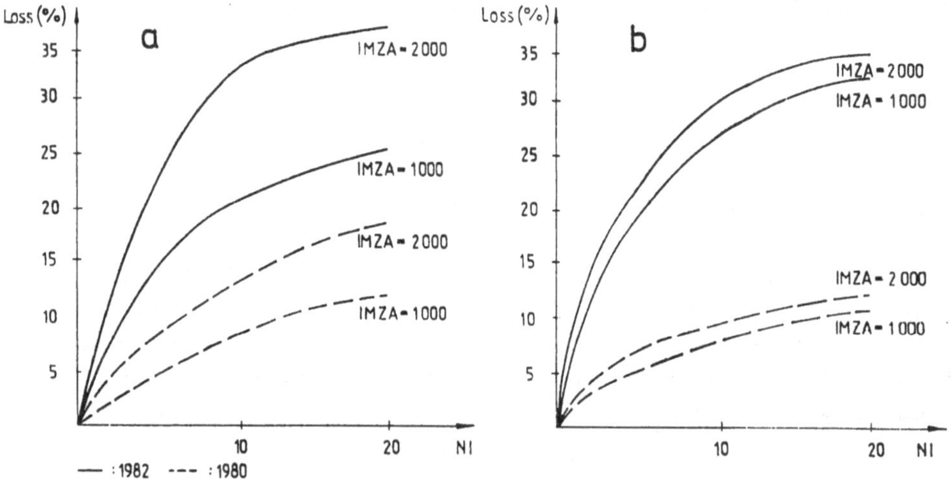

Fig. 6 Relative loss of a) grain and b) biomass in dependence on different immigration abundances of Macrosiphum (IMZA) and Oulema (NI) and different weather data

A much more impressive behaviour of yield loss for two pests (beetle and aphid) is demonstrated in Fig. 7.

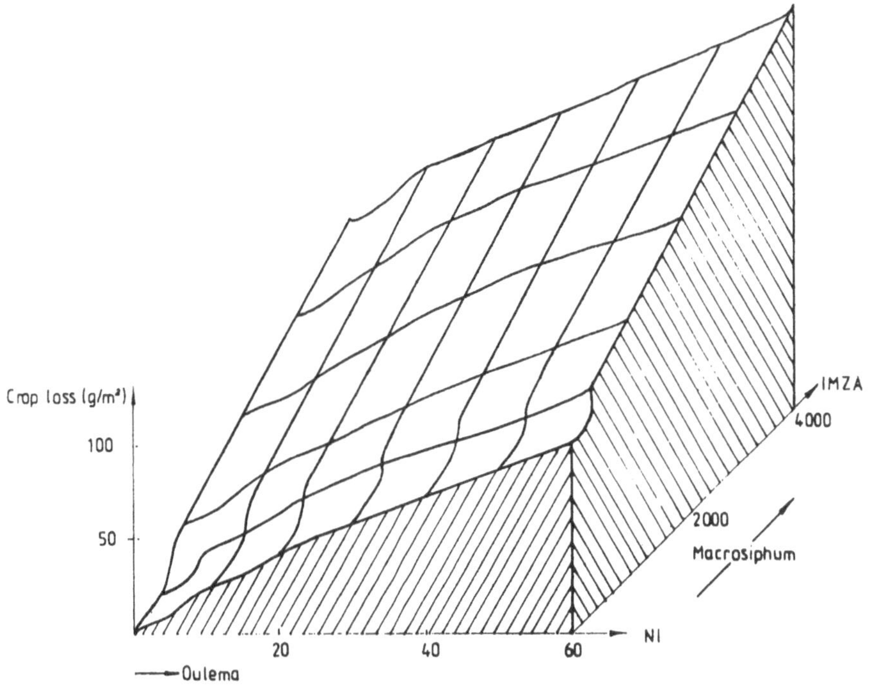

Fig.7: Common infestation of Oulema and Macrosiphum. Grain yield loss in dependence on different immigration abundances of Oulema (NI) and Macrosiphum (IMZA). Weather data for 1980.

By a lot of simulation runs for different pest immigration rates a fine structure with clear nonlinear, threshold like characteristics of pest to yield damage can be pointed out, which can be used as first ansatz for a dynamically based spraying standard. Complex pest damage of species with totally differing interaction pattern, like in this case, can be calculated by a linear superposition of the single pest crop damaging effects.

Model aided derivation of pest-management strategies was also done by means of the scenario-technique of the simulation system SONCHES varying the date of pesticide application. Spraying is mapped as a mean dependent additional mortality. The results for a short-term operating mean (killing efficiency per day: 1. day 50 %, 2. day 30 %, 3. day 10 %) are shown in fig. 8 and fig. 9, presenting the yield loss and the abundance maximum of the pest as function of the spraying date. From the practical point of view the optimal killing of the beetle is much easier than killing the aphids optimally, because the optimal spraying

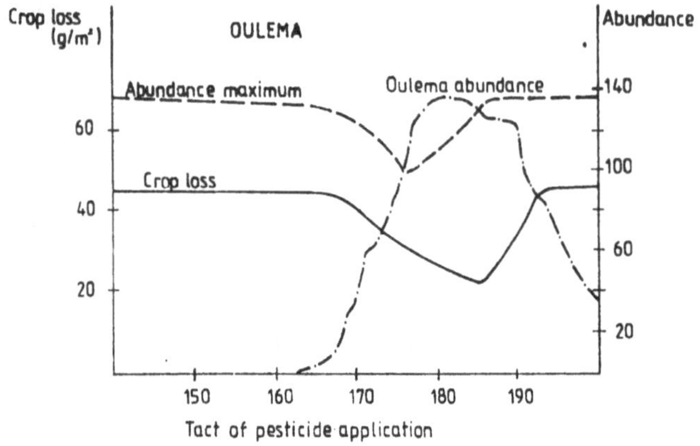

Fig. 8 Grain yield loss by Oulema infestation (full line) in dependence on tact of pesticide application; abundance maximum of Oulema (dashed line) in dependence on tact of pesticide application and abundance of Oulema without any pesticide application (dashed dotted line)

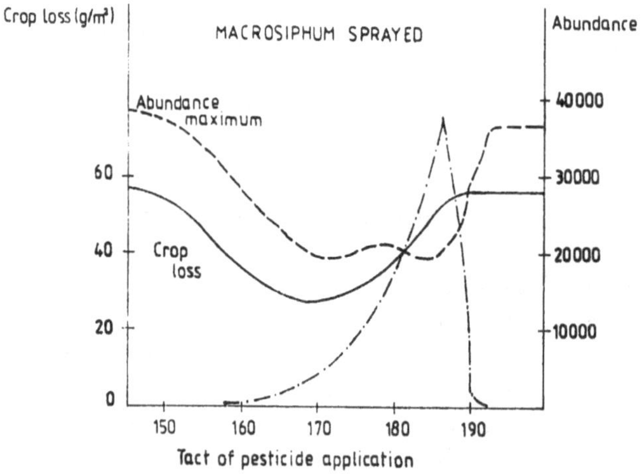

Fig. 9 : Grain yield loss by Macrosiphum infestation (full line) in dependence on tact of pesticide application, abundance maximum of Macrosiphum (dashed line) in dependence on tact of pesticide application and abundance of Macrosiphum without any pesticide application (dashed dotted line)

date for the beetle pesticide (185. day) coincides nearly with the date of maximal beetle abundance (180. day) which is observable. For the aphid pesticide application a model aided advisory system is nessecary because the optimal date lays in an early developmental stage with low population density (168. day).
In dependence on the used pesticide a 50 percent reduction of yield loss is possible in the case of a single pest and a single pesticide application. In the case of two pests but only one cocktail spray a reduction of yield loss by 38 percent is attainable and the optimal date is the 180. day indicating the stronger damaging effect of the beetle under the weather condition of the present example.
This example underlines the necessity of species specific pesticide application. By means of further pesticide application a better pest killing is on principal attainable, but in real situations a cost-benefit analysis limits the number of pesticide applications. The presented model results and model aided statements are experimentally obtainable only with enormous effort, if generally possible. This approach of making general conclusions from computer experiments of special examples underlines the necessity and the advantage of modelling crop-pest interactions within an agroecosystem framework.

References

- Bellmann,K., A. Knijnenburg
Simulationssystem SONCHES zur Modellierung von Ökosystemen-Konzept und Anwendung auf Agroökosysteme. IN: "Analyse, Optimierung und Automatisierung techn. und nichttechn. Prozesse" A1, Heft 2 S. 15-218 (27. Internationales Wissenschaftliches Kolloquium TH Ilmenau DDR, 25. - 29.10.1982
- Bellmann,K., A. Knijnenburg, E. Matthäus, V.Wenzel
Concept and Usage of the interactive simulation System ecosystems SONCHES (with application to agroecosystem winter wheat)
ZKI-Informationen 1/83. Hsg. Zentral Institut für Kybernetik und Informationsprozesse der AdW der DDR, Berlin, S. 61-100
- Knijnenburg, A., E. Matthäus, V. Wenzel
Concept and Usage of the interactive simulation system for ecosystems SONCHES
Ecol. Modelling, 26, 51-76 (1984)
- Matthäus,E., M. Flechsig, K. Bellmann, V. Wenzel
Use of task facilities of the simulator "SONCHES" for agroecosystem model investigations and control In: Proceedings 2. International Symposium on Systems Analysis , Berlin, 1985, Vol 2, pp 124-131
- Tag.Ber., Akademie der Landwissenschaft der DDR, Berlin 1985 (in press)
Computer aided Modelling, Simulation and Pest-Management of Winter-Wheat agroecosystem AGROSIM-W
- Wenzel,V., K. Bellmann, E. Matthäus, M. Flechsig
SONCHES - an interactive simulation system for design, validation and usage of ecosystem models In: Proceedings 2. International Symposium on Systems Analysis, Berlin 1985, Vol. 2, pp.120-123

OPTIMISATION IN SIMULATION PACKAGES AND SIMULATION LANGUAGES FOR CONTINUOUS PROCESSES

F. Breitenecker, Technical University Vienna*

The contribution deals with the optimisation features in continuous simulation languages. First compiler- and interpreter- oriented languages are compared using macro features for implementing optimisation (supermacro- concept). Large CPU- times and difficulties in state- event- handling lead to a comparison of digital and hybrid simulation. The outlined aspects are demonstrated in simulating the optimal control of a crane- crab, the calculation of optimal trajectories for a metro network and the optimal strategy in a hydro- energetic system.

1. Introduction

The aim of simulation of continuous processes is to get insight into the system behaviour (usually in the time domain). Variation of model parameters allows to detect and analyse a lot of characteristical phenomena of the system. Soon the question arises with which values of the parameters the system (the real process modelled by the dynamical system) works best- optimisation turns out to be an essential feature in analysing systems. Simulation of continuous processes is supported by simulation languages (SL's)- general purpose one or specialised one. All these SL's usually "work" only in the time domain so that optimisation is supported insufficiently. On the other hand optimisation packages offer no features for simulation of the system in the time domain. Consequently optimisation should be implemented in SL's so that simulation and optimisation can be performed.

2. Optimisation features in compiler- and interpreter- oriented SL's

Interpreting optimisation as 'experiment' with the model (basic experiment:simulation run, s./3/) then there exist two possibilities for implementing optimisation in a SL; the first one is using macro features within the runtime interpreter (RTI) of the SL, the second one is to program the optimisation procedure within the model description; but also a mixture between these both methods is possible. The second method turns out to be well suited for compiler- oriented SL's (the model description is compiled to an object, which can be executed within the RTI), the first and the mixed one for interpreter-oriented SL's (the model description is interpreted, so that the model can be changed within the RTI). This is now demonstrated with the compiler- oriented simulation language ACSL (/1/) and with the interpreter- oriented SL HYBSYS (/7/).
Due to the CSSL- standard (/3/) the model description in ACSL consists of four sections, namely INITIAL SECTION (calculation of initial values for parameters, etc), DYNAMIC SECTION ('loop' for integration over a communication interval and data recording for documentation) with embedded DERIVATIVE SECTION (description of dynamic equations) and TERMINAL SECTION (calculation of terminal values, for instance cost functions). A possibility for optimisation is now to define an 'iterative' model: jumping from TERMINAL into INITIAL SECTION and changing the parameters corresponding to the values of the cost functions for the previous iteration. The great disadvantage is that the optimisation algorithm has to be programmed by the user in INITIAL or TERMINAL SECTION, because usual optimisation procedures from program libraries use subroutines for evaluating the

* Technical University Vienna, Wiedner Hauptstrasse 6-10, A-1040 Wien

cost functions (which need a simulation run). Consequently this method is useful only up to three parameters.

A much more efficient way (proposed in /2/, /4/) is to implement the optimisation in the main program (FORTRAN), which is generated by the precompiler of ACSL. ACSL generates a FORTRAN main program consisting of three large subroutines, namely ZZDLOC (initialisation of files, etc.), ZZSIML (execution of the compiled model description) and ZZEXEC (runtime- interpreter).

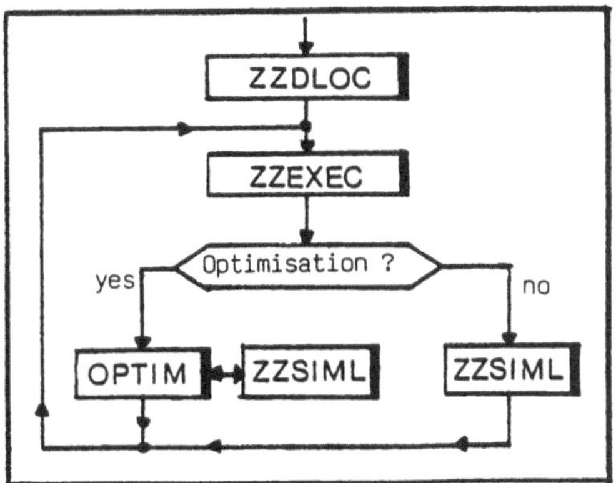

Fig.1: Optimisation in ACSL main program

Now the call of ZZSIML can be replaced by a call of an optimisation program 'OPTIM' of a library, which itself calls several times ZZSIML for cost function evaluation. Data from optimisation to simulation are exchanged via a COMMON block representing the user dictionary of the SL. Figure 1 shows the changed main program, extended by a flag which allows choice between simulation run and optimisation.

HYBSYS- developed at the Hybrid Computation Center of TU Vienna- is an interpreter-oriented SL which allows model decription and problem investigation in an efficient and hardware- independent manner using either analog or digital integration for solving the system governing differential equations (/7/). A hybrid time sharing (multi- user) system supports simulation in HYBSYS, in case of analog integration automatic patching and automatic scaling is available.

HYBSYS offers a very efficient macro feature. Each standard HYBSYS command has an equivalent FORTRAN subroutine call (HYBSYS subroutine), and a FORTRAN subroutine which may contain 'HYBSYS subroutines' or not can be made a new HYBSYS command; data communication is performed via the model data base generated by HYBSYS which is not only a user dictionary but also a data base characterising the structure of the model. This recursive macro feature allows now to implement optimisation in an efficient way: a subroutine (=HYBSYS command) reads initial data from the model data base and starts an optimisation procedure which itself calls the RUN- subroutine, the equivalent of the simulation run in HYBSYS.

HYBSYS offers standard features for optimisation, a macro 'ZERO f_1 $f_2,..,f_n$ BY p_1, $p_2,..,p_n$' which calculates the zeros of the cost function $f=(f_1,f_2,..,f_n)^T$ subject to the parameter vector $p=(p_1,p_2,..,p_n)^T$ and a macro 'OPTI, method $f_1,f_2,..,f_n$ by $p_1,p_2,..,p_m$' which optimises f subject to p using the algorithm 'method' (zyclic parameter variation, (conjugate) gradients, variable metric, Monte- Carlo method). A macro 'POLY, method $f_1,f_2,..,f_n$ BY $p_1,p_2,..,p_m$' which calculates (and plots) the efficient set of parameters p subject to the vector (poly-) optimylity of f, will be available soon.

The outlined methods for implementing optimisation have a common denominator: extended macro features of the SL are used.

Consequently these methods can be seen implemented in the so- called supermacro- concept (/3/). This concept- developed in correspondence with old and new CSSL- standard (/3/) - is based on the three basic elements of a SL, namely model, method and experiment; there an experiment has to be seen as performance of a method with certain models (simulation

run, sensivity analysis, etc). Optimisation consequently can be interpreted as a very complex experiment with iterative simulation runs, solution of additional differential equations (sensivity functions), etc.
In (/3/) it is proposed to use 'supermacros', extended special experimental macros for implementing complex methods for problem investigation.
A supermacro consists of three (groups of) subexperiments: generation or change of model equations, generation of methods for analysing the models, performance or initialisation of the generated methods with the generated models. Thus, implementation of optimisation in supermacro- technique combines the following subexperiments: change of model equations (time transformations,..) or generation of additional equations (sensitivity functions,..), choice and/or modification of optimisation procedures (gradient methods, polyoptimisation, stochastic methods), performance or initialisation of optimisation (iterative simulation runs). Because choice of the algorithm can be supported by software too, this supermacro concept can be seen as first stage of an expert system.
It is evident, that an essential requirement for supermacros, the possibility for change and/or generation of model equations is only available in interpreter- oriented SLs because the model has to be changed at run time. A new release of the fore- mentioned HYBSYS macro OPTI will be a 'real' supermacro which is able to generate the sensitivity functions which extend the model.

3. Digital versus hybrid simulation in optimisation

The well known advantages and disadvantages of both methods shall not be mentioned here. Only two aspects are to be noted which make hybrid simulation very interesting. First, the amount of simulation runs cause large CPU times usually exponentially growing with the complexity of the system- in case of hybrid simulation a run takes a fixed (short) time independent from growing complexity. Second, state events which often arise in technical systems cause difficulties in numerical integration which may lead to wrong values of the cost functions- in hybrid simulation state- depending changes (although discontinuous one) are treated without any difficulties because of the full parallelism. Consequently also only numerical algorithms with integrated state event handling (like integration in ACSL) are useful.

4. Applications

The first example is the time optimal control of the crane crab with the dynamics

$$d^2w/dt^2=-a.\sin w+b.u, \quad d^2s/dt^2=-c.w+d.u$$

where w(t) and s(t) denote angle and position of the crane crab, u(t) the control which is of bang- bang type with the unknown switching times t_1, t_2, t_3 and unknown terminal time T. Moving the crane- crab from one point to another the time instants t_i and T have to be determinded so that $w(T)=\dot{w}(T)=\dot{s}(T)=0$, $s(T)=s_T$ holds. This boundary value problem was solved with optimisation- using ACSL and HYBSYS. Optimisation with the standard macro ZERO in HYBSYS takes about 20 simulation runs, each about 0.8 milliseconds. Digital simulation and optimisation with ACSL takes also about 20 simulation runs, but the time for each run naturally depends on integration algorithm, stepsize and 'complexity' of the model (nonlinear or linear).
Table 1 shows normalized CPU- times for one simulation run. Extending the model (1) by bounds for w(t) and s(t) the necessary state event handling causes numerical difficulties and/or CPU times five times as long as before.

Stepsize / Algorithm	linear 0.01	linear 0.001	nonlinear 0.01	nonlinear 0.001
Euler	0.19	1.80	0.25	2.18
RK 2nd order	0.21	1.79	0.48	4.57
RK 4th order	0.65	6.19	0.66	8.06
Adams-Moulton	0.80	2.03	0.85	2.14
Gears Stiff	1.29	3.26	1.33	3.41

Table 1: Integration-CPU times, example 1

Stepsize	Energy.10^7	CPU-time
1 m	4.3458788	1.295
0.5 m	4.3457845	2.177
0.1 m	4.3457687	4.503
0.05 m	4.3457678	18.978

Table 2: Integration-CPU time, RK 4th order, example 2

and/or CPU times five times as large as before.

A more realistic example is the time optimal control of a metro train with the dynamics (/6/)

$$dv/dx=(u-y-\int_{x-L/2}^{x+L/2}(a.\sin w(s)+b.\cos w(s))ds)/v-cv, \qquad (2)$$

where v(x), L denote velocity and length of the train, u(x) and y(x) forces accelerating and braking and w(s) the slope of the section. Again the control is of bang- bang type, additionally bounds for v(x) and u(x), y(x) cause state events. Unknown parameters are switching time instants in u(x), y(x) and the terminal time T, cost functions are boundary values due to the stop of the train in a station. Hybrid simulation and optimisation work within technical accuracy. Digital simulation and optimisation depend extremly on stepsize, also the terminal values for consumed energy differ (table 2, normalised CPU times). An optimisation takes about 13 simulation runs. This fact becomes important if in a second step of optimisation the slope of the section w(x) is to be determined so that the consumed energy becomes minimal (dynamic programming, literature see (/5/)).

As last example a hydro- energetic system consisting of three reservoirs s_i with the dynamics (/4/)

$$ds_1/dt=y_1-u_1,\ ds_2/dt=y_2-u_2-e,\ ds_3/dt=u_1+u_2+r.e(t-t_v)-u_3 \qquad (3)$$

is considered (y_i natural inflows, u_i control outflows, e(t) water users with delay time t_v). To be optimised are energy production, flood control and recreational use. As the costs are contradictory one, vector (poly-) optimisation is used. Again the computation time is of interst because for determining the efficient set about 200 simulation runs are necessary.

5. Literature

/1/ ACSL User Guide/Refernce Manual. Mitchell & Gauthier Ass., Concordia, MA.

/2/ Bausch- Gall I.: Parameteroptimierung bei technischen Modellen mittels einer kontinuierlichen Simulationssprache. Informatik- Fachbericht 56, Springer.

/3/ Breitenecker F.: The concept of supermacros in today's and future simulation languages. Mathematics and Computers in Simulation XXV, pp279.

/4/ Breitenecker F., Schmid A., Peschel M.: Simulation and optimisation of a multipurpose hydro- energetic system using standard simulation software. Proc. Int. Workshop "Hydroenergy-Optimisation", Linz, Juni 1984.

/5/ Breitenecker F.:Optimierung in kontinuierlichen Simulationssprachen: Aspekte bei Modellen technischer Systeme. Informatik-Fachbericht 85, Springer, pp.656.

/6/ Hoang H.H.,Polis M.P.,Haurie A.: Reducing energy consumption through trajectory optimisation for a metro network. IEEE Trans.Autom.Con. AC20, Nr.5.

/7/ Solar D.,Berger F.,Blauensteiner A.: HYBSYS- interactive simulation software for a hybrid multiple- user system. Informatik- Fachbericht 56, Springer, pp 257.

On Numerical Methods in Simulation Systems

Krug, W., Jurisch, R.[1)]

Summary:

Simulation as a problem solution method developed into one of the most important aids of the modern engineer in the last few years. Central problems in the development and evaluation of simulation systems have been, on the one hand, the questions of the adaptation to the user's requirements and, on the other hand, the robustness of the system. These problems are closely associated with the simu-lation language and the numerical modules implemented. The present contribution is based on the experience and generalizations resul-ting from the work performed on the simulation systems DISIP 1/2 at the Engineering College of Koethen during the last ten years. It is an attempt to deal with the interrelations between numerical methods and simulation on the basis of system theoretical approaches from a uniform aspect.

Numeric, Simulation, Systemtheorie

1. The position of numerical methods within simulation

The approaches are based on the following abstract mathematical model:

$$F(u) = 0 \;,\; u \in X \;,\quad 0 \in Y$$

$$F:\; X \rightarrow Y \;,\quad X,\; Y \text{ infinite dimensional Banach spaces} \qquad (1)$$

In the special case, (1) may be, for examples a system of partial differential equations which (including the corresponding boundary and initial conditions) was formulated in the simulation language. As a direct solution of (1), i.e., the determination of $u \in X$, in the sense of the digital simulation on the computer, is not possible, in the usual way a discretization of the problem (1) is made which can be represented in the form of the following discretization and approximation scheme (2):

$$\begin{array}{ccc} X & \xrightarrow{F} & Y \\ \updownarrow & & \updownarrow \\ X_n & \xrightarrow{F_n} & Y_n \end{array} \qquad X_n,\; Y_n \text{ finite dimensional Banach spaces} \qquad F_n : X_n \rightarrow Y_n \qquad (2)$$

As an approximate solution of (1) in the sense of (2) a $u_n \in X_n$ is to be found for which $F_n(u_n) = 0_n \in Y_n$ is valid.

If (2) is considered only from the aspect of numerical analysis, this leads to the concepts of convergence and stability of the scheme (2) /1/generally used there.

---

1) Prof. Dr. sc. Wilfried Krug, Dipl.-Math. Ronald Jurisch, Ingenieurhochschule Köthen, 4370 Köthen, Bernburger Str. 52 - 57

However, it must be taken into account that these are asymptotic propositions (e.g., for $n \to \infty$ and $\mathcal{l}h \to 0$, respectively, vector of discretization parameters). This approach is to be rejected as insufficient or unsuitable in the sense of simulation for the following reasons:

- Error estimations for ensuring the convergence in real problems are hardly accessible and sometimes they are not accessible at all.
- Practically, approximate solutions u_n in the sense of (2) can be calculated only for few values of n from which hardly any asymptotic propositions can be obtained.
- Because of the requirements of computing time and storage space n is practically restricted in its upper limits.

If (2) is now regarded as an independent simulation problem which for the reasons mentioned seems to be quite reasonable, then the problem of the validization of the simulation model $F_n(u_n) = 0$ with regard to the original system $F(u) = 0$ arises immediately. As such a validization using means and methods of mathematics is hardly possible, the question of other possibilities and their combination with the usual numerical methods arises.
In the sense of such a simulation problem the discretization parameters $\mathcal{l}h$ and their interrelations play a central role. The next chapter will deal with this problem in some more details.

2. Evaluation of numerical methods from the aspect of simulation

If at first the problem (1) is considered in the sense of system theory (or the theory of determined signals), then X, Y assumes the role of system function interconnection of input and output signals). The determination of u in (1) corresponds to the signal theoretical problem of the determination of the input signal if output signal and system function[2] are given.
In this sense the simulation model (discretization scheme) (2) is a simple transfer of the problems to the finite-dimensional case. The structure of the simulation model (in the sense of the triple (X_n, Y_n, F_n) must be given a priori by corresponding discretization parameters so that a feedback in the sense of a correction of the simulation model through the possible degrees of freedom of the simulation problem (the discretization parameters) cannot be realized. To a simulation experiment the determination of an approximate solution u_n corresponds where the evaluation of different experiments is left to the user (mainly empirical experiences are utilized).

In the following reference is made to possibilities of a feedback and, associated with it, a control of the simulation experiments in the sense of numerical methods.
Historically, here first the techniques with time step width control for time dependent problems are to be mentioned whose theory is nowadays already developed to a relatively high level. However, much more difficult is the problem of space discretization. In this field only in the last few years serious successes were achieved which find thei expression, on the one hand, in the theory of adaptive refinement processes and, on the other hand, in the development of multigrid algorithms. Then the evaluation of these trends of development in the sense of the above problems is dealt with in some detail. The basic idea of the adaptive refinement processes can be represented by the following model:

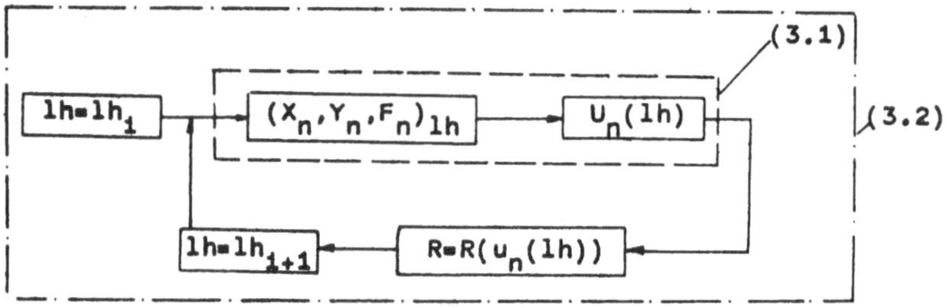

(3)

The control R is based on the principle of equalprobability distribution of local errors developed by Lentini-Pereyra |7| and Babuska |3|.

However, as this principle is of static nature the control R is performed in the practical realization of (3) by a learning process (adaptive method) where (3) is a non-autonomous structure scheme in the sense of Zypkin |4| (in each learning step must be made available u_n in the sense of (2)). However, also the following fact is interesting in this conjunction. If (3) is considered as a simulation model, then (3) represents a process of discrete structure optimization (in each step from a discrete quantity of structure possibilities corresponding to the possible refinements of the grid the optimum is selected in the sense of the control R), i.e., the original model (1) is approached by a more complex problem (simulation model) in the sense of (3). A critical assessment of (3) shows two main disadvantages. One is the non-autonomy which expresses itself in a high expenditure (calculation of u_n in each step), the other is the control principle R itself which through the a posteriori error estimations is indirectly related to the unknown simulation objective u.

Moreover, in each step the complexity of the model (expressed by n) is increased.

Quite another idea provides the basis for the multigrid method |5|. By giving a lot of grids, i.e., a system of structures in the sense of (2) that are coupled with each other (for example, by the principle of "nested iteration") the solution process of (2) is considerably intensified. However, in the sense of the simulation problem there is again the problem of validization as this is a simple transfer of the original model to the finite-dimensional case as it was described already at the beginning of chapter 1. Although there are already systems connecting the two principles with one another, the disadvantages associated with scheme (3) remain |5|.

3. Conclusions and prospect

From the above-mentioned, the question for the possibility of autonomous structure schemes in the sense of Zypkin results. Necessarily, such a scheme must combine the advantages of both principles (i.e., adaptive structure and multigrid idea) and avoid the disadvantages described in chapter 2. (according to (3) transition from (3.1.) to (3.2.)

Proceeding from the method of finite elements, in the last few years the basic ideas of such a system were developed |6| at the Engineering College of Koethen. For reasons of space, however, here only a general description can be given. The starting point for this system are function spaces above variable grid structures. On such spaces minimization problems for functionals are investigated resulting directly from (1) (for potential operators). These functionals directly depend on the discretization parameters so that from the necessary optimality conditions ($grad_{lh}\ \hat{F}\ (lh) = 0$) an operator can be derived which makes possible a direct (continuous) structure optimization. This structure optimization is performed by local structure adaptation where the complexity of the structure (expressed by n) is not increased. Only in the case of sufficient structure adaptation (in the sense of $grad_{lh}\ \hat{F}\ (lh)$) a corresponding increase in the complexity takes place (by refinement of the grid). First test calculations of simple model problems showed very favourable results so that in further investigations in this field an enrichment of the existing theory can be expected.

4. Literatur

1 Vainikko,G.: Funkt.-Anal. d. Diskret.-Meth./Teubner,Leipzig 1976

2 Peschel,M.; Wunsch,G.: Meth. u.Prinz. d.Syst.-Th./Verl.Technik, Berlin 1972

3 Babuska,I.: Anal. of opt. f. e. m. in R^1/Math.Comp.33(1979)435 ff.

4 Zypkin, J.: Grundl. d. Theorie lern. Syst./Verl. Techn. Berlin 1970

5 Lecture notes in mathem., Nr. 964, Springer Verlag 1982

6 Jurisch, R.: Diss. A, einger. im Sept. 1984 an d. TU Magdeburg

7 Lentini, M.: An adapt. f. d. solver... /SIAM J. Num. An.14(1977) 91 - 111

CONSTRUCTOR OF DECLARATIVE REPRESENTATION
OF COMPLEX SYSTEM SIMULATION MODELS

I.V. Maksimey

Gomel State University, Department Of Control, Sovietskaya Street, 104,246019, USSR

Résumé

A constructor possessing declarative type system describing language is suggested to be used as an instrument of complex system simulation models construction. The models are represented in terms of Queueing Network Systems (QNS) by means of graphic mode. The constructor is built on the basis of CASM modelling complex, which provides automatic transition of graphic models to the internal language of processors representation in a model and realizes network description on Common System of Computers. Key Words: CONSTRUCTOR,GRAPHIC SIMULATION MODELS

1. Graphic Representation Of Complex System Models

The process of complex systems functioning may be represented in terms of states change sequence, and the duration of these system components presence in these states is defined either by behaviour algorithm of the elements themselves or by the influence of external interferences. The functioning of any complex system components is represented by Queueing Network Device (QND), and its behaviour is graphically represented in figure 1.

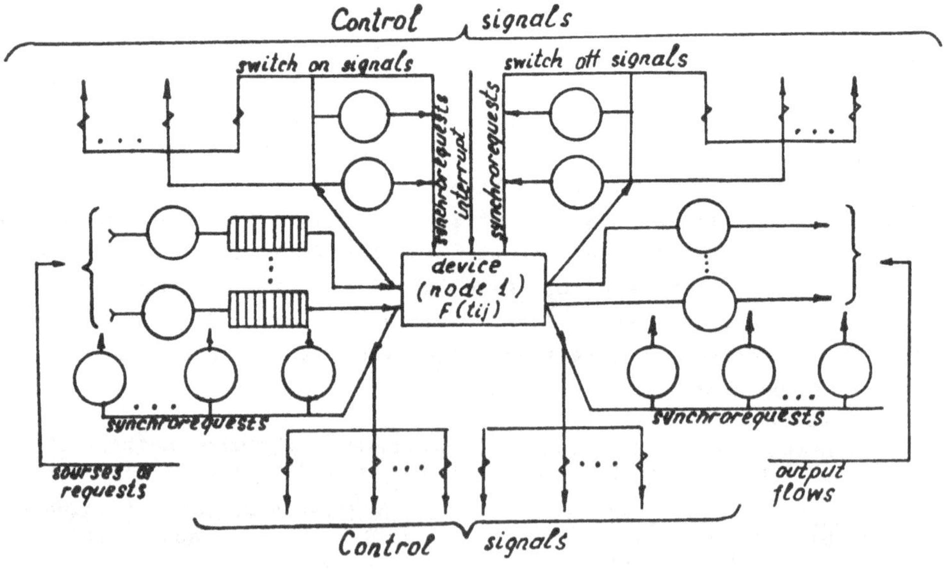

Figure 1 Common Node Of QNSM

The possible behaviour of this device is essentially different from the conventional QND. It is possible to represent any number of transacts queues at the device input and output. Any QND can serve any number of transacts according to the individual law of time distribution of - i transacts by F(tij) complex system j-element. Each QND possesses three control inputs:"switch on","switch off","interrupt". These control inputs can receive the corresponding control signals. If the device is "switched off"then its input receives the corresponding queues in spite of the fact the device can be free of service. This state may last until the signal"switch off" is received. On QND output the further root of i-transact receiving is marked after the end of its service by j-QND. Either the probability choice of the direction of I-transact receiving according to the probability Pij, or the determined choice of its movement according to i-transact index is possible. On QND output the following transitions are assumed; i-transact multiplication by the number of transact copies mentioned before; the formation from i-transact of any previously said number of control signal of all the three types ("switch on","switch off" and "interrupt"). The alteration of control signals structure and their generation places at the QND constant composition leads to the alteration of logics of complex system simulation model's functioning. The absence of logic control elements in a modelling object leads to the typical models of Queueing Network Systems (QNS).

For the complex system models experiment organization the two-stage mode of graphic model transition to program simulation models (PSM) is suggested. At first, the complex system graphical representation is recorded by declarative statements, which reflect all the aspects of PSM functioning. The device composition of the modelling object belongs to them and also, relations between them,functioning logics and complex system elements relationship. Then the received complex system elements description are automatically transitted to a computer simulation model by means of CASM modelling complex [1] .

2. CASM modelling Coplex Language

Complex system elements which are graphically represented here are described by six basic sentences: DEVICE, INPUT FLOW, QUEUE,SERVICE MECHANISM, SERVICE TIME, OUTPUT FLOW. The first sentence identifies the network's node, all the rest describe its characteristic features. The group of all nodes description in accordance with the set syntactic rules of CASM modelling complex represents the model's commong text which is accomplished by MODELLING TIME sentence.

The sentence DEVICE identifies each node by name pointer of the device TYPE and priority pointer of the device INDEX.

The sentence INPUT FLOW characterizes the process of transacts arrival at the input to QND. Its description consists of the set of sources which give the transacts for service. Transact sources are divided into the sources out of system and the sources in system. For each transact source acertain class and priority according to the provided service disciplines are assigned. For the regulation of transact entering process in time by means of the construction THE STARTING PERIOD, the transact source active and passive periods are set. The sources in system set the arrival order of transacts from the other nodes of the network. Any source in system is characterized by the unique number and network's node identifier from which the transacts arrive. For the simulation of systems with internal control it is possible to transit QND from the state "ready" to the state "occupied". QND transition from one state into another is realized by the signals "switch on" or "switch off". These signals generation may be realized by any in-system transact flow at the moment of their arrival to the QND input. The signal generation properties setting is realized by the construction "WITH DIVISION ON TYPE SIGNALS".

The sentence SERVICE MECHANISM gives the set of rules which define the processing order of transacts, arriving at QND.The sentence contains the service discipline descriptors. By setting of the construction WITH LOSSES or RE-SERVICE the transacts processing order is pointed out after the interruption of its service because of the arrival of more prioritive one.The character of re-service is set by the constructions WITH NEW TIME and WITH PREVIOUS TIME.

The sentence QUEUE describes the input queues composition to QND. The set of three types of queues to the node is possible with the help of pointers: COMMON, SEPARATE and MIXED. The place that the arriving at the queue transacts occupy is defined by service mechanism. The queues formation may be overlapped by limitations, queue's length, the time of being in the queue and probability of transacts putting into the queue. For the set of these limitations there are pointers in the sentence. They are: LENGTH,TIME, PROBABILITY.

The sentence"service time" describes the QND transacts service duration. The descriptor DISTRIBUTION sets the type and parameters of duration and distribution law of transact service. The setting of service mode with quantization is possible with the construction WITH QUANTIZATION MODE.

The sentence OUTPUT FLOW is used in three cases when after service stage at QND the transact is the source of new transacts or control signals. The division process is set by the construction WITH SPLITTING IN REQUESTS. New transacts rules of planning for further processing are different from standard ones. Their routes may be determined routes, as well as stochastic ones and are set by the combinations of the constructions DEVICES LIST and WITH PROBABILITY. Control signal setting is realized by the construction PLANNING ACCORDING PROBABILITY. The sentence MODELLING TIME sets time interval during which it is necessary to simulate QNS. There may be a number of instructions for managing the experiment in the sentence. The construction STATISTICS GATHERING IN STATIONARY MODE provides automatic searching for the controlling parameter entering the stationary mode by queue's mean length value.

3. The Methods Of Models Construction Usage

For CASM modelling complex usage in the given research it is suggested to use the graphic submodels library. The submodels are described in CASM modelling complex declarative language. It is suggested to use the methods of models constructor, using complex system modelling, as an example. There is a library of submodels hardware components and computer systems software. Due to the known computer-aided modelling systems for considered types of computer system simulation models, it would take an order more calendar time and computer resources which are related to the expences on the organization of complex system model debugging.

Computer experiment organization by models constructor includes the following stages: drawing up the complex system description, modelling tasks definition and the formulation of complex systems variants quality, simulation models formal description in the form of QNS, simulation model realization in CASM language, carrying out experiments by CASM modelling complex, the interpretation of modelling results. The main concept in the basis of simulation model realization in CASM input language lies in complex system decomposition into separate QND-s and their further independent description in arbitrary order. As a result of complex system decomposition, the detailed graphic schemes of all nodes of the network or even complex system submodels are being built (or have already been built). The transition from a detailed block-diagram to simulation language description in CASM modelling complex language does not give rise to difficulties for the researcher. The possibility of parametric setting of composition and QNS models structure is provided by means of the constructor.

REFERENCE

I. Максимей И.В., Семишин Ю.А. Технологические возможности имитационного моделирования сетей массового обслуживания.-Электронная техника. Сер.9.Экономика и системы управления.I983,вып.I(46), с.45-50.

SIMULATION - A HELPFUL TOOL OF SOFTWARE DEVELOPMENT

Koeppe, Reinhard 1)

Problems of the development of real-time software

There arise many problems in the development of real-time software:

- Real-time systems must satisfy strong reliability requirements. They can only be satisfied, if both hardware and software will do so.
- The processing of programs is inferior to time conditions.
- The randomness of input data causes a randomness of processing of control programs.
- The treatment of a requirement can be interrupted by other requirements.
- The hardware and software functions are closely connected with each other.
- Real-time systems are very complex.

These problems can be obviated only by an effective support of software development. By that, simulation can be a valuable tool.

Stages of software life cycle, supportable by simulation

The first stage is requirement analysis. A main task of it is the determination of response time. If it is too long, there can occur so many events of environment process during it, that response becomes worthless. If it is too short, there will be an increase in the expense for realization of the whole system.
The result of requirement analysis is the requirement specification.
The next stage is design. It is a classical field for the application of simulation. Here hard- and software structures, both closely connected, are specified. Often the data needed to design the system may not be available until the system is implemented. Estimating procedure will in most cases be an iterative process. Besides that, effective control strategies must be developed. Their efficiency has to proved under different loading conditions. The functions of the whole system have to be allocated to various modules.
The design is finished by validation, where simulation can help to avoid mistakes.
The result of design is documented as the design specification.
The next stage supportable by simulation is the dynamical testing. Test data must be generated in the right logical and temporal order, the program has to be processed with them and the results checked. Testing causes nearly half of the whole expense of software development.

---

1) Technische Hochschule Magdeburg
DDR - 3010 Magdeburg, PSF 124

One reason for this fact are the costs of software testing tools. The test examines whether the program satisfies the demands established by design and requirement specification. This shows the close connection between requirement analysis,design and testing.

The connection between requirement, design and testing

Advantage of this connection can be taken by making the specifications machine-readable. In this way, error-prone manual work of creating software testing tools can be reduced. For instance, this can be done by special notations.
Another possibility is given by simulation. Investigation of environment and control process causes their simulation at the stage of requirement analysis and design. In a sense test data reflect the environment process. The control process is reflected by the programs to be tested. The simulation model used in design reflects (one aspect of) design specification. So the simulation model, including both environment and control processes, is a possible form of machine-readable design specification and requirement specification respectively. It represents an opportunity of connecting requirement analysis, design and testing.
For taking advantage of this connection, the creation of the simulation model during requirement and design has to be done in a way allowing further use at the stage of testing. That requires especially the division of the simulation program into modules and a well defined interface between environment and control processes in the simulation model. For testing, the modules of the simulation model representing the control process are replaced by coded programs.

Implementation

For creating the simulation model, the GPSS-like Simulation System SIMDIS is used. To apply the idea of the preceding part to the software development for microcomputers, there was built a cross simulator which permits correct connection to SIMDIS in reference to events and time.
The cross simulator offers powerful debugging tools:
All registers, storage, gates and interrupt chain can be changed and listed.
Assertion controlled actions can be carried out.
For assembler coded programs, symbolic access to data and coverage are planned.
Debugging can be done interactively.
There is a library for simulation of various combinations of periphery and processor units.
The present program can be easily adapted to special requirements by chosing special parameters or fitting additional programs. So

simulation of the whole configuration is possible.
The flexibility of the cross simulator allows experimentation while testing.

Summary

Simulation is a powerful tool for developing reliable control software. It helps to avoid errors in requirement and design specification and represents an opportunity of connecting requirement analysis, design and test. In this way, error-prone manual work of creating software testing tools can be reduced. For realization of this concept, there was built a special cross simulator for microcomputers which permits correct connection to SIMDIS in reference to events and time.

References

/1/ Boehm,B.W.: Software engineering.
IEEE Trans. Comput. C-25 (Dez 1976), 1226-1241

/2/ Fairley,R.E.: Software testing tools.
Computer program testing, Proc., Amsterdam 1981

/3/ Giddings,R.V.: Accommodating uncertainity in software design.
Communications of the acm, 27(1984)5, 428-434

/4/ Lorenz,P.;Ziems,D.: Konventionen fuer SIMDIS-Quelltextmoduln.
edv-aspecte 3(1984)2, 22-25

/5/ Roberts,J.W.;Talparu,R.: A cross test system for PEARL.
Real-time data handling and process control, Proc.,
Amsterdam 1981, 191-194

MESSIL - SIMULATION PROPERTIES OF MEASURING SYSTEMS

Jacek Nalepa [1/]

SUMMARY: The paper discusses the problem of computer simulation of measurement systems. The operations appearing in the system are defined. Application possibilities of the MESSIL program package created for simulation analysis and design of measuring systems are described. The range of possible applications of the presented modelling methodics of measuring system simulation is specified.
KEY WORDS: measurement system, computer simulation.

1. INTRODUCTION

The measuring system is a system of interconnected measuring and computing modules which perform the functions of measurement of many quantities, of recording and processing of measuring data, and of outputting the final results of measurements and computations.
The principal aims of the measuring systems are:
- evaluating parameters of models of objects for the purpose of identification or control
- determining parameters of object quality models for diagnostics.

Computer simulation of measuring system models can be an effective tool for determining their metrological properties.

2. MODELLING OF MEASURING SYSTEMS

The measuring part of the system equipment includes a set of measuring converters such as: analog, analog-digital and digital converters /controlled or not/ together with auxiliary equipment. The assembly of the computing equipment includes the hardware and the software.
The general model of a measuring system has the form [1] :

$$[m_r] = [R_p]\ [m_q] \qquad /1/$$

where:

$[m_q]$ - vector of the measured real quantities /the analog ones/
$[m_r]$ - vector of measuring results and computations
$[R_p]$ - system of operations and algorithms representing the processing in the measuring system

In order to built a useful model of the measuring system it is necessary:
- to determine the properties and the shape of signals $[m_r]$ and $[m_q]$
- to determine the properties and the shape of the operation and algorithm systems $[R_p]$.

The form and the properties of signals $[m_q]$ depend on the properties of the measured objects or on the type of the measuring properties determined for the system /e.g. response to unit step/. As the $[m_r]$ signals are generated by the measuring system, their shapes and properties depend on the properties of the $[R_p]$ model. The model of a measuring system $[R_p]$ can be in the general case a system of sequences of complex operations of measuring processing and computation algorithms which is a representation of the physical structure of the system /e.g. subdivi-

1/ Academy of Mining and Metallurgy in Cracow
Institute of Electric Machines and Power System Control
30-059 Cracow, al. Mickiewicza 30, bl. B-1, Poland

sion of the system into channels/.
Depending on the type of the metrological properties which we will determine for the system, models of the system or its part with various degree of insight into their structure and into the physical phenomena occurring in them can be built.
However, in the created models there must appear the related engineering parameters of this equipment or those being directly the parameters of given metrological properties.
For instance: sensors, electromechanical converters, filters, analog electric transducers can be modelled for dynamic investigations by differential equation sets, integral equation sets or transmittances. The same elements, for statical investigations, or with their dynamic properties being negligible, can be modelled by their static characteristics of conversion. The pulse type elements can be modelled by sets of difference or differential equations, separating the impulsators and continuous parts in the structure of these elements.
The digital elements, after omitting their dynamic properties, can be modelled by their static characteristics of processing of numeric sequences. The dynamics of these elements can be accounted for by introducing such an equivalent parameter as e.g. delay time. The computing parts of the system can be modelled by their algorithms with consideration given e.g. to the computation time /delay time/.
The example mentioned above illustrate the characteristic feature of the presented structural method of modelling i.e. the modelling of the individual elements of the system's structure at the level of measuring signals /carrying the measuring information/.
Thus, the system model represents its structure and the tool for simulation investigation of such models have to ensure the simulation of functions representing measuring signals and operations appearing in $[R_p]$.

3. THE PROGRAM PACKAGE FOR SIMULATION OF MODELS OF MEASURING SYSTEMS

Computer simulation is the only effective tool for investigation of structural models of measuring systems[2,3] .
For simulation analysis and design of measuring systems, the MESSIL program package /Measurement Systems Simulation Language/ was created, which includes the simulation language GODYS-4 and two supplementary libraries: MAKRO and OPROG. The GODYS-4 language is intended for simulation of continuous dynamic systems, where discrete events can occur, causing changes of its structure and/or of the eigenvalues of its mathematical model. The language has 52 standard operations /single-output/ and optional operations which can be defined in FORTRAN by the user. The user can also define the multioutput macro-operations /submodels/ constructed of the language standard operations. The macro-operations are used in the model in the same manner as the standard operations. The macro-operations can be defined in a given program or/and call from the MAKRO library.
The expressions occuring in the model description are formed of variables, constants and standard operations connected by arithmetical or relation type operators or/and logically.
Taking advantage of the standard operation file, the models of linear and nonlinear elements of systems can be built, described by differential equations as well as models of typical pulse type and digital elements.
The macro-operations are used as models of complex digital elements /e.g. LSI/ or as submodels of elements are often used in the model of the whole system. The simulation parameters /e.g. step, time, error/

and the forms of results /print-outs/ are declared by the user.
In the MESSIL system, the GODYS-4 language is equipped with an parametric optimization module. This allows us to determine an optimum /in the meaning of a certain feature function/ value of indicated model parameters.
The objective function is formed from standard operations of the language. The optimization occurs by linear simulation experiments using /at present/ two optimization algorithms: MGS for unidimensional problem and DSC for a multidimensional problem. The parameters of the optimization module /e.g. method, time, error/ are declared by the user. The MAKRO library includes models of complex digital systems such as multiplexer, converters of codes, summators, counters etc. The OPROG library contains subroutines /in FORTRAN/ serving for: modelling elements of systems described by integral equations and serving for processing of the simulation results.
The MAKRO and OPROG libraries can be supplemented by users. The MESSIL program package is available at present on the computers of the ODRA 1300 /ICL1900/ series.

4. SIMULATION INVESTIGATIONS OF MEASURING SYSTEM MODELS

The simulation of the system model is carried out for concrete forms and values of the input functions m_q and model parameters. The aim of such investigations is to determine the metrological properties of the system under investigation in order to analyse its functioning or for design purposes.
In the practice, this activities are reduced to:
a/ investigation of the time behaviour of functions representing the measuring signals in the system models,
b/ determining the static and dynamic characteristics of measuring systems, their channels and elements,
c/ investigation of the effect of internal and external random and deterministic interferring factors,
d/ testing of the programs controlling and processing the measuring data,
e/ optimization of the metrological characteristics of the system through choice of the model structure and parameters.
The limitations concerning the classes of the investigated models and the types of performed simulation experiments result from the possibilities of the MESSIL package and the parameters of the used computer.
A particular feature of the simulation model investigations of measuring systems is the possibility of determining the error of the whole system as a multidimensional object, the possibility of examining the effect of any interferrence and the possibility of optimization of the model parameters and its characteristics.
The simulation investigations carried out so far included among others: determining the static and dynamic errors of systems channels, determining frequency responses and parameters of response to a step function for channels and elements, investigation of the effect of interferences of signals and element parameters /e.g. temperature/ on the measuring error, optimization of measuring filter parameters.
The gained results confirm the effectiveness of the assumed method of modelling and simulation of measuring systems.

LITERATURE

1. Szyper, M.: The Structural Method in the Theory of Measuring Converters. Zeszyty Naukowe AGH, z. 114, Kraków 1979.
2. Szyper, M., Nalepa, J., Wojtasik, A.: Measuring System Analysis with Digital Simulation of Structural Model. Acta IMEKO VII, Londyn-Budapeszt 1976.
3. Nalepa, J., Szyper, M.: Modelling of Measuring Systems by Computer Simulation. Simulation of Systems'83, Praha 1983.

ADVANCED SIMULATION, USING WORLD-VIEW OF TRANSACTION FLOW

Heinz Stahn[1]

"Under the transaction flow world-view, the operation of a system is described from the viewpoint of the behaviour pattern experienced by units of flow (transactions) as they move through the system." /2/. We agree to HENRIKSEN: The substantial content of simulation is the description of the behaviour pattern of the operation of a system. Therefore /5/: the behaviour pattern in the relation behaviour pattern - system is the primary one. And we must ask: How we may describe the behaviour pattern and how we may implement the description of the behaviour pattern?

TWO ALTERNATIVE WAYS TO BEHAVIOUR PATTERN DESCRIPTION

Without restrictions of generallity we may assume, that a behaviour pattern description i a structure S_{BU} of behaviour pattern description units. A structure S_{BU} is /4/:

$$S_{BU} = (BU, N_{BU}, F_{BU})$$

with BU - set of behaviour pattern description units

N_{BU} - evaluation set

$F_{BU} : BU \times BU \longrightarrow N_{BU}$ - behaviour structure transformation.

There are two alternative ways I and II to describe such a structure S_{BU} on a set BU of behaviour pattern description units. Everybody in the simulation field knows the common way I:

(I) A behaviour pattern description is a structure on a set of system units.

Let system structure S_{SU}

$$S_{SU} = (SU, N_{SU}, F_{SU})$$

with SU - set of system units

N_{SU} - evaluation set

$F_{SU} : SU \times SU \longrightarrow N_{SU}$ - system structure transformation.

Only if the behaviour structure S_{BU} is isomorphic to the system structure S_{SU}, the common way I is well.

This singularity exists e.g. in the standard, elementary single

---

1 Technische Universität Dresden, Sektion Informationsverarbeitung, DDR-8027 Dresden, Mommsenstr. 13

serving problem /6/. In general such isomorphism does not exist. Therefore, specially the application field of flexible automation of production processes demands a new, problem solving way II.

(II) A behaviour pattern description is a structure on a set of behaviour pattern description units, which in general are not system units.

In our ARS-technology /1/ we call such a behaviour pattern description unit 'process unit', using the term process in sense of PETRI, KAAZ and DIN. A process unit bu is the set of all possible realizations of the transformation bu

$$bu: X_{bu} \times Z_{bu} \longrightarrow Z_{bu} \times Y_{bu} \qquad bu \in BU.$$

Every process unit may recursively decomposed into a structure of other process units.
Other authors use the term process in sense of one sequence of realizations or in sense of one realization $(x_{bu}, z'_{bu}) \longleftrightarrow (z''_{bu}, y_{bu}) \in X_{bu} \times Z_{bu} \times Z_{bu} \times Y_{bu}$. Often there is no distinction between process and system. For example MESAROVIC is calling our process an (abstract) system: "By the term 'system' we simply mean a transformation of the input data into outputs."

DESIGN OF A BEHAVIOUR PATTERN DESCRIPTION

Therefore process unit is synonym to behaviour pattern description unit. Process units may be, e.g.:

process = structure of subprocesses
subprocess = structure of tasks
task = structure of events
event = unsharable process unit.

The type of such process units determines a hierarchical stratum. We demand:

In every stratum: for every process unit of a stratum the transition rules (of PETRI) must be valid.

From this grant follows, that the process structure transformation F_{BU} is determined by the direct-preceding-relation:

$$a \longrightarrow b \Longleftrightarrow a \text{ is a direct preceding process unit of } b \quad (a,b \in BU),$$

which we also call "activating-relation" (see 'Aktivierungsplan' in /1/). A process unit is in the state 'activated', if all direct preceding process units are in the state 'realized'. The result is the flow of "activations" ('activation-transaction') on the process structure. This activation-transaction flow dominant determines the operation characteristics of a system /1/. If we design a structure S_{OD} on a set OD of operand units using the direct-needed-up-stream-relation, then there exists a direct relation between S_{OD} and S_{BU}: between the skeleton of S_{OD} and S_{BU} exists a morphism. Every process unit bu for its realization must be assigned to a system unit su. Using the unambiguous transformation alg: BU ⟶ SU we may demand

$$alg(F_{BU}) = F'_{BU} \stackrel{!}{\leq} F_{SU}.$$

That means, that operand-transaction flow in the system structure represents only in special cases (e.g. /6/) the operation characteristics of a system.
We still must mention, that great processes and great systems must hierarchically decomposed not only into different strata, but also into layers and into echelons.

IMPLEMENTATION OF A BEHAVIOUR PATTERN DESCRIPTION

The GPSS language was first made available 1961 from GORDON. In 1967 from PETRI was given his fundamental paper about the description of discrete processes. The main concept of GPSS was the concept of transaction flow. Based upon this concept it no surprise, that a GPSS-transaction[1] may be a 'operand-transaction' in sense of queuing theory or may be a 'activation-transaction' in sense of PETRI-nets. Therefore it is not correct, to speak about possibilities of GPSS only in the sense of queuing theory. The term flow describes a relative situation: one thing is immovable, another thing is movable. By application of duality principe may be the same one thing movable and the other thing immovable. In a structure S_{BU} are activation-transactions movable to process units. By duality principe we may implement either activation-transaction in GPSS-transactions or process units in GPSS-transactions. Both implementation of a structure S_{BU} to be realized on S_{SU} are correct and are comparable by criteria of effectivity /3/.
But the knowledge, that the behaviour pattern description is an activation structure on a set of process units is not exclusively connected with GPSS. An activation process structure is a logical structure and becomes by its implementation a physical structure. This is valid for every software system, also for simulation software systems, for example for an universal simulation tool, implemented on

1 Also in GPSS-SIMDIS-2 from ROBOTRON (GDR).

microcomputer, for simulation hierarchically structured control processes, which are elements of the lowest echelon of a coordination type control process structure for flexible manufacturing.

REFERENCES

/1/ Engelien, M.; Stahn, H.: Software-Engineering - ARS-Technologie. Akademie-Verlag, Berlin 1984.

/2/ Henriksen, J.O.: State-of-the-art GPSS. Proceedings of the Summer Computer Simulation Conference, Cancouver, July 1983, 918 - 933.

/3/ Kieszling, M.: Simulation flexibler, gesteuerter Produktionsprozesse. Dissertation, TU Dresden 1985.

/4/ Stoschek, E.: Angewandte Strukturtheorie. Akademie-Verlag, Berlin 1981.

/5/ Stahn, H.: Ein Beitrag zur Methodologie der Simulation. ZfR-Informationen Nr. 80.11, Berlin 1980, 3 - 9.

/6/ Stahn, H.: Diskontinuierliche Simulation des Single-Server-Problems mit prozeßabhängiger Bedienzeit. rechentechnik/datenverarbeitung 20(1983)6, 29 - 30.

/7/ Stahn, H.: Simulationstechnik als rechnerunterstütztes Hilfsmittel zum Entwurf und zur Implementation von diskreten Steuerungsprozessen. Tagungsberichte der Akademie der Landwirtschaftswissenschaften. In preparation.

SISNET 3 - A SIMULATOR OF SWITCHING NETWORKS WITH EFFICIENT DATA STRUCTURES

Janusz Kleban, Jerzy Kubasik [1/]

INTRODUCTION

Connecting networks play an important role in various systems composed of many functional modules, e.g. telecommunication exchanges, multiprocessor systems and measurement complexes. They provide interconnecting routes from a set of network inputs, through a set of switches, to a set of network outputs. In most practical networks the switches are arranged in groups called stages. The stage is composed of x switches each having y inputs and z outputs. Two adjacent stages are connected by a set of links. Along with the switches, the link patterns are responsible for the distributive characteristics of the network. In some networks blocked calls /requested connection cannot be completed immediately/ are lost, in others they are delayed.

In order to properly design a connecting network, the probability of loss or delay for a given network should be determined. A purely analytical approach for the solution of teletraffic problems frequently leads to unmanageable mathematics. Then simulation is the only alternative. The simulation program SISNET 3 /SImulator of Switching NETworks/ is used for traffic investigations of switching networks. It is the improved version of the simulator SISNET 2 [2]. SISNET 2 is not enough effective. In this simulator the data structures are redundant and the computation time is relatively long, i.e. cost of simulation is big. Because the size of the simulated network is limited by the available memory, in SISNET 3 the special data structures, which save required memory and computation time have been applied. The new algorithm for the selection of the next event has also been used.

SISNET 3 is applicable to two-sided networks having practically any structure. It makes it possible to investigate loss, delay and combined delay and loss systems.

INPUT DATA

In order to run a simulation experiment a set of punched cards should be prepared. This set consists of two subsets: the cards describing structure of the connecting network, and the cards describing various parameters of the simulation experiment.

The network description language makes it possible to prepare data for

---

1/ Technical University of Poznań, Institute of Electronics, ul. Piotrowo 3a, 60-965 Poznań, Poland

users not having programming experience. Most of the cards of input data contain the command and a set of parameters. The command specifies the kind of parameters. The description of a network's structure consists of the following commands: NOSTAGES, STAGE, SWITCH, INPUTS, OUTPUTS, and NEWNET. The NEXTS, MULTSW, MULTIN, and MULTOUT cards facilitate the description of switching networks having regular structures.
The set of cards describing the parameters of an experiment contains the following six cards: ALGOR, PARAM, FLAGS, STARTGEN, PARAMGEN, and NEWPARAM [2] .

SIMULATION ALGORITHM

In the model of connecting network there are two types of events directly dependent on time: a subscriber's arrival and a termination of a call. The simulation algorithm in SISNET 3 is based on the event scheduling approach [1] [3]. This approach emphasizes a detailed description of the steps that occur when an individual event takes place. The main feature of this approach is maintenance of a set of event notices. Each event notice contains time at which an event is scheduled to occur and event type. The event set is organized in this simulator as a linked lists of event notices. The list is ordered by event times.
The simulation algorithm performs three main functions: scheduling future events, executing events in the proper order and updating the clock variable which represents simulated time. SISNET 3 employs the variable time advance method to update simulated time, i.e. the simulated time is advanced to the time of the next event.

DATA STRUCTURES

The structure of a switching network is represented in SISNET 3 by following tables: STAGES, LINKS, CODES, ADDRESS, SUBA, and SUBB.
The STAGES(A,B) table, where A indicates the number of the stage, and the variable NS describe the size of a switching network. The entries in the table denote:

- for B=1, the number of switches in stage A,
- for B=2, the number of inputs per each switch in stage A,
- for B=3, the number of outputs per each switch in stage A.

This table contains 3×NS elements, where NS indicates the number of stages in the simulated network.
The LINKS(C) table, having

STAGES(1,1)×STAGES(1,3) + ... + STAGES(NS,1)×STAGES(NS,3)

elements, describes the state of an interstage link and the state of a called subscriber, i.e. the states of all outputs of the switches which are used in the switching network. An interstage link can be in one of the following four states: free /0/, busy /1/, temporarily busy during path hunting /-2/, faulty /-1/. The called subscriber can be in any of the following two states: free /5/, and busy /6/. C denotes here the number of a link.

The CODES (C) table, has the same size as LINKS. Each element of the CODES table indicates the stage and the switch at which the succeeding outputs of the given switch are connected /by an interstage link/. The number of stage and the number of switch are coded. The value of the element of the CODES table is calculated in the following way: the number of stage times 100 plus the number of switch. For example, if CODES (1) is equal to 201 then it means, that the first output of the first switch of the first stage is connected to the first switch in the second stage. The entries of the CODES table also define the points at which the called subscribers are connected to the network.

The ADDRESS (D) table, where D indicates the number of switch, contains the addresses of the elements of the LINKS table, which denote the states of the first output of the D-th switch. The table ADDRESS plays an auxiliary role. It is used to shorten the computation time. This table have STAGES (1,1) + ... + STAGES (NS,1) elements.

	LINKS	CODES
1	1	201
2	1	202
3	0	203
4	0	201
5	0	202
6	0	203
7	0	201
8	0	202
9	0	203
10	1	301
11	0	302
12	0	303
13	0	301
14	0	302
15	1	303
16	0	301
17	0	302
18	0	303
19	6	301
20	5	301
21	5	301
22	5	302
23	5	302
24	5	302
25	5	303
26	5	303
27	6	303

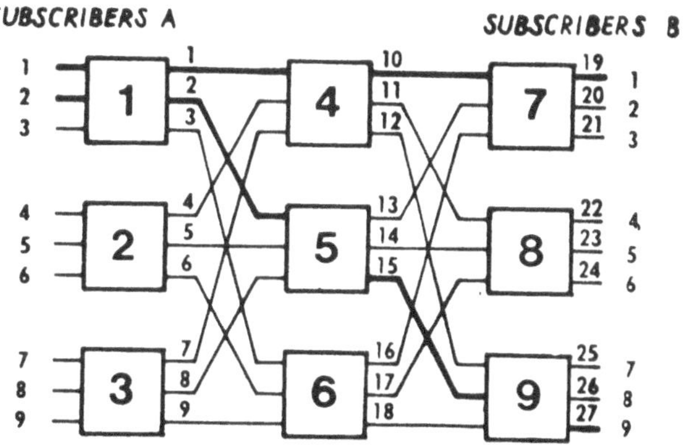

	ADDRESS	SUBA (X,+)		SUBB
1	1	101	1	19
2	4	101	2	20
3	7	101	3	21
4	10	102	4	22
5	13	102	5	23
6	16	102	6	24
7	19	103	7	25
8	22	103	8	26
9	25	103	9	27

Fig. 1. The exemplary structure of a switching network and the values of elements of tables LINKS, CODES, ADDRESS, SUBA, and SUBB.

The SUBA(X,F) table contains information about the calling subscribers. For F=4, entries in the table denote the point at which the X-th calling subscriber is connected to the network. Elements of this table are coded in the same way as in the table CODES. For example, the code 101 means that the subscriber is connected to the input of the first switch in the first stage. For other values of F entries of this table denote parameters of calling subscribers, which are used in the simulation procedure. The table SUBA contains 8×STAGES(1,1)× STAGES(1,2) elements.
The SUBB(Y) table, where Y is the number of the called subscriber, indicate the addresses of the elements of the CODES and LINKS tables, concerning to the Y-th subscriber. These entries denote the states and the codes of the called subscribers connection points. This table have STAGES(NS,1)×STAGES(NS,3) elements.

CONCLUSION

SISNET 3 has been programmed in FORTRAN and run on an ODRA 1305 computer /compatible with ICL 1900/. It has been used to examine several practical networks for private automatic branch exchanges and for rural exchanges. For a five-stage network having 256 inlets and 256 outlets SISNET 3 requires 15 K memory words /SISNET 2 requires 26 K memory words/ and works five times faster than SISNET 2.

REFERENCES

1. Fishman, G. S.: Concepts and methods in discrete event digital simulation; John Wiley and Sons, New York 1973.

2. Jajszczyk, A., Kubasik, J., Rajski, J.: A universal simulator of telecommunication switching networks; Proc. 6th European Conf. on Electrotechnics EUROCON'84, Brighton 1984, pp. 409-413.

3. Jajszczyk, A., Rajski, J.: Discrete simulation of connecting networks - sequencing algorithms; Proc. Int. AMSE Conf. "Modelling and Simulation", Paris-Sud 1982, vol. 13, pp. 85-89.

A HYBRID SIMULATION APPROACH FOR QUEUEING NETWORK MODELS

F. Burghardt [1)]

Hybrid technique combines simulation and analytical methods in order to overcome the disadvantages of these both methods. This paper presents a program supporting hybrid method, especially for analyzing queueing network models. It is based on decomposition and aggregation of submodels by load dependent servers.

keywords: hybrid method, extended queueing networks

Introduction

Queueing network models play an important role in performance evaluation of computer systems and communication networks. There are three major methods by which queueing network models can be analyzed. Analytical methods are very efficient for certain model classes. However, we have to consider that some complex models would require expansive effort in developing an analytical algorithm. Then we prefer simulation, but we have to pay with high simulation time for analyzing the model. The selected solution method can also depend on the aim of investigation. If we are interested in time-dependent behaviour or in distributions, then we have to decide on simulation. In general, analytical methods only provide mean values.

Hybrid methods combine the two other approaches in order to overcome their disadvantages. For instance, succesful applications are possible for models consisting of submodels with strongly different event frequencies. In pure simulation approach we would have to simulate a sufficient large number of events in submodels with low event frequency, i.e. a lot of events would be in submodels with high event freouency in the simulated time. We can assume that these submodels reach a quasi-equilibrium before there is an interaction between the submodels. So we can apply analytical methods to compute model characteristics such as mean residence times or throughputs in order to simplify the simulation procedure. Hybrid method has been proved to be cost-effective in analyzing such model classes. /1/

A program supporting hybrid method for queueing network models

A SIMULA program with following properties was implemented:

1. The description of extended open and closed queueing networks is supported in an interactive manner, including manipulations like reading, saving and changing a model description. An extended class of queueing networks was introduced in order to use more realistic models of computers and distributed systems.

1086 Berlin, Humboldt-Universitaet, PSF 1297

So we admit
- simultaneous resource possession (by population constraints in submodels),
- blocking and polling as special synchronisation mechanisms.

2. There are various solution methods for analyzing queueing network models:
   * simulation for any admitted queueing network model
   * exact and approximate analytic methods for various classes of queueing networks
   * special iterative analytical algorithms for special configurations.
3. Decomposition and aggregation of queueing networks is supported by commands creating modified model descriptions. For execution of decomposition the well known conditions (local balance or weak interactions between the submodels) must be valid in order to obtain sufficient exact results.

This tool allows to combine various solvers by offline evaluation of certain submodels. That can provide some advantages. (To simplify matters in the following we suppose that there are only two submodels, one inner and one outer model. See fig. 1)

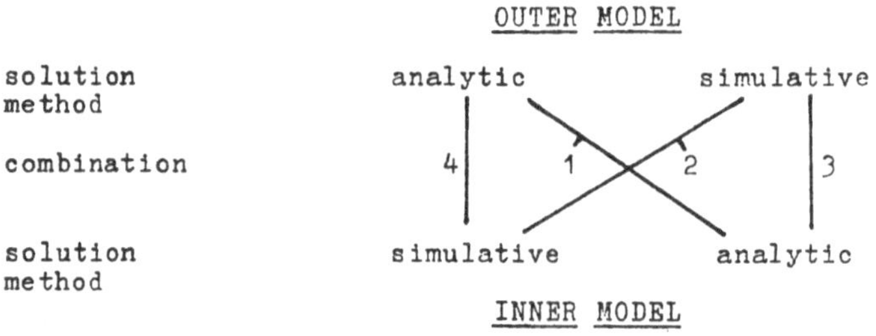

Fig. 1: Combinations of solution methods

Combinations 1 and 2 from fig. 1 are not hybrid, but they are useful in analyzing large models because of their reduced memory request. Further more, good results were obtained with combination 2 in reducing simulation time against simulation of the total model./2/
Most of the known models solved in hybrid manner correspond to combination 3. In some cases there are dramatic reductions of computer time and improvements of confidence intervals.
Finally, we consider models, which we can analyze only by simulation because of "bad" properties of a submodel. Solving this submodel by simulation, we have the possibility to analyze the remainding model by an analytical method (combination 3).

Load dependent server - an essential component of the hybrid approach

We have supposed that we can analyze the inner submodel offline by any technique. Which information do we need from this system? How can we embody the results of this analysis into the outer model?
Obviously, we can assume that the behaviour of the inner model depends on the customer population in this model. Therefore a load dependent server is used as a surrogate for the inner model. A load dependent server has a service rate which is a function of the customer population. The service rates correspond to the load dependent throughputs of the model, which we want to replace.
This is a powerful concept in parametric analysis of queueing network models, hierarchical modelling and approximation of submodels. Representing population constrained submodels is an other important application of load dependent servers./3/,/4/

Now we have to estimate the troughputs of the inner model for any customer population. In combinations 1 and 3 we get these values automatically by estimating the troughput for the total population. In combination 2 and 4 we only investigate the subsystem for some populations. All other throughputs are derived by interpolation.

Since load dependent servers are compatible with local balance queueing network models, the resulting outer model in combinations 1 and 4 can be evaluated efficiently. In combinations 2 and 3 we need an entity representing a load dependent server in the simulation program. Customer arrivals and completions effect changes in the service rates for all customers present at this server.
From analytical methods it is known that the processor sharing server and infinite server disciplines are special cases of the load dependent server. We make use of this fact in the simulation program. It is also possible to define a m-server node as a load dependent server.

There is another interesting property of simulation program component of load dependent server. We can imagine that some customers get service simultaneously, i.e. in general some customers are delayed at this server together. The delay times correspond to the customer residence time of the subsystem represented by this server. If we assume that there is an analytical expression or algorithm for evaluating the delay time, we can combine simulation method and analytical methods in an online approach. In this case we use the load dependent server as a "delay node" (infinite server node).Further more, we can include this component, representing a queueing network model for instance, into a general SIMULA - simulation program. So we have created a general connection between simulation and queueing network theory. This hybrid technique has proved to be a efficient method in performance evaluation of computer systems and packet-switching networks.

Ref.:

/1/ Schwetmann, H.D.
Hybrid Simulation Models of Computer Systems
Communication of ACM, Vol.21, No. 9, Sept.1978

/2/ Blum, A.; Donatiello, L.; Heidelberger, P.;
Lavenberg, S.; Mac Nair, E.
Experiments with Decomposition of Ext. Queueing Networks
Intern. Conf. on Modelling Techniques and Tools for
Perf. Analysis, Preprints, Paris, 1984

/3/ Chandy, C.H.; Sauer, K.M.
Computer Systems Performance Modelling
Prentice-Hall, Inc., Englewood Cliffs, New Jersey, 1981

/4/ Zahorjan, J.; Lazowska, E.D.
Incorporating Load Dependent Servers in Approximate Mean
Value Analysis
Perf. Eval. Rev., Vol.12, No. 3, Aug. 1984

A DIALOGUE COMPUTER MODEL AS AN AID TO THE OPERATOR FOR TAKING DECISIONS

Cl. Vasilev [1)]

ABSTRACT. A method for constructing computer program for algorithmic and high level language is considered. It is built-up for editing and processing of documents recorder of a two dimensional array. The method enables such a program to be written easily and faster than other known approaches. It may be applied for information activities, for example, in railway, motor or air transport, place reservations, time-tables and other domestic information processes.

KEYWORDS. Programming techniques, Editing, high level language.

The object of the paper is building an editing program enabling the users. When fulfilling information activities. The decision making means here that the user with the help of an in-instruction repertoire.

Let us S_1 denote the set of element-symbols by means of the algorithmic language works. The elements are one byte long for most of the minicomputers. Let $S_2 = S_1 \times S_1$ where x denotes a Descartes multiplication. Then S_2 represents the set of the feasible two-bytes symbols. If the number of the elements in S_1 is N, the same number for S_2 will be N^2 .Let also $P \subset S_2$, $Q \subset S_2$, $R \subset S_2$ be denoted. We propose the editing command to have the following format:

$\langle p \rangle \langle q \rangle \langle r \rangle$ with $p \in P$, $q \in Q$, $r \in R$.

We suppose also that $P \cap Q = P \cap R = Q \cap R = \langle a,b \rangle$ is fulfilled where a is a two-byte symbol which indicates an error; b is a two byte symbol which indicates end of the editing.

Other notations used are:

$S_a = \langle a \rangle \times Q \times RUP \times \langle a \rangle \times RUP \times Q \times \langle a \rangle$

$S_b = \langle b \rangle \times Q \times R\,U\,P \times \langle b \rangle \times R\,U\,P \times Q \times \langle b \rangle$

S is the set of editting commands:

$S = P \times Q \times R \setminus (S_a \cup S_b)$

The symbol $\setminus$ stands for the difference of the sets. Obviously, $S \subset S_2 \times S_2 \times S_2$. The block diagram of this editing programm is given on the Figure 1. The error message can be of seven types mostly. For example, the error message $\langle a \rangle \langle a \rangle \langle a \rangle$ means that the symbols $\langle p \rangle \langle q \rangle \langle r \rangle$ are not correctly composed: $(p \in S \setminus P) \wedge (q \in S \setminus Q) \wedge (r \in S \setminus R)$.

A program in FORTRAN for PDP-11 was written with the document as a two-dimensional array (see the Table).

1) Central Laboratory of Control Systems, 1113 Sofia, Bl.1o7

Table

№	A1	B1	C1	D1	E1
01					
02					
03					
04					
05					

1 - line arrangement
2 - column arrangement

A command is denoted in the Bacchus language by <p> <q> <r>

<p> ::= <operation>
<q> ::= <columns>
<r> ::= <number of the line>
<operation> ::= <adittion> V <delete in a few lines> V
V <delete> V <add in a few lines>
<column> ::= <A1> V <B1> V <C1> V <D1> V <E1> V
V (<A1> Λ <B1> Λ <C1> Λ <D1> Λ <E1>)
<line> ::= <01> V <02> V <03> V <04> V <05> V
V (<01> Λ <02>) V (<01> Λ <02> Λ <03>) V
V (<01> Λ <02> Λ <03> Λ <04>) V
V (<01> Λ <02> Λ <03> Λ <04> Λ <05>)

The program is recorded as (TSK.FILE) according the accepted terminology for PDP-11 and is called with the aid of RUN RED by a command of PDP-11, after which the user works with it aided by the described editing commands. Such editing programs are adequate in use in the folowing cases:

a) When for some reason , the utilisation of the program products for dialogue and editing is not needed.

b) When a speedy action for answering by editing is not requested.

c) When it is not expedient to work with other more powerful program systems.

The advantage of the editing program are as follows:
It is written with eax and fast by an operator.
A user not familiar with programming technique learns quickly to work with the aid only of a short instruction.
The program can be applied in creating real and simulating models , on purpose of research and control of different processes in the railway, motor and air transport,place reserva-

tions for a vehicle, and checu-up and time-table, and others.

REFERENCES:

1. Кузнецов, И.П. Кибернетические диалоговые системы, М.Наука,1977.
2. Павлов Р. Математическа лингвистика,Народна просвета,1982.
3. Хуторецкий, Б.И.,И.А.Медведь, Взаимодействие человека с ЭВМ в сфере управления,М.Статистика, 1976.

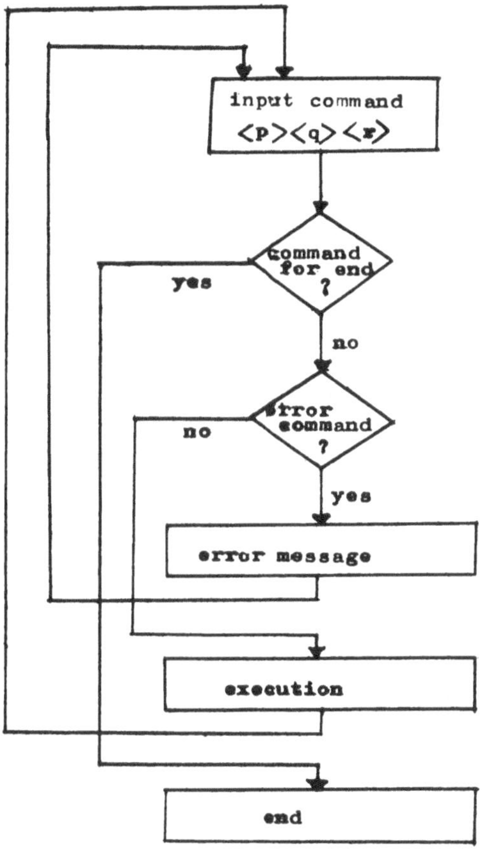
input command
<p><q><r>
command for end ?
yes
no
error command ?
no
yes
error message
execution
end

Fig. 1

KOMBINAT -A MULTILEVEL GAME FOR SIMULATING COMPLEX INDUSTRIAL PROCESSES

Hans Gernert [1)]

Management simulation games increase efficiency in management education and research. The game KOMBINAT is a multilevel simulation wherein participants take on the roles of managing large and complex socialist enterprises over several periods, representing 6 month of real time each. The negotiations between the teams of players are of special value in this exercise.
The comprehensive simulation model BES 4 is used to simulate the behaviour of the car producing factories in accordance with the decisions made by the teams of participants in fulfilling the plan targets. The program is written in FORTRAN and implemented on main frame computers. The use of video terminals is recommended.

Management game, multilevel game, simulation, management education, simulation model

Hierarchical multilevel management simulations are important in research on different economic strategies. They are an exellent tool for teaching students and senior management various concepts of managing enterprises efficiently.
The game KOMBINAT has been developed to simulate complex structures within the socialist industry. It aims at simulating several passenger car producing factories being organized under the joint management of the Kombinat.
As a multilevel management simulation KOMBINAT includes the following levels of responsibility:

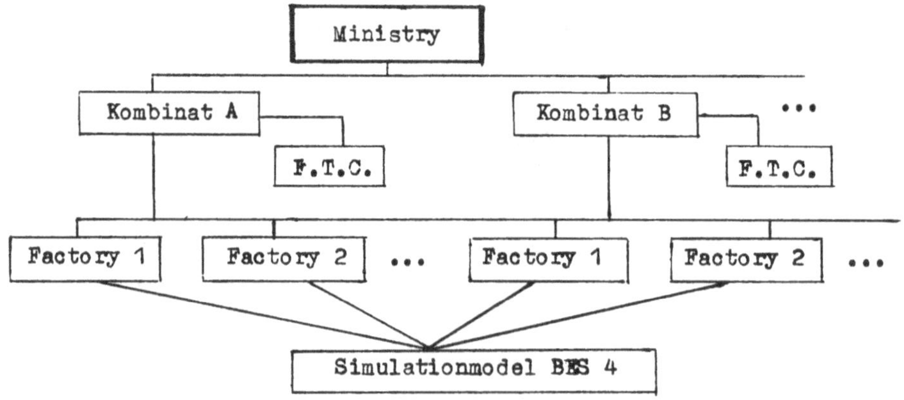

Doz. Dr. sc. H. Gernert
Humboldt-Universität zu Berlin, Sektion Wirtschaftswissenschaften, Bereich Statistik, 1020 Berlin, Spandauer Straße 1

The structure of this game corresponds to three levels of management in the real world.

1. Level of the Ministry of Industry
At this level the umpires represent the centralized planning authorities. All influences from the national economy of the GDR as a whole (plan targets, general orientation for future development, resource constraints etc.) are elaborated by the team of umpires that acts in the role of a minister of industry. Therefore they provide all data and information to run the game. The participants have to report to them on the results of their efforts to manage the factories in accordance with the given plan targets.

2. Level of Kombinat management
On the Kombinat level several player teams are in charge of running large enterprises consisting of one or more factories each. These factories are independent within the general frame of the Kombinat. The Kombinat management therefore has to break down all plan targets to the next lower level, to the factory management. In doing so the recent economical and technological situation of these factories has to be considered.
Special responsibility goes with in the Foreign Trade Corporation of the Kombinat. This enterprise has to carry out all negotiations as far as the import and the export business in concerned.

3. Level of factory management
The lowest level of management in the game is concerned with all the problems of developing the individual factories as highly efficient enterprises.
In order to fulfil the given plan targets decisions are necessary covering areas like production, marketing, fixed and current assets, manpower, working conditions, research and development, finance, etc.

To simulate the behaviour of these factories the computerized simulation model BES 4 (Gernert/Messerschmidt, Humboldt University Berlin, 1984) is used.
The algorithm consists of 19 modular parts that have to be calculated for each factory and each period.
To simulate <u>one</u> factory and period the program uses the following data-sets (in brackets are given the № of data items).

- Data - Set	:	starting situation	(175)
- -"-	:	plan targets	(10)
- -"-	:	decisions	(66)
- -"-	:	parameters	(267)
- -"-	:	results of simulation	(903)

In accordance with his specific objectives of running the game the user of BES 4 may generate as many Kombinat's resp. factories within each Kombinat necessary.
The simulation exercise can last up to 6 periods, each of them representing 6 month in reality.

There are several alternatives to run the game.
It is possible to use only the model BES 4 (without the administrative framework of the Kombinat) in order to simulate the behaviour of the factories alone.

The KOMBINAT game can be used as part of an international trade exercise GDR - Great Britain as well.

The model BES 4 is programmed in FORTRAN for main frame computer (EC 1022 etc), approx, 4500 statements.
The usage of video terminals is advisable but not essential.

The KOMBINAT model has been used in Universities and Polytechnics and in several courses for senior management since the 10. International Seminar on Computer Games, Humboldt University Berlin, December 1984.

RISK-1: A COMPUTERIZED SIMULATION GAME FOR FIRM VALUATION

E.Naumienko, L.Czarny [1/], B.Naumienko [2/]

Abstract. The purpose of this paper is to present structural features of a computerized management game RISK-1. The game combines together some principles of such disciplines as operations research, statistics, managerial accounting, decision-making processes, feedback processes, structured design and simulation techniques.

RISK-1 simulates firms competition in the field of crediting and placing their product on an unstable market. The internal structure of each competing enterprise is in the form of interconnected feedback loops combining together flows of material, orders, production,manpower, investment and finance. Choice of taxation, cost and profit determination methods out of different options is provided. Some synthetic financial ratios serving as criteria for firm valuation are also built-in the model.

RISK-1 has been structurally designed and fully programmed on ZX SPECTRUM.

Keywords: computerized management game, interconnected feedback loops, structured design, simulation, profitability evaluation, risk index.

INTRODUCTION

One way of gaining new insights into the behavior of specific socio-economic system is to create a game representing some of its important aspects. In the game people take responsibility for making decisions. The results of their decisions are controlled by a model or a set of verbal rules that represents the same cause-effect mechanisms that govern the behavior of the real world. In this way individuals can personally experience and experiment with the forces at work in determining system behavior, including the impact of delays, time pressures, conflicting goals, bad information and related forces [1].

Teaching courses in systems analysis, information systems design and dynamics of socio-economic systems or talking to people in companies top management we notices that most of them, while making analysis

1/ Dept. of Information Systems,University of Lodz,90-214 Lodz,Poland

2/ Institute of Systems Research ASG, 00-910 Warsaw, Poland

of existing systems tend to think in terms of separate events rather then in terms of structure and behavior patterns as results of certain structures. When people focus on events they tend to feel they are victims of outside forces or of circumstances outside their control. When they recognize the behavioral patterns within which their particular events occured they begin to undrstand the results. When they see the structure that causes the behavior, they begin to feel some sense of control.

Therefore, taking an advantage of having a microcomputer we designed a game that combines together some basic principles of such disciplines as operations research, managerial accounting, control theory, feedback processes, structured design and simulation techniques. This particular game serves as a tool for teaching systems thinking not only students but mainly companies executives.

DESCRIPTION OF THE GAME

RISK-1 is an on-line, multiplayer game in which the competition among firms operating in an unstable market conditions is simulated. Game participants share the same market with its own instabilities. Growth of each competing firm is assured by investments based on own or external sources. Bank crediting policy is based not only on classical criteria like balance sheets or profit-and-loss statements, but mainly on specially designed synthetic financial ratios, e.g. risk index to predict the probability of firm's bankruptcy within the next three to five years.

The winning company is the one with the lowest total risk index for the entire period of time and highest productivity and profitability trends.

Customer demand is generated as a random number and shared among competing firms in a reverse proportion to their market prices. Customer orders are placed in the backlog and fulfilled either from inventory or directly from production based on FIFO principle. There are certain costs associated with orders and inventories, like cost of holding inventory and cost of inventory shortages. Other groups of costs represent capital and labour utilization costs and carrying costs. The choice of cost determination and taxation method based either on proportional or an uneven taxation of profit elements is provided for each game. The internal structure of each competing enterprise is in the form of interconnected feedback loops combining

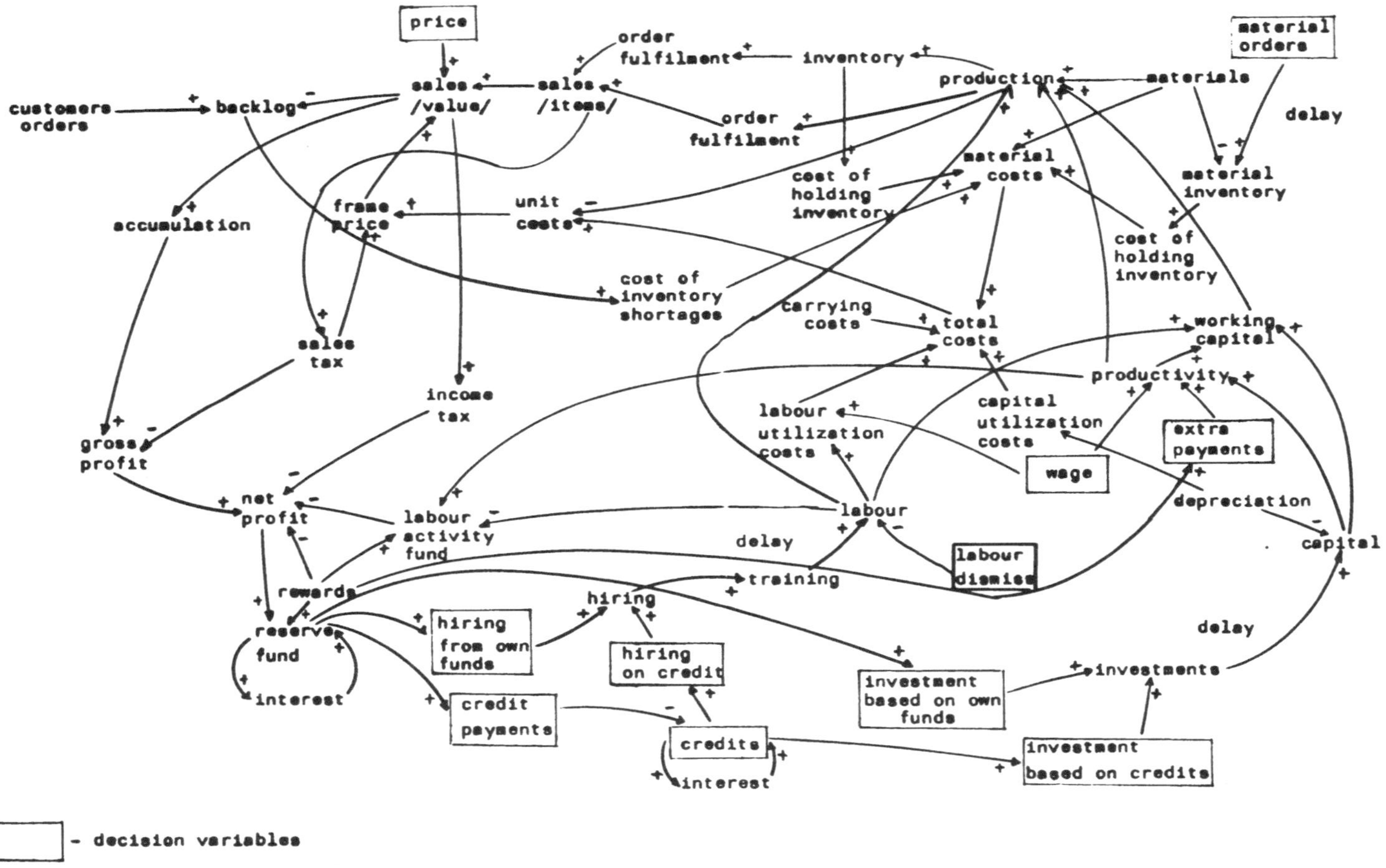

- decision variables

Fig. 1 Main interrelationships in RISK-1 firm model

together flows of material, orders, production,manpower and finance as shown on fig. 1.

We should notice, that some of interdependencies, like wage-productivity or profit-taxation rate are given in the form of table functions capital life-time is limitted with straight-line depreciation. We also assume that labour and material are not limited, credits are given based on risk index up to the value of equity capital. A firm goes bancrupt if its total liabilities, including interest exceed equity capital.

PLAYING THE GAME

While initiating the game, number of participants, number of runs, cost and taxation determination option and some other input parameters must be given. Each competing firm starts with the same initial conditions concerning number of workers, productivity, capital, price, wage, interest, customer orders and material orders.

Each run begins with making decisions for a given period of time /hiring or dismiss, wage and price level/ or the next one, e.g. level of investment from own or external sources, material orders, extra payments etc. After simulation of company's behavior with a given data the results concerning main variables, like number of workers, capital, production, labour productivity, profits, liabilities, rate of interest, and unit cost are printed. Based on certain financial ratios like equity capital and reserves, working capital, total liabilities, profit trend, current ratio, value of production and inventory, sales and receivables, all the companies are ranked and risk index is computed. According to this index and also profitability and productivity, the position of each firm on the market is established and banking decisions are made. The only information about other competitors which is obtained by a given firm concern their product price and market position. It is possible to form alliances with other players, f.ex. with a purpose to knock the most succesful competitor out of the no 1 slot, but not in the form of a fussion.

When the game is over, final results in tabelar and graphic forms are printed.

RISK-1 has been structurally designed and fully programmed on ZX SPECTRUM.

CONCLUSIONS

Games are great teaching tools. But it's almost impossible to build a game that teaches itself. You learn discussing. Therefore, if games are to be used in education, it is crutial that they be followed by a systematic debriefing, which consists of five parts [3] :

1. Summarize what happened during the game, what did people do, how did they feel, what did they pay attention to, what were the results for the outcome of the game?
2. What causal mechanisms, what structures were responsible for the outcomes and events reported under ≠ 1?
3. What would be an appropriate strategy, or useful heuristics for playing the game better next time?
4. How are the results in ≠ 1 and causal mechanisms in ≠ 2 related to what we observe in the real system?
5. In view of our conclusions in ≠ 4, what would be a set of effective strategies or heuristics for use in the real system.

REFERENCES

[1] Meadows,D.: "Production-Distribution Game Instructions", Dynamics of Sustainable Resources Workshop, Hungary, 1984

[2] Naumienko,E.: "Alternative Taxation Strategies", Organization and Management Quarterly, 1985 /in polish, in print/

[3] Senge,P.: "Debriefing of the Production-Distribution Game", Dynamics of Sustainable Resources Workshop, Hungary, 1984

[4] Tamari,M.: "Financial Ratios as Means of Forecasting Bankruptcy", Trends in Financial Decision-Making, Martinus Nijhoff Social Sciences Division, Leiden, 1978

[5] Wierzbicki,A.:"Critical Essey on the Methodology of Multiobjective Analysis", Regional Science and Urban Economics, vol. 13 , 1983

CASAM - AN INTERACTIVE PACKAGE FOR COMPUTER AIDED SYSTEM ANALYSIS AND MODELLING

A. VARGA#

Abstract. This paper presents an interactive package - CASAM, for the computer aided analysis and modelling of multivariable systems described either by state equations or by transfer-matrices. CASAM is based on several powerful, portable Fortran subroutine packages (BIMASC, BIMAS, EISPACK, LINPACK, ODEPACK) which implement the latest advances in numerical algorithms. All functions are performed by the means of a command language. The CASAM package is implemented on the Romanian family of minicomputers INDEPENDENT and CORAL, as well as on the DEC PDP-11 systems.

Keywords: System analysis; Modelling; Computer aided design; Numerical methods; Simulation.

INTRODUCTION

In the last few years, major developments have been achieved in elaborating efficient and reliable algorithms for most computational problems in the analysis, modelling and simulation of linear and nonlinear systems. The existence of high performance linear algebra packages LINPACK [2] and EISPACK [5],[8] as well as the ordinary differential equations solvers package ODEPACK [6] contributed decisively to reliable computer implementation of these algorithms. Recently, two powerful packages of portable Fortran subroutines, BIMAS [13] and BIMASC [14] have been developed for the computer aided system analysis and design. These packages implement the latest advances in numerical algorithms using the highest quality available numerical software.

CASAM is an interactive Computer Aided System Analysis and Modelling package which is mainly based on BIMAS and BIMASC packages. Most of modelling and analysis functions are provided for linear continuous or discrete models in standard state-space form. However several programs can deal with linear models in generalized state-space form. Simulation programs are provided for both linear and nonlinear systems. Graphical facilities are included for plotting simulation results. All functions are performed by the means of a command language. A flexible data organization allows an easy communication among various CASAM programs.

LINEAR SYSTEMS MODELLING

The purpose of modelling primarily consists of determining those forms of systems models which are appropriate for the analysis of systems properties or which can be used in the simulation. Some model transformations are provided in order to obtain better conditioned state-space parameterization of models or reduced order approximate models. The use of

Central Institute for Management and Informatics
Bd.Miciurin, No.8-10, 71316 Bucharest, Romania.

$$A' = Q^{-1}AQ,\ B' = Q^{-1}B,\ C' = CQ,\ D' = D \qquad (6)$$

where Q is a given or computed transformation matrix. TSO can be used to reduce by orthogonal transformations, the state-matrix A to Hessenberg or ordered real Schur forms. TSI uses non-orthogonal transformations to reduce A to condensed forms (balanced, Hessenberg, Schur, block-diagonal [1]). TCF uses orthogonal similarity transformations to compute standard controllability-observability forms of the system [11]. TBAL and TRED perform system balancing transformations [7]. TRED can also be used to compute reduced order approximate models.

For GSSM, two programs GTSO and GTCF perform orthogonal similarity transformations of the form

$$E' = Q^TEZ,\ A' = Q^TAZ,\ B' = Q^TB,\ C' = CZ,\ D' = D \qquad (7)$$

where Q and Z are orthogonal transformation matrices. GTSO reduces the pair (E,A) to triangular-Hessenberg or triangular-ordered real Schur forms. GTCF computes controllability or observability canonical forms [16].

SYSTEM ANALYSIS

The CASAM package provides several programs for the analysis of properties of linear state-space systems. The programs SMA and GSMA can be used for the analysis of stability, controllability-stabilizability and observability-detectability of SSSM and GSSM, respectively [11],[9]. Both programs also computes the system poles. MZE and GMZE are programs for computing the multivariable systems zeros using the algorithms in [3],[9].

The simulation of systems serves as a powerful tool for analysis. SDS is a simulation program for discrete linear systems, while SCS and SSCS are programs for simulating continuous linear systems. SSCS is appropriate for systems described by "stiff" differential equations or for high accuracy requirements. GSSC can be used for the simulation of linear systems described by GSSM. SCS uses the Runge-Kutta-Fehlberg method [4], while SSCS and GSSC are based on implicit Adams and backward difference methods [6]. All these programs exploit the linearity of the system, by reducing the number of operations required to evaluate derivatives [15].

Nonlinear systems described by the explicit system of differential equations

$$\dot{x}(t) = f(x(t), u(t), t)$$
$$y(t) = g(x(t), t)$$

or the implicit system of differential equations

$$A(x(t),t)\dot{x}(t) = f(x(t),u(t),t)$$
$$y(t) = g(x(t),t)$$

can be simulated using the programs SNS and GSNS, respectively. This programs are based on the LSODE and LSODI packages which belong to the ODEPACK collection of ordinary differential equation solvers [6]. Simulation results can be plotted using the program PLOT.

CASAM modelling programs assumes the availability of the linearized models of the physical systems which are studied. These models are determined either from model building using the basic physical principles or through system identification.

The CASAM modelling programs deal with linear time-invariant systems described by standard state-space models (SSSM) of the form

$$\begin{aligned} \lambda x(t) &= Ax(t) + Bu(t) \\ y(t) &= Cx(t) + Du(t) \end{aligned} \qquad (1)$$

or by generalized state-space models (GSSM) of the form

$$\begin{aligned} E\lambda x(t) &= Ax(t) + Bu(t) \\ y(t) &= Cx(t) + Du(t) \end{aligned} \qquad (2)$$

where x, u and y are the state, input and output vectors, respectively, A, B, C, D, E are constant coefficient matrices, and where λ is the differential operator d/dt for continuous systems or the forward shift operator $\lambda x(t) = x(t+1)$ for discrete systems. An alternative system description accepted by some CASAM programs is the input-output description given by

$$Y(\lambda) = G(\lambda)\ U(\lambda) \qquad (3)$$

where Y and U are the transform output and input vectors, respectively. In continuous case, the Laplace - transform is used, while in discrete case the Z - transform is used. The matrix $G(\lambda)$ is the system transfer matrix and generally is a rational matrix.

We list below the modelling facilities implemented in the CASAM programs.

a. Discretization

TMCD performs the discretization of continuous input-output models using the matrix exponential method [10]. TSCD performs the discretization of SSSM using the matrix exponential method with Padé approximations [10], combined with a very efficient block-diagonalization procedure [1].

b. Transformations between input-output and state-space descriptions

These two system descriptions are related by the relations

$$G(\lambda) = C(\lambda I-A)^{-1}B + D \qquad (4)$$

for SSSM, and

$$G(\lambda) = C(\lambda E-A)^{-1}B + D \qquad (5)$$

for GSSM. The programs NMR and GNMR construct state-space realizations for proper and non-proper transfer matrices, respectively. These programs determine generally non-minimal order state-space models and therefore they are usually followed by MNR and GMNR, respectively, which extract minimal order modells from non-minimal ones, using orthogonal similarity transformations [11],[16]. The programs TSM and GTSM computes the transfer matrix corresponding to a SSSM or GSSM, respectively [13], [16].

c. Similarity transformations

Several CASAM programs performs similarity transformations of the form

CASAM OPERATION AND IMPLEMENTATION

All functions of CASAM are activated by the means of a simple command language. Each program, when it is activated, fetches its input data from the disk and puts its results on disk. The outputs of most programs can be used as inputs to other programs of the package. A flexible data organization allows an easy communication ammong various CASAM programs. CASAM can solve problems with 35-40 state variables, all computations being done in double precision. The programs are written in Fortran, excepting a few routines written in MACRO-11 language. CASAM is implemented on the Romanian family of minicomputers INDEPENDENT and CORAL, and on a DEC PDP-11/34 system.

REFERENCES

[1] Bavely, C.A., and G.E. Stewart: An algorithm for computing reducing subspaces by block diagonalization, SIAM J. Numer. Anal. 10(1979), 359-367.

[2] Dongarra , J.J., J.R.Bunch, C.B.Moler, and G.W.Stewart: LINPACK User's Guide, SIAM, Phyladelphia 1979.

[3] Emami-Naeini, A., and P. Van Dooren: Computation of zeros of linear multivariable systems. Automatica. 18 (1982), 415-430.

[4] Forsythe, G.E., M.A. Malcolm, and C.B. Moler: Computer methods for mathematical computations. Prentice-Hall, Englewood Cliffs,1977.

[5] Garbow, B.S., J.M.Boyle, J.J.Dongarra, and C.B.Moler: Matrix eigensystem routines - EISPACK guide extension, Lect. Notes in Comp. Scie., vol.51, Springer Verlag, Berlin 1977.

[6] Hindmarsh, A.C.: Large ordinary differential equation systems and software. IEEE Control Systems Magazine. 2 (1982) 4, 24-30.

[7] Moore, B.C.: Principal component analysis in linear systems: controllability, observability and model reduction. IEEE Trans. Autom. Control. AC-26 (1981) 17-32.

[8] Smith, B.T., J.M.Boyle, J.J.Dongarra, B.S.Garbow, Y.Ikebe, V.C. Klema, and C.B. Moler: Matrix eigensystem routines - EISPACK guide. Lect. Notes in Comp. Scie., vol.6, Springer Verlag, Berlin, 1976.

[9] Van Dooren, P.: The generalized eigenstructure problem in linear system theory. IEEE Trans. Autom. Control. AC-26 (1981), 111-129.

[10] Van Loan, C.F.: Computing integrals involving matrix exponentials. IEEE Trans. Autom. Control. AC-23 (1978), 395-404.

[11] Varga, A.: Numerically stable algorithm for standard controllability form determination. Electronics Letters. 17 (1981), 74-75.

[12] Varga, A., and V.Sima: A numerically stable algorithm for transfer function matrix evaluation. Int. J.Control. 33 (1981), 1123-133.

[13] Varga, A. and V.Sima: BIMAS - A basic mathematical package for computer aided analysis and design. Preprints of the IFAC 9th World Congress, Budapest, 2-6 June, 1984.

[14] Varga, A., and A.Davidoviciu: BIMASC - A package of Fortran subprograms for analysis, design and simulation of control systems. Preprints of 3rd IFAC Symp. on CAD in Control and Engineering Systems, Copenhagen, July 31 - August 2, 1985.

[15] Varga, A., V.Sima, and C.V.Varga: On numerical simulation of linear continuous control systems. Preprints of SIMULATION'83 Symp., Prague, June, 1983.

[16] Varga, A.: Transfer functions evaluation of generalized state-space models. Report ICI, TR-05.83, 1983.

PACSID - A SOFTWARE PACKAGE FOR DYNAMIC SYSTEMS MODELLING, DESIGN AND SIMULATION

I.Dumitrache[1)], M.Dumitru[2)], C.Vasiliu[2)], C.Oprică[3)]

Abstract. PACSID is an integrated software package for dynamic systems computer aided modelling, design and simulation. All necessary interface between user's application and the package is provided by a specialized data base. Simple or complex dynamic structures can be handled by a nucleus for symbolic manipulations, which enables the user building models with any degree of complexity and also operations in order to simplify or obtain canonical forms. The simulation module is fully interactive and includes more than 10 integration methods, some of the most performant published algorithms; various graphic facilities are also provided. PACSID contains also a parametric optimizer for dynamic systems with given structure.

Keywords. Computer aided design; symbolic computation; computer graphics; digital simulation.

1. INTRODUCTION

During last years a special attention has been paid to CAD projects for control systems. A number of software tools (program packages) with strictly dedicated goals were created, in some cases combined into interactive software systems.

This paper describes an integrated medium, PACSID, for the problems appearing in the field of dynamic systems modelling, design and simulation. It is intended to provide the interested users a concise survey on the organization and the main functions/subsystems of PACSID; consequently, in the next sections, special emphasis is put on the distinctive features of PACSID viewed as a general CAD tool, a few remarks concerning its operation and potential applications being also added when appropriate.

2. SYSTEM ORGANIZATION

The PACSID system has a modular organization. Reflecting the functions implemented up to the present state of development, the following subsystems (modules) are defined (Fig.1): modelling, simulation, graphics, optimization and data base manager. In the following essential information is provided for each of these modules.

2.1. The modelling subsystem

This subsystem allows the interactive description of the structure of dynamic systems, in a specially designed language. The models may be

1) Polytechnical Institute of Bucharest, 77206 Bucharest, Splaiul Independenţei No.313, ROMANIA
2) Central Institute for Management and Informatics, 71316 Bucharest, Bd. Miciurin No.8-10, ROMANIA
3) I.C.L. Tehnometal, Str. 30 Decembrie No.2-4, Bucharest, ROMANIA.

linear or nonlinear, with concentrated or distributed parameters [1]. The user can describe simple or complex dynamic structures with the aid of a symbolic manipulations nucleus (SMN), which enables building structures with any degree of complexity and also operations in order to simplify or obtain canonical forms. The SMN provides the necessary internal processing of symbolic data, the editing of symbolic forms and the I/O interface with the user.

The main components of the modelling subsystem are (see Fig.2) :

a) I/O monitor, for the dialogue with the user. Its functions are: decoding input strings in internal format, checking for coherence all symbolic expressions, printing out in 1D or 2D symbolic expressions and editing (creating, updating or deleting) symbolic expression from the system data base.

b) Basic functions executive, the module which provides the specific symbolic operations (at system level) [3]:

- arithmetic operations in the field of rational numbers (in order to eliminate round off errors) ;
- basic operations with polynomials;
- symbolic differentiation of an expression respective to a certain variable ;
- basic operations with symbolic matrices ;
- simplification of symbolic expressions according to a predefined set of rules ;
- total or partial evaluation of a symbolic expression and substitution;
- automatic generation of complex models (aggregation) starting from their structural chains.

The editor of the modelling subsystem provides various facilities for symbolic data manipulations: definition of individual or group characteristics, definition of global and local options, editing symbolic expressions, saving/deleting symbolic expressions into/from user context area, substitution of variable with symbolic expressions, partial or total evaluation of symbolic expressions, trace-back of user activity during an editing session.

Summarily, the available commands and facilities enable the user to construct symbolic models step-by-step. The symbolic expressions, internally stored in prefixed Polish format, are manipulated by their names and can be used to produce new expressions.

Consequently, for complex models only the structure (i.e. the links between components) is required by the aggregation module in order to generate the global symbolic model.

2.2. The simulation subsystem

The simulation subsystem performs the integration of dynamic models, either generated via the modelling subsystem or specified from outside the PACSID medium (but observing PACSID conventions).

The simulation subsystem consists of :

- the interactive input data-file editor, to specify actual parameters of the model, integration parameters and I/O options ;
- the object modules library, which contains the necessary routines to build the executable image of user's applications, starting from the model. This library contains also predefined non-linear blocks that can only be invoked in user's model, in order to reduce the modelling effort;
- the on-line interrupt processor, a module intimately related to the operating system. Both user's interrupts from the application terminal and operating system synchronous interrupts (such as arithmetic overflows) are treated.

Pertaining to the simulation subsystem, the following functional details should also be noted :

a) user's models are written in FORTRAN IV; no check whether the equations are correct or not is provided.

b) there is no precompilation of user model to generate special code for the FORTRAN compiler, like in CSMP/360 job oriented language program; however the user can compile and correct models separately, during the initial development without having to load the larger simulation subsystem ,

c) the simulation subsystem can deal with either ordinary differential equations (ODE) or partial differential equations (PDE), providing automated integration of ODE and automated spatial discretization for PDE. Partial differential equations are converted into sets of ordinary differential equations by replacing the partial derivatives with appropriate discretization formulae;

d) a number of 12 integration methods based on Runge-Kutta-Fehlberg and Gear algorithms [2] are available.

The Gear methods use the Reid sparse matrix routines in the Hindmarsh-Gear algorithm. They are divided into two categories: Adams formulae for general equation sets and Gear backward difference formulae for stiff equation sets. All of these methods use predictor-corrector formulae of variable order, chosen as to optimize the step size. The convergence of thepredictor-corrector iteration determines the maximum allowable step size; to accelerate convergence, it is important to perform the matrix manipulation involving the Jacobian matrix in as accurate and efficient manner as possible.

The Runge-Kutta-Fehlberg methods are robust, economical in storage and efficient in many cases.

The output files produced by the simulation subsystem are processed by the graphic subsystem; so, if the results don't fit with user's objectives an interrupt can be generated from the terminal causing the input menu to be activated in order to make appropriate modification in the model or the integration parameters.

2.3. The graphis subsystem

This subsystem gives the user the ability to obtain simulation results

in various forms, depending on the output device facilities: alphanumeric, semigraphic or full-graphic. The output terminal can be a simple line-printer or a CRT - terminal. We selected the GKS standard for the implementation of the graphic nucleus; depending on user's available terminal appropriate software interfaces are provided. To work with PACSID only a few of the graphic nucleus facilities are needed.
Common graphics in 2D with default abscisa the independent variable time is available but any two model variables can be correlated in a such 2D diagram.
For PDE equation sets, a 3D module is provided for terminals with graphic or semigraphic facilities. Gridded surfaces will be traced and details of such diagrams can be expanded on the whole screen for better analysis.

2.4. The optimization subsystem

For static models, a library of optimization algorithms is available, which can be linked with user routines (either provided by the modelling subsystem, or obtained externally to PACSID).
The algorithms are well known in the scientific literature; for reference purposes we point out some of them: SUMT (the Fiacco-Mc. Cormick procedure for nonlinear programming with penalty functions), FLEXI (the Himmelblau-Paviani algorithm with the Nelder-Mead hyperpolyedron method), GRACOMIN (the conjugated gradient algorithm), COMPLEX (the improved Box algorithm) and so on.

2.5. The data-base manager subsystem

The data-base of PACSID system has two parts: kernel and user, and provides all necessary interfaces between the system components.
The Kernel part of data-base contains the predefined symbolic models available to the user and the configuration parameters for all the subsystems (established at the installation phase according to the hardware and software options of the host system).
The user part of data-base contains all symbolic expressions and models described by the user during his terminal sessions, the trace-back files for automatic generation of a particular model and all the temporary working files which make possible the communication between the subsystems of PACSID in activity.

3. CONCLUDING REMARKS

The PACSID integrated software system is operational on DEC PDP-11 compatible minicomputers. It is an efficient CAD tool for such applications as: design of industrial control systems, design of industrial robots, modelling and simulation of nonlinear processes (diffusion, thermal reactions, nuclear power plant), sensitivity analysis.

References:

1. Dumitru, M. and I.Dumitrache: PROSIM-A package for design and simulation of dynamic system,Proc. A.I. Symp., Lille, 1983.
2. Dumitru M.: SIO - A software system for high-level process control functions Proc. MIMI Symp., Davos, 1982.
3. Dumitru, M. and C.Vasiliu: Applications of artificial intelligence techniques to control system design, Univ. of Galaţi, 1982.

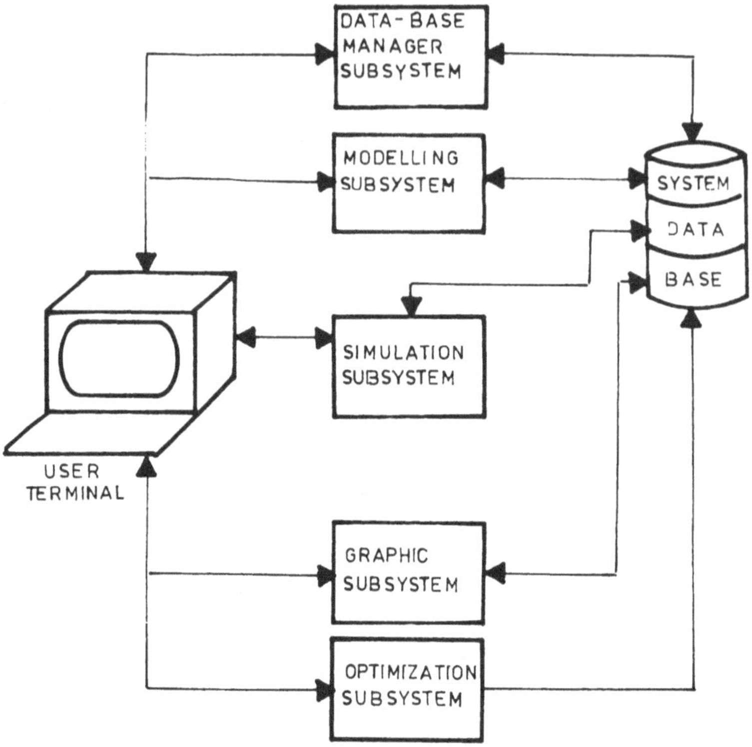

Fig 1 PACSID System structure

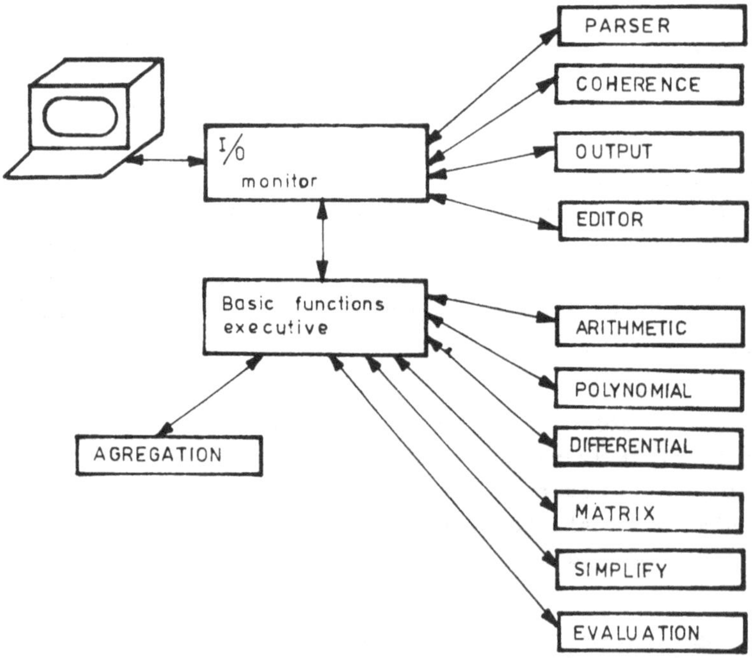

Fig. 2 Modelling subsystem components

SONCHES - AN INTERACTIVE SIMULATION LANGUAGE FOR DESIGN, VALIDATION AND USAGE OF ECOLOGICAL MODELS

Wenzel, V., Bellmann, K., Matthäus, E., Flechsig, M. [1)]

Abstract: A simulator for ecological models is presented. Modelling language is shortly characterized. The main components of the simulation system are mentioned in connection with corresponding services. A full list of ecosystems modelled in SONCHES-calculus and running at present time is given.

Key words: model building, simulation, language, ecological system

1. Language characteristics

SONCHES (completely: Simulation of Nonlinear Complex Hierarchical Ecological Systems) is a tool for computer aided experimental investigations of the dynamical behaviour of ecological systems. Subject of the investigations may be agro-, aquatic or forest ecosystems as well.
The core of the simulation system is the modelling language based on a comprehensive ecosystem prototype. Above the structure of a net rate equation model in terms of a general (non-branch-specific) simulation language, more-specifically ecological structure can be expressed and recognized according to the genuine structuring principles for this kind of systems (trophy, ontogenesis, food uptake dominance etc.).
A special semantics is realized by distinction of suitable types of variables and identification of ecological objects. Most important, however, is the classification of elementary ecological processes into the types listed in figure 1:

DD	driving forces dynamics	AG	abundance dependent growth
SO	self-organization	EG	environment dependent growth
SA	self-adaptation	OG	output dependent growth
EN	compartmental environment	AM	abundance dependent mortality
CO	input coupling	EM	environment dependent mortality
DE	compartmental demand	OM	output dependent mortality
OU	compartmental output	AD	ontogenetic advance
TO	trophic output	MF	migration/fluxes
HC	habitat control	ST	state change
TS	Trophic supply		
OT	competition		

Fig. 1

1) Central Institute of Cybernetics of the Academy of Sciences of the GDR, 1086 Berlin, Kurstr. 33

blish its most efficient form for usage.

Further advantages:

- unification, simplification and shortening of model building procedures
- comparison of different models concerning the same matter
- initiation of systematical measurement programms
- increased efficiency in exploitation of disposable data
- computation of forecasts
- design and optimization of control procedures for minimization of damages, losses and desages as well as for maximization of yields.

3. Applications

The following systems have been modeled in SONCHES-calculus and are running to solve certain simulation tasks:

PILOT - agroecosystem of crop and 3 pests for development of combined pest management; component models are

WHEAT - growth model of winter wheat (triticum)

OULEMA - population model of a cereal leaf beetle (outlema lichenis)

LAUS - population model of a grain aphid (macrosiphum avenae)

EMMO - eutrophication model of lake Müggelsee

BEM - eutrophication model of lake Balaton

TUNDRA- food chain model of grass-lemmings-polar foxes-hunters for determination of optimal fox hunting rate

BIOBENZ-toxicokinetic model of benzen metabolism and distribution in human tissues

BETA - sugar beet growth model

FOREST- forest yield table model

SUCC - model of secondary succession of a fallow cropland

PHARM - pharmacokinetic demonstration model

EPIDEM- dynamics of susceptible, sick and cured man according to an epidemics

FORA - simplest Volterra-lotka-model for demonstration

4. Some technical details

SONCHES is running on PDP-like computers (PDP-11/40, PDP-11/70, CM-4/20, I102F, A 6402) under RSX-11M and similar operating systems.

Memory requirement: 128 K Bytes (using of VIRTUAL)

Basic language: FORTRAN 77

SONCHES is developed at the Central Institute of Cybernetics and Information Processes in Berlin

Here are some of the languages characteristics:
- time-discrete dynamics (difference- and recurrence equations)
- spatial structure to be mapped as logical or functional relations between homogeneous habitats
- elements are compartments bearing some ecologically relevant funtion, measured by individual units free for users determination
- processes are ecologically interpreted terms of equations
- flexible aggregation of partial rates
- dynamical self-organization
- processes with different time steps
- macro-processes (and-variables) for efficient mapping of age class structure
- automatic rate balances (multiple balance within one time step)
- referable rate variables
- support of an unique data base (time series and function tables)

2. Simulation system

The modelling language forms the entrance to the simulation system, interactively supporting all user activities. Figure 2 gives a block structure diagram of the systems main components:

Model input is followed by structure analysis and translation, compilation and linkage of model equations, detection and correction of syntactical and semantical arrors.
The simulation experiments are governed by the command language and the corresponding language interpreter to determine the contents of the current scenario. Solving of complex multi run simulation tasks, especially in connection with model validation purposes, is supported by various facilities like parameter behaviour and sensitivity analysis, parameter optimization, Monte-Carlo-studies. Further, for debugging a one step regime with full information about process performances can be activated.
Details of the users model are made transparent by the component of model self-documentation.
Model and data are changeable. Above a recalibration of numerical values the change of process functions and of the topological model structure itself is possible.
Seperate models stored in the model library can be linked for getting a more comprehensive one. The latter options are useful for model synthesis. One starts with simplest structure adds successively more structure to investigate more and more complex models with respect to relevance and sensitivity of certain variables, processes and parameters. By means of these informations the model is then to be reduced to esta-

Further installations:
Institute of Plant Protection Research in Eberswalde
Institute of Water Resources in Berlin

At present a transfer onto a Honeywell Bull computer of the Hungarian State Computer Service (AS_zS_z) in Budapest is performed.

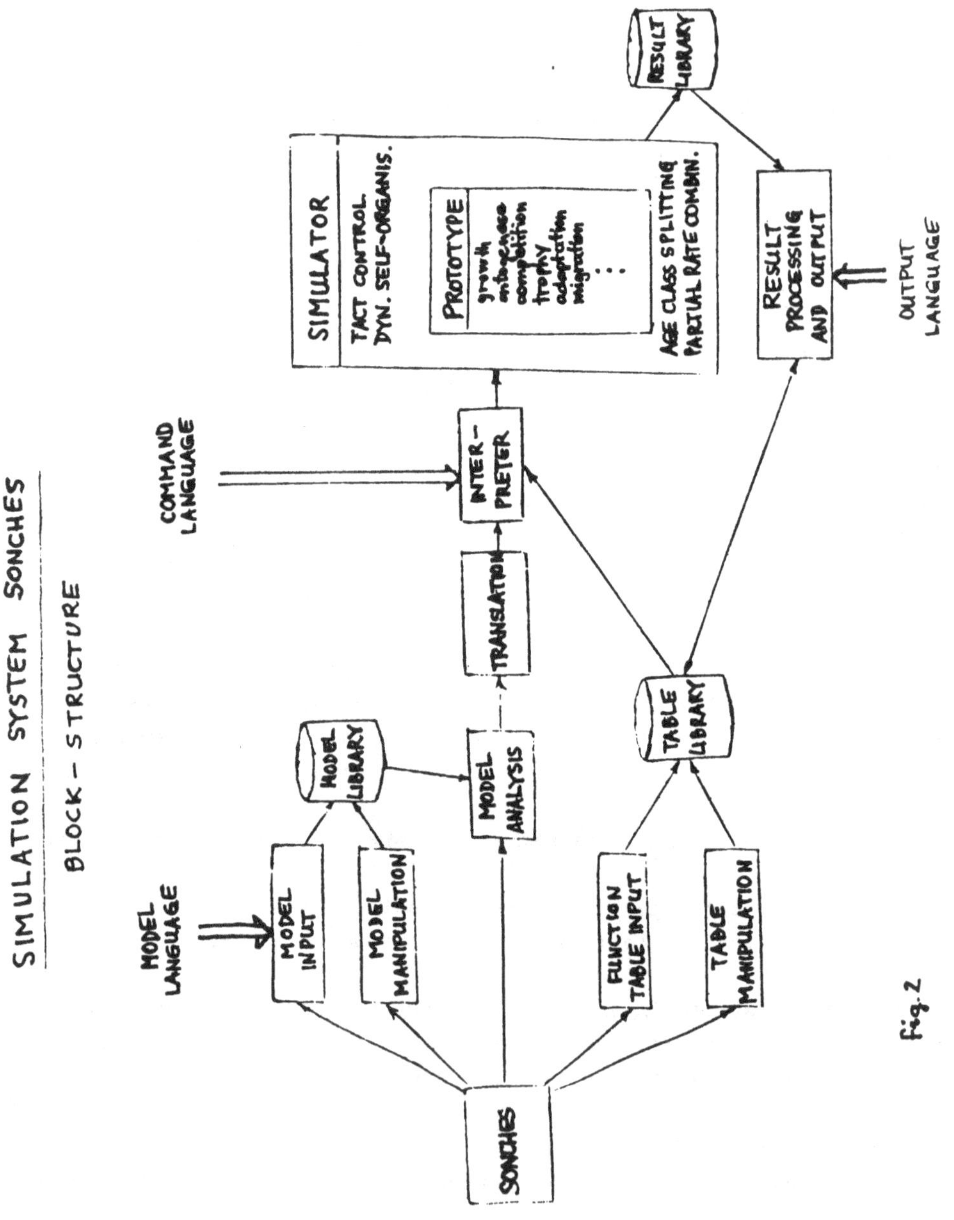

Fig. 2

USE OF TASK FACILITIES OF THE SIMULATOR "SONCHES" FOR AGROECOSYSTEM-MODEL INVESTIGATIONS AND CONTROL

E. Matthaeus; M. Flechsig; K. Bellmann; V. Wenzel [1)]

Abstract: Multi-run facilities of the simulation system SONCHES are introduced and their use for investigation of an interdisciplinary agroecosystem model with special view to pest-management are demonstrated. Despite of the general limitations of simulation results strategic recommendations for pest control can be given.

Keywords: Simulation, Simulation system, Model evaluation, Model use, Agroecosystem control policy design

The simulation system SONCHES /1, 2/ has been improved recently to support the systems analysis of complex ecological systems by an user-friendly command language and command interpreter for multi-run options. These enlargement of the interactive system helps the modeller not well-trained in mathematics and programming to concentrate on testing and using his model. The dialogue is based on menue-technique for selecting different tasks and a language calculus for specifying targets and operations within the task. The following tasks are available for model experimentation besides the usual single run performance

(i) behaviour analysis
(ii) sensitivity analysis
(iii) optimization
(iv) Monte Carlo simulation
(v) debugging

Model error analysis /5, 6/ can be performed either by behaviour and sensitivity analysis or by Monte Carlo simulation. Long term simulation is supported by a command file technique and dialogically prepared batch jobs.
Model results can be presented on video terminal, graphic terminal and line printer. Time series are presented as tables, curves versus time or phase portraits with automatic scaling. Besides time series for variables and referable rates, extreme values and moments defined over time intervals (aggregated variables) can be specified. Results can be stored in the SONCHES library for later use.

Behaviour analysis comprises the presentation of model behaviour /3/ as function of parameters, initial values or driving force dynamics. Finite sensitivity analysis comprises trajectory, output or performance sensitivity function calculation /4/. Different sensitivity functions

---

[1)] Central Institute of Cybernetics and Information Processes, Academy of Sciences, GDR, Berlin like linear, squared or relative sensitivity coefficients can be selected. Linear sensitivity coefficients and maximal deviations can be used for error analysis based on small perturbations /5/. Monte Carlo simulation comprises a stochastic simulation of moments of model variables for parameter or initial value distributions or for instantaneously fluctuating process rates. Monte Carlo simulation of the first kind is used for an error analysis in the large /6/ taking into account also parameter correlation. Different moments and extreme values, correlation among variables and between variables and parameters can be specified. Distribution of variables and rates can be displayed for any time step.

All task and result performance commands are checked syntactically and are supported by a help service. For repetitive tasks in routine-jobs the commands can be stored in a command file and the system performs automatically all calculations up to the result presentation.

The interdisciplinary research project for systems analysis of the agroecosystem winter-wheat /7/ has been led to a complex agroecosystem model containing crop and pest model prototypes - a grazing pest (leaf beetle), a sucking pest (grain aphid) and a fungous disease (mildew) - pest management control possibilities for spraying pesticides and for biological control by aphid's predators, and yield programming control possibilities like irrigation and fertilization. Mapped interactions are matter flows from crop to pest and both secondary effects of pest to crop photosynthesis and senescence, and secondary effects of crop to pest mortality and reproduction in the case of uptake limitation. Spraying is mapped as additional mortality of the pest.

In the following the paper concentrates to behavioural studies of the crop damaged by different pest infestation strengths. For the time discrete agroecosystem model M_A

$$M_A = \langle X, Z, \delta, Y, \lambda \rangle$$

X input space $X \subset R^n$
Z state space $Z \subset R^m$
Y output space $Y \subset R^p$
δ time discrete state transition function
$z_t = \delta(x_t, z_{t-1}) \qquad z \in Z$
λ output function
$y_t = \lambda(z_t) \qquad y \in Y$

we only regard the feasible subset of states of the real system. The behaviour B is defined as collection of mappings β_z of input sequences on output sequences for different states. $B = \{\beta_z\} \quad \forall z$

$$\beta_z : [x_{t_0}, \ldots, x_{t_n}] \longrightarrow [y_{t_0}, \ldots, y_{t_n}]$$

For demonstration we restrict the approach to one input sequence (special year), i.e., we are interested in the behavioural subspace which reflects the state dependence of the output sequence. From a practical point of view the most interesting variable of the crop model is the grain yield (G) depending on the leaf beetle (Ouelema (O)) and grain aphid (macrosiphum (M)) infestation strength determined mainly by the corresponding immigration. To demonstrate the use of task facilities the user's dialogue with the simulation system is sketched.
From the task menue

```
* *    SIMULATION TASK OPTIONS    * *
---------------------------------------


R    -    SINGLE RUN
D    -    SINGLE RUN WITH DEBUGGING
P    -    TEMPORARY PARAMETER CHANGE
I    -    TEMPORARY INITIAL VALUE CHANGE
---------------------- MULTIRUN OPTIONS---

B    -    BEHAVIOUR  ANALYSIS
SP   -    PERFORMANCE SENSITIVITY ANALYSIS
ST   -    TRAJECTORY SENSITIVITY ANALYSIS
A    -    PARAMETER ADAPTION
MP   -    MONTE CARLO EXPERIMENTS WITH
          PRE-RUN PERTURBATIONS
MT   -    MONTE CARLO EXPERIMENTS WITH
          TACTWISE PERTURBATIONS
-------------------------------------------

YOUR OPTION:
```

you have to choose the option B. After entering the information for run terminations the system prompts

```
ENTER COMMAND LINES FOR BEHAVIOUR ANALYSIS (PRESS  ^HELP FOR SYNTAX
                                            PRESS  ^BREAK FOR INTERRUPT)
------------------------------------------------------------------------
```

The command syntax has the following structure, the task specificator "BEHAV:" is always prompted at the beginning of a new line

BEHAV: <target> <operation> <specification list>

The target variables are initial values (immigration) of the pest (NI, IMZA), the operation can be an additive, multiplicative or relative change of the target variable corresponding to the values contained in the specification list either as list of increments or loop specification. For example

```
BEHAV: NI ADDITIVE 5 TIMES 12
BEHAV: IMZA A    200 T    20
BEHAV: COMB: (NI * IMZA)
```

These commands lead to a change of NI by 5 units 12 times and of IMZA by 200 units 20 times. The combination command lines allow a sequential change (separator +), simultaneous change (separator ,) and a combinative change (separator *). The combination command line leads to 273 runs including one run with the nominal initial value configuration. After specifying the result variables grain yield (G) at harvest date (time step 200) using the interval specification keyword _I

```
ENTER COMMANDS FOR RESULT STORAGE       (PRESS   ^HELP FOR SYNTAX
                                         PRESS   ^BREAK FOR INTERRUPT)
---------------------------------------------------------------------
BEHAV:      _I = 200, G
```

the simulator is active.
After finishing the results can be represented correspondingly to the menue

```
               * *   RESULT OPTIONS   * *
          ------------------------------------
               V  -  SWITCH ON VIDEO MODE
               M  -  SWITCH ON MEMORY MODE
               G  -  SWITCH ON GRAPHIC MODE
          ------------------------------------
               P  -  PRINT TIME SERIES
               S  -  PLOT CURVE SHAPES
               A  -  PRINT AGGREGATED VARIABLES
               AS -  PLOT AGGREGATED VARIABLES
               D  -  PRINT AGGREGATION RESULTS OF OUTPUT VARIABLES
               DS -  PLOT AGGREGATION RESULTS OF OUTPUT VARIABLES
          -------------------------------------------------------

YOUR OPTION:
```

Choosing G and S the results are displayed either as curve family or pseudo 3-dimensional plot. Specifying as result the difference G∅ - G_I = 200 with G∅ as yield without pest interaction the yield loss is displayed as function of NI and IMZA (see Fig. 1).
A fine structure can be pointed out with clear nonlinear, threshold-like characteristics of pest to yield damage for low infestation. This result can be used as first approach to a dynamically based spraying standard. Complex pest damage of species with totally differing interaction pattern like in this case (see /7/, /8/) can be calculated by a linear superposition of the single nonlinear pest-crop damaging effects.

Model-aided derivation of pest strategies can also be done with the behavioural analysis technique varying the date of pesticide application. The application date for the Oulema pesticide and Macrosiphum pesticide (DO, DM resp.) are parameters. The killing efficiencies of the two pesticides are stored in tables. Short-term, strong acting pesticides are used with a killing efficiency of 50% at the first day, 30% at the second day and 10% at the third day. Specifying the simulation task we are interested in the behaviour of the yield loss and the abundance maximum (aggregated variable) of the pests as function of the spraying date. Starting at the 160th day of vegetation with spraying and choosing option B (as task) the target variables are now DO and DM.

The command BEHAV: DO, DM A 1 TIMES 40
BEHAV: COMB: DO + DM

leads to a sequential change of the spraying date for Oulema and Macrosiphum pesticide within the time interval t = [160, 200] .
For calculating the maximum of Oulema and Macrosiphum abundance as output variable the abundance state variables of all mapped development stages have to be specified in the result storage command

BEHAV: IO, FO, L13O, L4O, PUP (immigrants, femals, larvae 1-3, larvae 4 pupae)

BEHAV: IM, L1M, L2M, NM, A1M, A2M (immigrants, larvae 1, larvae 2, nymphs, alatae, apterae of Macrosiphum)

BEHAV: _I = 200, G (yield at harvest date)

During the result processing the system calculates the abundance maximum, minimum, mean value and variance per time interval entering the option DS and the command

BEHAV: IO+FO+L13O+L4+PUP resp.
BEHAV: IM+L1M+L2M+NM+A1M+A2M

The results are displayed in Fig. 2 and 3 as function of DO and DM, resp. The figures contain additionally the time course of the Oulema abundance and Macrosiphum abundance, resp., which were taken from the result library.

From a practical point of view killing of the beetle is much easier than killing the aphids effectively because the optimal spraying time coincides nearly with the beetle abundance maximum which is easy to observe. Aphid killing needs a model aided calculation of spraying date because these species dynamics require an early killing on principle, independent on the special weather conditions. Weather conditions have a modifying influence because they determine abundance maximum. The presented model results are of great importance, experimentally they can be gained, if generally complete, only with giant effort.

/1/ A. Knijnenburg; E. Matthaeus; V. Wenzel
Concept and Usage of the Interactive Simulation System for Ecosystems SONCHES
Ecol. Modelling, 26 (1984), p 51-76

/2/ V. Wenzel; K. Bellmann; E. Matthaeus; M. Flechsig
SONCHES - an interactive simulation system for design, validation and usage of ecosystem models
In Proceedings of 2. International Symposium on Systems Analysis, Berlin 1985, Vol. 2, pp. 120 - 123

/3/ B. P. Zeigler
The Aggregation Problem. In: B. C. Patten (Ed.)
Systems Analysis and Simulation in Ecology, Vol. IV,
Academic Press, New York, 1976, pp. 299-312

/4/ P. M. Frank
Introduction to Systems Sensitivity Theory
Academic Press, New York, 1973

/5/ H. Stehfest
Decision Theoretical Remark on Sensitivity Analysis
IIASA Report RR-75-3, 1975

/6/ R. V. O'Neill; R. M. Gardner; J. B. Mankin
Analysis of parameter error in a nonlinear model.
Ecol. Modelling 8 (1980), pp. 297-311

/7/ Computer-aided Modelling, Simulation and Pest-Management of the Winter-wheat Agroecosystem AGROSIM-W, Tag. Ber., Akad. Landw. d. DDR, Berlin, 1985, in press

/8/ K. Bellmann; W. Ebert; E. Matthaeus; A. Schultz; V. Wenzel
Modelling and Simulation of Agroecosystems - the Winter-wheat Agroecosystem Model AGROSIM-W
In Proceedings of 2. International Symposium on Systems Analysis and Simulation, Berlin 1985, Vol. 1, pp. 64 - 77

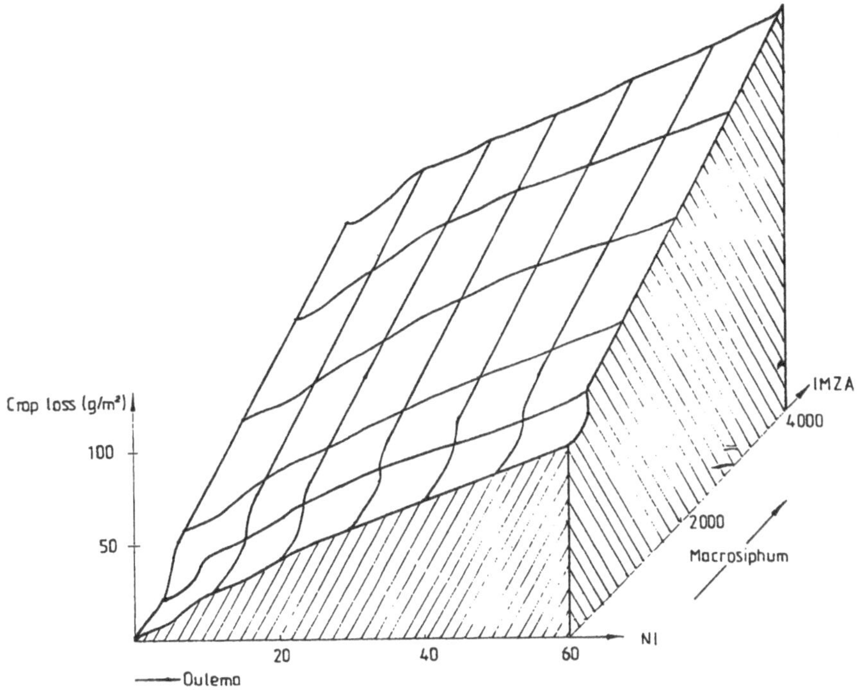

Fig.1: Common infestation of Oulema and Macrosiphum. Grain yield loss in dependence on different immigration abundances of Oulema (NI) and Macrosiphum (IMZA). Weather data for 1980

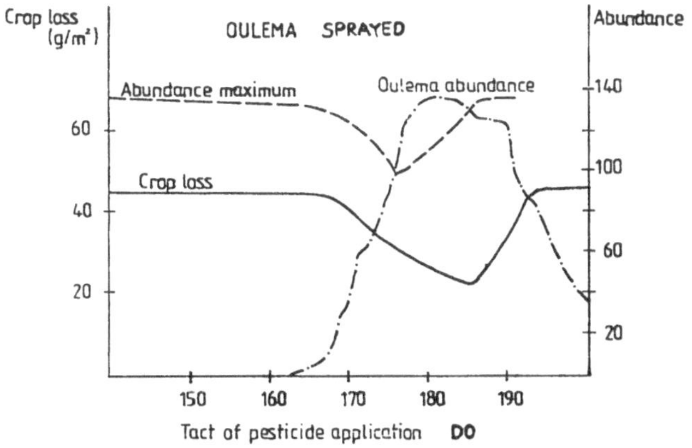

Fig.2: Grain yield loss by Oulema infestation (full line) in dependence on tact of pesticide application; abundance maximum of Oulema (dashed line) in dependence on tact of pesticide application and abundance of Oulema without any pesticide application (dashed dotted line)

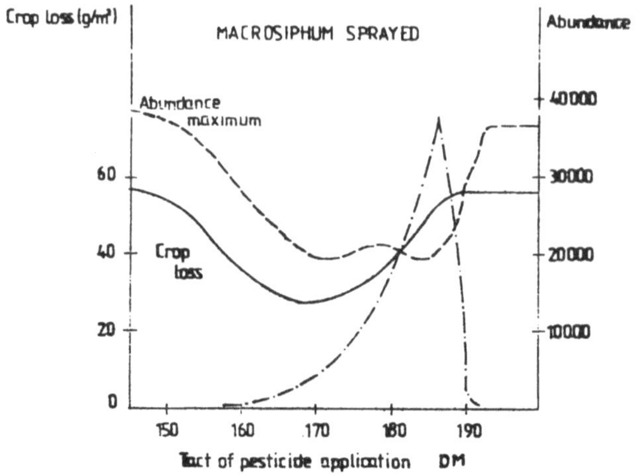

Fig. 3 : Grain yield loss by Macrosiphum infestation (full line) in dependence on tact of pesticide application, abundance maximum of Macrosiphum (dashed line) in dependence on tact of pesticide application and abundance of Macrosiphum without any pesticide application (dashed-dotted line)

THE INTENSITY OF ENERGY FOR CONSUMPTION INVESTIGATED WITH INPUT - OUTPUT MODELS

Edith Biebler [1)]

On the basis of input-output models coefficients of energy intensity for consumption and for some complexes of needs have been calculated and investigated for a period of 10 years.Despite of the existent inaccuracies these coefficients permit to calculate the amount of energy demanded in regard to the further shaping of the socialist way of life. By that,it is possible to create a methodical enlargement for prognostic investigations because these coefficients can be estimated outside of the input-output model for instance with econometric functions,valuations by experts and others.

1. Introduction

The further shaping of the socialist way of life and the increased scientific-technological development in the sphere of consumption have also influenced the consumption of energy. In regard to the comprehensive intensification of the reproduction process questions of energy intensity and economic use of energy get increasing significance also in the sphere of consumption.

At the Central Institute of Economics of the Academy of Sciences of the GDR these relations between consumption,production and the input of recources - mainly of energy - have recently been increasingly investigated with input-output tables.As indirect costs are involved in applying input-output models it is possible to calculate the needed total gross product of energy both for increase and structural changes of consumption.Investigations of the consumption structure in connection with the demand of energy are based on the following complexes of needs

- nutrition
- housing
- education/culture/recreation
- clothing
- health care
- transport/communication
- general public needs

For calculating the consumption structure a constant matrix of the complexes of needs was used.The complexes of needs 'housing','health care','education/culture/recreation' and 'general public needs' include only the consumption of goods and material services. As the nonproductive investments were not considered these complexes of needs -regarding to the total economic costs - are not complete and therefore too low.

The investigations of energy relate to the kinds of energy as electroenergy, heat energy,hard coal,brown coal, mineral oil and natural gas.

2. The intensity of energy for consumption and its complexes of needs

2.1 The direct intensity of energy

The direct intensity of energy for consumption or for complexes of needs $\mathcal{E}$ is calculated as the consumption of energy in Mio M in the sphere of consumption needed for 1 Mio M final demand of consumption or com-

1) Dr. E. Biebler ,Zentralinstitut für Wirtschaftswissenschaften der AdW derDDR, 1100 Berlin , Prenzlauer Promenade 149/151

plexes of needs :

$$\mathcal{E}_k = \frac{c_k^h}{c_k} \quad (1)$$

where

$\mathcal{E}_k$ the direct intensity of energy for the complex of needs k

c_k the final demand of consumption for the complex of needs k,

$$c_k = \sum_{i=1}^{n} c_{ik}$$

c_{ik} goods and services of industrial branch i for the final demand of the complex of needs k

i number of industrial branches, i = 1,..,h,..,n;

k number of complexes of needs, k=1,...,m;

c_k^h consumption of energy h in the complex of needs k, $c_k^h = \sum_{i=1}^{h} c_{ik}$

From 1972 to 1982 the direct intensities of energy $\mathcal{E}$ rose for both consumption and all complexes of needs, whereas the growth of $\mathcal{E}$ was lower in the period 1977/72 than in 1982/77. As in the sphere of consumption the prices for energy remained constant the growth of $\mathcal{E}$ was especially caused both by structural changes between and within the kinds of energy for instance increase of consumption of electroenergy resulting from construction of new flats and week-end houses and also through a better equipment of housholds with durables and the improvments of services. The complexes of needs 'transport/communication' (consumption of benzine), 'housing' (durable goods) and 'health care' have the greatest coefficients $\mathcal{E}$.

2.2 The total intensity of energy

The total intensity of energy for the consumption γ expresses the total gross product input of energy needed for 1Mio M final demand of consumption. Therefore, the total gross product input of energy includes not only the consumption of energy and the direct costs of energy for the consumption, but also the indirect costs of energy required by interweaving the different branches in the production sphere. Using static input-output models the total gross product input of energy for consumption and for the complexes of needs was calculated:

$$X = (E - A)^{-1} \hat{c}_k \quad (2)$$

$$r_k = \sum_{i=1}^{h} x_{ik} \quad (3)$$

where

$X = (x_{ik})$ matrix of the gross production of branch i for the final demand of the complex of needs k

E identity matrix

A matrix of input-output coefficients

r_k total gross product input of enrgy for the final demand of the complex of needs k

h number of sectors of energy

The total intensity of energy γ_k for the complex of needs k was calculated as follows :

$$\gamma_k = \frac{r_k}{c_k} \quad (4)$$

As the total intensity of energy γ is calculated on the basis of gross product of energy some factors influence γ which have not any or even contrary effect on ε . Thus, the rise in prices of energy in the production sphere has also an influence on the total intensities of energy; on the other hand the production of consumer goods or the construction of houses with an economic use of energy but possibly with high energy costs in their production sphere can increase γ and decrease ε .
In the period 1972-1982 the total intensities of energy γ for consumption and for all complexes of needs rose considerably, especially for the complexes of needs 'transport/communication', 'health care' and 'nutrition'. This is mainly caused by increasing the prices of mineral oil and natural gas which influenced clearly the complexes of needs with great input of these kinds of energy. The ranks of complexes of needs in regard to γ are changed only unessentially for the time from 1972 to 1982. The complexes of needs 'transport/communication', 'health care' and 'housing' have the greatest coefficients γ , whereas the complex of needs 'nutrition' holds the rank 5 within 7 complexes of needs.

3. Relation between the total gross product input of energy and the total intensity of energy for consumption and for complexes of needs

It is not possible to draw conclusions from the total intensities of energy for the complexes of needs γ_k to the total gross product input of energy and therefore to the necessary production of energy, because the structural development of consumption (α_k) has a great influence on it:

$$\beta_k = \frac{\gamma_k \cdot \alpha_k \cdot c}{r} = \frac{r_k}{r} \quad (5)$$

where

β_k the share of the total gross product input of energy for the complex of needs (r_k) in the total gross product input of energy for consumption (r)

α_k the share of the final demand for the complex of needs (c_k) in the final demand of consumption (c) : $\alpha_k = c_k / c$

The ranks of β_k are clearly variant from those of γ_k because the values of complexes of needs are different. Thus, the complex of needs 'nutrition' includes the main part of the final demand of consumption ($\alpha > 40\%$); therefore, the coefficient β of this complex of needs has the rank 1. Though, the complexes of needs 'health care', 'transport/communication', 'housing' have a greater intensity of energy and a higher growthrate than the complex of needs 'nutrition', the greatest demand of energy will also in future be caused by the complex of needs 'nutrition'.
As β_k is distorted by the influence of α_k, β_k was related to α_k :

$$\delta_k = \frac{\beta_k}{\alpha_k} = \frac{\frac{r_k}{c_k}}{\frac{r}{c}} = \frac{\gamma_k}{\gamma} \qquad (6)$$

where

δ_k indicates the relation between the total intensity of energy for the complex of needs k and the total intensity of energy for consumption

$\delta_k < 1$ means that the total intensity of energy for the complex of needs k is smaller than the total intensity of energy for the consumption, i.e. $\gamma_k < \gamma$ and $\beta_k < \alpha_k$

$\delta_k > 1$ means that the total intensity of energy for the complex of needs k is greater than the total intensity of energy for the consumption, i.e. $\gamma_k > \gamma$ and $\beta_k > \alpha_k$.

From 1972 to 1982 δ_k is greater than one ($\delta_k > 1$) for the complexes of needs 'housing','health care','transport/communication' and 'general public needs',i.e. these are the complexes of needs with the greatest total intensity of energy γ_k.The calculation of δ_k allows to conclude mediately from the structural shares of complexes of needs both in the total gross product input of energy (β) and in the final demand of consumption (α) to the total energy intensities of the complexes of needs (γ) though these total energy intensities are related by the total energy intensity of consumption.

4. Final remarks

With these coefficients of enrgy intensity for the complexes of needs calculated on basis of input-output models it is possible

- to investigate the development of demand of energy more differentiated than only on the basis of the total consumption despite of the existent inaccuracies. Thus,the amount of energy demanded in regard to the structural changes of consumption can relatively easy be computed;
- to estimate these coefficients outside of the input-output models for instance with econometric functions,valuations by experts and others.

By this it is additionally possible to enlarge prognostic investigations methodically.However it also became evident that investigations on the structure are considerably influenced by the prices.Even if the consumer prices for energy remained constant for years,the price changes in the production sphere would influence the results as they are reflected in the total intensity of energy by the calculation of the total gross product input.Therefore,besides further work on the structural matrix of complexes of needs future research must involve the influences of price changes.

References

1 See U.Ludwig/E.Biebler/M.Kraft,Modelling the Structural Change of Consumption and the Input of Primary Resources in the GDR Economy by a Time-Series and Input-Output Approach,in:Input-Output Modelling, Proceedings of the Third IIASA Task Force Meeting,Laxenburg,1983

2 See G.Knobloch,Bedürfnisbefriedigung-Produktionsstruktur-Grundproportionen,Berlin 1974(Forschungsberichte des ZIW der AdW der DDR,Nr.10)

COMPUTER MACROECONOMIC MODELS FOR SIMULATION EXPERIMENTS

MARTCHEV A.A. and M.R. MOTZEV [1]

This paper discusses the design of macroeconomical models in the form of a system of simultaneous equations. A new approach in this field is described. The main idea is that the synthesis of the model is based on a multi-stage selection procedure (MSSP). Characteristic of this procedure is that at each stage of selection are generated a variety of hypothesises about the wanted model. Each generated hypothesis is verified and estimated, after which by inconclusive limit choice of a given number, only few of them are selected as "best" in the sense of a predefined selection criteria. MSSP combines in itself the principle of grouping of variables [1], which lets to remove the consequences of multicollinearity and undersized samples of data, the principle of externally addition "Cross validation", which is asymptotic equivalented with the Akaike's criterion for choice of model [4] etc.

Keywords: simultaneous equations systems, multi-stage selection procedure, macroeconomic simulation models for Bulgarian economy.

INTRODUCTION

This paper represents a part of research done in the scientific laboratory for systems' analysis and management in the Highere Institute of Economy "Karl Marx" - Sofia. The main result of this research is a detailed methodology for automated design of simulation models of economic systems with a great number of algorythms and programs organized in a few packets of annexed programs.

MSSP FOR SYNTHESIZING SIMULATION MODELS IN THE FORM OF SIMULTANEOUS EQUATIONS

In the MSSP which is designed for syntesizing simulation models in the form of a system of simultaneous equations (SSE)[3], we can differentiate three main parts:

A. In the first part, on the basis of existing practical data, grouped in a table of observations of researched system, we make a first choice of significant factors, which will be included in the model. Then:

A.I. The generating of the variety is done by including non-linear transformations and taking into account the pre-history of the factors through euristic considerations by the researcher and automatically using a computer program.

A.II. Each competing hypothesis represents a hypothesis of inclusion or non-inclusion of an additional factor in the table, that is, a hypothesis on the fact is this factor potentially important for

1) Highere Institute of Economy "Karl Marx" - Sofia - Bulgaria

the mathematical description of the researh system.

A.III. The limit choice and the limiting of the variety is done using the criteria "correlation with a dependant variable", but can be supplemented with euristic solutions given by the researcher. In all cases the preservation of the initial factors - "protection of the variables" is guaranteed.

B. The second part represents a procedure for multi-stage selection of each equation participating in the model. The form of the wanted equation is not predetermined. The task of finding the form and coefficients of each equation is divided in many tasks for determining the coefficients of equations with two variables - the principle of grouping of variables. In this way, we make certain of obtaining statistically important coefficients, based on a limited number of observations (undersized samples of data). The form of the equation is determined by a few consequitive stage of selection:

B.I. The generation of the variety is done by introducing new intermidiate variables for each stage and the creation of a new generation of intermidiate equations function with two variables which include indirectly more and more complex combinations of the initial factors.

B.II. Each generated equation is considered as potential description of a given connection in the model, which competes with the other possible descriptions "fighting for survival".

B.III. After each selection stage, there is no choice of a unique equation, but a limit choice of a predetermined number of good equations is done - the principle of non-finilized solutions. This gives us the possibility to obtain a set of alternative good equations.

B.IV. The estimation of the coefficient in each intermidiate equation is done using the criteria of the average square error. The existing set of data is divided into two parts: a teaching set which is used for estimating the coefficient in the intermidiate equations, and the control set used for evalueing the adequacy of the obtained equations - the principle of "Cross validation" choice of good model.

B.V. The chosen equations in a given generation are used for generating new, more complex equations of the next generation. From these predetermined number are chosen as good etc.

B.VI. The selection procedure ends when certain conditions, limits or results are achieved, which the researcher wants to obtain based on a given number of selections, achievement of an average minimum for the generation error and others.

B.VII. At the end of the second part of the MSSP automatically the full form of the best equations is restored. Each equation describing a given connection in the model has a given number of alternative variations.

C. In the third part of the procedure from the obtained equations a synthesis of the simulation model is done in the form of SSE:

C.I. The variety of alternative variants of the model, which represent SSE is obtained by combined, already chosen, best equations.

C.II. Each of the competing hypothesises represents a hypothesis for inclusion in the model of a given variant of the separate equation.

C.III. Each generated SSE is considered as a potential model which imitates the analyzed system.

C.IV. The evaluation of these competing models is complexly done, using a great number of statistical ctiteria - coefficient of correlation, variation coefficient, average relative error, precission in prognosis and others.

C.V. The final choice of the best model is made by the researcher, who has the possibility to introduce some additional non-formalized considerations, but after having the guarantee of evalueting a great number of possible models and the choice of a small numbers of good ones.

We have to remark, that in difference to existing traditional methods I2I, when the structural and parametrical identification of SSE is done separately, in the described MSSP, this is done in a unified, highly automated procedure. In it, we have the possibility for imposing predetermined limits on SSE coefficients as in traditional methods. The diference is that an additional structural identification is done during multi-stage selection.

During this structural identification automatically some of the variables are removed and in the synthesized model remain only those, which are important. In this way, simultaneously the form of each equation is determined as well as the important variables participating and an estimation of the coefficients is done.

In addition the SSE coefficients can be obtained as time functions, which makes the MSSP applicable to nonstationary systems, describing most economic processes.

SIMUR - A FAMILY OF SIMULATION MACRO-MODELS

Together with the design of a procedure for automated synthesis of simulation models, we designed a family of macroeconomic models called SIMUR. The models from this family can be considered as a result and at the same time as a base for this procedure.

Each model from the SIMUR family is designed with a special purpose - experimentation and improvement of a given part of the MSSP for synthesis of SSE. The conclusions obtained from each experiment with a model are used on one hand for designing the next model and on the other for improving MSSP. In this way we achieved a parallel development and interaction between the family of models SIMUR and the MSSP for synthesis of SSE.

Short information for the SIMUR models is presented in table 1.

THE FAMILY OF MODELS "S I M U R" Table 1.

Model (Year of design)	Main purposes of design	Specific characteristics
SIMUR-I (1978-1980)	Analysis of possibilities for automatic synthesis of SSE during the run of multi-stage selection procedures.	A one-product macroeconomic model in the form of SSE with 5 equations. Contains 5 endogenous, 5 lag and 1 exogenous variables.
SIMUR-II (1981-1982)	Design and experimenting of a programming system for automatic holding of simulation experiments with SSE. Analysis of different criteria for the precision's estimating of SSE	Agregated macroeconomic model in the form of 12 interdepending simultaneous equations. Contains 12 endogenous, 5 exogenous and 30 lag variables with lag of up to 3 years.
SIMUR-III (1983-1985)	Improving the MSSP for synthesis of SSE with many equations. Simulating and forecasting of main macroeconomic indexes.	It is being designed. Contain 39 equations, 39 endogenous, 7 exogenous and about 100 lag variables with a timé lag of up to 5 yers.

CONCLUSIONS

This paper presents a new and perspective line in research and modelling of economic systems - using a multi-stage selection procedure for synthesis of simulation models. The accumulated experimental data [1,3] shows, that MSSP has great advantages over the existing methods - system dynamics, econometric approach, agregated modelling etc.

In this connection, the presented approach has a large field of application and needs further development in the following main directions:

A. Further design of algorythms and programs for automated

elaboration of simulation models.

B. The design of new simulation macroeconomic models. In this direction, special interestrepresents the design of the macro-models for different socialist countries using the same set of variables. On the basis of this, it wil be possible to make valuable comparisons and analysis of the common and specific characteristics reflected in the different models.

LITERATURE

I1I Ivachnenko, A. and J. Müller: Selbst-organisation von Vorhersage-modellen. VEB Verlag Technik, Berlin, Verlag Technika, Kiew, 1984.

I2I Malinvaud, E.: Methodes Statistiques de L'econometrie. Dunod, Paris, 1969.

I3I Marchev, A. and M.Motzev: Simulation Macroeconomic Models Designed in the Form of Systems of Simultaneous Equations Synthesizing During the Run of Multi-stage Selection Procedure. Scientific papers of the Highere Institute of Economy "K.Marx", vol.I, Sofia 1983.

I4I Stone, M.: An Asymptotic Equivalence of Choice of Model by Cross-validation and Akaike's Criterion. J.R.Statist. Soc. (1977) B39, 44-47.

A NEW FORECASTING AND SIMULATION MODEL FOR THE POLISH ECONOMY - "SAPO"

Jerzy Pawilno-Pacewicz[1)]

The model, named "SAPO" for the abbreviation of the Polish description of related research project, is a large-scale, computer simulation model of economic change in Poland. It is designed for economic policy analysis on the national level and for short- and medium-term forecasts of essential macroeconomic indicators. The model is also intended to help the planning authorities to control the feasibility of central plans and to asses the degree of their fulfillment.

1. General description

The model is intended to represent the underlying structure of the economy for the purpose of projecting one to five years in the future, although technically it is possible to extend projections for periods up to ten years. The model treats the interrelated issues of productive capacity, employment, production, exports, consumer supplies, investment and market equilibrium. The economy is aggregated to the size of 13 major sectors and 11 basic groups of commodities exported and imported to and from both socialist and non-socialist countries. The consumer market covers demands and supplies of three groups of products: (1) food and alcoholic beverages, (2) non-food products, (3) services.

2. Structure of the model

The general aim of the SAPO Model is different from that of earlier models for the Polish economy by Barczak [1], Maciejewski [3], and Juszczak and Welfe [2]. Figure 1 decribes the general framework of the model. The submodels I-VII can also be regarded as being parts of the computer algorithm developed to solve the model. First, the set of input data is transformed from its original form, familiar to the central planners, to a modified form, more useful in the process of computations. Second, expected values of initial variables are calculated for the year (called "basic") which directly preceeds the first year of the forecast. This is necessary because statistics for the basic year are usually not yet known at the moment when projections for the following years are mostly needed.

[1)] The model was developed at the Central School of Planning and Statistics in Warsaw. The members of the research team were: M.Garbicz, T. Kuszewski, K.Mazur, W.Misiąg and J.Pawilno-Pacewicz (the leader of the team).

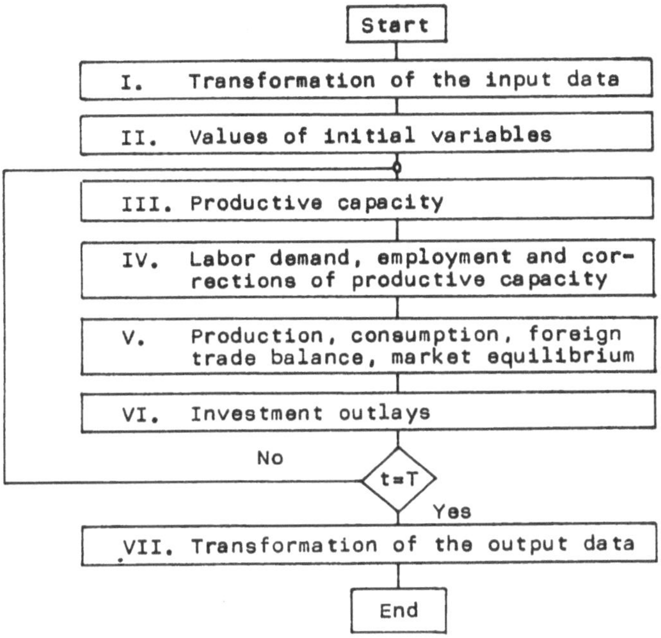

Figure 1. General structure of the SAPO Model

Next, there are four sub-models numbered III-VI. They create the real economic "heart" of the model. The sub-model III determines productive capacity as a function of gross fixed capital. However, it is assumed that in order to be utilized the installed capital must be operated by a necessary amount of labor. If the available labor is not sufficient (which tends to be a common practice in most CPE's) the sub-model IV brings the necessary corrections of earlier estimates for productive capacity. The submodel V has a form of a mathematical programming problem with exchangeable objective functions. The objective functions may represent broad macroeconomic targets (such as maximization of consumption or national product) or they may reflect an expected behavior of the economic organisations on the micro-level (a function maximizing the minimal level of productive capacity utilization is an example). The set of constraints contains the standard input-output model with a section for imports determination, a group of constraints reflecting the existing limits for exports and imports increases, and acceptable levels of the foreign trade balance and of the inflationary gap on the consumer market. The sub-model IV is a set of equations for sectoral distributior of disposable investment supplies. Shares of selected sectors in the total value of disposable investment may be treated as instrumental variables.

The sub-models III-IV are solved recursively with the output of earlier models entering the input for the following submodels, and with the output of one-year simulation run affecting the initial conditions for the following years in the simulation period. In this regard the model resembles its earlier version developed in 1979 [4]. Finally, the submodel VII generates the numerical results for the whole forecasting period and transforms them into a set of tables which are designed to suit appropriate planning documents.

3. Equations

The following set of relationships describes the main features of the model:

$$k_j = F_{1j}(k_j^{t-1}, g_j^{t-1}, g_j^{t-2}, \ldots, g_j^{t-\overline{t}}) \qquad j \in S \qquad (1)$$

$$z_j = F_{2j}(k_j, l_j, h_j) \qquad j \in S \qquad (2)$$

$$\sum_{j \in S} l_j = l^s \qquad (3)$$

$$x_j \leqslant z_j \qquad j \in S \qquad (4)$$

$$m_i = F_{3i}(x_1, \ldots, x_n) \qquad i \in S \qquad (5)$$

$$f_i = F_{4i}(x_1, \ldots, x_n) \qquad i \in S \qquad (6)$$

$$f_i \geqslant s_i f_i \qquad i \in S \qquad (7)$$

$$e_i = F_{5i}(f_i) \qquad i \in S \qquad (8)$$

$$e_i \leqslant \overline{e}_i \qquad i \in S \qquad (9)$$

$$m_i \leqslant \overline{m}_i \qquad i \in S \qquad (10)$$

$$\sum_{i \in S'} p_i^e e_i - \sum_{i \in S'} p_i^m m_i \geqslant b \qquad (11)$$

$$c_i = F_{6i}(f_i) \qquad i \in S \qquad (12)$$

$$d = F_7(x_1, \ldots, x_n) \qquad (13)$$

$$d + r^{t-1} - \sum_{i \in S} p_i c_i = r \qquad (14)$$

$$r \leqslant \overline{r} \qquad (15)$$

$$g_i = F_{7i}(f_i) \qquad i \in S \qquad (16)$$

$$v_j = F_{8j}\Big(\sum_{i \in S} g_i\Big) \qquad j \in S \qquad (17)$$

$$f_i \equiv c_i + g_i + e_i + u_i \qquad i \in S \qquad (18)$$

$$u_i = F_{9i}(x_i - x_i^{t-1}) \qquad i \in S \qquad (19)$$

where:

k - gross fixed capital

g - investment outlays by sector of destiny

z - productive capacity

l - employment

h - number of working hours per week

l^s - labor supply

x - input (production)

m - imports

f - output

s - lower bounds for output changes

e - exports

$\overline{e}$ - upper bounds for exports

$\overline{m}$ - upper bounds for imports

b - lowest acceptable foreign trade balance

p^e - foreign prices of one unit of exported goods

p^m - foreign prices of one unit of imported goods

p - current consumer prices

c - consumption supplies

d - current money demand for goods and services

r - inflationary "overhang"

$\overline{r}$ - upper bounds for the inflationary "overhang"

v - investment by sector of origin

u - inventory increase

t - time variable

j - number of sector

i - number of product

$S = \{1,\ldots,13\}$, $S' = \{1,\ldots,11\}$

$F_{1j},\ldots,F_{9j}$ symbolize function-relationships between variables.

For lack of space, this set of equations is not a comprehensive account of the model, but it does show the principal relationships between variables. The number of instrumental (decision) variables is 196 for every year of the forecast. The number of constant parameters appearing in the model is 587. Most of them (exactly 338) are the import coefficients and the technical coefficients in the input-output model. They are available from current statistics. A group of parameters is obtained by means of econometric estimation. This especially concerns parameters of the F_{1j} and F_{2j} functions in equations (1) and (2). There is also a set of parameters with no specific economic interpretation. They are necessary to remove some of the existing price differences and other incompatibilities in the available data series.

Basically, all variables are expressed in constant prices. They are the prices of the latest input-output table (i.e. the 1982 prices at the moment). However, some variables are additionally calculated in current prices. This is the case for the foreign trade and the consumer market segments of the model.

In order to make the implementation of the model easier and less costly a package of user-oriented computer programs was designed and developed for the computer Univac 1100. The programs enable the user to manage the data basis, to reestimate parameters, and to obtain a variety of simulation runs quickly, easily and with no need for extensive konwledge of detailed model structure.

4. Applications

The SAPO Model was developed in 1980-1983. It was completed in June 1983. Since then the authors presented several studies to diagnose changes in the growth rate for the polish economy. In July 1983 the first conditional forecast for 1983-1985 was prepared. Its updated version, covering the 1984-1985 period, appeared at the end of 1983. In June 1984 the authors presented a detailed forecast of plan fulfillment for the 1984 central plan. This study was followed by the consistency analysis of the blueprint for the 1985 central plan. An attempt to extend the forecasting horizon was made in a separate paper on the long-term structural problems of the Polish economy. A two-year experience is certainly not enough to justify any far-reaching conclusions on the model usefulness. However, the preliminary results and the policy recomendations obtained from the model seem to be encouraging.

References

[1] Barczak A, B.Ciepielewska, T.Jakubczyk, Z.Pawłowski: Model ekonometryczny gospodarki Polski Ludowej. Warszawa 1968.

[2] Juszczak W., Welfe W.: A Conditional Forecast for Development of the Polish Economy in the Period 1983-1985. Prace Instytutu Ekonometrii i Statystyki Uniwersytetu Łódzkiego, Łódź 1983.

[3] Maciejewski W., M.Opara, J.Zajchowski: Ekonometryczny model gospodarki Polski. Gospodarka Planowa 6(1973).

[4] Pawilno-Pacewicz J.: Simulations to Diagnose Changes in the Growth Rate of the Polish Economy. State University of New York at Stony Brook, Department of Economics, Working Paper No. 218, October 1979

TAX REGULATION DESIGN AS A STRATEGY SELECTION IN A SYSTEM WITH CONFLICTING GOALS

Jacek Stefański and Wojciech Cichocki 1)

ABSTRACT. The problem of analysis and designing tax regulations influencing the enterprise's activity is considered. The enterprise is described as a coalition of workers and managers; it is taken into account that they have different goals. On the other hand the government, being in a position to influence the situation inside a firm, has different utility function. The situation is modelled as a three-level system with government at the highest and workers at the lowest level. Analysis is focused on the ways in which tax regulations influence situation inside a firm. Examples refer to the legislative framework describing the activity of firms in Poland.

KEYWORDS: microeconomic systems; conflicts in organization; hierarchical games.

1. INTRODUCTION

In this paper we study the ways in which three decision-makers influence the enterprise's activity. Usually it is assumed that the firm's goal is set by its management; here we assume that the workers have also noticeable effect on the formulation of a firm's policy [3]. On the other hand the existence of the upper level (say, the government) is taken into account. There are same other models in which the government's decisions play an important role [1], [2], [4]. It is evident that the government's utility function differs considerably from the firm's objectives. There is also no reason to suspect that the management's and worker's goals are consistent. The disagreement of interests suggests an approach based on game theoretical considerations. An important fact is that the roles and statuses of the three parties are different which is reflected by the organization of the decision making process. The situation is modelled as a three-level system with government at the highest and labour at the lowest level.

An enterprise is free to act within some legislative framework which contains also tax regulations. Some regulations of this kind can be used to influence, to a certain extent, the behaviour of a firm. In the model formulated and investigated in this paper two types of the taxes concerning labour income are used as upper level's decisions. Analysis in focused on the ways the situation in a firm depends on these decisions. Examples refer to the legislative framework describing the activity of firms in Poland. The problem of choosing optimal tax regulations is formulated as a strategy selection in a three-level game.

---

1) Systems Research Institute, Polish Academy of Sciences, Newelska 6, 01-447 Warsaw, Poland.

2. CONFLICTING GOALS OF THE PARTIES INFLUENCING AN ENTERPRISE ACTIVITY

As already mentioned we take into account three groups of people the activity of an enterprise heavily depends on. Clearly the main decision makers is the firm's management and the other two are the workers and the upper level (government). Each of the parties uses different decision variable and tries to maximize different objective function. We will distinguish three types of decisions: production rate q, wage fund w and two types of tax regulations, described by the functions T_w and T_b, bound up with the labour income. Taxes of this kind have been introduced in Poland in order to reduce inflation.

It is reasonable to assume that the workers, treated here as a group, want to maximize their income

$$J_2(T_w,T_b,w,q) = w + B(T_w,T_b,w,q) \tag{1}$$

which consists of the wages w and the bonus (conected with the attained profit) $b=B(\cdot)$. According to the Polish regulations each of these two parts is individually taxable, which means that the taxes $T_w(w,q)=T(d_2w-d_1q-d_o)$, $d_2,d_1,d_o \geq 0$, and $T_b(b)$ must be paid by the firm. Usually only a part of the net profit, i.e. $\alpha\Pi(w,q)$, $\alpha \in [0,1]$, is reserved for the bouns and the mentioned income taxes:

$$\alpha\Pi(w,q) = b + T_w(w,q) + T_b(b). \tag{2}$$

Calculating b from the above equation we obtain the function B which appears in (1). The net profit Π is computed taking the profit tax rate $t_p \in [0,1]$ into account:

$$\Pi(w,q) = \begin{cases} (1-t_p)\Pi_1(w,q) & \text{if } \Pi_1(w,q) > 0 \\ \Pi_1(w,q) & \text{if } \Pi_1(w,q) \leq 0 \end{cases} \tag{3}$$

where the profit:

$$\Pi_1(w,q) = (1-t_o)qp(q) - C(q) - (1+t_w)w, \tag{4}$$

where, in turn, $t_o \in [0,1]$ is the rate of turnover tax, p is the inverse demand function, C is the production cost function (excluding wages) and $t_w \in [0,1]$ is the wage tax rate.

The managers want high net profit and growing market share which can be formalized by the management's utility function as a linear combination of the two goals:

$$J_1(w,q) = a_1\Pi(w,q) + a_2q, \tag{5}$$

$a_1, a_2 \geq 0$.

The government interests are more complex; in the model under consideration maximization of tax revenue and reduction of inflation is incorporated in the objective function:

$$J_o(T_w,T_b,w,q) = \beta_1[t_o\, q\, p(q)+t_p\Pi(w,q)+t_w w] - \beta_2 J_2(T_w,T_b,w,q)/q\,, \qquad (6)$$

where $\beta_1, \beta_2 \geq 0$. The first part con tains the tax revenue connected with turnover, profit and wages. The additional income taxes, i.e. T_w and T_b are not taken into account here becouse the money received is reserved for special purposes. The second part in (6) determines the labour income per production unit - its minimization mirrors the aim to reduce inflation (since it leads to lower labour income and production rate increase which ought to result in lower prices).

3. TAX REGULATION DESIGN AS A STRATEGY SELECTION

It is evident that the roles of the decision makers whose goals we have described in the previous section are not the same. As a matter of fact we deal with a hierarchical system in which the government is at the highest and labour at the lowest level. The hierarchy is reflected by the following decision sequence: the government declares its strategy first, i.e. announces tax regulations T_w and T_b, this is followed by the management announcing the wage fund w to the workers, who act last working with the chosen productivity which means that they determine the production rate q. We assume that the parties know their own as well as partners objectives.

Let us look at the outlined economic system from the government's point of view. The natural question of interest concerns the ways in which the enterprise reacts to the tax regulations T_w and T_b. If we assumed, as it is usually done, that the firm is profit (and market share) maximizer, the change of regulations T_w and T_b would not influence the firm's behaviour at all (see (5) - the objective J_1 does not depend on T_w or T_b). But, as reality suggests, changes of T_w and T_b influence the enterprises' behaviour very much. The reason is that employees recognize their own objective (depending an T_w and T_b - see (1)) and act accordingly. Thus, the decisions taken in a firm follow from the labour-management game. The result can be, to a certain extent, influenced by the third party, the government, taking part in the game and playing the role of the leader.

The management, determining its decision, takes into account the workers reaction:

$$q_r^R(T_w,T_b,w) = \begin{cases} \arg\max\limits_{q \in \Psi(T_w,T_b,w)} J_2(T_w,T_b,w,q) & \text{if } \Psi(T_w,T_b,w) \neq \emptyset \\ q_o & \text{if } \Psi(T_w,T_b,w) = \emptyset, \end{cases} \qquad (7)$$

where the mapping $\Psi: R \to 2^R$ determines motivational conditions ($q \in \Psi(\cdot) \neq \emptyset$ means that the workers have financial incentive to produce q):

$$\Psi(T_w,T_b,w) = \{q \geq 0 \mid \frac{J_2(T_w,T_b,w,q)}{f(q)} \geq r \text{ and } \frac{\partial J_2(T_w,T_b,w,q)}{\partial q} \Big/ \frac{df(q)}{dq} \geq r\},$$
$$r \geq 0 \qquad (8)$$

In the above expression f is a function associating the workers' effort with production rate, and r is the motivational threshold. The first part in (8) introduces constraint concerning average (per effort unit) income, while the second part is bound up with the marginal average income. Thus, the management determines its decision in the following way:

$$w_r^R(T_w,T_b) = \arg \max_{w \in \Omega(T_w,T_b)} J_1(w,q_r^R(T_w,T_b,w)), \tag{9}$$

where, as in (7), the subscript r indicates the motivational threshold taken into account in (8), and $\Omega(\cdot)$ is the set of admissible management decisions:

$$\Omega(T_w,T_b)=\{w \mid (w,q^R(T_w,T_b,w)) \in A(T_w), \Pi(w,q^R(T_w,T_b,w)) \geq$$
$$\geq \pi_o, J_2(T_w,T_b,w,q^R(T_w,T_b,w)) \geq i_o\} \tag{10}$$

The set $A(T_w)$ which appears in the first constraint is defined as follows:

$$A(T_w) = \{(w,q) \mid w_o \leq w \leq \overline{w}(T_w,q),\ 0 \leq q \leq \overline{q}\}, \tag{11}$$

where $\overline{q}=p^{-1}(0)$, w_o is the minimal admissible wage fund, while the function $\overline{w}$ we obtain solving (2) for $b=0$, i.e. the equation $\alpha\Pi(w,q)=T_w(w,q)$ with respect to w. Thus, $\overline{w}(\cdot)$ determines the maximal admissible wage fund since for $w > \overline{w}(\cdot)$ we have $\alpha\Pi(\cdot) < T_w(\cdot)$, i.e. there is not enough money for the wage tax $T_w(\cdot)$. Note, that the decision of the management (9) heavily depends on T_w, T_b although the managements objective does not.

The optimal decisions of the leader of the game can be determined in the following way:

$$(T_w^*,T_b^*) = \arg \max_{(T_w,T_b) \in \Theta} J_o(T_w,T_b,w^R(T_w,T_b),q^R(T_w,T_b,w^R(T_w,T_b))). \tag{12}$$

The set Θ which appears in the above expression introduces reasonable constraints concerning the income tax functions:

$$\Theta=\{(T_w,T_b) \mid T_w \text{ and } T_b \text{ are non-negative, continuous and monotonic; the functions } w^s \text{ and } q^s \text{ are continuous}\}, \tag{13}$$

where the functions w^s and q^s are defined as follows:

$$w^s(r) = w_r^R(T_w,T_b)$$
$$q^s(r) = q_r^R(T_w,T_b,w_r^R(T_w,T_b)), \tag{14}$$

where, in turn, r is the motivational threshold and w_r^R, q_r^R are given by (9),(7). Continuity of w^s and q^s eliminates the possibility of making great mistakes by the management in the case when the value of the workers' motivational threshold differs from the value which has been taken into account when solving (9). The first condition in (13) prevents the government from applying a threat strategy, e.g. in the form:

$$T_w(w,q) = \begin{cases} C & \text{if} \quad w \neq w^G \\ \tau^G & \text{if} \quad w = w^G, \end{cases} \tag{15}$$

where C is a great "discouraging" value (a threat) and w^G, τ^G are the values preferred by the government.

Let us now come back to the second condition in (13). It turns out that it eliminates all the tax functions with discontinuous derivative. Therefore we will focus on the quadratic forms of T_b and T_w (or T since $T_w(w,q)=T(d_2w-d_1q-d_o))$, where $T_b(0)=0$ and $T(0)=0$.

Let us take into account the pair of decisions (14). It determines the Stackelberg equilibrium point in the game between the two lower levels of our system. Actually, this point may be treated as a status quo during the labour-management negotiations since it turns out that these both parties have incentive to cooperate. It appears that such collaboration is advantageous to consumers (because it results in higher production) but not always advantageous to the government.

Note, that the management's decision (9) depends on the tax functions T_w and T_b in a specific way (J_1 does not depend directly on T_w or T_b). First, changing the parameters of T_w and T_b the government can change the shape of the set Ω of admissible management's decisions and second, it can influence the workers' objective J_2 and, consequently, the reaction q^R. What makes this problem interesting is that the changes of Ω and q^R are not simple and they take place at the same time.

The next aspect we would like to mention concerns the possibilities of influencing the workers' objective by the government. It turns out that if the ratio between the part of the net profit $\alpha\Pi(\cdot)$ (see (2)) and the wage fund is within a certain range then the government is in a position to change the shape of J_2 in a special way. Namely , if $(w_m,q_m)= = \arg\max_{(w,q)\in\Omega(T_w,T_b)} J_2(T_w,T_b,w,q)$ then (w_m,q_m) is an internal point of $\Omega(\cdot)$. Thus, for instance - if the workers were free to chose any wage fund w, they would not choose the maximal admissible one.

The main aim of the model which has been roughly outlined here is to provide insight into the nature of the interactions among government, management and workers. In order to study various aspects of the relevant problems an interactive computerized model has been built. Most of the results can be nicely interpreted graphically and therefore the appropriate procedures have been incorporated in the model.

REFERENCES

[1] Salman M.A., Cruz J.B.Jr.: An Incentive Model of Duopoly with Government Coordination. Automatica 17 (1981).

[2] Stefański J.: Control of the Firm Activity - a Two Level Game Model. "Measurement and Control 83" IASTED Symp. Athens, Greece, 1983.

[3] Stefański J., Cichocki W.: Disagreement and Consistency of Interests in an Enterprise. "Systems Theory and Math. Economics" 6th Italian-Polish Symp. Rome, Italy, 1984.

[4] Takayama T., Simaan M.: Multilevel Interactions of Goverment and Private Sectors in Economic Development. 4th IFAC/IFORS/IIASA Conf. on Modelling and Control of National Economies, Washington DC, USA, 1983.

MODELS IN THE ECONOMY OF INFORMATION

Dorin Rădulescu[1)]

ABSTRACT. This paper deals with some models considering information as an economic ressource as well as capital goods, raw materials, energy and manpower. This approach is not only possible, but quite necessary, because of the increasing weight of information processes at each level of socio-economic activities. Some specific peculiarities of the information as an economic ressource are to be taken into account in order to introduce information and the industry of information in macroeconomic models.

KEYWORDS: Information, industry of information, interaction between information and the other economic ressources, quality of information processes.

1. Introduction

The very rapid increasing of the weight and complexity of information processes in the socio-economic activities is generating studies and research dealing with the industry of information (viewed as an economic branch) and even with the economy of information. This approach is involved in "some extents of the economic theory beyond its classical framework" [1]. So, considering the informational features of the human work, the new concept of <u>social energy</u> has been developed. This concept generalizes the more approximate concept of human work viewed only as worktime, involving various degrees of complexity of the work as well as some specific informational features of the work: creativity, skill, talent and initiative. More considerations on this subject are to be found in [2, 4, 10].

Another basic concept is that of <u>industry of information</u>. According to [1, p. 7] the structure of the industry of information may be described as follows:

---

1) Central Institute for Management and Informatics, 8-10 B-d Miciurin, 71316, Bucharest - 1, ROMANIA

INDUSTRY OF INFORMATION	(1)	education, communications, mass-media, culture and art, administrative processes, etc.	industry of information in its limited meaning
	(2)	informatic industry	
	(3)	production of equipments and materials for (1) and (2)	
	(4)	production of components for (3)	

The most concentrate and direct intervention of information in the work processes is the substitution of human work by robots. The new science of robotics is not only a technical one. Its implications concern the economic theory. In [6] the new relationships between work and value are discussed under the circumstances of substitution of work by a special class of capital goods: the robots.

2. Econometric models for the informatic industry

Using the data presented in [3], the informatic industry in Romania has been described in [7] by simple Cobb-Douglas dynamic models. In this way, it became possible to calculate the productivity of labour and of capital goods, as well as the marginal productivities in the informatic industry. It is to be emphasized that all these productivities have higher values than in the other branches of the economy.

The connection between the informatic industry and the whole national economy has been analized by the means of an regressional model:

$$p = aL_1 + bL_2 + cL_3 + d \qquad (1)$$

in which: p = medium productivity of labour in the whole national economy; L_1 = labour in directly productive work; L_2 = labour in office work; L_3 = labour in the informatic industry. The identification of the coefficients a, b, c, d (medium error $< 2\%$) demonstrates that to increase 1% the medium productivity of labour, the contribution of L_3 is 2,4 times, respectively 4,2 times larger than the contributions of L_1 or L_2 [7].

Obviously, such simple models cannot have a considerable explanatory power, but they emphasize convincingly the main trends of the role of informatic industry in the whole system of national economy: 1) the informatic industry is, itself, a lucrative branch, not an auxiliary one; 2) its role is, first at all, to increase the general productivity of labour and to provide structural changes within the labour.

3. An input-output model: information as an economic ressource

This model has the following structure:

$$X^k = AX^k + Y^k \tag{2}$$

$$X^{k+1} = X^k + \Delta X^k = X^k + BY^k \tag{3}$$

which is reducible to the recurrence:

$$X^{k+1} = I + B(I-A)\ X^k \tag{4}$$

The variables of the model are the following : X_1 = the general national product (GNP); X_2 = the "production" of information (in conventional units); X_3 = the total amount of social energy (conventional units). The Y-vector does not mean the final demand as in many classical I/O models, but has the significance of specific accumulations. So, this model generalizes the well-known hypothesis that accumulations lead to self-reproduction, from the "material" production field to those of information and social energy. To exemplify the structure of the A-matrix: $x_{13} = a_{13}X_3$ = material production embodied in the social energy (domestic consumption, domestic equipments, housing, public transports, etc.); $x_{32} = a_{32}X_2$ = social energy embodied in the "production" of information (or in the industry of information). The B-matrix describes the policies of transforming the accumulations, Y, into the growth ΔX.

The A and B matrix have been identified [8] using statistical yearbook data and assuming some qualitative hypotheses.

Another model [8], based on the previous one, identifies the contributions of X_i (i = 1, 2, 3, i $\neq$ j) to X_j as follows:

$$\begin{aligned} X_1 &= a_{11}X_1 + c_{IP}(a_{21}X_1) + c_{EP}(a_{31}X_1) \\ X_2 &= c_{PI}(a_{12}X_2) + a_{22}X_2 + c_{EI}(a_{32}X_2) \\ X_3 &= c_{PE}(a_{13}X_3) + c_{IE}(a_{23}X_3) + a_{33}X_3 \end{aligned} \tag{5}$$

The coefficients $c_{IP} = 1/c_{PI}$, $c_{EP} = 1/c_{PE}$, $c_{EI} = 1/c_{IE}$, which "transforms" material production (P), information (I) or social energy (E) into one another, once identified, give some very interesting conclusions. So 100% material production comprises 52,3% previous material production, 25,9% information and 21,8% social energy. The considerable weight of information requires that this matter be considered separately, not as a component of P or E.

4. Qualitative models

In [5] and [9] two qualitative models were described. The first, based on the "catastrophe" theory of Réné Thom, concerns the relationship, $V = f(m, x_1, x_2)$, between: V = the economic efficiency of informatic industry, m = the optimal level of utilization of the informatic industry, x_1 = economic potential (expressed as GNP); x_2 = "degree of informatization" of the economy.

Another, more complex model introduces the information as a specific economic ressource in a classical model of the economic growth. The structure and the variables of the model suggest the complex implications introduced by considering information as well as the features differencing the social energy from the simple work time.

5. Final remarks

This paper tried to put together some Romanian contributions in the field of the Economy of Information. (A more detailed presentation of the models worked out in this field can be found in the research reports (EI series) issued by the Control Institute for Management and Informatics, Bucharest.)

However it must be emphasized that such models, no matter about their type, demonstrate the fact that industry of information has its own identity as a branch of the national economics. Moreover, the important weight of information and of information processing in the whole economic system gives information the role of a primary ressource of economic growth and of socio-economic development.

REFERENCES

(1) Drăgănescu, M.: Economy of Information (in Romanian), preprint, Central Institute for Management and Informatics (CIMI), Bucharest, 1983

(2) Drăgănescu, M.: Work and Economy, Political Publishing House, Bucharest, 1974

(3) Drăgănescu, M.: Achievements and Prospects in Informatics, preprint, CIMI, Bucharest, 1983

(4) Dumitru, N.S.: Social System, Praxis, Experiment (in Romanian), Scientific and Encyclopaedic Publishing House, Bucharest, 1974

(5) Guran, M., Popescu, I.A., Rădulescu, D.: A Qualitative Model for the Development of Informatics, preprint, CIMI, Bucharest, 1980

(6) Nistorescu, Gh.: Problems of Value in the Context of Production Cybernetization, Econ. Computation and Econ. Cybernetics Studies and Research, vol. XV, nr. 2/1981, p. 65-73

(7) Rădulescu, D.: On Some Indicators Concerning the Economic Efficiency of Informatics (in Romanian), Romanian Bulletin of Informatics, nr. 2/1983

(8) Rădulescu, D.: On the Structure and Functions of a Model in the Economy of Information, Research report EI-01/83, CIMI, Bucharest, 1983.

(9) Rădulescu, D.: An Econometric Model with Considering the Role of Information as Ressource of Development, lecture held at the 3-rd monthly seminary "Economy of Information", CIMI, Bucharest, 1984

(10) Toffler, A.: The Third Wave (translated in Romanian, Political Publishing House, Bucharest, 1983)

SIMULATION MODEL FOR CONTROL OF AIR POLLUTION AND ANALYSIS OF ITS IMPACTS TO BIOSPHERE (EXAMPLE FOREST)

Bellmann, K.[1)], P. Lasch[1)], W. Niefnecker[2)]

Abstract: The interactive simulation model supports the analysis of immission depending forest state dynamics and productivity with regard to fertilization and forest management, and allows to evaluate policies in implementation of antipollution technologies by calculation of site specific emission, immission, and biosphere state transitions. The user is aided in conflict analysis in environmental protection.

Keywords: simulation model, air pollution control, environmental impacts, forest.

In order to get more insights into the linked dynamic behavior of pollutant emission(SO_2), its transport from the source to a receptor area(immission), and state changes of different elements of the bio - sphere in such an area due to immission(combined with climatic stress), a simulation model must consist of subsystems (i) production/emission, (ii) pollutant distribution/immission, (iii) impacts to biosphere and environmental impact assessment.

1. The subsystems

1.1. Subsystem 1: Production/Emission

for calculation of site specific pollutant emission(SO_2) and quantities of production

Definitions :

Production matrix $V = \|v_{i\ j}\|$; $m \times n$ matrix

$i=1,\dots,m$ indices of materials(products, pollutants, raw materials)

$j=1,\dots,n$ indices of technologies $\{T_1,\dots,T_j,\dots,T_n\}$

Emission sites $S = \{s_1,\dots,s_q\}$ are grid squares describing pollutant source sites

Technological innovation matrix ($m \times n$), as control variables for technological changes

1) Central Institute for Cybernetics and Information Processes of the Academy os Sciences of the G.D.R., DDR-1086 Berlin, Kurstr. 33

2) Institute of Forest Sci., DDR-1300 Eberswalde-Finow 1, Alfred-Möller-Straße

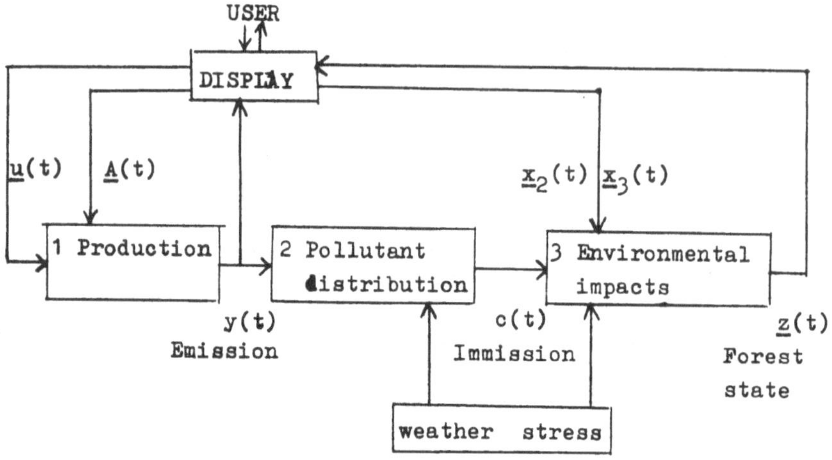

fig. 1 : Scheme of the simulation model
3 subsystems : Production(using V),Pollutant distri - bution (using P), Environmental impacts(using FOREST)

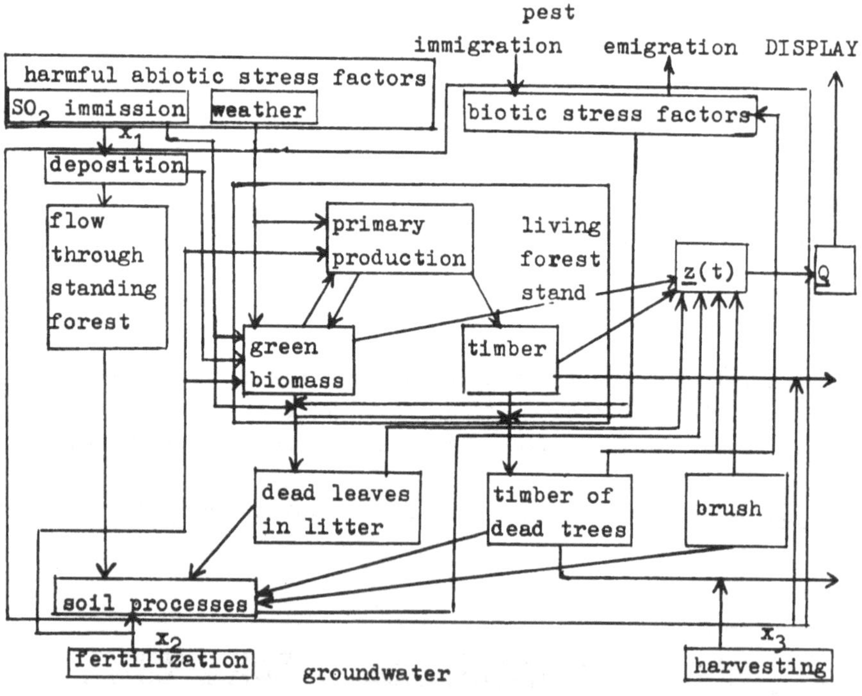

fig. 2 : Scheme of model FOREST of subsystem 3

$$A(t) = \|a_{ij}(t)\| \qquad a_{ij} = \begin{cases} \neq 0 & \text{iff } v_{ij} \text{ is changed at time } t \\ 0 & \text{else} \end{cases}$$

Planned main product figure (1×n vector) as control variable for amount of production and (qualitative) technological innovations

$$\underline{u}(t) = \begin{bmatrix} u_1(t) \\ u_j(t) \\ u_n(t) \end{bmatrix} \qquad u_j(t) = \begin{cases} \neq 0 & \text{iff } T_j \text{ works at time step } t \\ 0 & \text{else} \end{cases}$$

$u_j(t)$ is the planned quantity of main product per year produced by T_j in year t

Site and time depending production matrix (m× n)

$$\bar{V}^s(t) = \bar{v}^s_{ij}(t) \qquad \bar{v}^s_{ij}(t) = \begin{cases} v_{ij} & \text{iff } s \text{ is site of } T_j \text{ and } u_j(t) \neq 0 \\ 0 & \text{else} \end{cases}$$

$$V^s(t) = \bar{V}^s(t) \cdot A(t)$$

Emission $y(t,s)$ of site s and time step t is

$$y(t,s) = V^s(t) \cdot u(t) \qquad y - \text{a } (1\times m) \text{ vector}$$

$$y(t,s) = \left[v^s_1, \ldots, v^s_j, \ldots, v^s_n\right] \cdot u(t) \qquad \text{,considering only } SO_2 \text{ emission}$$

1.2. Subsystem 2 : Pollutant Distribution/Immission

For calculation of immission concentrations $c(t,s)$ of SO_2 for all receptor grid squares $S = \{s_1, \ldots, s_b, \ldots, s_B\}$,describing the receptor area

Standardized immission matrices (r×r) $P_h = \|p_{kl}\|$; $k,l = 1, \ldots, r$ for different stack hights $h=1,\ldots,5$ are used.

p_{kl} is the standardized immission concentration at site s_{kl},related to a given emission standard figure,due to a source T_j of hight h lo - calized at grid square s_{kl},as the central element of P_h. The immission concentration $c(t)$ at site $s_b = s_{kl}$ is calculated by linear super - position :

$$c(t, s_{kl}) = \sum_{j \in J_{kl}} c^j(t, s_{kl})$$

with J_{kl} - index set of T_j's which influence s_{kl}.

$c^j(t, s_{kl})$ - contribution of T_j into the considered grid square s_{kl} calculated by setting the coordinates of T_j as central element of matrix P .

1.3. Subsystem 3 : Impacts to biosphere (forest)

By the forest model FOREST consequences of SO_2 immission (combined with climatic stress) to state changes of a coniferous forest eco - system under specific fertilization and harvesting management can be investigated.

1.3.1. Input and control variables of the model FOREST

- environmental stress : SO_2 immission (x_1), climate, weather stress[1)]
- biotic stress (pest immigration)[1)]
- nitrogen fertilization ($\underline{x}_2$)
- management control input (harvesting) ($\underline{x}_3$)

1.3.2. Subprocesses/state variables

a) growth and mortality : - dynamics of biomass ($\underline{z}_1$)
- dynamics of forest stand variables ($\underline{z}_2$)

b) damage state of biomass/toxicity ($\underline{z}_3$)

c) dynamics of net assimilation (carbohydrat pool)[1)]

d) soil processes/brush[1)]

e) pollutant flow through the ecosystem[1)]

1.3.3. Dynamic equations

ad a) $\underline{z}_1 = (z_{11}, z_{12}, z_{13})$ with z_{11} - height z_{12} - diameter
z_{13} - stem volume

- state variable height z_{11} :

$z_{11}(t) = z_{11}(t-1) + \Delta z_{11}(t)$ with the stem elongation rate

$\Delta z_{11}(t) = A_{11}(AGE(t), IH(AGE(t), z_{32}(t)), FH(AGE(t), z_{32}(t), \underline{x}_2(t)))$

IH - table function, response of stem elongation to toxicity

FH - table function, response of stem elongation to fertilization

A_{11} - state transition algorithm

- state variable diameter (bhd) z_{12} :

$z_{12}(t) = A_{12}(z_{12}(t-1), AGE(t), ID(AGE(t), z_{32}(t)), FD(AGE(t), z_{32}(t), \underline{x}_2(t)))$

ID - table function, response of diameter growth to toxicity

FD - table function, response of diameter growth to fertilization

A_{12} - state transition algorithm

- state variable stem volume z_{13} :

$z_{13}(t) = r_1 \cdot z_{11}(t)^{r_2} \cdot z_{12}(t)^{r_3}$, r_i - parameters

- forest stand variables $\underline{z}_2$ (basal area, density, ...) :

$\underline{z}_2(t) = A_2(\underline{z}_2(t-1), \underline{z}_1(t), \underline{x}_2(t), \underline{x}_3(t))$

A_2 - state transition algorithm

ad b) $\underline{z}_3 = (z_{31}, z_{32})$ z_{31} - toxicity of needles
z_{32} - damage level of green biomass

- state variable toxicity z_{31} :

$z_{31}(t) = z_{31}(t) + \Delta z_{31}(t)$

1) in preparation

$$\Delta z_{31}(t) = f_1(t) \cdot AGE(t) \cdot f_2(x_1(t), x_1(t-1))$$

f_1 - toxicity accumulation function

f_2 - toxicity delay function

- state variable damage level z_{32} :

$$z_{32}(t) = f_3(z_{31}(t))$$

f_3 - scoring function of toxicity

2. Model output

a) emission figures $y(t,s)$ for SO_2
b) immission concentrations $c(t,s)$ for SO_2
c) forest state variables $(\underline{z}_1, \underline{z}_2)$ for a pine forest as timbermass (living, dead, standing, harvested), density, basal area, etc.
d) damage level of green biomass $(\underline{z}_3)$
e) Furtheron, it is possible to calculate
(i) comfort functions (Q) with free parameters for the user in order to evaluate the forest state of a given area in the sense of environmental quality functions and
(ii) the loss of biomass gain per year due to SO_2 immission with special regard to treatments with nitrogen fertilizer and to evaluate these losses monetary.

3. Scenario example

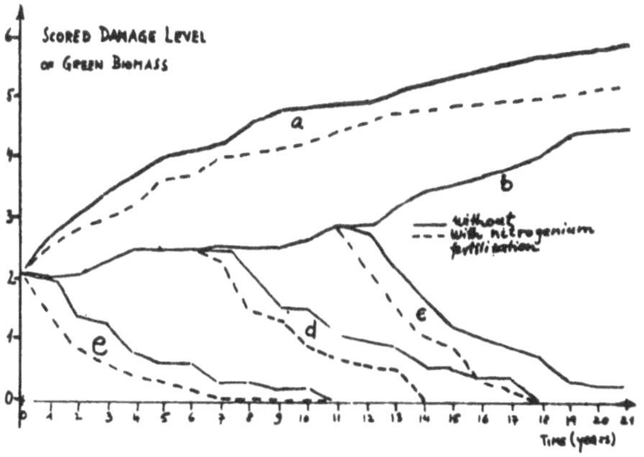

fig.3 Forest damage dynamics depending on emission reduction and fertilization : 0 - without visible damage, 6 - fully damaged (dead) green biomass; (a),...,(e) scenarios (see text)

In figure 3 simulated dynamics of the mean damage level of green biomass of all(living and dead) trees(scenario (a)) and of living trees(scenarios (b),...,(e)) in a pine forest area is shown,depending on different innovation years of antipollution technologies in the pollutant emittents.The technological scenarios are

(a),(b) - without any emission reduction(ER) within the 20 years interval

(c),(d) - 0.5 ER from 10-th year and 5-th year resp.

(e) - 0.9 ER immediately introduced(1-st year).

The experiments for analysis of fertilization effects on damage dynamics of green biomass were performed by the Institute of Forest Science Eberswalde.

Reference

Lembke,G.;Knapp,E.;Dittmar,O. : Die neue DDR -Kiefernertragstafel 1975,Beiträge für die Forstwirtschaft,2/1981,55-64

A COMPUTER MODEL FOR SHORT-TERM FORECASTING AND CONTROLLING AIR QUALITY IN A CITY

Piotr Holnicki and Antoni Żochowski [1)]

Abstract. An urban-scale computer model for short-term prediction of city air quality is presented. The physical process of pollutant dispersion is described by distributed parameters advection-diffusion equation. The model, utilizing meteorological forecast , calculates spacial and temporal distribution of pollutant concentration in complex urban area. Some test computations have been conducted for Warsaw case. Basing on prediction model, a computer program for real-time control of emission redistribution in urban area is elaborated.

Keywords. Air pollution, distributed parameters systems , computer model, optimal control.

1. THE FORECASTING MODEL

This paper describes the urban-scale computer model URFOR predestined for short-term prediction of air pollution in a large city. The horizontal

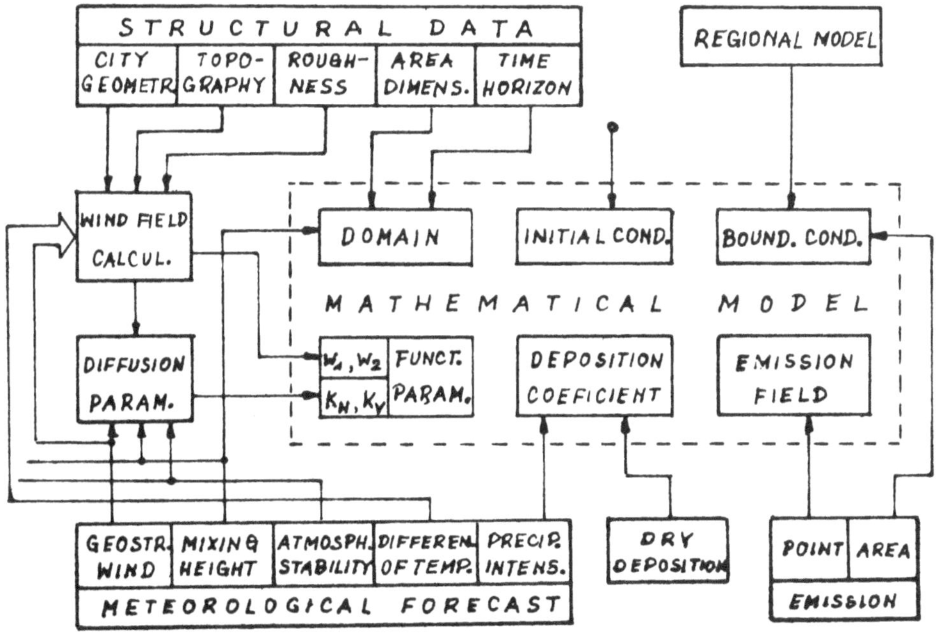

Fig. 1. The block diagram of the model

1) Systems Research Institute of the Polish Academy of Sciences
01-447 WARSZAWA, ul. Newelska 6, POLAND

scale of simulated dispersion process is 20 ÷ 40 km, while the time horizon of model's prediction - T = 1 ÷ 3 days. The input data can be divided, in general, into the following three groups (compare Fig.1 for details): i) independent of time structural data, ii) meteorological forecast, iii) characteristics of emission sources.

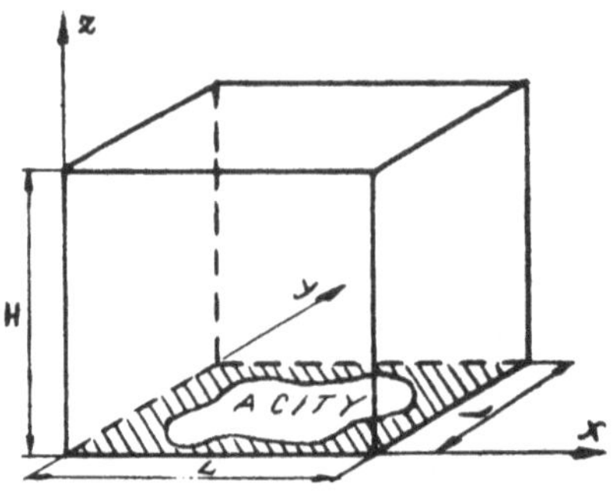

Fig.2. Domain of simulation

The block MATHEMATICAL MODEL represents the physical process of pollutant dispersion in the atmosphere, which is considered in a square domain $\Omega = L \times L$ (see Fig.2). It is described by two-dimensional, averaged over the mixing height - H, advection-diffusion equation of the form [3,4]

$$\frac{\partial c}{\partial t} + \underline{w} \cdot \nabla c - K_H \Delta c + \gamma c = \bar{Q} + (E - v_d c)/H \quad \text{in} \quad \Omega \times (0,T), \tag{1}$$

along with the boundary conditions

$$\begin{aligned} \frac{\partial c}{\partial \underline{n}} &= 0 \quad \text{for} \quad \underline{w} \cdot \underline{n} \geq 0 \\ & \qquad\qquad\qquad\qquad \text{on} \quad \partial\Omega \times [0,T], \\ c &= 0 \quad \text{for} \quad \underline{w} \cdot \underline{n} < 0 \end{aligned} \tag{2}$$

and the initial condition

$$c(0) = c^{o} \quad \text{in} \quad \Omega. \tag{3}$$

Here c - denotes a pollutant concentration in $[\mu g/m^3]$, $\underline{w} = [u,v]$ - the wind vector in $[m/s]$, K_H - horizontal diffusion coeficient in $[m^2/s]$. $\bar{Q}$ - averaged over H pointwise emission field in $[\mu g/m^3 s]$, E - area emission field in $[\mu g/m^2 s]$, v_d - dry deposition coeficient $[m/s]$, γ - wet deposition factor depending on precipitation intensity.

The values of the wind field vector $\underline{w}(x,y,t)$ in domain Ω are generated by a special procedure which calculates succesively: i) the averaged over Ω value $\underline{w}_o$, basing on meteorelogical forcast, ii) topographical correction $\underline{w}_t$ depending on the ground topography and aerodynamical roughness, iii) thermal correction $\underline{w}_\theta$ depending on a city "heat island" effect. Finally

$$\underline{w} = \underline{w}_o + \underline{w}_t + \underline{w}_\theta \,. \tag{4}$$

The model generates a sequence of forecasts of pollutant concentration within a period - T, which is discretized with time interval of the model - DT. Its length is determined by the frequency of introducing meteorological data (in our case DT=6 hrs). Each time interval is segmented with

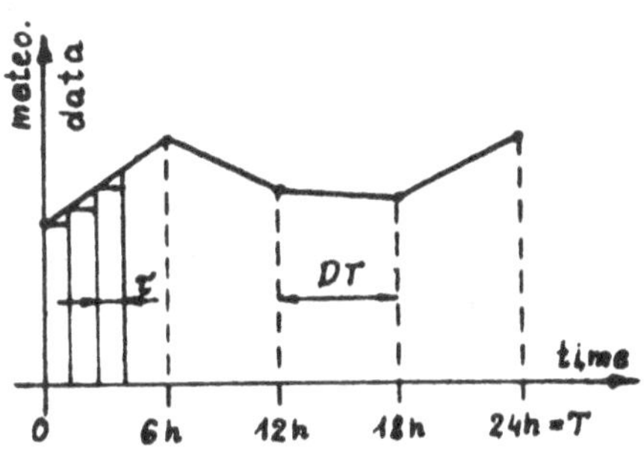

Fig.3. Time discretization

time discretization step - τ of numerical procedure solving advection-diffusion equation (see Fig.3). All time-dependent data at the inner points of the interval DT are linearly interpolated. The initial-boundary value problem (1) -(3) is numerically solved by an effective combination of the methods of characteristics with F.E. procedure [2].

The model has been implemented on SM-4 computer and tested on real data for Warsaw Metropolitan Area. The computing time for one 24-hrs forecast on 20 × 20 space discretization grid does not exceed 10 min [1].

2. THE EMISSION CONTROL MODEL

Basing on the forecasting model URFOR, the real-time emission control problem for the system of sources covering the area was formulated. A general idea of controlling consists in minimizing an enviromental damage by redistributing the production (emission) among a set of selected sources, according to the meteorological situation. Formally, the problem is stated as follows:

Minimize the cost functional

$$J(\underline{u}) = \alpha_{11} \int_0^{T_h}\int_\Omega r \cdot \max^2(0, c(\underline{u}) - c_d)\,d\Omega dt + \alpha_{12} \int_0^{T_h} \sum_{i=1}^{N} \beta_i (u_i - u_i^o)^2 dt , \qquad (5)$$

$$\underline{u} = [u_1, \dots, u_N]^T ,$$

where $c(\underline{u})$ is the solution of the state equation

$$\frac{\partial c}{\partial t} + \underline{w} \cdot \nabla c - K_H \Delta c + \gamma c = Q + \sum_{i=1}^{N} \chi_i \cdot F_i(u_i) \quad \text{in} \quad \Omega \times (0, T_h), \qquad (6)$$

with the boundary-initial conditions (2)-(3), under the constraints

$$g(\underline{u}) = \alpha_{21} \int_0^{T_h}\int_\Omega r \cdot \max^2(0, c(\underline{u}) - c_d)\,d\Omega dt + \alpha_{22} \int_0^{T_h} \sum_{i=1}^{N} \beta_i (u_i - u_i^o)^2 dt \le K , \qquad (7)$$

$$\underline{u}_i \le u_i \le \bar{u}_i \quad \text{and} \quad -D_i \le \frac{du_i}{dt} \le D_i , \quad i=1,\dots,N , \qquad (8)$$

$$\sum_{i \in N_j} a_{ij} u_i \ge b_j , \quad j=1,\dots,M ; \quad \bigcup_{j=1}^{M} N_j = N . \qquad (9)$$

The functions $\chi_i(x,y)$ in (6) describe the location of controlled sources; $F_i(u_i(t))$ relate emission to production level. The factor $r(x,y)$ in (5), (7) is a region weight function, $c_d(x,y)$ - the admissible level of pollution. The second term in (5) constitutes the cost of deviation of production levels u_i from the economical values u_i^o. The inequalities (8) represent technological constraints, while (9) reflects demand requirements b_j, imposed on homogeneous groups of plants - N_j.

The parameters α_{ij} in (5), (7) make it possible to formulate a variety of optimization problems, ranging from minimization of enviromental damage under limited resources ($\alpha_{ii}=1$, $\alpha_{ij}=0$) to minimization of outlays with enviromental constraint ($\alpha_{ii}=0$, $\alpha_{ij}=1$).

The problem is solved by a quickly convergent version [2] of linearisation algorithm [5]. The numerical experiments involved a case of five controlled sources, based on modified data for Warsaw. An illustrative result ($\alpha_{ii}=1$, $\alpha_{ij}=0$) is shown in Fig.4. Since the exact solutions are

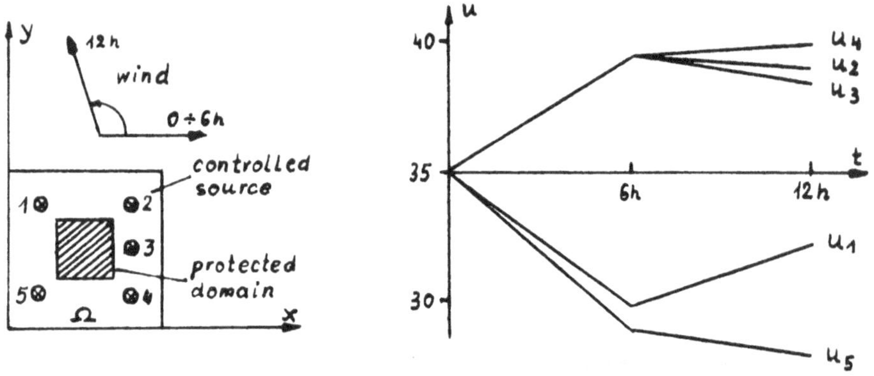

Fig.4. The results of numerical experiment

not known, the result must confirm physical intuition. In our case the transitions of production from sources {1,5} to {2,3,4} in the first 6 hrs, and then from {5} to {1}, satisfy this condition.

REFERENCES

[1] Holnicki P. et al.: An urban-scale computer model for short-term prediction of air pollution. Arch. Autom. i Telemech. (to appear).

[2] Holnicki P. and Żochowski A.: Numerical methods in forecasting and controlling air pollution problems. Systems Research Institute, Raport No. ZTS-15-7/84, Warsaw 1984 (in Polish).

[3] Fronza G. and Melli P. (eds.): Mathematical Models for Planning and Controlling Air Quality. Proceedings of an IIASA Workshop. Pergamon Press 1982.

[4] Martchuk G.I.: Mathematical Modelling in Enviromental Problems. Nauka 1982 (in Russian).

[5] Pshenitchny B.N.: Linearisation Method. Nauka 1983 (in Russian).

REPRODUCTION MODEL FOR A FOREST MANAGEMENT UNIT

D. Gerold [1)]

The forest fund is a hierarchically structured dynamic system subdivided in management units. In order to get more insights into the dynamic behaviour a simulation model for long and intermediate forecasting of forest fund development and for designing management strategies and tactics was built.
Its application is shortly described by means of scenario techniques.

Keywords: Simulation of forest stand funds, model-aided strategy design, scenario techniques

1) Technische Universität Dresden, Sektion Forstwissenschaft, DDR - 8223 Tharandt, Pienner Str. 8

LONG TERM PREDICTION OF FOREST RESOURCES

H. Kurth; B. Anders; G. Lucas [1)]

A macromodel is described for men-controlled dynamics of forest stand funds depending in different options of management strategies.
In this simulation model a functional connection between the accumulated variable stem volume/ha, annual growth rate of stem volume, and harvesting rate is used to design strategy variants of forest stand funds in order to support decision making with regard to forest growth enhancing treatments.

Keywords: Forest stand dynamics, simulation model, computer-aided decision making

[1)] Technische Universität Dresden, Sektion Forstwissenschaften, DDR - 8223 Tharandt, Pienner Str. 8

DISPERSION OF POLLUTION IN RESERVOIRS AND LAKES

Ivo Chytil [1)]

Summary: Reservoirs and lakes are classified in the several groups. Each group is characterised. The basic characteristic features of dispersion of pollution for every type of reservoir or lake are discussed.

Keyworde: SIMULATION, SURFACE WATER, LAKES, RESERVOIRS, POLLUTION

Introduction

Reservoirs and lakes accumulate important storages of fresh surface water. Therefore it is important to simulate the dispersion of pollution in them as to predict the influence of new sources of pollution or of new reservoir on the quality of water.

Mixing processes in two different lakes or reservoirs can be very different. Therefore it is convenient to classify lakes and reservoirs in the several groups (table No 1). The lagtime and the shape of the lake or of the reservoir are important parameters of this classification. The lagtime T of the reservoir or the lake of volume V is defined:

$$T = \frac{T}{Q} \qquad (1)$$

where Q is the average flow.

The boarders between groups are not strong. Very often the character of the reservoir or lake is on the boarder of two groups. Therefore it is necessary to find the best solution separatetly.

Through - flow reservoirs

Dispersion of pollution in through - flow reservoirs is the same as in the running waters. Due to the slower water

1) Research Institute of Fuel and Energy Complex, dep. 133 Vladislavova 4, CS - 113 72 Praha 1, Czechoslovakia

velocity and the larger depths the total mixing is formed later than in the former channel (overflow of weir is important for depth mixing not for transversal mixing). The lagtime is given by the equation (1).

Reservoirs with temperature stratification

Excluding the input part the whole reservoir has vertical density stratification. Except some special cases (e.g. floods) density and temperature stratification corresponds. This stratification changes during the year.

Due to the stratification characteristic parameters (velocity, dispersion coefficients and others) differ with the depth.

Therefore as for the simulation of the density (temperature) stratification as for the simulation of the dispersion of the pollution it is usually convenient to divide the reservoir in two parts: 1) inflow part, 2) stratificated part.

Extremly inflow part can be described as part with total mixing water [3] or as a part of slowly flowing river (through - flow reservoir). The real situation is somewhere between these two extreme cases (in the dependance on the flow and temperature of input water as the function of the time, on the shape of input part and on stratificated part) and it is experimentally investigated.

Usually the stratificated part is separated in horizontal layers [2], [3], [4]. The thickness of each layer can be constant or variable. It is assumed tha t in each layer characteristic parameters are constant (velocity, coefficient of dispersion). The influence of the shape of the bottom, bays, important tributary streams, operation of hydro-electric power plant on homogenity of layers is now experimentaly investigated.

In many cases the layer is taken homogenous i.e. totally mixed. This assumption is not exactly correct and it can influence the lagtime and real mixing of pollution.

Deep lakes

Wind streams are typical for deep lakes (due to the shape of the lake and the lagtime the stream input - output is unimportant). Therefore the direction and the intensity of the stream depent on the wind. The wind produce in the surface layer the streams in the direction of the wind and in the layer under the surface layer in the opposite direction. Therefore the distribution of the streams during the year in the surface layer is in the correspondence to the wind rose and in the layer under the surface layer is opposite.

Due to the surface inlet and outlet, long lagtime, wind streams and the depths tens meters it is possible to divide the lake in two main horizontal layers: 1) surface layers with important wind mixing and important influence of inlet and outlet, 2) deep layer with minimum mixing and relatively restricted exchange with surface layer (the lagtime is significantly larger than the average lagtime of the lake (1). Very often the exchange between layers is restricted by temperature difference.

Shallow lakes and reservoirs

The direction of the streams is significantly influenced by the shape of the bottom (in the reservoirs the main stream very often traces the former channel) and by wind. Water in some parts is nearly exchanged only by wind streams.

Repumped reservoirs

Due to the great difference of the relative quantity of the repumped water (1 ÷ 80 %) there are important differences in mixing processes in separate reservoris. Our first experimental results show that extreme possibilities are: 1) totally mixed reservoir and 2) density stratificates reservoir. Under the surface layer there is situated a thick layer without stratification with the temperature of repumped water. The surface layer is not homogenous. In the part near the dam it is influenced by repumped water. The stratification is signigicantly influenced by the shape of the bottom during the operating of the hydro - electric power plant.

Literature:

1 Chytil I.: Dispersion of Radioactive Pollution in Surface Water, Ecol. Modelling, in print
2 IAEA: Methods for Cost Effectiveness Analysis of Risk Reduction in Energy Production, Draft, Vienna, 1984
3 Markofsky, M., D.K.F. Harleman: A Predictive Model for Thermal Stratification and Water Quality in Reservoirs, Water Pollution Control Research Series, 16 130 VJH 01/71, Environmental Protection Agency, Water Quality Office, Department of Civil Engineering, Massachussets Institute of Technology, Cambridge
4 U.S. Nuclear Regulatory Commission: Regulatory Guide 1.113: Estimating Aquatic Dispersion of Effluents from Accidental and Routine Reactor Releases for the Purpose of Implementing, Appendix I, Revision 1, Washington D.C., 1979

Table No 1 - The main types of reservoirs and lakes

Type of reservoir or lake	Characteristic features
Through - flow reservoirs	The lagtime is shorter than 10÷20 days. The shape of the reservoir is nearly the same as the shape of the former channel
Reservoirs with temperature/density/ stratification	The lagtime is tens or hundreds days. The shape of the reservoir is elongated in the direction of the former channel. The length is in order larger than the width. Important depth (tens meters). The main outflow is under the level. Typical depth density stratification (created by temperature stratification).
Deep lakes	The lagtime is tens or hundreds days. "Circular" shape (the length is not significantly longer than the width). The outflow is on the level. Significant depth (tens or hundreds meters).
Shallow lakes or reservoirs	The lagtime days or tens days. In the most parts the depth less than ten meters. Very often: Significant relative differences in the depth. A lot of islands.
Repumped reservoirs	The water is repumped between two reservoirs (repumping part of water can be very different 1 ÷ 80 %). The lagtime tens or hundreds days.

NASH MODEL OF WATER RESERVOIR POLLUTION

Maciej Krawczak [1] and Andrzej Ziółkowski [1]

Abstract : The paper describes some concept of water quality control in a reservoir. Two factors assessing the water pollution is considered, viz: dissolved oxygen and biochemical oxygen demand. Several polluters can be adjacent to the reservoir. Each polluter minimizes his own costs of treatment of discharged waste . Each polluter is treated as a player. The Nash feedback solution concept is proposed. A numerical example illustrates the paper.

Keywords. Control theory, differential game theory, modelling, water pollution.

I. Introduction

Each water system should contain three subsystems, namely : water distribution control, water retention control and water quality control, Little attention has been paid to the last problem, however this problem no longer can be neglected. Water pollution control of the reservoir is done by proper treatment of sewage by each polluter. After the treatment the sewage is discharged into the reservoir. Each polluter (e.g. factory, city) has his own cost of waste treatment in order to keep pollution level close to predefined standards.

In related papers, several authors tried to clarify the nature of the problem. Young and Beck [7] have identified parameters of the one-reach model, Singh [4] has compered hierarchical to suboptimal optimization of the multi-reach model. Ozguner and Perkins [3] as well as Krawczak and Mizukami [1] , [2] applied game theory concept to multi-reach models. In all above mentioned works it was assumed that one polluter is adjacent to one reach.

Here, in this paper we consider a reservoir (or one reach) on a river and several polluters allocated around it. The reservoir is on the river and water flows in and out. Each polluter is treated as one player. We shall applied the Nash feedback solution concept in order to find strategies for each player-polluter.

II. The Reservoir Model.

Water in the reservoir is used for several purposes, e.g. for potable

---

1) Systems Research Institute, Polish Academy of Sciences, ul. Newelska6, 01-447 Warsaw, Poland.

supply or for recreation. It has been shown [7] that water quality can be considered in terms of the dissolved oxygen (DO) concentration and the biochemical oxygen demand (BOD) concentration. The most often used models describing balance between DO and BOD are based on the classical work of Streeter and Phelps [6] . The general model of DO-BOD balance in the reservoir can be written |4| , [7] as follows

$$\text{BOD:}\quad \dot{z} = \left(-K_1 - \frac{Q + \sum_{i=1}^{N} Qe_i}{V}\right) z + \frac{Qo}{V} zo + \frac{1}{V} \sum_{i=1}^{N} Qe_i m_i \tag{1}$$

$$\text{DO:}\quad \dot{q} = -K_1 z + \left(K_2 - \frac{Q + \sum_{i=1}^{N} Qe_i}{V}\right) q + \left(K_2 q^s + \frac{Qo}{V} qo - \frac{\eta}{V}\right) \tag{2}$$

here:

z is the BOD concentration of flowing out water,

q is the DO concentration of flowing out water,

N is the number of polluters,

m_i is the BOD in the effluent of polluter i,

Q is the volumetric rate,

Qe_i is the flow of effluent of polluter i,

V is the volume of reservoir water,

K_1, K_2, q^s, η - are some parameters,

zo is the BOD of flowing water into reservoir,

qo is the DO of flowing water into reservoir.

The main purpose of water quality control is to keep levels of BOD and DO in the output of the reservoir as close as possible to the predefined levels. There predefined levels of BOD and DO are steady-state standards. Due to the assumed fact that each polluter has his own cost criterion

$$J_i = \int_{t_o}^{T} \left[(z - z^*)^2 F_i + (q - q^*)^2 G_i + m_i^2 H_i \right] dt \tag{3}$$

where z^*, q^* are standards, F_i, G_i, H_i are proper weights - the differential game approach seems to be very natural.

III. Nash Strategy for Pollution Control.

Here we will give some definition of the nonzero-sum differential game. Consider the system.

$$\dot{x} = Ax + \sum_{j=1}^{N} B_j m_j , \qquad x(to) = xo \tag{4}$$

with the cost functions,

$$J_i = \frac{1}{2} \int_{to}^{T} \left(x' Qi\, x + \sum_{j=1}^{N} m_j R_{ij} m_j \right) dt \tag{5}$$

the Nash feedback strategy is well known [5] to be

$$m_i^* = - R_{ii}^{-1} B_i' P_i x \tag{6}$$

where P_i is found from

$$\dot{P}_i = - P_i A - A' P_i - Q_i - \sum_{j=1}^{N} (P_j B_j R_{jj}^{-1} R_{ij} R_{jj}^{-1} B_j' P_j - P_i B_j R_{jj}^{-1} B_j' P_j \tag{7}$$

$$- P_j B_j R_{jj}^{-1} B_j' P_i) , P(T) = 0.0 ; \qquad i = 1, 2, \ldots, N.$$

According to the definition , the strategy set $\{m_1^* , m_2^*, \ldots, m_N^*\}$ is a Nash equilibrium strategy set if , for $i=1,2,\ldots,N$

$$J_i (m_1^*, \ldots, m_i^*, \ldots, m_N^*) \leqslant J_i (m_1^*, \ldots, m_i, \ldots, m_N^*) \tag{8}$$

where m_i is any admissible strategy for player i.

The model of water pollution of the reservoir (1),(2) with the cost criteria (3) correspond accurately to the description of the Nash game (4),(5).

IV. Example

The only reported data [7] of water pollution were collected at the River of Cam (England). Here, we use synthetic data which are based on these of the River Cam. We consider two polluters associated with a reservoir. The matrices in (4) and (5) are

$$A = \begin{bmatrix} -1,32 & 0,0 \\ -0,32 & -1,2 \end{bmatrix}, \; B_1 = \begin{bmatrix} 0,1 \\ 0,0 \end{bmatrix}, \; B_2 = \begin{bmatrix} 0,05 \\ 0,0 \end{bmatrix}, \; Q_1 = Q_2 = \begin{bmatrix} 2 & 0 \\ 0 & 1 \end{bmatrix}$$

$$R_{11} = R_{22} = 1,0 \quad , R_{21} = R_{12} = 0,0$$

Nash solutions were obtained using (6) , (7) in order to obtain the trajectories of Fig. 1 and strategies of Fig. 2 . Initial values are $q_0 = 10.0$, $z_0 = 0.0$, $z^* = 2.5$, $q^* = 8.0$, $m_1^* = 19.25$, $m_2^* = 27.5$, $z(0) = q(0) = 5.0$

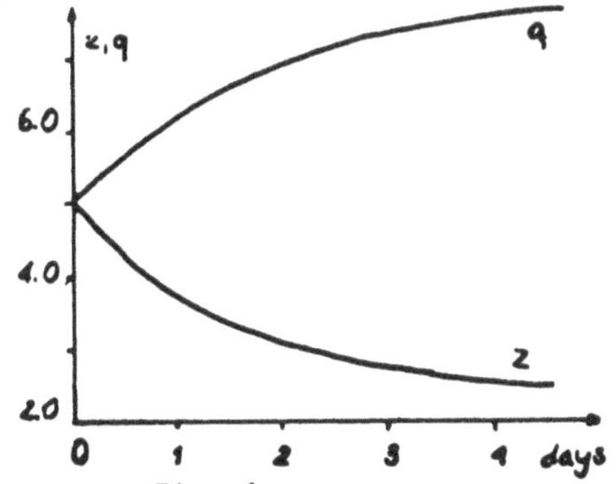

Fig. 1.

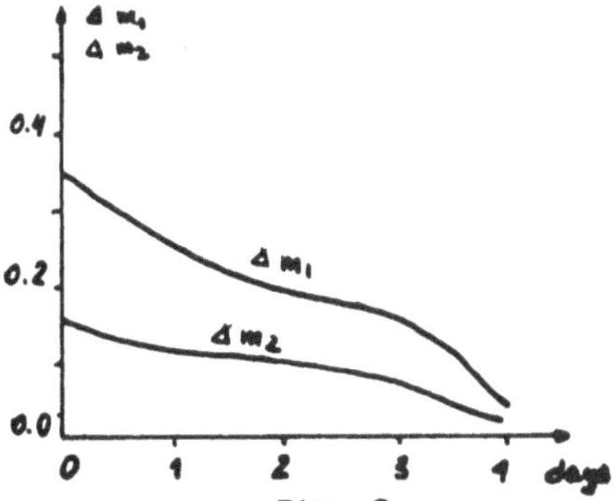

Fig. 2.

V. Conclusion

A strategy for reservoir pollution control based on the Nash differential game concept has been introduced. Polluters around the reservoir are in a conflict and can be treated as players. Differential game theory gives another tool for optimization problem.

References

[1.] Krawczak,M., and K.Mizukami : On Stackelberg approach to river pollution control, Memoirs of the Faculty of Integrated Arts and Sciences III, Hiroshima University, Vol.7 , 1983 .

[2.] Krawczak,M., and K.Mizukami : Game - theoretic approach to river pollution control : discrete case , in Proc. 3 rd IFAC/IFORS Symposium LSSTA, (A.Straszak ed.) , Warsaw 1983 .

[3.] Ozguner,U., and W.R.Perkins : A Nash feedback approach to control of river pollution, IEEE Transactions, vol. SMC-9, no.5,May 1979.

[4.] Singh,M.G.: River pollution control, Int.J.Syst.Sci. vol.6,no.1, 1975.

[5.] Starr,A.W., and Y,C,Ho.: Nouzero-sum differential games,J.O.T.A., vol.3,no.3, 1962.

[6.] Streeter,H.W., and E,B.Phelps: A study of pollution and natural purification of the Ohio river, Biulletin No 146,U.S. P.H.S., 1925.

[7.] Young,P., and B.Beck:The modelling and control of water quality in a river system, Automatica, vol.10, no.5, Sept.1974.

SET PROBABILITY IDENTIFICATION IN FOREST FIRE SIMULATION

T.N.Ivanilova[1)]

Abstract. Average measure simulation of forest fire spread is one of the applications of set probability theory. Probability spread calculations are carried out using set identification methods of a model of random spread ,taking into account initial data of fire condition in a present moment of time we get prognosis - average measure fire contour - in any post coming moment of time.
Keywords: set probability identification, average measure prognosis, local spread probabilities, experimental and model divergence.

Introduction

The paper presents methods of set identification and average measure prognosis of random spread. Forest fire is spontanious spreading process. Methods of probability set simulation are used for the forest fire description. As a model of fire spread process of random spread is chosen [1] .

Prognosis of geometrical fire development to be done it is necessary to carry out an identification of random spread model. In present methods of plane and three-dimensional identification problems are worked out, identification methods of are researched for homogeneous (rules of fire spread are unique for every point of space) and nonhomogeneous spread. The main idea, on which identification methods are based consists in calculations of local spread probabilities depending of geometrical shape of fire [1] . Set identification algorythm is an iterative procedure for spread probabilities and ends in approach of a given exactness.

Initial data for fulfilment of random spread model identification and further for average measure prognosis are the two fire contours in fixed moment of time and a map of homogeneties. Burning edge of forest territory will be call fire contour, if may be tied or untied. A map of homogeneties is a laying - out forest territory in parts homogeneous in relation to burning. A plane square lattice is used for an input in computer a map of homogeneties and a map of forest territory. As a result of using it becomes possible to consider fire spread in every point in eight directions.

1) Computer Center, SB AS USSR, Krasnoyarsk, 660036, USSR

Short description of mathematical model of random spread process (RSP)

Let environment of x , including itself eight the nearest heibours, be called an environment of local spread. And accept the assumption that do not contradict experimental observations:

- local spread from each pair of points happens independently,
- local spread in each direction happens independently,
- every homogenety possesses stationary qualities in relation to burning.

Then denote:

$$f(x) = \{f(1,j), \dots, f(8,j)\} \tag{1}$$

a set of independent probabilities of local spread from point x in one of eight directions in unit of time on unit of length, $g(y)$ - probability of burning of point y from environment of point $x, (y \in O_x)$; x belongs to j-th homogenety.

Set-recurrent coorelation, tieing the present K_t and forthcoming K_{t+1} conditions of a process in field of random finite sets of local spread is

$$K_{t+1} = \bigcup_{x \in K_t} {}^tS_x^{f(x)} (g(y), y \in O_x) \tag{2}$$

where ${}^tS_x^{f(x)}(g(y), y \in O_x)$ random finite set of local spread from point in moment t , given by probabilities $f(x), g(y), y \in O_x$.

Then distribution of local sets spread is completely denoted by spread and ignition probabilities. For any point $y \in O_x$, laying in i-th direction and belonging to j-th homogenety, probability of its covering by local random set ${}^tS_x^{f(x)}(g(y), y \in O_x)$ is determined by the formula

$$\pi(y) = f(x) \cdot g(y) \tag{3}$$

The set of probabilities of covering (3) for all $y \in O_x$ - the distribution of the local random set, since the probability of any its value calculates by the formula

$$P({}^tS_x^{f(x)} = A) = \prod_{y \in A} \pi(y) \prod_{y \in O_x \setminus A} (1 - \pi(y)) \tag{4}$$

Set probability identification

Let's consider the method and the algorythm of identification of random spread model on geometrical criterion. As it has been said above, experimental fire condition are determined by a map of contours, and simulation fire condition dependes on values of spread probabilities $f(x)$ and ignition probability $g(y)$.

Let designate Y_1 - fire condition in moment t_1 , Y_2 - fire condition in moment t_2, $M = \bigcup_{j=1}^{m} M_j$ - laying out of forest territory in homogeneous parts. Problem of average measure identification consists in calculations for its probabilities (1) such, that process of random spread (2) determined by these probabilities having begun in condition $K_0 = Y_1$ from $\tau = t_2 - t_1$ moments of time became in average measure condition $\int K_\tau$ [2], that is the least deviating from Y_2.

As criterion of proximity of experimental and average measure conditions of sets deviations of area of these sets are used in every of eight cone of directions. Deviation $\int K_\tau$ from Y_2 in i-th direction on j-th homogenety M_j is defined as difference

$$S(i,j) = S_+(i,j) - S_-(i,j) \qquad (5)$$

where $S_+(i,j)$ -area of a set on which $\int K_\tau$ bigger that Y_2 in limits $C_i \cap M_j$, and $S_-(i,j)$- area of a set on which $\int K_\tau$ less that Y_2 in limits $C_i \cap M_j$.

So, for any collection of spread probabilities

$$f = \{ f(i,j),\ i=1,\dots,8;\ j=1,\dots,m \} \qquad (6)$$

one must determine a collection of deviations

$$S = \{ S(i,j),\ i=1,\dots,8;\ j=1,\dots,m \} \qquad (7)$$

characterizing a deviation Y_2 from $\int K_\tau$, where $\int K_\tau$ - average measure fire condition in moment τ, and probabilities of fire spread are equal (6). Coorelations for calculations of probability collection on $(n+1)$ step is an iterative formula.

This coorelation allows carry out search of f for which $S \neq 0$ with help of linear interpolation of deviation dependence S from probabilities f.

Components, which reaccount $f^{(n+1)}$ is depended on are: $S^{(max)}, f^{(max)}, S^{(min)}, f^{(min)}, f^{(n)}$

Initial probability values $f^{(max)}, f^{(min)}, f^{(n)}$ are set. Corresponding to collections of probabilities given above deviation $S^{(max)}, S^{(min)}, S^{(n)}$ are estimiated by Monte-Carlo method by means of RSP simulation on these probabilities. Then the linear iteration is repeated for $f^{(n+1)}$ to be found.

In iterative algorythm of reaccount sliding boards of the interval $f^{(min)}$, $S^{(min)}$ and $f^{(max)}, S^{(max)}$ are used the values of which are changed depending on coorelations between the value of deviation $S^{(n)}$ and parameter of δ that is set. The first stage of spread model identification is the assumption that forest territory is a whole homogeneous one. Probabilities are used as initial probabilities for the next step - identification of model of nonhomogeneous spread. Further definition of spread probabilities for these directions, which zero probabilities have not been done are held depending on its values in homogeneous case.

Algorythm of identification is finished as soon as for every homogenety

M_j the maximal deviation do not become less than number ε that have been got beforehand:

$$\max_{i=1,\dots,8} S^{(n)}(i,j) \leq \varepsilon \qquad (8)$$

Average measure prognosis of random spread

The method of average measure prognosis of fire spread is a method of average measure set simulation [2] on the basis of an equation (2). This necessary to set an initial fire condition K_0, a map of homogeneties $M=\bigcup_{j=1}^{m} M_j$ and collections of probabilities $f(x), g(x), \forall x \in M$. Method of average measure prognosis allows to get an average measure set $\int K_T$ in any given moment T, which is a geometrical RSP condition in moment T. If fire condition in moment T is known, one may define the exactness of prognosis, using the value of deviations of prognosed and real sets.

Application of methods and numerical computes calculations

Algorithm of set identification and average measure prognisis, which have been considered above are realized in programme software system for computers and micro-computers. Software is intended for average measure prognosis of great forest fire spread. Micro-computer exploitation allows to carry out operative fire forest prognosis. For this purpose forest fire contours ficsation is made by means of airophoto in consequent moments of time. This fhotos are processed in special way for computer input. All processes of forest fire spread may be simulated in computer. It allows to organize dialogue system "operator-computer".
On the basis of this software there have been carried out calculations for test experimental forest fires, they give satisfactory results. Practical application of given methods are possible not only in pyrology, but in any field where fenomenon dynamics depends on random geometry evolution of a dimensional spread which is under research. They are: processes of cell formation in biology, spread of epidemies in medicine, processes of crystalization ets.

References

1 Vorobiev, O.Yu.: Average measure simulation. Nauka, Moscow 1984.

2 Ivanilova, T.N.: Programme software of random spread prognosis. Preprint 22, Computer center of SB AS USSR, Krasnoyarsk 1982.

PRODUCTION OF PHOTOOXIDANTS IN THE ATMOSPHERE BY THE ULTRAVIOLET RADIATION

Heinz Schilling[1)]

Epidemic tree dying appears in heights between 500 and 1200 m, especially at the southern slopes of mountains. In the industrial concurbations no epidemic tree dying takes place.
After Mohr's hypothesis [1] for increasing height the increasing intensity of ultraviolet radiation and therefore the increasing production of photooxidants, as e.g. atomic singulet oxygen and peroxiacetyl nitrates (PAN) gives the reason that trees die. These vegetable poisons are produced in provable quantity, if the relative portion of affine nitrogen gases increases to values of 1 ppm and a strong insolation of ultraviolet light takes place. The production of PAN supplementary requires the existence of hydrocarbon.

Ultraviolet radiation of wavelength 313 nm decomposes ozone in molecular oxygen and highly stimulated singulet oxygen $O\,(^1D)$ [2] :

$$O_3 \xrightarrow{313\ nm} O_2 + O(^1D) \;. \qquad (1)$$

It reacts with water to hydroxil-radicals:

$$O\,(^1D) + H_2O \longrightarrow 2OH \;. \qquad (2)$$

The production of PAN takes place under ultraviolet light in a chain of four parallel going reactions [3] : influence of hydroxil-radicals on hydrocarbon (e.g. propane) and on the product, oxidation and reaction with nitrogen dioxide

$$C_3H_6 \underset{}{\overset{uv1,\ OH}{\rightleftarrows}} CH_3\ CHO + HCHO \;, \qquad (3)$$

$$CH_3CHO \overset{uv2,OH}{\rightleftarrows} CH_3CO + H_2O \;, \qquad (4)$$

$$CH_3CO + O_2 \overset{uv3}{\rightleftarrows} CH_3COO_2, \qquad (5)$$

$$CH_3COO_2 + NO_2 \overset{uv4}{\rightleftarrows} CH_3COO_2NO_2 \;. \qquad (6)$$

Decisive importance is due to the intensity of the ultraviolet radiation. For weak concentrations of the reacting substances the reaction velocity is proportional to the intensity of the ultraviolet light in the spheres of eigen-wavelength for each of the four reactions. If these spheres cover each other the output of PAN is proportional to higher power of the radiation intensity, maximum up to fourth power.
Weak changes of the radiation field thus lead to strong displacement of the chemical equilibrium and the reaction velocity.

---

1) Academyc of Sciences of the GDR, Central Institute of Kybernetics and Information Processes, 1086 Berlin, Kurstr. 33

The Field of the Ultraviolet Radiation

Radiation flux density in any point ot the space is composed of primary radiation ϕ_0, reflection from the ground, Rayleigh - and Mie-scattering [4]. Therefore Radiation flux density is represented by

$$\phi = \phi_0 R e^{-\tau \cos \chi} \qquad (7)$$

χ characterizes the zenith-solar-angle. Amplitude R considers primary flux and reflection from the ground. The effective optical depth τ is the sum of the optical depths of Mie-scattering, Rayleigh-scattering and absorption.

Mie-scattering is caused by the diffraction of large particles as dust and water drops. Optical depth of Mie-scattering increases proportional to the concentration of dust. Thus with increasing concentration of dust the intensity of the ultraviolet field decreases after an exponential function. Therefore by Mie-scattering the pollution in the conurbations suppresses the production of the photooxidents. That leads to a preserving of plants which by smaller concentrations of pollution in greater height die off. It is a problem, whether it will be economic to shelter strongly irradiated planes of first-rated plants during peak hours by the immission of dust , e.g. by agricultural airplanes.

Rayleigh-scattering is caused by the stimulated radiation of the neighbouring air molecules. Near the visible sphere Rayleigh-scattering and absorption V depend only weakly on the height. There is no argument for the growth of the production of photooxidants with increasing height. But below the wavelength 340 nm the intensity of the radiation field grows distinctly due to the increasing height. With decreasing height of sun the growth of the radiation field exposes in some hundred metres height, e.g. for 300 nm wavelength under a zenith-solar-angle of sixty degrees to the ninefold of the ground in the height of 2 km, to the fivefold of 1 km, to the threefold of 500 m (s. Fig.1).

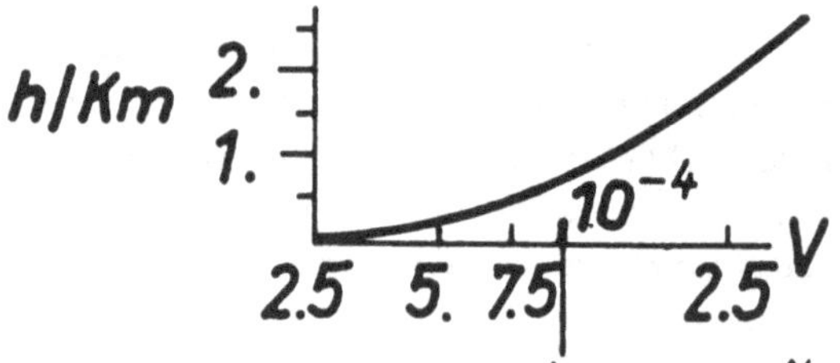

Fig.1

Ultraviolet radiation for λ =300 nm, χ = 60° in dependence of height

Increasing height for eigen-wavelength below 340 nm thus leads to increasing energy sources. The output of secondary pollutants therefore does not change proportional to the raw materials, but due to increasing height, if their concentrations exceed decided critical values. Corresponding to the quantity of parallel reactions for heterodyned spheres of

eigen-wavelength the accretion factor rises to higher power;that means e.g. in the height of 1 km under a zenith-solar-angle of sixty degrees efficiency
of ultraviolet radiation 300 nm for the power two rises to the 25-fold respective to the ground.

A Fuzzy Model for the Production of Photooxidants

The measurement of the primary pollutants as well as the radiation field fluctuates within large spheres. On the other hand the efficiency of a pollutant is determined only for a few markings of density, e.g. stimulus threshold, compatibility, weak damage, strong damage. Therefore only a fuzzy model for the production of secondary pollutants at any place is necessary.
Each of the primary pollutants is characterized by its mean density ϱ at the observation place, its spheres of eigen-wavelength and its effective cross section for the stimulation by the radiation field. The output of the stimulated material is determined by

$$\dot{\varrho}^{*} = \int \varrho \sigma(\lambda) \phi(\lambda)\, d\lambda, \qquad (8)$$

where $\phi(\lambda)$ is the flux density of the radiation field.

For singulet oxygen after (1) the reaction velocity is proportional to the intensity of the ultraviolet field for 313 nm and to the concentration of ozone. This concentration increases due to the concentration of nitrogen dioxide.
For PAN after (3) and (4) the reaction velocity is proportional to the second power of the concentration of hydroxil-radicals, which increases due to the concentration of affine nitrogen oxides,after (6) to the concentration of nitrogen dioxide, after (3) to the concentration of stimulated hydrocarbon.
Hydrocarbon consist of a multitude of components with diverse variable properties which are not yet completely investigated. Therefore in dependence on quantity and structure of raw materials a fuzzy function $n(\lambda)$ is defined as concentration of pollutants above the eigen-wavelength.
With this function the integral

$$\dot{\varrho}^{*} = \int n(\lambda) \sigma(\lambda) \phi(\lambda)\, d\lambda \qquad (9)$$

approximately determines the output of stimulated hydrocarbon at the observation place (see Fig.2)

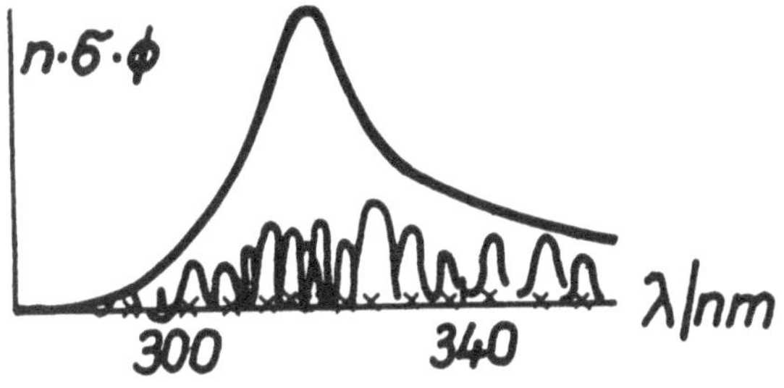

Fig.2 Fuzzy model for the production of stimulated hydrocarbon

Below 300 nm wavelength the intensity of the radiation field increases abruptly [4]. Thus the production of photooxidants can be globally suppressed. For this primary pollutants, which can be raw materials, before they are emitted in the atmosphere, are to be reduced in such a way that they dissociate at radiation below 300 nm or at dissociation energy greater than 400 kJ per mol. That means the peak in Fig. 2 is to shift to smaller wavelength below 300 nm. Decisive importance moreover comes up to the reduction of affine nitrogene oxides.

Literatur

[1] Mohr,H.: Biologie in unserer Zeit 13 (1983),3, 74
[2] Kümmel,R.: Zeitschrift für Chemie, 23 (1983) 8, 287
[3] Penkett,S.A.: Nature 24 (1983),
[4] Meier,R.R. and D.E. Anderson: Planet opace Sci. 30 (1983),9, 923

ON SUPERPOSED SYSTEMS (I)
---AN ITERATIVE PROCEDURE FOR ECOSYSTEM MODELLING

He Shan-Yu*

Abstract---In this paper, a straight approach, the Superposition Procedure, is suggested for modelling multicomponent ecosystems. The advantages of this procedure are: (1) the information provided by subsystem models can be utilized to a greatest extent, (2) the models thus obtained are suitable for numerical prediction, and (3) the superposition procedure is also applicable in the identification problem of other complex systems.

Keywords---ecosystem modelling, identification, complex systems

I. Motivation

Because of the existence of intricate mutual interactions, it is usually impossible to model a multicomponent ecosystem by intuition. Furthermore, most simple ecosystem models are not suitable for practical numerical prediction. Even the famous Lotka-Volterra equation [1,3] can be used only as a start point of rough qualitative analysis.

To obtain accurate quantitative model of complex ecosystems, the author introduced the concept of Superposed System and suggested a Superposition Procedure. The fundamentals of this are: (1) A whole ecosystem is thought to be formed by superposing its subsystems one after another, (2) we have already satisfactory separate models for all subsystems when they were in isolated state.

II. Outline of Superposition Procedure

Assume that the ecosystem S is composed of N subsystems(populations), we denote this fact by $S = S_1 \vee S_2 \vee \cdots \vee S_N$ For each isolated subsystem S_i, $i = 1, 2, \cdots, N$ we have already satisfactory model

$$M_i : \qquad \dot{X}_i = f_i\,(X_i, u_i) \qquad (1)$$

where $X_i = \left(x_{i1}(t), x_{i2}(t), \cdots, x_{in_i}(t)\right)^T$ is the state vector of S_i,

$u_i = \left(u_{i1}(t), u_{i2}(t), \cdots, u_{im_i}(t)\right)^T$ the control vector used to represent actions from outside, in evolution case $u_i = 0$, and f_i a vector function with the same dimension as X_i, i.e., n_i.

* Institute of Automation, Academia Sinica, Beijing, People's Republic of China

We start the Superposition Procedure with two subsystems, say, S_1 and S_2.

Usually we may take the union vector $(X_1^T \vdots X_2^T)^T$ to describe the superposed system $S_{12} = S_1 \vee S_2$. No loss of generality, we can write the model M_{12} of S_{12} as following

$$M_{12}: \quad \begin{aligned} \dot{X}_1 &= f_1(X_1, u_1) + h_1(X_1, X_2, u_1, u_2) \triangleq \varphi_1(X_1, X_2, u_1, u_2) \\ \dot{X}_2 &= f_2(X_2, u_2) + h_2(X_1, X_2, u_1, u_2) \triangleq \varphi_2(X_1, X_2, u_1, u_2) \end{aligned} \tag{2}$$

We may naturally assume that the first state variable x_{i1} is the "size" (population) of S_i, so that if at some instant t_0, $x_{i1}(t_0) = 0$ the subsystem S_i will disappear for all $t \geq t_0$, (extinction!) and the corresponding φ_i in (2) should be reduced to f_i i.e., $h_i = 0$ for $t \geq t_0$. This is the Submodel-Fitting Requirement for φ_i.

When possible, we may, under certain initial conditions, apply a test control $(u_1^T \vdots u_2^T)^T$ (e.g. impulse sequence, pseudo-random code, etc.) to S_{12}. With some identification technique (e.g., Volterra series approach) [2], we can obtain some experimental data which may be fitted by approximate closed form $X_1 = \tilde{X}_1(t, u_1, u_2)$, $X_2 = \tilde{X}_2(t, u_1, u_2)$.

Based on $\tilde{X}_1$ and $\tilde{X}_2$, we can obtain a set of differential equations by elimination

$$\begin{aligned} \dot{X}_1 &= \tilde{\varphi}_1(X_1, X_2, u_1, u_2) \\ \dot{X}_2 &= \tilde{\varphi}_2(X_1, X_2, u_1, u_2) \end{aligned} \tag{3}$$

$\tilde{\varphi}_1$ and $\tilde{\varphi}_2$ are, generally not submodel fitting, although they can fit the observed data quite well, i.e., data fitting.

Let Ω be a set of vectors $(\bar{\varphi}_1^T \vdots \bar{\varphi}_2^T)^T$, where $\bar{\varphi}_1$ and $\bar{\varphi}_2$ are submodel fitting and have rational components. If we choose a suitable norm $\|\cdot\|$ (e.g., Euclidean) in the set Λ of all vectors $(\varphi_1^T \vdots \varphi_2^T)^T$ where φ_1 and φ_2 are piecewise continuous functions.

By computation, we can finally find out a vector $(\varphi_1^{*T} \vdots \varphi_2^{*T})^T$ such that

$$\|(\varphi_1^{*T} \vdots \varphi_2^{*T})^T - (\tilde{\varphi}_1^T \vdots \tilde{\varphi}_2^T)^T\| = \min_{(\varphi_1^T \vdots \varphi_2^T)^T \in \Omega \subset \Lambda} \|(\varphi_1^T \vdots \varphi_2^T)^T - (\tilde{\varphi}_1^T \vdots \tilde{\varphi}_2^T)^T\| \tag{4}$$

Then

$$\begin{aligned} \dot{X}_1 &= \varphi_1^*(X_1, X_2, u_1, u_2), \\ \dot{X}_2 &= \varphi_2^*(X_1, X_2, u_1, u_2). \end{aligned} \tag{5}$$

will be a submodel fitting and best data fitting model of $S_{12} = S_1 \vee S_2$.

Continue this procedure by iteration, we can finally arrive at a good model $M_{12\cdots N}$ of the whole S suitable for numerical prediction.

III. Remarks

1° In this paper only the procedure itself is given. Rigorous mathematical discussions are left for later papers.
2° Computers should be used for superposition procedure.
3° The theory of Superposed System could be developed in different (including algebraic and geometrical) ways.

Literature

[1] Christiansen, B., Fenchel, M.: Theories of Populations in Biological Communities, Sec. 2.2, Springer-Verlag Berlin, 1977.
[2] Eykhoff, P.: System Identification, Chap. 4, John-Wiley, 1977.
[3] Volterra, V.: Variations and Fluctuations of the Number of Individuals in Animal Species Living Together. J. Conseil, 3 (1928), 1-51.

SIMULATION MODELS AS TOOLS IN ECOTECHNOLOGY

Milan Straškraba[+]

Summary. Simulation experiments with possible controls of eutrophication of water bodies suggest new ecotechnological measures. Experiences with the dynamic optimization model suggest, that by combined effects of several measures it would be feasible to control eutrophication in different deep water-bodies in an economic way.

Key words: simulation, ecological systems, optimization models, management.

By ecotechnology I understand the use of technological means for ecosystem management, based on deep ecological understanding, in respect to minimize the costs of measures and their harm to the environment. From ecosystem theory [4] follows, that natural ecosystems are open disipative systems far from thermodynamic equilibrium, differentiated in function and structure. Feedback mechanisms are typical features leading to domination of indirect effects. Ecosystems are hierarchically organized and so is their internal control. They are selfadaptive and selforganizing nonlinear systems, coherent with their environment.

Therefore, not only understanding but also management of ecosystems is not an easy task. Simulation models appear to be a corresponding tool for studying, understanding and managing such systems. Simulation models provide a simplified description of the major features of the ecosystem, understanding of their dynamic reactions and sensitivity to perturbations. New methods of control can be developed when the reactions are understood, and combined effects of different measures evaluated. Because indirect effects are dominating in ecosystems, most methods of brute force technology like chemical treatments have many secondary effects, which have to be carefully evaluated.

Inclusion of costs of different measures and performance criteria based on the corresponding use of water, like the supply of healthy and tasty drinking water into verified simulation models leads to management (optimization) models.

The eutrophication problem

This is demonstrated here for one management problem of aquatic ecosystems, management of eutrophication. Eutrophication is a burden to many lakes and reservoirs. Organic matter formation by aquatic primary producers (particularly phytoplankton - microscopic algae living in free water) rises as a consequence of increases nutrient inputs in such a way, that difficulties with uggly appearance, for recreation and water use appear. Particular harm is caused to drinking water supply, with consequences not only to treatment costs and water tastes and odors, but often inability to keep hygienic and water quality criteria. As the

[+]Hydrobiological Laboratory, Institute of Landscape Ecology, Academy of Sciences, Na sádkách 702, 370 05 České Budějovice, Czechoslovakia.

critical nutrient leading to increased primary production in most inland waters phosphorus was recognized. A different situation is known from the sea and coastal waters with nitrogen limitation dominating.

Simulation models of eutrophication

Models describing in a simplified way aquatic ecosystems with respect to their eutrophication have been recently reviewed in [8] . Only few of them have been extended into management (optimization) models for eutrophication control.

One of the models somewhat adequate for such purposes is based on simulation models of the series AQUAMOD, with the hydrodynamic simplicity corresponding to AQUAMOD 1 (one completely mixed layer) but biologically more elaborated as in AQUAMOD 2 and 3. The model is far from adequately including all the theoretical features of ecosystems which we have recognized above. Particularly, it lacks the selfadaptive and self-organizing features of the ecosystem. Although attempts to construct models including the internal control mechanisms are made [1,5] , their methodology is not yet developed adequately and their numerical sollution is rather difficult. Nevertheless, in other respects the model seems to represent gross features of reality. The management model structure was described in a few previous papers [7,3,6] .

We now examine some of the reactions of the modelled ecosystem to parameters possibly subject to human control. One problem of this is the nonlinearity of the ecosystem reactions. Examining nonlinear systems sensitivity to a given control parameter u_i in one (nominal) point of the parameter space p^* (where p are the uncontrolled system parameters) does not provide a complete answer about its behaviour during the same changes of u_i for other values of control parameters u_j nor for other values of p^* . Here p^* is kept constant to represent an average Czechoslovak reservoir and values of both u_i and u_j are always indicated.

Classical eutrophication control is based on reductions of phosphorus inputs. In the model, this is represented by the parameter u_i=PRFOS, concentration of phosphorus in the inflows to the reservoir. The reactions on PRFOS were simulated for an intensively throughflowing upper layer (Q/V=0.1 which corresponds to retention time 10 days) with fairly clean water and not very deep mixing (EPS = 0.1 and ZMIX = 4m) and average zooplankton mortality of 7.5% per day (MORTZ = 0.075). The reaction is very strong, and is evident for all three state variables: phytoplankton algae, zooplankton and phosphorus. For the extremely high phosphorus concentration of 500 mg/m^3P a very high spring peak of phytoplankton is formed, which is heavily reduced afterwards by high growth of zooplankton. Phosphorus concentration in the waterbody remains high and almost unaffected by algal uptake (close to 100 mg/m^3) and does not limit phytoplankton growth. For the second PRFOS value studied, 80 mg/m^3 the phytoplankton does not reach such a high peak, which is a consequence primarily of the phosphorus limitation. Phosphorus is rapidly exhausted (taken up by phytoplankton) up to very low residual values. Zooplankton still grows well and acts as a reducing force. For values of PRFOS decreasing further up algae decrease and so does zooplankton, which becomes negligible at PRFOS = 30 and 10. Phosphorus concentrations decrease and the period of exhaustion is prolonge, so that phytoplankton is continuously limited by phosphorus and its annual pattern is regular.

Other parameters which can be changed artificially represent the extinction depth EPS.ZMIX (u_2 = EPS, u_3 = ZMIX). Investigation of the reaction of system suggests that also during fairly high inflow phosphorus concentrations (PRFOS = 100) and throughflow and zooplankton mortality conditions as above the phytoplankton can be reduced considerably by increasing the value of $u_2 \cdot u_3$. In both instances the effect is caused by reducing light available to the phytoplankton within the upper mixed zone. In the model the value $u_2 \cdot u_3$ determines the limits of integration of photosynthesis over depth - in a deep layer if relatively well iluminated (light drops with depth approximately exponentially) average phytoplankton production is the same as in a shallow mixed layer (low ZMIX), but with water colored or turbid (high EPS). The multiplicative nature of the effect of both controls shows that the effect is identical if the one or the other are changed proportionally.

Based on these simulation experiences we suggest two effective ecotechnological controls of eutrophication. One is the increase of EPS, which corresponds to the extinction coefficient of water for light. This can be done by increasing the color or turbidity. Experiments with dyes in Scandinavian lakes [2] and Czechoslovak ponds suggest that the use of dyes is feasible, although large scale experiments have not yet been performed and problems with coloured water and harmless dyes arise. For the other control, increase of mixing depth, a long tradition exists in shallower water bodies: London supply reservoirs which are mixed by hydraulic guns until the bottom. For deep lakes we suggest a modification of the air mixing used for destratification and/or hypolimnic aeration. This modification consists of releasing compressed air from a variable depth, determined by the model to minimize algal growth.

The other controls, u_4 and u_5 are given by regulating the relative throughflow of the productive layer, Q/V, and regulating the zooplankton populations and their effect on reducing phytoplankton. The first can be realized in deep reservoirs by the so called multiple outlet structures, which means that the same inflow, Q, is mixed to a varying degree into the upper productive layer. Positive effects on phytoplankton growth (increased phosphorus load) are coupled with negative ones - washing out of algae at high Q/V. The last is represented in our model by varying MORTZ which is not fully equivalent to complex changes of zooplankton due to biomanipulation. In any case, the realization is by artificially changing fish populations of the water bodies, which causes the desired changes of zooplankton. Simulation experiments similar to the above suggest, that both controls are feasible.

New experiences were obtained by using the optimization model as modified for the Želivka Reservoir in Czechoslovakia (drinking water supply of Praha). Several situations were simulated, corresponding to hydrological conditions in different years, and to present and predicted future phosphorus inflow concentrations. Optimal management solutions were obtained under two assumptions, either with building of the purification plant (phosphorus reduction) or without. In Želivka Reservoir the biomanipulation measures (regulated fish populations) are already in use. The cheapest measure is always regulation of Q/V, because the opening of other outflow gates is practically at no cost. Therefore, according to optimization results this control is always set first, and only then u_2 and u_3 are used. Because a dynamic optimization procedure using extended Lagrangian computation is applied, the computed controls start to operate before the values of CHA (measure of phytoplankton algae) rise above the critical level. By changing the controls in 20 day intervals according to the computed optimal trajectory, the critical value is not exceeded (by more than 0.5%) at minimum cost.

Literature

[1] Fedra, K.: Modeling Biological Processes in the Aquatic Environment (With Special Reference to Adaption). IIASA WP-79-20, Laxenburg 1979.

[2] Jørgensen, S.E.: Lake Management, Pergamon Press, Oxford 1979.

[3] Kalčeva, E., J.V. Outrata, Z. Schindler, and M. Straškraba: An optimization model for the economic control of reservoir eutrophication. Ecol. Modelling 17 (1982), 121-128.

[4] Patten, B., S.E. Jørgensen, M. Straškraba: Eco-Systems Emerging. Systems Principles in Ecology. 1984, Prospectus for a book.

[5] Radtke, E., and M. Straškraba: Selfoptimization in a phytoplankton model. Ecol. Modelling 9,4 (1980), 247-268.

[6] Schindler, Z., and M. Straškraba: Optimal control of reservoir eutrophication. Vodohosp. časopis SAV 30,5 (1982) 536-548 (In Czech with English Summary)

[7] Straškraba, M.: The application of predictive mathematical models of reservoir ecology and water quality. Can. Water Resourc. J. 7 (1982) 283-318.

[8] Straškraba, M., and A. Gnauck: Aquatische Ökosysteme. Modellierung und Simulation. Fischer, Jena 1983 (In English Elsevier, Amsterdam 1985).

DEVELOPMENT OF METHODS FOR MODELLING COMPLEX ECOLOGICAL SYSTEMS

V.I. Belyaev [1)]

In modelling complex ecological systems (i) a new theory of diffusion of non-conservative particles, (ii) procedures for aggregation and averaging of ecosystem components with subsequent hierarchical decomposition and for (iii) joint analysis of dynamic and logical informational models were developed and shortly described. Numerical procedures support the elaboration of prognoses about the behaviour of men-controlled ecological systems. Some results in modelling the North-West Black Sea shelf ecosystem are demonstrated.

Keywords: Dynamic equations, aggregation, decomposition, prognosis of ecosystem dynamics

---

[1)] Institute of Marine Hydrophysics of the Academy of Sciences of the Ukrainian SSR, 33500 Sevastopol, Lenin Street 28

WHAT IS STABILITY? A MATHEMATICIAN'S AND ECOLOGIST'S POINT OF VIEW

Pavel Kindlmann and Jan Lepš[1)]

Summary. The term "stability" is frequently used both in ecology and mathematics. Whereas in mathematics exist a lot of rigid definitions, the understanding of this term in ecology is often vague. The stability concept commonly used in mathematical models of ecological systems is the Liapunov stability. However, it does not completely fit the needs of ecologists. In this paper, different stability concepts are discussed. A new mathematical definition of stability, which fits better the demands of ecologists is proposed.

Key words: stability, persistence, resistance, resilience, perturbations.

1. Introduction

The term stability is frequently used in the contemporary ecology. Despite of a lot of definitions in literature, its understanding in ecology is more or less intuitive and often vague. Moreover, we may distinguish many aspects of stability of ecological systems. In field studies of ecological systems, the "operational definition" of stability is always needed (cf. Pimm 1984). In mathematical analysis of ecosystem models their stability properties are usually studied. The rigid mathematical concepts of stability (usually the Liapunov one) are applied (e.g., May 1973). The stability of the model is then considered to characterize the stability of its real counterpart. However, the suitability of the mathematical stability concept for solution of given ecological problem is often neglected. In this paper we would like to show, what does the term "stability" mean for an ecologist in various situations and compare it with adequate stability definitions in mathematics.

2. Stability in ecology

The first decision to be made is the choice of variables under consideration. Each ecological system may be characterized by many various variables and it is impossible and impracticable to measure all of them. In mathematical models the state variables (usually population sizes, characteristics of abiotic environment) describe the system completely. In field investigations, only restricted numbers of variables may be used. The omitted ones are considered to be either constant or to have no significance from the point of view of the study. The population size may be characterized not only by the number of individuals, but also by its biomass, cover or some characteristics of its activity (e.g. respiration, photosynthesis). Similarly, various characteristics

1) Department of Biomathematics, Biological Research Centre, Na sádkách 702, 370 05 České Budějovice, Czechoslovakia

of the whole community or of any part of the community (e.g. total plant biomass, total respiration) or the whole ecosystem are used (e.g. nutrient retention). The choice of variables has often crucial effect on evaluation of the stability. For example, the change of species composition need not be necessarily accompanied by the change of total biomass and vice versa.

In ecology, two approaches to the study of stability can be distinguished. In the first approach, the lack of changes in the course of a sufficiently long period is considered to be a sign of stability. No action of particular disturbing force is assumed. In the second approach, stability is considered to be the ability to remain in "equilibrium" when faced to some disturbing force and to return toward this equilibrium after having been displaced.

The first approach is used by Woods and Whittaker (1981). They provide the definition: "A community in which several species coexist, with relative importances (and population structures) remaining fairly constant for some significant amount of time, is stable." They stress the relativity of the definition with respect to perceptual scales (particularly temporal and spatial ones). Connel and Sousa (1983) suggest that one of the community characteristics may be the area, on which the community can be considered to be stable. Obviously, the developing successional stages are not stable according to this definition.

The second approach (see Harrison 1979, Rejmánek 1979) is based on the study of the system behaviour when faced to some disturbing force. It supposes that community is in equilibrium under normal conditions. Afterwards, particular disturbing force is applied for a limited period of time (stress period). The behaviour of the system during and after this period is investigated. Following characteristics are then derived: resistance - the ability to remain in equilibrium when faced to disturbing force; resilience - the ability to return toward equilibrium (or the rate of return); persistence - the ability to remain within some arbitrarily predetermined range.

The definition of normal conditions and the type of disturbing force play a crucial role in such studies. Each community has developed under certain conditions and is "adapted" to them. The conditions we consider normal for one community may be a disturbing force for another one and vice versa.

We suppose that under normal conditions the system remains in equilibrium. However, the environmental conditions are never constant in nature and the endogeneous dynamics of each community causes fluctuations in values of particular variables. The quantitative determination of equilibrium usually causes difficulties in the field. However, all stated stability concepts may be considered too simplified and schematic from the field ecology point of view.

3. Stability in mathematics

In mathematics, there exist more than 30 various definitions of stability. As already stated, Liapunov stability is commonly used in mathematical analysis of models of ecological systems. However, there are several points of this stability concept, which may be criticized from the ecologist's point of view:

1) This type of stability assumes that the system is in some steady state - in equilibrium - under normal conditions. But many ecological systems and even many models of ecological systems exert periodical motion caused either by their own dynamics (e.g. the well known Lotka-Volterra systems), or by periodicity of their coefficients (e.g. due to seasonality). Other ecosystems are able to exist only far from equilibrium. As a nice example may serve snow avalanche paths (Rejmánek 1984), where the great species diversity is caused by avalanches which disturb the system from time to time. If these avalanches were not there, many species would die out.

2) The type of Liapunov stability used is a local property of the system. It assures only the existence of some neighbourhood of the equilibrium, in which the system is stable, without saying anything about the size of this neighbourhood. In ecology, however, we are interested in the consequences of fairly great perturbations.

3) In biological systems we meet perturbations which act permanently (climatic changes, biological variability etc.). Neither this fact is incorporated in the Liapunov stability concept.

How to overcome these problems? We can assume that the system remains in some invariant set G under normal conditions rather than in equilibrium. The behaviour within this set is not of any interest to us - it may be periodical, almost periodical or even chaotic motion. The stability of this invariant set is then studied. Local stability may be replaced by global one. We define the set H, within which the system is stable, i.e., the system, after having been displaced by some external force to some point $x \in H-G$, never leaves the set H and finally reaches the set G. The existence of permanently acting perturbations leads either to stochastic models, or to some perturbed differential or difference equations. None of various stability concepts of perturbed equations in mathematics (see, e.g. Hahn, 1967) is satisfactory for the use in ecological applications. All of them assume that the "perturbation function" is small in some sense: that its norm approaches zero as the state of the system approaches equilibrium, or that there exists some constant (maybe very small), such that the norm of the perturbations is smaller than this constant. But in ecology the norm of the "normal" perturbations may be fairly high. Therefore, a new stability concept is to be defined in the following way: Let us have a constant K, two sets H, G, $G \subset H \subset R_m$, and a system

$$\dot{x} = f(t,x) + g(t,x), \qquad (1)$$

$x \in R_m$, $f: R_{m+1} \to R_m$, $g: R_{m+1} \to R_m$, where the function f is assumed to be known (it may be some of the classical models) and the function g performs the "normal" - permanently acting perturbations (variations in climate, biological variability etc.). We do not know exactly the form of the function g, the only information available about it is that $\|g(t,x)\| < K$. Then we define that G is K-attractive in H for the system (1), if and only if for each g satisfying $\|g(t,x)\| < K$ each motion which starts in G remains there and each motion which starts in H finally enters G. The difference between this stability concept and the common ones is that the constant K is a priori defined (and therefore it may be fairly great) and does not depend on x. This stability concept distinguished two kinds of perturbations: "normal", which are bounded and permanently acting (performed by the function g) and "extreme" - catastrophes like fire, some human activities etc., which may displace the system outside the regions of its "normal" motion - outside the set G.

There are several questions which may be answered by means of this stability concept only:

1) For given K - how large is H? This question can be interpreted in ecological terms: What can we afford without any drastic consequence for the system?

2) Which is the maximal value of K, for which all components of G are positive? In biological terms: How constant must be the conditions for the whole system to survive - i.e., all species to survive?

3) Which is the maximal value of K, for which $G \subset H$? Of course, if K exceeds this value, the predictability of the model is not assured.

This stability concept is much more realistic from an ecologist's point of view than the Liapunov stability. Nevertheless, it does not overcome all the problems mentioned in the first part of this paper.

Literature

[1] Connel, J.H., and W.P. Sousa: On the evidence needed to judge ecological stability or persistence. Amer. Nat. 121 (1982) 6, 789-824.

[2] Hahn, W.: Stability of Motion. Springer, Berlin 1967.

[3] Harrison, G.W.: Stability under environmental stress: resistance, resilience, persistence, and variability. Amer. Nat. 113 (1976) 5, 659-669.

[4] May, R.M.: Stability and complexity in model ecosystems. Princeton University Press, Princeton 1973.

[5] Pimm, S.L.: The complexity and stability of ecosystems. Nature 307 (1984), 321-326.

[6] Rejmánek, M.: Stability and complexity in biotic communities: theoretical and empirical approach. In: M. Růžička ed. Fifth Intern. Symp. Probl. Ecol. Land. Res., Bratislava, pp.65-72, 1979.

[7] Rejmánek, M.: Perturbation-dependent coexistence and species diversity in ecosystems. In: P. Schuster ed. : Stochastic phenomena and chaotic behaviour in complex systems. Springer, Berlin, pp. 220-230, 1984.

[8] Woods, K.D., and R.H. Whittaker: Canopy-understory interaction and the internal dynamics of mature hardwood and hemlock-hardwood forests. In.: D.C. West, H.H. Shugart and D.B. Botkin eds. : Forest succession. Springer, Berlin, pp. 305-323, 1981.

BIMODALITY IN SINGLE EVEN-AGED POPULATION: A SIMULATION STUDY

Jan Lepš[1)]

Summary. The simulation model of the self-thinning even-aged population was constructed and calibrated on the basis of field data. The aim of the simulation study was to find and biologically interpret the conditions leading to bimodality in distribution of plant heights. These are: strikingly asymmetric competition influencing the growth of individuals and high initial density. The results of simulations were compared with generalized 3/2 power law.

Key words: simulation, even-aged plant population, bimodality, self-thinning, 3/2 power law.

1. Introduction

The single even-aged populations are those of the simplest ecological systems. They seem to be very suitable for studying some processes and phenomena that may be obscure in more complicated systems. In pure even-aged populations of sedentary organisms (usually plants), self--thinning is the process very often studied. Our simulation study is based on the field data of Prach (1982) from self-thinning populations of ruderal annuals.

2. Some features of natural populations

The self-thinning populations usually follow the 3/2 power law (Yoda et al. 1963), i.e. the plant weight-density trajectory in the course of self-thinning may be described by the formula

$$w = k \cdot \rho^{-3/2} \qquad (1)$$

where w is the average weight of an individual and ρ is the density. Among the generalizations, that of Hozumi (1977) seems to be of great ecological significance.

The other interesting feature is the bimodality in distribution of plant heights and/or weights. It appears in some even-aged populations in the course of their self-thinning (e.g. Ford 1975, Prach 1982), it does not in others (e.g. Mohler, Marks and Sprugel 1978). The aim of the present study is to elucidate some possible causes of the appearance of bimodality.

3. The models

Munro (1974) distinguished three modelling philosophies of forest growth models. They may be directly applied on models of population growth. The resolution of models decreases from the first to the third. For purposes of the present study, a simple model with the second modelling philosophy was used and calibrated. Two other models, one with the first and the other with the third modelling philosophy were exami-

[1)] Department of Biomathematics, Biological Research Centre, Na sádkách 702, 370 05 České Budějovice, Czechoslovakia ned for their ability to simulate the appearing of bimodality. However, only the model with the second modelling philosophy is described here, the other two are briefly mentioned.

The model is based on ecologically feasible assumptions about the growth of a single individual within the canopy of its neighbours:

1) The length growth of each individual is described by logistic equation with an added term for competition of other individuals.

2) Other characteristics of the individual are considered an allometric function of its height. The plot occupied by a particular individual is considered to be proportional to the square of its height.

3) The strength of competition within the plot is a function of the ratio of the sum of plots occupied by all individuals to the area of the simulated plot.

4) The reduction of the growth of a particular individual depends on the strength of competition and on the status of the individual (i.e. on its height in comparison with the average height of the canopy). The competition is asymmetric - the taller individuals are influenced less than the smaller ones.

5) The probability of survival of a particular individual for one time unit depends on the competition strength and on the status of the individual.

The mathematical formulation of the model is given only very briefly. The model deals with a set of n individuals growing in the simulated plot. Their location within the plot is not considered. The length growth of the i-th individual ($i=1,\dots,n$) is described

$$L_{i,t+1} = L_{i,t} + r_i \cdot L_{i,t} \cdot \left(1 - \frac{L_{i,t}}{Lmax_i}\right) \cdot RED_{i,t} \qquad (2)$$

where $L_{i,t}$ is the length of i-th individual at time t, r_i is its growth rate, $Lmax_i$ is its maximal possible length. Both r_i and $Lmax_i$ were generated as random numbers with normal distributions N (μ_1, σ_1^2) and N (μ_2, σ_2^2) respectively to express the natural variability. In both cases: $\sigma \ll \mu$ were chosen (usually $\sigma = 0.05 \times \mu$). The difference equations were used. The suitable time-step for ruderal annuals is one day. The term $RED_{i,t}$ describes the influence of competition on the growth

$$RED_{i,t} = f_1 \left(\sum_{j=1}^{n} L_{j,t}^2, \frac{L_{i,t}}{Lmean_t} \right) \qquad (3)$$

where $Lmean_t$ is the average height of the canopy. The f_1 is a non-negative function, decreasing in the first variable and increasing in the second one. Different functional forms were examined. The probability of survival to the next time point, $P_{i,t}$, is expressed

$$P_{i,t} = f_2 \left(L_{j,t}^2, \frac{L_{i,t}}{Lmean_t} \right) \qquad (4)$$

The function f_2 decreases again in the first variable and increases in the second. $P_{i,t} \in \langle 0,1 \rangle$ However, different functions (or the same functions with different parameters) are used to enable characterizing different sensitivities of growth and survival towards competition stress.

The Monte Carlo method was used to decide whether the individual survives to the next time. Initial lengths were generated as random numbers with normal distribution, with the standard deviation corresponding to the original data of Prach (1982).

The model with the first modelling philosophy used for comparison is based on similar assumptions as the described one. Similar modelling techniques were employed. It differs in taking into account the horizontal structure. Each individual is located in a particular point within the plot. The competition strength (influencing both growth and survival) depends on joining neighbours only. We have used this model for simulating changes in spatial pattern during the course of self-thinning (Lepš and Kindlmann 1984). As a model with the third modelling philosophy, the adapted Leslie model of population growth was used. The population was divided into height classes. Transition and survival probabilites were variable and computed in each step in order to fit the above listed assumptions.

4. Results of simulations

The comparison of results of simulation with original data of Prach (1982) is given in Fig. 1. The sensitivity analysis and simulation experiments show that the appearance of bimodality depends on the parameter values and on initial conditions. The necessary conditions are the striking dependence of the RED_i value on the status of individual and sufficient initial density. The mortality of undercanopy individuals must not be very high. These conclusions hold roughly also for the two models

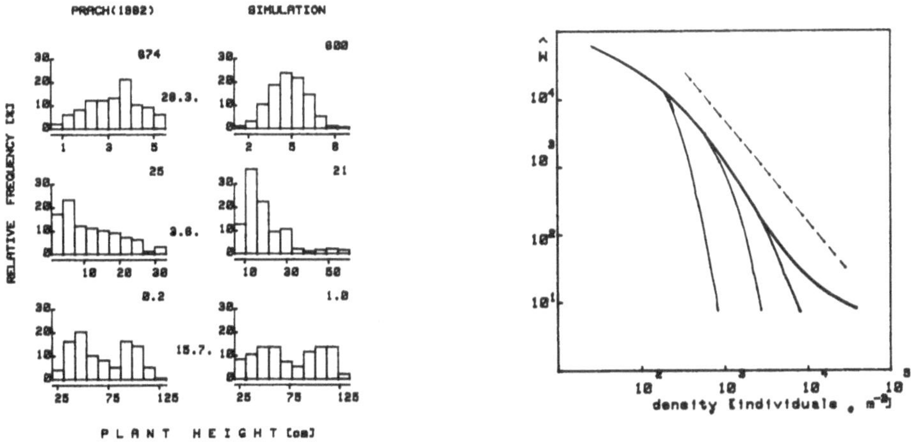

Fig.1 Results of simulations of the Atriplex nitens population model. Left: Comparison with original data of Prach (1982) - distribution of plant heights at dates 28.3., 3.6.,15.7. Densities are in the upper right corner of each histogram. Right: Simulated weight - density trajectories for different initial densities. The broken line corresponds to -3/2 slope.

used for comparison. The mean plant weight - density trajectory (the weight is considered to be proportional to the 3-rd power of the length) corresponds well to the Hozumi s (1977) generalization of the 3/2 power law (Fig.1). Note that originally the 3/2 power law is intended for overcrowded populations only.

5. Discussion

There are several models (both analytical and simulational) trying to explain the bimodality in even-aged populations (Diggle 1976, Gates 1978, Aikman and Watkinson 1980, Ford and Diggle 1981). All of them are based on the data of Ford (1975). Although the modelling techniques differ considerably, all the models including those presented here agree that the necessary condition for the appearance of bimodality is asymmetric competition. In sparse growths the bimodality does not appear. This conclusion is in agreement with field experience (Ford 1975, Mohler, Marks and Sprugel 1978). Extremely high initial densities observed by Prach (1982) resulted in distinct bimodality. However, at very high initial densities the growth may be slowed and the appearance of bimodality postponed. The intensity of bimodality cannot be used as measure of competition strength or asymmetry (cf. Rabinowicz 1979). However, there are some other reasons that may cause the bimodality in natural populations, which we consider to be even-aged. The most common one is seed germination in two or several waves (e.g. due to weather) which obviously results in bimodality even at early stages of development (cf. Prach 1982).

Literature

[1] Aikman, D.P., and A.R. Watkinson: A model for growth and self-thinning in even-aged monocultures of plants. Ann. Bot. 45 (1980), 419-427.

[2] Diggle, P.J.: A spatial stochastic model of inter-plant competition. J. Appl. Prob. 13 (1976), 662-671.

[3] Ford, E.D.: Competition and stand structure in some even-aged plant monocultures. J. Ecol. 63 (1975), 311-333.

[4] Ford, E.D., and P.J. Diggle: Competition for light in a plant monoculture modelled as a spatial stochastic process. Ann. Bot. 48 (1981), 481-500.

[5] Gates, D.J.: Bimodality in even-aged plant monocultures. J. Theor. Biol. 71 (1978), 525-540.

[6] Hozumi, K.: Ecological and mathematical considerations on self-thinning in even-aged pure stands. II. Growth analysis of self-thinning. Bot Mag. Tokyo 93 (1980), 149-166.

[7] Mohler, C.L., P.L. Marks, and D.G. Sprugel: Stand structure and allometry of trees during self-thinning of pure stands. J. Ecol. 66 (1978), 599-614.

[8] Munro,D.: Forest growth models - a prognosis. In: J. Fries ed. : Growth models for tree and stand simulation. Royal College of Forestry, Stockholm, pp. 7 - 21, 1974.

[9] Prach, K.: Self-thinning processes in selected ruderal species populations. In Czech. Preslia 54 (1982), 271-275.

[10] Rabinowicz, D.: Bimodal distributions of seedling weight in relation to density of Festuca paradoxa Desv. Nature 277 (1979), 297-298.

[11] Yoda, K., T. Kira, H. Ogawa, and K. Hozumi: Self-thinning in overcrowded pure stands under cultivated and natural conditions. J. Biol. Osaka City Univ. 14 (1963), 107-129.

FORECASTING THE FUTURE OF A LAKE USING STOCHASTIC AND DETERMINISTIC MODELS

Tibor Kutas[1/]

One of the most important tasks and at the same time a serious test for a model is to forecast the future of the system to be modelled. In lake ecosystem modelling a usual question for modellers is what the response of the lake on changing /mostly decreasing/ nutrient load is. Both statistical and dynamical models are used to answer the question. The first question of forecasting is what the water quality will be in the future, but the second question arising soon after the first is what the reliability of estimation is, to which a comdifence interval or estimation of standard deviation has to be given. In most cases this problem has been solved by randomizing the forcing functions /nutrient loads, meteorological functions/. This approach can answer some questions, but it is not the real solution of the problem as the system itself has a stochastic feature and this can only be modelled by using a stochastic model.

The approach described here is only one way in constructing stochastic models, but hopefully this can stimulate modellers to find other ways too.

The Balaton Eutrophication Model /BEM/ versions

In the case of Lake Balaton, where the process of eutrophication threatens the ability of the lake for recreation and measures have been done and will be done, the forecast has great practical importance.

First a deterministic model /a system of nonlinear differential equations/ was constructed. The model has 10 state variables. The first eight: winter-spring phytoplankton, summer phytoplankton, autumn phytoplankton, blue-green algae, bacterioplankton, dead organic matter, dissolved inorganic phosphorus, dissolved inorganic nitrogen describe the pelagic zone of the lake, but as the lake is shallow /the average depth is 3.2 m/ the sediment plays an important role in open water processes too, therefore the remaining two state variables: organic matter in the sediment and exchangeable phosphorus in the sediment are referred to the sediment. Since the lake is large it has been divided into four basins, the model with ten state variables refers only to one

1/ Computer and Automation Institute, Hungarian Academy of Sciences
H-1111 Budapest, XI. Kende u. 13-17. Hungary

basin, that is why it is multiplied four times for modelling the whole lake. Here only the model of one basin will be discussed. The matters circulate in two cycles: in the first one phytoplankton uptakes inorganic nutrients and produces organic matter, the mortalized phytoplancton is decomposed by bacterioplancton revealing inorganic nutrients; in the second cycle a part of all particulate matters settles down into the sediment, where it is decomposed. The inorganic nitrogen gets back directly into the water, the inorganic phosphorus first gets into a pool in the sediment and from this it gets back to the water.

As an example for the formulation of processes the primary production of summer phytoplankton is shown:

$$R_2(A_2, P, N, t) = PMAX \cdot TEMP(t) \cdot U(t) \cdot A_2(t)$$

where PMAX is the maximal production rate, TEMP is the temperature limitation depending upon water temperature which is one of the forcing functions of the model, U is the joint limitation of light, dissolved inorganic phosphorus and nitrogen, the light is also a forcing function. A full description of the deterministic model can be read in [2].

Having succeeded in reproducing measured values with the deterministic model a stochastic counterpart of it has been constructed to describe inherent fluctuations /not measurement errors/ without introducing further parameters. The stochastic model is continuous time, discrete state, Markovian pure jump process having the same state variables and subprocesses as the deterministic one has. The same subprocess, the primary production of summer phytoplancton will correspond to the following state transition:

$$a_2 \rightarrow a_2 + 1 \qquad p \rightarrow p - pr \qquad n \rightarrow n - nr \qquad (1)$$

where a_2, p and n are the state variables of the stochastic model - they are summer phytoplankton, dissolved inorganic phosphorus and nitrogen respectively -, pr and nr are the phosphorus and nitrogen equivalent of organic matter. The assumptions are as follows

a/ the probability of state transitions type /1/ are proportional to their deterministic rates, e.g.

$$P((1)) = R_2(a_2, p, n, t) \cdot h + o(h) \quad \text{if} \quad h \rightarrow 0$$

b/ the state transitions are independent of each other

c/ the probability of no change in /t,t+h/ is

$$1 - \sum_i R_i \cdot h + o(h) \quad \text{if} \quad h \rightarrow 0$$

It can be proved that in the linear case, when the system of differential equations is linear in state variables the expected value of the stochastic model equals to the deterministic one. Defining the conditional expected velocity

$$D_1(t, \underline{n}) = \lim_{h \to 0} \frac{1}{h} \cdot E(\underline{x}(t+h) - \underline{x}(t) \mid \underline{x}(t) = \underline{n})$$

and using the notation for the deterministic model

$$\dot{\underline{X}}(t) = \underline{f}(\underline{X}, t)$$

the following equation holds

$$D_1(t, \underline{X}) = \underline{f}(t, \underline{X})$$

/see [1] /.

There is an alternative definition forthe same stochastic process, which is more convenient for simulating it:

1. All state variables remain constant for a random exponentially distributed period of time. Its expected value is the reciprocal of $\sum_i R_i$ and the subsequent intervals are independent of each other.
2. After this period of constancy one of the subprocesses selected at random and the state variables are adjusted according to the changes caused by the subprocess in question. The probability of choosing a subprocess is proportional to its deterministic rate. After this jump the first step will be performed again.

Based upon this second definition a computer program has been written in SIMULA 67 for studying the stochastic model.

Simulation results of the models

Having succeeded in describing measurement values with deterministic and stochastic models /these can be read in [3], [4]/ the forecast came to the limelight. Several simulation runs were carried out using the four basin model /[3]/, but here only the one basin version will be discussed as the stochastic model needs too much computer time. A less eutrophicated basin /Szemes basin/ has been chosen for examining the coincidence and difference between stochastic and deterministic models. The nutrient input of a basin has two sources: nutrient flowing from the preceeding basin /the lake can be considered to be a slow river/ and nutrient coming from the watershed of the basin. This second one can be controlled by measures, so this was reduced to the half and to the one fourth of the original nutrient load in simulation runs.

As it is impossible to look over realizations of the stochastic model and values obtained from deterministic model only some indices have been chosen for describing them. Table 1. shows the maximum value of summer phytoplankton obtained from the deterministic model, the maximum of the average, standard deviation and maximum of the stochastic realizations in case of: 1. no change in nutrient load; 2. reducing it to the half; reducing it to the one fourth. Table 2. shows the same values for the blue-green algae. The deterministic and mean values decrease as the load was reduced, this shows the strength of the theorems mentioned above, but the maximum values don't follow this tendency in any case showing the importance of stochastic models. In case of blue-green algae the maximum can increase in case of decreasing load revealing that water blooms can occur even in less eutrophicated basins, of course the probability and the height of it are lower than in a hypertrophic basin.

	determin.	average	stand.dev.	maximum
1.00	0.21	0.26	0.17	0.59
0.50	0.17	0.20	0.18	0.63
0.25	0.15	0.16	0.17	0.54

Table 1. Maximum of summer phytoplankton biomass / mg/l /

	determin.	average	stand.dev.	maximum
1.00	0.15	0.41	0.41	1.03
0.50	0.11	0.31	0.39	1.40
0.25	0.10	0.25	0.24	0.82

Table 2. Maximum of blue-green algae biomass / mg/l /

References

[1] Érdi,P., Tóth,J.: Stochastic reaction kinetics

[2] Herodek,S., Kutas,T., Csáki,P.: Simulation of phytoplankton dynamics in Lake Balaton. ISEM Journal, volume 4, Nos.3-4 /1982/

[3] Kutas,T., Herodek,S.: BEM, a complex model for simulation Lake Balaton ecosystem. in "Eutrophication of Shallow Lakes: Modelling and Management - The Lake Balaton Case Study" ed. by L.Somlyódy, S.Herodek, J.Fischer, IIASA, Laxenburg, Austria /1983/ pp 257-272

[4] Kutas,T., Tóth,J.: A stochastic model of phytoplankton dynamics in the Lake Balaton. J. of Statistical Computation and Simulation /in press/

MODELLING OF INTERPLANT COMPETITION

P. Ya. Grabarnik[1)]

A pure plant community with a simple structure of the arrangement of individuals is considered. The stochastic model represented by a Markov random field (MRF), describes a productivity property of the plant community. The parameters of the model may be interpreted as a strength of the competition between plants. The problem of the interplant competition modelling is to choose the probability structure and parameters of the model and to simulate a Markov random field. The synthesized image patterns may be applied to a study of real data.
Key words: plant community, Markov random field, simulation algorithm, image pattern.

It is known that the average weight of a plant drops when the density of a population increases. This fact is believed to be a consequence of competitive relations between plants. It seems natural to assume that the effect of competition should result in a specific pattern of individual weights (sizes) of plants in the community. We consider the pattern of productivity characteristics of plants as a realization of a two-dimensional random process, the parameters of which are interpreted in terms of the competitive relations between plants. The object of this paper is to present a Markov random field applicable as a model of a plant community. Indeed, plant within the community while consuming soil nutrients, water and light interacts with the nearest neighbours. It is thus possible to describe the community as a system of interacting random variables.

Let us assume that the plants are arranged in a simple structure, for example in a square lattice $I^2=\{s=(i,j),1\leq i,j\leq M\}$. In this case the neighbours of a plant in a site (i,j) will be the plants located in the sites $(i-1,j),(i+1,j),(i,j-1),(i,j+1)$. This description corresponds to a first-order MRF. A condition of markovianity means that

$$P\{x_{ij}|\text{all other values}\}=P\{x_{ij}|x_{i-1j},x_{i+1j},x_{ij-1},x_{ij+1}\}.$$

Following Besag [1] we consider a subclass of MRF, which has a probability structure satisfying

$$\frac{P\{x_s|\ x_r,r\in\Lambda\}}{P\{0_s|\ x_r,r\in\Lambda\}} = \exp\{x_s(\phi(x_s)+\sum_{r\in\Lambda}\beta_r x_r)\},$$

where Λ is a neighbour set, ϕ is an arbitrary function and β_r are arbitrary constants. By analogy with the models of statistical physics (the Ising model) the interactions between neighbouring plants are characterized by the parameters β_r, which may be considered as a measure of competition. In order to achieve a correspondence between the model and real data one should choose an appropriate functional form of the conditional distribution $P\{x_s|\cdot\}$ and fit parameters. Thus, the problem of the interplant competition modelling is to choose a probability structure and parameters of the model and to simulate the MRF.

A simulation of MRF is performed by an operator that operates on a set of independent random variables $\{u_s\}$. A linear model on a finite lattice I^2 may be represented by an equation

$$x_s-\alpha = \sum_{r\in\Lambda}\beta_r(x_r-\alpha) + \sqrt{\nu}\, e_s\ ,$$

1) Institute of Soil Science and Photosynthesis, Pushchino 142292, USSR

where the noise sequence e_s may be either correlated or uncorrelated.

A simulation algorithm for linear MRF models [2] is:

Step 1: Generate an array of independent random variables $\{u_s, s \in I^2\}$ with zero mean, unit variance and specified probability density.

Step 2: Find its discrete Fourier transform (DFT) $U(t)$. Find the corresponding DFT $X(t)$ by the model equation.

Step 3: Obtain the given image $\{x_s\}$ by means of inverse DFT of $X(t)$.

Using linear models we cannot achieve a full correspondence when we attempt to simulate a pattern exhibiting sharp boundaries. Note that linear models are more suitable for continuous variables than for discrete ones. In the case of integer variables nonlinear models arise . We will concentrate on a binomial model when a site variable changing in the range of $\{0,\dots,n\}$ has a binomial distribution, the parameter of which is controlled by the site's neighbours:

$$\theta = \exp(f)/(1+\exp(f)),$$

where $f = \alpha + \sum_{r \in \Lambda} \beta_r x_r$

A simulation algorithm for Binomial model [3] is:

A matrix of transition probabilities $\{P_{kl},\ k,l=0,n\}$ corresponding to conditional probabilities of the MRF $q_m = P\{x=m \mid \text{neighbours}\}$ is formed.

Step 1: Generate an array of independent random variables distributed binomially with parameters $\frac{1}{2}$ and n.

Step 2: Randomly choose a site (i,j) of the lattice and perform a transition $x_{ij} \to x'_{ij}$ according to the matrix of transition probabilities $\{P_{kl}\}$.

The described process converges to an equilibrium distribution with conditional probabilities $\{q_m\}$.

Below we present several examples of images simulated with different parameters of MRF.

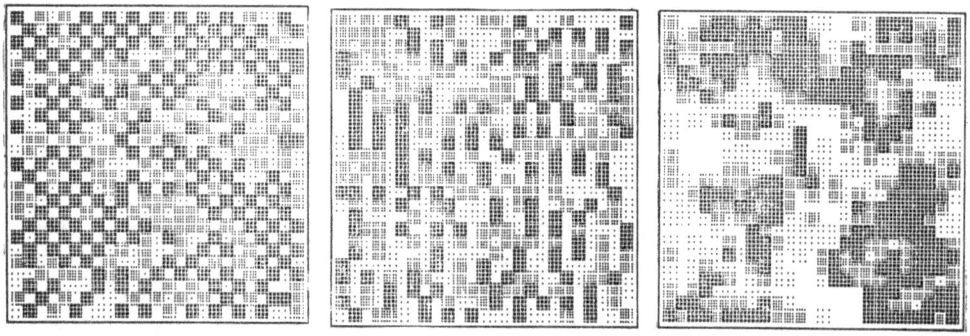

a) b) c)

Fig.1 Realizations of the first-order MRF with parameters a) $\alpha=3$, $\beta_1=\beta_2=-0.5$, $n=3$; b) $\alpha=1.2$, $\beta_1=-0.6$, $\beta_2=0.2$, $n=3$; c) $\alpha=-3$, $\beta_1=\beta_2=0.5$, $n=3$.

A case of highly inhibitory interaction corresponding to large negative values of parameter β is presented in Fig.1a. Fig.1b illustrates an effect of anisotropy when the interactions for different directions are not equivalent. In Fig.1c an example of cluster formation corresponding to positive values of parameter β is given.

These examples are not completely identical to the real data, nevertheless a good fit may be achieved by a reasonable combination of the model's parameters.

In the present paper we leave apart the problem of estimation of the parameters. We want to emphasize that in majority of cases the simulated patterns will be the only source of information on the properties of these estimates.

References:

[1] Besag, J.: Spatial interaction and the statistical analysis of lattice systems. J. Royal Statis. Soc., ser.B, 36 (1974), 192-326.

[2] Kashyap, R.: Analysis and synthesis of image pattern by spatial interaction models. Progr. in Pattern Recogn. 3 L.Kand & A.Rosenfeld, Eds. North-Holland (1981) 149-186.

[3] Cross, G., Jain, A.: Markov random field texture models. IEEE Trans. Pattern Anal. Machine Intell., 5 (1983) 1, 25-39.

NUMERICAL MODELLING OF EVEN-AGED PLANT POPULATION DYNAMICS. TWO-DIMENSIONAL MODEL

V.V.Galitsky, A.A.Krylov [1)]

Vegetation cover. Modelling is in rather wide use when studying plant objects and in the first place their production process. The main attention in it is paid to mathematical modelling on the level of a single plant and a vegetation cover in the whole. As usual, the models of a single plant are of compartmental character and reflect the existing conceptions of variety of the plants life processes. In analysis of cover in the whole the same set of conceptions is directly used, actually considering the cover as one big plant or, to be more exact, as a homogeneous film of plant substance. In this way, a certain part of biological point under consideration is lacking, since in reality the vegetation cover consists of separate interacting plants.

Horizontal structure and thinning of plant community. In any vegetation cover there is the so-called horizontal geometric structure, i.e. a certain scheme of distribution on the territory, occupied by interacting plants, which is reffered to as plant community. In this paper the simplest type of pure plant community of even age is dealt with. The effect of thinning, i.e. gradual dying out the coexisting plants is the consequence of the presence of horizontal structure of community and, correspondingly, of local competitive interrelation of plants. Thinning is a natural though usually neglected factor for agricultural crops as well. Dying out the plants in the community caused by their interaction with other plants is a tragical end of many but not all of them. Consideration of history of separate plants using the model described below or through the analysis of data of century-long observations made at the experimental forest station TSHA (Plotnikov,1979) shows that "the ways of development" of single plants which survived till present can be rather original: during their history they can repeatedly be oppressed and, vice versa, in conditions favorable for them and fatal for their neighbours they can hold on and go growing. Thus, the plants making community by the very essense of the process of thinning in reality can not be similar. Their characteristics are the functions of coordinate of plant distribution on the territory and therefore the idea of a cover as a homogeneous film of plant substance widely used in models of productive process ought to be connected with the presence of a real usually nonhomogeneous horizontal geometric structure of the corresponding plant community.

The model of plant community. The modelling of plant community dynamics is specified by actual account of the role of geometric factor in development of separate plants and the whole community. A method of numerical modelling (numerical experiment by Fedoseev et al., 1978) with computers or - to make it more concrete - a method of molecular dynamics (Bird, 1960) - appears to be most adequate to the task of modelling such objects. This method consists in investigating the dynamics of the system by computer direct integration of equations describing the dynamics of the elements of system.

The submodel of plant. The equation describing an element of our system - a plant - is presented in some works (Galitsky, Komarov, 1979; Galitsky, 1981)

$$\dot{B} = \kappa(T) \cdot (\alpha B^k + (B/B_f)^{\gamma} \cdot \dot{B}_f) - \alpha B^k \qquad (1)$$

where $\kappa(T) = \min\{A/A_f(T), 1\}$ - i.e. the income item of balance is proportional to the value of relative provision of plant with the necessary territory (resources); A - a measure of the territory, available for the growth of the given plant; A_f is a measure of of the territory needed for the free growth; B_f - the biomass of the freegrowing plant; T - age; αB^k - the expenditure item of balance, i.e. the expences for maintenance of biomass, is expressed by the power function of biomass B , which is weaker then the linear one ($0<k<1$). Analysis of the data related to the experiments on influence of removing some part of biomass of plant or its organ on respiration (Kuperman,Khitrovo, 1977) leads to the conclusion that in case when the growth of the plant is restricted by insufficiency of territory (resources) - $0<\gamma<k$ (Galitsky, 1984).

Free-growing plant. From analogy with some tasks of physics, for example, of solid state (Ziman, 1960) it is possible to say that A (T) and B (T) are self-consistent characteristics of the plant community and reflect the habitat conditions modified by community for the plants of the given species.

Biomass and phytomass of plant. Equation (1) describes the dynamics of plant biomass leaving that of phytomass beyond. It is biomass that is an active substance of

1) Institute of Soil Science and Photosynthesis of Academy of Sciences, USSR, Pushchino

the plant responding directly to the conditions of plant existence including the provision with the growth area. On the other hand the phytomass of the whole plant or of its organs is easy to observe and usually is rather easy to measure. Phytomass F bringing about mechanical properties and geometric form of the plant somewhat determines the area currently occupied by it. Let us agree that some part of biomass permanently looses its activity and transforms into a passive portion of biomass $F = p/T_1 \cdot \int_0^T B(\theta)\, d\theta$, where p is a measureless constant.

Poligonal mosaic. The presence of a certain number of resources available with plants is a necessary condition for the existence of any plant and consequently of the whole plant community. On ecologically homogeneous area the resources are likely to be distributed uniformly and their quantity intended for a given plant is determined by some measure of the territory which can be adequate for this plant. In the model under cosideration a polygonal mosaic by Voronoy (Voronoi, 1908; Galitsky, Mironenko, 1981) is used as geometric tool for determination of this territory.

Mechanics of the model of community. The essence of the model functioning is that biomass of each plant placed in its own cell of the mosaic keeps changing according to equation (1) independent of other plants of community until either this particular plant or some of its neighbours die for lack of territory. At the moment of dying actual interaction of the neighbouring plants takes place as "sharing" the polygon cleared by the dead plant. The redistributed parts of plot are added to the appropriate polygons of the adjoining living plants whereupon the biomass of the living plants starts changing anew according to equation (1) with new values for A until some other plant of the community dies next. In such a formulation interaction of plants takes place, firstly, locally and, secondly, instantly, i.e. disappearance of the given plant "is felt" only by its neighbours and these "sensations" are immediately and fully (according to the size of the occupied area) reflected on the dynamics of their biomass.

Rearrangement of mosaic. Along with the dying off of individual plants the initial mosaic according to the starting plant distribution begins to change locally (roundabout the dying plants) and under certain conditions these changes cover the increasing areas of mosaic. Thus, within the frame of this model it is necessary to indicate the way of geometric division of the vacant polygon territory among the neighbours of the dead plant. The division is made by rearranging the appropriate polygons in such a mode as if the disappearing polygone did not ever exist. This way of rearrangement always results in Voronoy's mosaic consisting of convex polygons.

Plant dying off. The question of determining the moment of plant dying off is of significance. Restricting here to a "structureless" model of plant, we should supply it with the definition of the dying off moment T_d : $B_d(T) = \varepsilon B_f(T)$, where ε is some constant $(0<\varepsilon<1)$. The plant is to be considered dead if biomass determined from equation (1) became less than the corresponding value of B_d.

On realization of the model on computer. In some of its parts a model is necessarily a pure imitation, i.e. it imitates possible situations and their results without reflecting appropriate processes in mechanisms. So, the initial plants distribution on their territory is initiated here by scattering points on the rectangular section of the plane and an element of randomness is introduced into the definition of plant coordinate, as required. The presence of stokhasticity in the model makes statistical treatment of data necessary when obtaining integral characteristics of plant community. The so-called technique of accumulation was applied for obtaining integral characteristics of the community. For this purpose, a predetermined number of model realizations differing only in chosen random values is made. In each case the characteristics of the community under study are summarized in a certain way, which makes provision for a considerable increase in the efficient number of the modelled community and in stability of its integral characteristics. Some particularities in programme realization of the model are also connected with the fact that the plant community gradually decreases in number and along with it the dimension of the system of differential equations describing the community decreases and its coefficients change.

Some results of the work of model. As seen from the above, even with the essential simplicity of the system element model there are extremely many variants of its geometric construction against the background of possible variations of parameters of the plant model. It is quite natural that full and thorough parametric study of the model can not be given here. We shall give only some example of its functioning which allow to judge more concretely about the possibility of such technique of modelling and about some particularities of the object. Fig. 2 shows data for the case of cluster sowing location of 3 plants per a hole sized 0.3 of square grid step. The rest parameters and functions $A_f(T)$ and $B_f(T)$ are the same as in Fig. 1. Parameter A_0 is supposed to be equal to 0.05

of an area unit, all the population being located on this area. The graphics of the dependences upon time are presented (normalized average values of population N, biomass B, phytomass P of community per unit area and the same per one plant).

Conclusions. The two-dimentional model of even-aged plant community described above demonstrates specificity of approach meant for investigating the influance of details of arranging an element (plant) and a geometric construction of the whole system (community) on dynamics both of all the community and its members. Principal methodological similarity of computer approach with a natural experiment which can not be completely substituted by the former one should be noted. Spreading this method is supposed to open a wide field of practically and theoretically useful research in geobotanics, forestry, phisiology of plant, plant breeding, etc.

Literature

[1] Bird G.A.: Molecular gas dynamics. Clarendon Press, Oxford 1976.
[2] Galitsky V.V.: On modelling a production process in plant community. In: Modelling biogeocenotical processes. Nauka, Moscow 1981.
[3] Galitsky V.V.: On influence of removing a part of biomass on plant growth. Izvestya Ac.Sc.USSR, Ser.Biol.,(1984) 6, 762-773.
[4] Galitsky V.V., Komarov A.S.: Dynamics Model of Plant Biomass. In: Proc.Internat.Sympos. "Stability of Spruce Forest Ecosystems". Ed. E.Klimo. Brno 1979.
[5] Galitsky V.V., Mironenko E.V.: Mosaic by Voronoy on the plane. Algorithmes of building. ECOMODEL, No 6, Pushchino 1981.
[6] Fedoseev D.V., Chuzhko R.K., Grivtsov A.G.: Heterogeneous crystallization from a gas phase. Nauka, Moscow 1978.
[7] Kuperman J.A., Khitrovo E.V.: Respiration gas exchange as an alement of plant production process. Nauka, Novosibirsk 1977.
[8] Voronoi G.F.: Nouvelles applications des parametres continus a la theoric des formes quadratiques. Deuxiems Memoire Recherches sur les paralleloedrus primitifs. J.reine angew. Math. 134 (1908) 198-287.
[9] Ziman I.M.: Electrons and phonons. Oxford at the Clarendon Press 1980

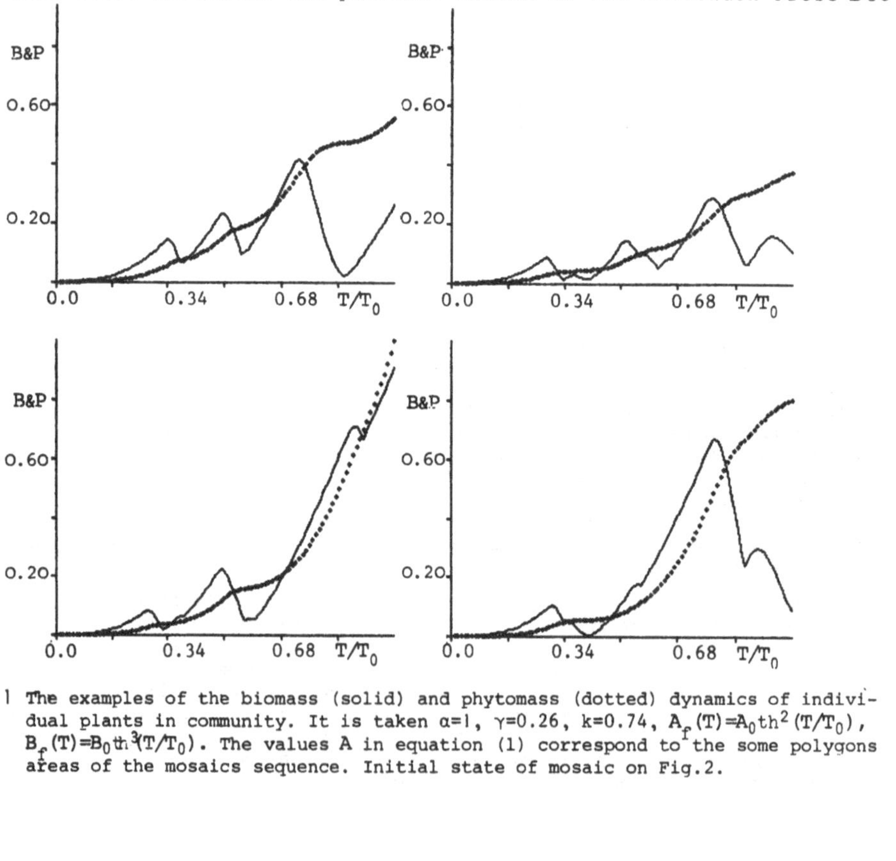

Fig.1 The examples of the biomass (solid) and phytomass (dotted) dynamics of individual plants in community. It is taken $\alpha=1$, $\gamma=0.26$, $k=0.74$, $A_f(T)=A_0 th^2(T/T_0)$, $B_f(T)=B_0 th^3(T/T_0)$. The values A in equation (1) correspond to the some polygons areas of the mosaics sequence. Initial state of mosaic on Fig.2.

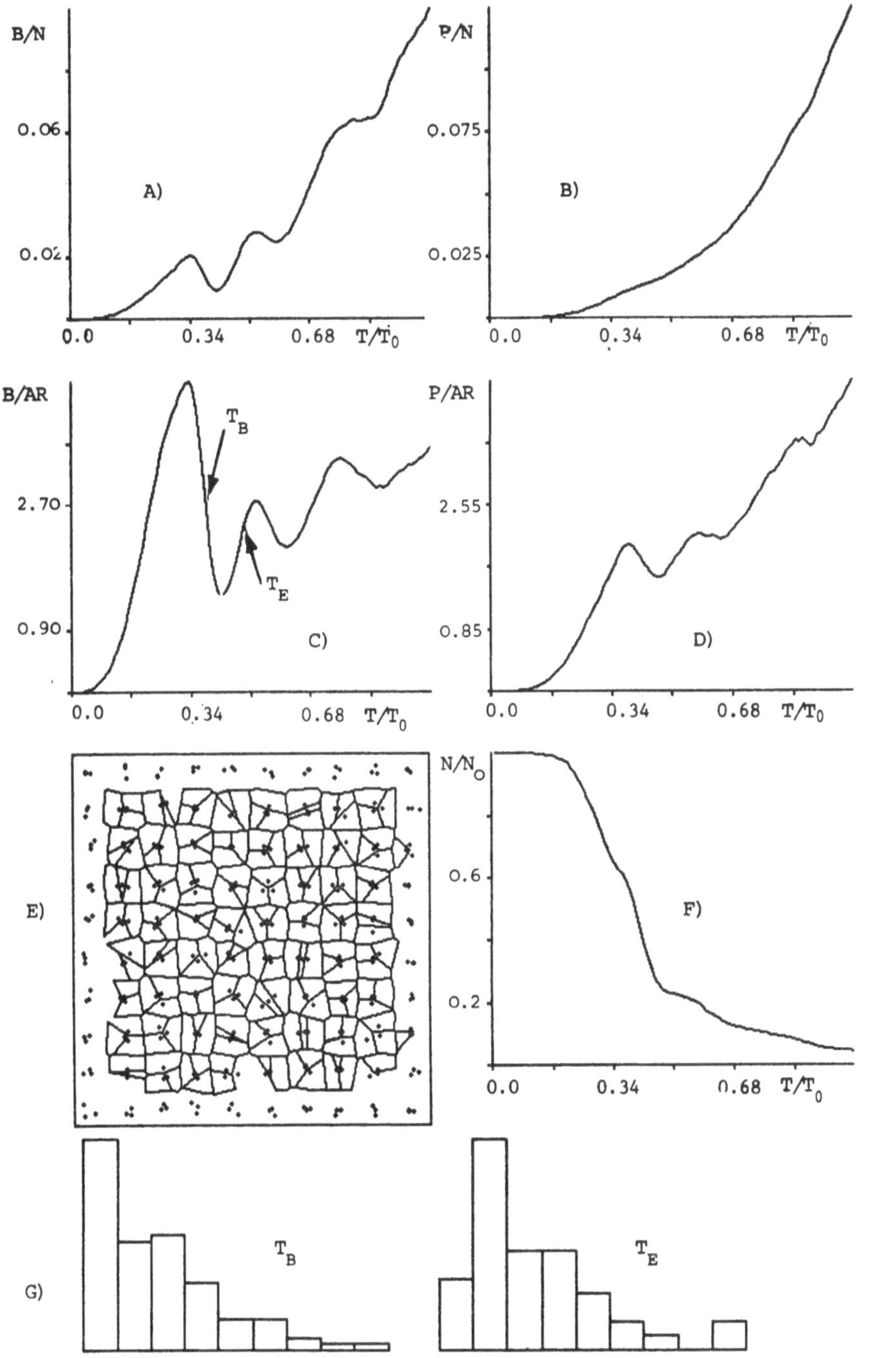

Fig.2 Some results of modelling of cluster sowing community.
A) dynamics of average biomass per individual plant; B) dynamics of average phytomass per individual plant; C) dynamics of the community biomass per area unit; D) dynamics of the community phytomass per area unit; E) Voronoy's mosaic for the initial pattern; F) dynamics of thinning; G) the individual biomass hystograms for moments corresponding to the arrows at T_B and T_E on C.

PRACTICAL USE OF ADAPTATION PRINCIPLES IN ECOLOGICAL SYSTEMS MODELLING

Peter Racsko [1)]

1. Introduction

The term "adaptation" is widely used in ecological, biological, economical systems research as a qualitative characteristic of complex systems. Generally speaking adaptation expresses the system's ability to reduce the amplitude of responses to the changes of the embedding system i.e. environmental conditions.

While the adaptation of biological objects is quite explicit and understood as a tool for survival of individuals and populations, for ecological systems with trophical structure, systems with external control - e.g. human impact - adaptation is treated as a type of behaviour oriented to preserve stability. The term stability is used in connection with the steady states of the system and stands for the system's ability to resist to not very significant perturbations.

We shall demonstrate a pair of examples, one of them illustrates the practical use of a hypothesized adaptation principle in a tree growth simulation, the other one proves the fact that a forced adaptation can destabilize a highly developed ecological production system.

2. A tree growth simulation model

The whole model description was published in [1], only a simplified version will be demonstrated below.

Let x_1, x_2, x_3 be the root, bole and leave biomass of a tree respectively. Let y be the new organic matter produced by the photosynthesis during the time period τ. The system's evolution can be described by the following system of difference equations:

$$x_i(t + \tau) = x_i(t) + e_i(t)\ y(t, \bar{\lambda}) \qquad (1)$$

$$i = 1, 2, 3$$

with the initial conditions $x_i(0) = x_i^o$, $(i=1,2,3)$, where

---

1) Computing Centre of the Eötvös Loránd University
1117. Budapest, Bogdánfy u. 10/b.

$e_i\ t$ - the distribution rate of the new assimilates between the parts of the tree; $\sum_{i=1}^{3} e_i(t) = 1;\ e_i(t) \geq 0.$

$\bar{\lambda}(t)$ - vector of the environmental conditions.

$y(t, \bar{\lambda})$ - is defined as follows:

$$y(t,\lambda) = y(x_1(t),\ x_2(t),\ x_3(t),\ \bar{\lambda}(t)) \qquad (2)$$

As one can see system (1) is not "closed", i.e. the $e_i(t)$ functions can not be obtained from the system itself. In order to "close" the system we postulate an adaptation hypothesis, that the system distributes the new assimilates in a way that maximizes the overall biomass production in the next time period supposing the environmental conditions do not change.

The functions $e_1(t)$, $e_2(t)$, $e_3(t)$ are defined from the expression:

$$y^*(t+\tau) = \max_{\{e_1(t),e_2(t),e_3(t)\}} y\{x_1(t)+e_1(t)y(t,\bar{\lambda})\tau,\ \ldots, \qquad (3)$$

$$\ldots,\ x_3(t)+e_3(t)y(t,\bar{\lambda})\tau,\ \bar{\lambda}\}$$

$$\sum_{i=1}^{3} e_i(t) = 1\ .$$

Now the system (1) with the adaptation principle (3) can be solved. A more complex but conceptually the same model was identified and investigated for pine forests of Carelia. The computer simulation results occured to be sufficiently adequate. The adaptation principle provides a numerical technique to investigate the limits of survival of a tree as a part of the forest biogeocenosis. The simulated system is very stable and tolerates relatively large changes of the exogenous parameters e.g. wheather conditions, nutrient and water quantity etc.

3. Ecosystem stability with external control

The second example demonstrates that an externally supported equilibrium state of an ecological system is unstable if the system is relatively well-developed and the output-control strategy is fixed.

In this case the system can not tolerate even small perturbations, i.e. it will never be adapted to the artificial equilibrium.

The ecological system is a two-level trophical chain (resource and consumer) with controlled resource input flow and biomass output.

The system's model:

$$\frac{dR}{dt} = Q(R) - mR - NV(R)\ \varphi(\bar{\mu})$$

$$\frac{dN}{dt} = \alpha NV(\bar{R})\ \varphi(\bar{\mu}) - S(N) \qquad (4)$$

where R - resource quantity, N - consumer (biomass or production) quantity, Q(R) - resource input function, S(N) - production output function, m - natural loss rate of resources, V(R) - trophical function, α - conversion parameter and $\varphi(\bar{\mu})$ - parameter expressing the impact of the external factors $\bar{\mu}$.

In real production systems the following conditions hold:

i) Q(R); S(N) and V(R) are sufficiently smooth functions in $\mathbb{R}^1$ $\varphi(\bar{\mu})$ is sufficiently smooth in $\mathbb{R}^n$

ii) $V(\dot{0}) = 0$; $V(R) \leq M$ for every $R \geq 0$

$\frac{dV}{dR} \geq 0$ for every $R \geq 0$

iii) $0 \leq \varphi(\bar{\mu}) \leq 1$

Assume the production output quantity has been selected and fixed, $S(N) = C$.

The necessary and sufficient conditions for the stability of a steady state (R^*, N^*) are:

$$\left.\frac{dQ}{dR}\right|_{R^*} - m > 0 \qquad (5)$$

$$\left.\frac{dQ}{dR}\right|_{R^*} - m + V(R^*)\ \varphi(\bar{\mu}) - \frac{c}{\alpha} \cdot \frac{1}{V(R^*)} \cdot \left.\frac{dV}{dR}\right|_{R^*} < 0 \qquad (6)$$

As far as Q(R) is an external control (5) can always be satisfied however (6) holds only for "small" $V(R^*)$ values, when $\left.\frac{dV}{dR}\right|_{R^*}$ is "large" i.e. when the system is in its early period of evolution.

During this period proper selection of Q(R) provides a stable equilibrium. For "large" V(R) values there are no control strategies providing stability. The changes of the environmental factor $\varphi(\bar{\mu})$ might shift the separating line between the domain of possible stable steady states and the part of the phase plane where stable steady states can not exist.

As we have seen from (6) in this type of forced ecological systems the stability strongly depends on the proportion $\frac{1}{V(R)} \cdot \frac{dV}{dR}$. When this value is small there is no way to maintain a stable equilibrium, i.e. the system loses the ability to adapt to small perturbations.

Consequently, computer simulations based on this type of models have to be performed very carefully.

Budapest 30. 01. 1985.

Bibliography:

[1] Рачко, П.: Имитационная модель роста дерева. Журнал Общей Биологии т. XXXIX/4. 1978. Москва.

[2] Svirezhev, Yu. M. - Logofet, D. O.: Stability of Biological Communities. 1983, MIR Publishers, Moscow.

GRADIENT DYNAMIC PROPERTIES OF COEVOLUTION OF SOME ECOLOGICAL COMMUNITIES

Passekov V.P.*

A model is presented for coevolution of an m-species community with polylocus genotypic structure of species and multiple symmetric individual interactions. It is shown that with weak genotypic differences and loose linkage for asymptotic dynamic equations there is a function of dynamic landscape. For some Riemannian metric the asymptotic equations are gradient.

Key words: biological communities, extremal properties, coevolution.

1. Consider a mathematical model of an m-species community. If genotypic structure is neglected the dynamics of m populations numbers is determined by n-fold ($n=1,2,\ldots$) random intra and interspecific interactions:

$$\dot{N}^i = N^i \sum_n \sum_{(k^n):k_1=i} a^{(k^n)}(N) N^{k_2} \cdots N^{k_n}, \quad i=\overline{1,m} \quad (1)$$

where $N=(N^1,\ldots,N^m)$, $a^{(k^n)} = a^{k_1,\ldots k_n}$ are the interaction coefficients, k_j equals species superscripts of interacting individuals. When $n=1,2$ and $a^{(k^n)}$ are constants (1) is the same as the Volterra model.
Let us change variables

$$\bar{N} = \sum_i N^i, \quad h^i = N^i/\bar{N}, \quad i=\overline{1,m} \quad (2)$$

Here h^i is the fraction of the i-th population in the community. Then

$$\dot{\bar{N}} = \sum_n (\bar{N})^n \sum_{(k^n)} a^{(k^n)}(\bar{N},h) \prod_{\ell=1}^n h^{k_\ell} = \sum_n (\bar{N})^n a(n), \quad (3)$$

$$\dot{h}^i = h^i \Big[\sum_n (\bar{N})^{n-1} \sum_{(k^n):k_1=i} a^{(k^n)}(\bar{N},h) \prod_{\ell=2}^n h^{k_\ell} - a(n)\Big] = h^i \sum_n (\bar{N})^{n-1} [a^i(n) - a(n)], \quad i=\overline{1,m}$$

where $h=(h^1,\ldots,h^m)$, $(\cdot)^n$ is the n-th power of $(\cdot)$.

2. Suppose that interaction coefficients $a^{(k^n)}$ in (1) are invariant to the superscript permutations. Then (allowing the dependence of $a^{(k^n)}$ on the random number of encounters for the corresponding type of individuals) (1)-(3) may be written as

$$\dot{N}^i = N^i \partial_{N^i} G(N), \quad i=\overline{1,m}; \quad (4)$$

$$\dot{\bar{N}} = \bar{N} \partial_{\bar{N}} G(\bar{N},h), \quad \dot{h}^i = (h^i/\bar{N})\Big(\partial_{h^i} G(\bar{N},h) - \sum_j h^j \partial_{h^j} G(\bar{N},h)\Big) \quad (5)$$

*) Computer Center, USSR Academy of Sciences, Moscow, USSR

where $\partial_{.}G=\partial G/\partial_{.}$. It is easy to see that for equations (4)-(5) $\dot{G}\geq 0$:

$$\dot{G}(\mathcal{N}) = \sum_i \partial_{\mathcal{N}^i} G(\mathcal{N})\cdot \dot{\mathcal{N}}^i = \sum_i \mathcal{N}^i(\partial_{\mathcal{N}^i} G(\mathcal{N}))^2 \geq 0; \quad (6)$$

$$\dot{G}(\bar{\mathcal{N}},h) = \partial_{\bar{\mathcal{N}}} G\cdot\dot{\bar{\mathcal{N}}} + \sum_i \partial_{h^i} G\cdot \dot{h}^i = \bar{\mathcal{N}}(\partial_{\bar{\mathcal{N}}} G)^2 + \sum_i (h^i/\bar{\mathcal{N}})(\partial_{h^i} G - \sum_j h^j \partial_{h^j} G)^2 \geq 0 \quad (7)$$

In (7) we took into account that under summation over i one may substruct the sum $\sum_j h^j \partial_{h^j} G$ (independent of i) from each of the multipliers $\partial_{h^i} G$ without changing the result.

Notice that $h^m \, (=1-\sum_{j=1}^{m-1} h^j)$ may be considered as a dependent variable. It can be checked that for independent variables the dynamic equations have the same form (5)-(6).

From (4)-(5) and (7) follows that the stationary points G (as function of independent variables) are the equilibria, that $\dot{G}\geq 0$ and equality holds only at the equilibrium, further on, there are no periodic trajectories, and local maxima of G correspond to the asymptotically stable equilibria, i.e. G determines the dynamical landscape for (4)-(5). Roughly speaking the community"tends" to reach equilibrium in the "nearest" peak of G .

These results are evident if one notices that the right hand sides of (4)-(5) are Riemannian gradients of G with the metrics g_{ij} in the phase space interior and with induced metrics on faces. For (4) $g_{ij} = \delta_{ij}/\mathcal{N}^i$, $i=\overline{1,m}$, for (5) matrix g_{ij} is block-diagonal with element $1/\bar{\mathcal{N}}$ (corresponding to $\bar{\mathcal{N}}$) and elements $\bar{\mathcal{N}}(\delta_{ij}/h^i+1/h^m)$ corresponding to $m-1$ independent coordinates h .

3. When autosomal diploid genotypic structure of species is considered, the dynamic equations have the same form as (1) but population numbers $\mathcal{N}^k$ are replaced by numbers $\mathcal{N}^k_{ij}$ of the corresponding genotypes ($i^k j^k$) of interacting individuals (i^k, j^k are gamete indices of k-th species). Interaction coefficients $a^{(k^n)}_{ij} = a_{i^{k_1}j^{k_1},\dots,i^{k_n}j^{k_n}}$ for individuals with genotypes $(i^{k_\ell} j^{k_\ell})$ are supposed to be invariant to permutations of gamete subscripts of individual and genotype subscripts. Let subscripts of gametes of individual be ordered. Then

$$\dot{\mathcal{N}}^k_{ij} = \sum_n \sum_{(k^n)} \sum_{i^{(k^n)} j^{(k^n)}} a_{i^{k_1}j^{k_1},\dots,i^{k_n}j^{k_n}}(\mathcal{N}) \prod_{\ell=1}^{n} \mathcal{N}^{k_\ell}_{ij}. \quad (8)$$

Consider only one locus. Then in the second sum $i^{k_1}=i^k$, j^{k_1} (or i^{k_2})$=j^k$. If $j^{k_1}=j^k$, then $a^{(k^n)}_{ij}$ includes the mortality component $m^{(k^n)}_{ij}$ of genotype $(i^k j^k)$ with n-fold interactions. When $i^{k_2}=j^k$, it is supposed that the first two interacting individuals may form a mating pair and the corresponding coefficient $a^{(k^n)}_{ij}$ includes its fertility component $b^{(k^n)}_{ij}$. Choose new variables:

$$\bar{\mathcal{N}} = \sum_k \sum_{i^k j^k} \mathcal{N}^k_{ij}, \; h^k = \sum_{i^k j^k} \mathcal{N}^k_{ij}/\bar{\mathcal{N}} = \bar{\mathcal{N}}^k/\bar{\mathcal{N}}, \; p^k_{ij} = \mathcal{N}^k_{ij}/\bar{\mathcal{N}}^k. \quad (9)$$

Here $\bar{N}$ is the total community number, h^k is the fraction of k-th species in the community, p_{ij}^k is the genotype ($i^k j^k$) frequency. Rewrite (8) as

$$\dot{N}_{ij}^k = \sum_n (\bar{N})^n \sum_{(k^n)} \sum_{i^{(k^n)} j^{(k^n)}} a_{ij}^{(k^n)}(\bar{N},h) \prod_{l=1}^n h^{k_l} p_{ij}^{k_l}. \qquad (10)$$

In new variables (9) from (10) we get:

$$\dot{\bar{N}} = \sum_n (\bar{N})^{n-1} a(n), \quad \dot{h}^k = h^k \sum_n (\bar{N})^{n-1}(a^k(n) - a(n)), \qquad (11)$$
$$\dot{p}_{ij}^k = (1/h^k)\sum_n (\bar{N})^{n-1}(a_{ij}^k(n) - a^k(n)).$$

Here $a(n)$ is the sum of n-fold interaction coefficients over all kinds of species and genotypes with weights $\prod_{l=1}^n h^{k_l} p_{ij}^{k_l}$, $a^k(n)$ differs from $a(n)$ by fixing the first individual superscript k and in $a_{ij}^k(n)$ i^k, j^k are fixed.

4. Further suppose there are weak genotypic differences between fertilities-mortalities, i.e.

$$a_{ij}^{(k^n)} = b^{(k^n)} + \varepsilon b_{ij}^{(k^n)} - (m^{(k^n)} + \varepsilon m_{ij}^{(k^n)}) \qquad (12)$$

where interaction coefficients are expressed in terms of fertilities - mortalities differences and $0<\varepsilon\ll 1$. Then

$$\dot{N}_{ij}^k = (\sum_n (\bar{N})^n \sum_{(k^n):k_1=k_2=k} b^{(k^n)} \prod_{l=1}^n h^{k_l}) p_i^k p_j^k - \qquad (13)$$
$$- p_{ij}^k h^k \sum_n (\bar{N})^n \sum_{(k^n):k_1=k} m^{(k^n)} \prod_{l=2}^n h^{k_l} + O(\varepsilon),$$

$$\dot{p}_{ij}^k = h^k (\sum_n (\bar{N})^{n-1} \sum_{(k^n):k_1=k_2=k} b^{(k^n)} \prod_{l=2}^n h^{k_l})(p_i^k p_j^k - p_{ij}^k) + O(\varepsilon). \qquad (14)$$

Here i^k-th gamete frequencies $p_i^k = \sum_{j^k} p_{ij}^k$ will be slow variables when $b^{(k^n)} \gg \varepsilon$:

$$\dot{p}_i^k = \sum_{j^k} \dot{p}_{ij}^k = O(\varepsilon). \qquad (15)$$

When slow variables $\bar{N}$, h, $p=\{p_i^k\}$ are frozen and $\varepsilon=0$ system (14) has a unique asymptotically stable equilibrium, $p_{ij}^k = p_i^k p_j^k$. That is why slow variable dynamics with small ε approximately satisfies the equations

$$\dot{\bar{N}} = \sum_n (N)^n a(n) = \bar{N}\partial_{\bar{N}} G,$$
$$\dot{h}^k = h^k \sum_n (\bar{N})^{n-1}(a^k(n) - a(n)) = (h^k/\bar{N})(\partial_{h^k} G - \sum_j h^j \partial_{h^j} G), \qquad (16)$$
$$\dot{p}_i^k = (p_i^k/h^k)\sum_n (\bar{N})^{n-1}(a_i^k(n) - a^k(n)) = (p_i^k/2Nh^k)(\partial_{p_i^k} G - \sum_j p_j^k \partial_{p_j^k} G);$$

$$G(\bar{N},h,p) = \sum_n \sum_{(k^n)} \sum_{i(k^n)j(k^n)} \int \prod_{\ell=1}^{n} \bar{N} h^{k_\ell} p_i^{k_\ell} p_j^{k_\ell} a_{ij}^{(k^n)}(x)dx, \quad (17)$$

a, a^k, a_i^k are calculated using weights $\prod_{\ell=1}^{n} h^{k_\ell} p_i^{k_\ell} p_j^{k_\ell}$

From (16) it follows that $\dot{G} \geq 0$, and the equality holds only at the equilibrium points of (16), i.e. G is the dynamical landscape for (16). If in (16) the dependent variables (h^m and last alleles frequencies $p_{n^k}^k$) are absent,then the right hand side is Riemannian gradient of G with metric g_{ij} in the phase space interior (and with induced metrics on faces). The block-diagonal matrix g_{ij} has non-zero elements $1/\bar{N}$ (corresponding to $\bar{N}$), $\bar{N}(\delta_{ij}/h^i+1/h^m)$ (corresponding to coordinates of h) and $2\bar{N}h^k(\delta_{i^k j^k}/p_i^k+1/p_{n^k}^k)$ corresponding to coordinates p_i^k (the independent allele frequencies of k-th species).

5. Considering the ℓ^k locus genotypic structure of k-th species $(k=\overline{1,m})$ in (13)-(14) p_i^k is replaced by $\sum_{U|V} r^k(U|V) p_{i_U}^k p_{i_V}^k$ (similarly p_j^k changes). Here $r^k(U|V)$ corresponds to probability of exchange of two gametes of an individual by two sets of loci U and V during sex cell formation, $p_{i_U}^k$ is the marginal frequency of U-locus gamete, containing the same alleles as the i^k-th one ($i^k = i_1^k, \dots, i_{\ell^k}^k$, i_j^k is i^k-th gamete j-th locus allele index). Then

$$\dot{p}_i^k = h^k \left(\sum_n (\bar{N})^{n-1} \sum_{(k^n):k_1=k_2=k} b^{(k^n)} \prod_{\ell=1}^{n} h^{k_\ell}\right) \times \quad (18)$$

$$\times \sum_{U|V} r^k(U|V)(p_{i_U}^k p_{i_V}^k - p_i^k) + O(\varepsilon).$$

From (18) we get that when $b^{(k^n)} \gg \{r^k(U|V)\} \gg \varepsilon$ the k-th species j-th locus allele frequencies $p_{m_j}^k = \sum_{i^k:\, i_j^k = m_j} p_i^k$ will be slow variables, i.e. $\dot{p}_{m_j}^k = O(\varepsilon)$.

As a result the community dynamics includes three stages. At the first one the genotype frequencies become equal to quasiequilibrium products of their gamete frequencies. The gamete frequencies quickly reach quasiequilibrium values which are equal to their allele frequency products. Finally the slow variable dynamic equations will be of the form (16) where interaction coefficients depend on all loci of individual genotypes, the calculation of G includes summation over all allele indices.

EXPERIMENTAL DESIGNS FOR MODELLING OF THE AGROECOSYSTEMS

Alexander N. Sadovski[1)]

The yield programming in Bulgaria is based on empirical and semiempirical mathematical models which identification is accomplished by the help of data from field experiments. An experiment for investigating the influence of water-supply, crop density and rates of nitrogen on the maize yield is described. The design varies those three factors on five levels and contains 36 variants. For construction of models including from 3 to 16 controlable factors also on 5 levels, a new family of almost saturated designs is developed.

KEY WORDS: Agroecosystems, experimental designs, models.

The yield programming investigations in Bulgaria proceed more then 10 years. Their future development and practical implementation are expressed by the agroecological approach. As a theory, this approach makes use of the definition of the agroecosystems as a plant ecosystems exploited and controlled by the man, which are determined by four sets of factors - biological, meteorological, soil and anthropogenical. As a practice, the agroecological approach means development and implementation of differentiated systems of cultivation, which permit the most effective use of the possibilities in different agroecological regions to get the expected obtainable or optimal yields

The control and management of agroecosystems are based on certain empirical and semiempirical mathematical models, identification of which is accomplished by the help of data from experiments conducted in a geographic set. The agricultural experiments are carried out in the presence of considerable drift both in time and space (Peregudov 1972) and they have to allow solving of two problems simultaneously: discrimination among several alternative models and optimal estimation of their parameters. Consequently, such designs must contain several factors varied on 4 or more levels (Klevtsov, Sadovski 1981).

In the experimental stations of N.Poushkarov Institute of Soil Science an experiment for investigation of the influence of water-supply, crop density and the rates of nitrogen fertilization on the maize yield is realized. During 10 years the experiment presents factorial design from 4^3 type in two replications or 128 plots in all. During the last three years, this design is replaced by a composite de-

1) N.Poushkarov Inst. of Soil Sci., 5 Shosse Bankya, BG-1080, Sofia

sign, which varies the same three factors on five levels and contains 36 variants in 3 replications, or 108 plots at total.

The last design is constructed from a factorial design 3^3 in respect to the levels codded by 0,2 and 4, factorial design 2^3 of the levels 1 and 3, and 1 central point added. The variants are as follow·

No. 1- 9: 000, 002, 004, 020, 022, 024, 040, 042, 044;
No. 10-18: 200, 202, 204, 220, 222, 224, 240, 242, 244;
No. 19-27: 400, 402, 404, 420, 422, 424, 440, 442, 444;
No. 28-36: 111, 113, 131, 133, 311, 313, 331, 333, 222.

The factor "density" S is quantitative and its levels have the following values: S_0 = 25 000; S_1 = 45 000; S_2 = 65 000; S_3 = 85 000; S_4 = 105 000 plants/ha.

The rates of nitrogen are: N_0 = 0; N_1 = 100; N_2 = 200; N_3 = 300; N_4 = 400 kg/ha.

The levels of the water-supply represent: W_0 - without irrigation; W_3 - almost optimal irrigation for maintenance 80 % of field capacity; W_4 - 4/3 from the optimal moisture; W_2 - 2/3 from the optimum; W_1 - 1/3 from the optimum.

As in essence, the water-supply factor W is qualitative, to express its 5 levels four dummy variables are introduced $w_i (i=1,\dots,4)$ in the following manner:

W	w_1	w_2	w_3	w_4
0	0	0	0	0
1	1	0	0	0
2	0	1	0	0
3	0	0	1	0
4	0	0	0	1

The experimental data permits identification of empirical models of the type y = f(w, S, N) for example

$$y = b_0 + \sum_1^4 b_i w_i + b_5 S + b_6 N + b_7 S^2 + b_8 N^2 + b_9 SN + b_{10} w_1 S + $$
$$+ b_{11} w_2 S + b_{12} w_3 S + b_{13} w_4 S + b_{14} w_1 N + b_{15} w_2 N + b_{16} w_3 N + b_{17} w_4 N.$$

The optimal conditions for the maize cultivation are determined from those models.

The existence of interactions among the factors is detected in many field experiments, but the high-order interactions are almost always statisticaly insignificant. In addition it is well known that with the increase of the number of factors and their levels, the practical realization of the full and fractional factorial designs becomes impossible.

A new family of almost saturated designs is developed for construction of models, including from 3 to 16 controlable factors (different mineral nutrients, irrigation, depth of tillage, crop density, etc.), varied also on 5 levels. Those experimental designs are composite and the manner of their construction may be illustrated in the case with number of factors k = 4.

The basic configuration consist of:

one zero point	(0, 0, 0, 0),	
one central point	(2, 2, 2, 2),	
and 2k star points	(0, 2, 2, 2),	(4, 2, 2, 2),
	(2, 0, 2, 2),	(2, 4, 2, 2),
	(2, 2, 0, 2),	(2, 2, 4, 2),
	(2, 2, 2, 0),	(2, 2, 2, 4).

This configuration looks like a kite and provides degrees of freedom for estimation of the constant term, four linear and four quadratic terms and for the error. For estimation of the two-factor interactions of first order, it is combined with $\frac{k(k-1)}{2}$ points, presenting a part of Rechtschaffner's (1967) saturated designs

(3, 3, 1, 1), (3, 1, 3, 1), (3, 1, 1, 3),
(1, 3, 3, 1), (1, 3, 1, 3), (1, 1, 3, 3).

The number of the variants n, depending on factors number - k, can be determined by the next formula

$$n = \frac{(k+1)(k+2)}{2} + 1.$$

The new family of asymmetric designs is intended for obtaining of regression models in the form of full second-order polynomial. The analysis could be accomplished easy after Gram-Schmidt's orthogonalization.

The same designs could be used successfuly and for indentification of other empirical and semiempirical models - linear or nonlinear, additive of multiplicative, with number of unknown coefficients not exceeded $\frac{(k+1)(k+2)}{2}$, for example

$$y = f(x_1, x_2, \ldots, x_k) = a \prod (n_i + x_i)^{b_i} \exp \left[\sum c_i (n_i + x_i) \right] ,$$

where a, b_i, c_i and n_i are regression coefficients. The described designs are one of the most economical solutions known in the statistical literature and appear exeptionally convenient for agricultural experiments.

The results of long-term field experiments conducted in a geographic set according to the indicated and other designs, are organized in a data base, which includes soil, climatic, biological, agro-technical, socio-economical and experimental data (Sadovski 1984). The same results are used for development of an Integrated computerized agriculture management system.

References

[1] Klevtsov A., Sadovski A.: Agricultural Science, XIX (1981), 4, 108 - 115, Sofia (in Bulg.).

[2] Peregudov V.: Agrochimiya, 6 (1972), 132 - 243. Moskow (in Russ.).

[3] Rechtschaffner R.: Technometrics, 9 (1967), 569 - 575.

[4] Sadovski A. - In: Soil Information Systems Technology (P. Burrough and S. Bie, Eds.), 73 - 78. Pudoc, Wageningen 1984.

GROUNDWATER QUALITY MANAGEMENT MODELING USING VIDEODIGITIZED DATA AND INTERACTIVE COMPUTER GRAPHICS

Daniel P. Loucks †

ABSTRACT

In the last few years, interactive modeling, digital cartographic analysis, and computer graphics have become important activities in many engineering and resource management organizations. Such computer-aided water resources planning systems are becoming increasingly powerful, both computationally and graphically, and are also becoming more portable and affordable. These systems will be important aids in the synthesis and analysis of resource management plans and policies. They also will provide an improved means of communicating technical data to professional engineers and planners as well as to citizen participation groups.

An important component of these interactive systems will be their use of digital, color-coded mapping. Since such mapping has long been regarded as an effective means of presenting spatially- and time-variant information associated with groundwater management problems, for example, its incorporation into the digital world of computer simulation analysis is an important step towards more effective groundwater and other resource planning.

This paper very briefly reviews some of our research in the development of interactive modeling-computer graphic systems applied to groundwater quality management problems, and describes the use of videodigitized maps and other images as a means of providing low-cost, cartographic information needed for effective interactive input of geographic and other model data and for the display of model results.

INTRODUCTION

During the last 10 years significant progress has been made in the fields of interactive groundwater quantity and quality modeling, digital cartographic analysis, and computer graphics. This progress has made it technically feasible to develop computer-aided planning workstations for examining groundwater management problems. These workstations can be designed to aid engineers and technicians in both quantitative and qualitative analyses. The incorporation of interactive computer graphic capabilities into such systems also can be an important aid in the difficult and time-consuming process of managing and displaying spatial- and time-varying data, and communicating this and other technical information to planners, managers, policy makers, and citizen advisory groups.

The basic research which has made it possible to develop such systems has occurred in a number of different fields. Besides the rapid development in modeling techniques and general computer technology, research and development in the areas of geographic information systems, remote sensing and image processing, large-scale data base management, and management information systems have all served as contributions to the design of computer-aided groundwater management modeling.

As computers and interactive computer graphic techniques were developed to handle

† Professor, School of Civil and Environmental Engineering, Cornell University, Ithaca, New York 14853, USA.

large geographic and other data bases, research began in the application of this technology to the planning and management of water and other natural resources. Since the mid-1970's systems of interactive simulation and optimization models coupled to computer graphics have been developed and applied to surface and groundwater and related land and environmental resource problems and issues. This research has demonstrated that substantial improvements in the ability to explore, analyze and synthesize alternative plans and policies can be achieved through the use of this technology [see references 1 through 9].

One of the accomplishments of this research has been the development of a prototype computer-aided groundwater management modeling system which incorporates models used by various government agencies and other organizations responsible for groundwater management. A single data base definition and management module interfaces with numerous groundwater quantity and quality models, including those commonly used by the U.S. Geological Survey and the Environmental Protection Agency. In addition to these models, the interactive modeling system contains many cartographic analysis capabilities which are used as aids in data definition, data management, and display and analysis of model results. The models are designed to help define, at any location in an aquifer, the probability distributions of various constituent concentrations at a given time, or of the time it takes for a given constituent concentration to occur. Optimization modeling is used to find how a given amount of money can be spent in obtaining data that will maximize the reduction in the uncertainty of these constituent concentration predictions.

USE OF COLOR-CODED MAPPING

An important characteristic of the interactive system is its use of color-coded mapping. Color-coded mapping has long been regarded as a very effective means of presenting spatially-variant information. For governmental agencies and private firms engaged in groundwater management and protection, large amounts of time and money are often allocated to procuring, storing, updating and manipulating model and other data. It is now feasible, and rapidly becoming economically necessary, for many of these model users to integrate electronic mapping into their operational procedures for inputing, editing, and displaying data.

It is certain that the next decade will see a major shift in how mapping is prepared and distributed. Already the major governmental agencies in the U.S. with mapping responsibilities have established digital cartographic groups, and others are attempting to integrate digital cartographic information into their operations. Based upon these activities and the many advances being made in the hardware and software aspects of computer science, an important opportunity is arising for smaller agencies and firms to adopt these techniques. This opportunity is being brought about by advances in several areas, including: (1) low-cost computer memory and associated processing devices; (2) dependable, high-resolution raster color displays used as information system/user interface devices; (3) "on-line" and "off-line" computer storage technology; and (4) digitizing technology and methodology.

Our work in both surface and groundwater management modeling has involved the use of computerized video base maps that serve as a versatile and effective medium for data presentation. Computerized base maps, created by videodigitizing aerial photography,

USGS 7.5-minute contour and land-use maps, or other suitable hardcopy, contain the complex "extras" - the high school track, the interstate's cloverleaf, parking lots, fields under cultivation - that provide for instant recognition and visual georeferencing. This type of mapping has been used for preparing model input data to resource evaluation models and for displaying the analytical results of those models.

Videodigitizing is a computer data entry technique that uses a video camera and specialized computer hardware. It can be used to convert (digitize) a representation of a photograph, map, or other image into a computer-compatible form. For color mapping using a non-color video camera, a triple-scan process is used. This process captures full color images by separately digitizing three monochrome images through color separation filters and constructing a composite full color image.

Using a videodigitizing camera and associated computer hardware and software, maps can be digitized and represented by a matrix of numbers in computer memory. Each matrix position, or pixel, corresponds to a small area on the scanned map. Each pixel is represented by three numbers which indicate the intensity of each of the primary additive colors (red, green, and blue) sensed by the video camera from that small area on the map. The resulting digital mapping can be displayed using standard color graphics equipment.

The digital mapping produced by videodigitizing can also be used as a source of information for a second step, interactive data classification. This is a process of creating a data base of site-specific attributes from the digitized image. Having this data base in the form of color-coded geologic, soils, and land-use maps, one is able to create computer-generated overlavs of various data types, including tabulations and color-coded maps of various groundwater model parameter values within a specified aquifer boundary. This capability is an important aid when estimating model parameter values such as the hydraulic conductivity, aquifer depths, etc.

Digital mapping is also useful in the display of results from computer groundwater simulation and optimization models. We have displayed spatially-variant results from simulation models (such as groundwater heads, flows, and concentrations of contaminants) as computer-generated contours or color-coded areas overlain by a transparent digitized map of surface features. Such displays are useful for readily identifying the location of particular aquifer conditions, and in understanding, for example, how different surface land-use practices or different hydrologic or geologic conditions affect groundwater quantities and qualities. One additional use of our interactive system has been in the combining and simultaneous display of model results and other data. For example, using the computer-generated display of areas having excessive toxic contaminant concentrations together with land use, demographic, or other data, it is possible to obtain a visual impression of the potential health risks in various areas.

Despite the benefits realized through the use of interactive modeling with videodigitized maps, agencies needing a large amount of "hardcopy" mapping currently may see the cost of digitizing as a disincentive to adopting digital mapping techniques. Since 1978 the writer and his colleagues have been developing and using videodigitizing techniques, and have examined the feasibility of using low-cost techniques for digitizing color imagery. The paper to be presented will review some of the promises and problems of using videodigitizing to develop a cartographic groundwater modeling data base. In the limited space available for this summary paper, a discussion of this more unique aspect of our interactive groundwater modeling work is not possible.

CONCLUSIONS

This paper has discussed very briefly the development of interactive computer-aided modeling systems for groundwater and other natural resources management. Undoubtedly in the future these systems will be based on small, portable, and affordable workstation designs. The use of color-coded digital mapping and the integration of that mapping into modeling activities will be prevalent. Videodigitizing techniques will continue to provide a cost-effective means of developing the required digital mapping and for converting hardcopy information into digital data.

In the talk, slides will be shown illustrating how interactive graphics has been used to define and calibrate groundwater quality management models, and how the results of such models for predicting aquifer characteristics can be displayed under transparent videodigitized maps or photographs of the surface for aiding in the exploration and synthesizing of appropriate groundwater quantity and quality management policies.

REFERENCES

[1] Fellows, J. "Real-Time Information System and Interface for Hydrologic Modeling." Proceedings, Joint Automatic Control Conference, June 1981.

[2] French, P.N., L.E. Johnson, D.P. Loucks and D.P. Greenberg. "Water Resources Planning Using Computer Graphics." J. Water Resources Planning and Management Div., ASCE, WR1, March 1980.

[3] Johnson, L.E. An Interactive Method for Developing and Evaluation of Reservoir Operating Policies. Ph.D. Thesis, School of Civil and Environmental Engineering, Cornell University, Ithaca, NY, January 1981.

[4] Kelly, J. "Graphics System Gives Advance Flood Warnings." Computer Graphics News, Vol. 3, No. 2, July/August 1983.

[5] Kleinstreuer, C. "Computer-Aided Analysis of Water Resources Systems." J. Technical Councils, ASCE, Vol. 107, No. TC1, April 1981.

[6] Loucks, D.P., P.N. French and M.R. Taylor. "Computer-Aided Resource Planning." Computer Graphics World, Vol. 5, No. 10, October 1982.

[7] Males, R.M. "Computer Applications in Water Resources Planning." J. Technical Councils, ASCE, Vol. 108, No. TC1, May 1982.

[8] Taylor, M.R., P.N. French and D.P. Loucks. "Interactive Graphics for Flood Plain Management." Proceedings, ASCE Water Forum '81, August 1981.

[9] Walski, T. and P. Schroeder. "Preliminary Design Tool for Corps Water Supply Study." Proceedings, ASCE Water Forum '81, August 1981.

APPLICATION OF METHODS OF SYSTEMS ANALYSIS FOR THE CONTROL OF GROUNDWATER WINNING PLANTS

G. Blasberg, H. Schwartz,[1] W. Vermeiren[2]

In this paper we propose three types of forecasting models which are useful for groundwater works of different size and with different regimes. These models are

a) analytical models originated from analytical solutions of the geofiltration process,
b) time series models originated from statistical analysis of time series,
c) state vector models originated from analysis of inputs and outputs of the spatial discretized aquifer in consideration of the interior state of the system (hydraulic heads at discrete node points) and as well additionally initial state determination by methods of state estimation.

These models are parts of a software package which is useful for seconding the regular regime of water winning plants of groundwater works by prognostic computation of the groundwater state and flow.

Keywords: forecasting and control of the regime of water winning plants in groundwater works,
analytical solution of groundwater flow,
time series models, model reduction,
state equation model, state estimation,
software for control in water management.

Introduction

The increasing requirements to the water supply have to be secured now preferably by an intensive exploitation of the opened water storages and an appropriate regime of the plants because of the already everywhere great degree of utilization of the water resources.

Groundwater as the most important base of the municipal water supply is extracted by well groups in an enough permeable aquifer in the catchment area of water works.

Manifold meteorological and hydrological influences and different quantities of demand cause more or less large oscillations of the actual water storage and the actual and prognostic possiblities of extraction, respectively.

Here the objective of the optimal process control is the forecasting of a demand appropriable drinking water supply also in difficult

---

1/ Institute of Water Management
DDR 1190 Berlin, Schnellerstr. 140

2/ Central Institute of Cybernetics and Information Processes of the Academy of Sciences of the GDR
DDR 1086 Berlin, Kurstraße 33; POB. 1298

conditions of resources at always economical effectively policy control of operation of the plants.
Therefore a control model is needed which computes the dynamics in the aquifer forecastly on recharge and extraction conditions above all of the wells.
So a distributed parameter system with complicated dynamics has to be controlled.
In the G.D.R. already of many years for simulation of fluid dynamics in aquifers simulation models have been used now above all by way of computer programs.
Following we are going to present the forecasting part of these water technological control systems. The forecasting part is their main element wich is based upon one of different simulation techniques for computation of groundwater flow /1,2/. We shall show procedures of model formation and the application of those models.
You can make use of three different types of forecasting models seconding the special case of process control of wells and artificial infiltrations.
At this above all the various size and the different regimes of the plants are taken into consideration.

Analytical Models

Analytical methods are used for plants if there is enough information about the hydrogeologic parameters and the flow conditions. They base upon linearization of the fundamental differential equation by use of schematizations and simplifications with reference to the geohydraulic time constant and to the hydrologic initial and boundary conditions.

Time Series Models

Further on forecasting procedures originated from time recurrent multivariable input-output models were successful used to small and medium-sized groundwater works. The model formation happens by statistical analysis of time series (multiple regression) of long standing measuring and operating data of the operating time of the water works. These models are structured as follows

$$\begin{aligned} A(k) = {} & a_0+a_1A(k-1)+a_2A(k-2)\dots \\ & +b_1E_1(k-1)+b_2E_1(k-2)\dots \\ & +c_1E_2(k-1)\dots \\ & \vdots \end{aligned}$$

The parameter $a_i, b_i, \dots$, the dimensionality of the model with reference to the number of inputs E_j, and time lag (k-n) as well as statistical test and performance measures are determined optimally by computers. Especially, the small effort for model formation is advantageous.

State Equation Modells:

A third procedure in forecasting is founded on the application of state vector equations (differential equation systems). Such state-space representations arise from the analysis of the hydrogeological conditions and the conditions of adits and of the spatial discretization of the aquifer in consideration of the interior state variables of the process which are hydraulic heads and amount of water at discrete node points. These models are capable of taking into consideration precise knowledge of the aquifer and various modes of operation. Their application for the explained control task requires reduced forms of that complex model which describe the flow dynamics in groundwater sufficiently. The degree of simplicity of these forms depends on the control objective. Models originated from step responses as well as convolution integrals are suitable only for some few input-output relations.

Reduced models originated from the method of momenta /3/ have wide possiblities of application. The model structure

$$\underline{\dot{z}} = \begin{bmatrix} A_1 & & & \emptyset \\ & A_2 & & \\ & & \ddots & \\ \emptyset & & & A_q \end{bmatrix} \underline{z} + \begin{bmatrix} B_1 & & & \emptyset \\ & B_2 & & \\ & & \ddots & \\ \emptyset & & & B_p \end{bmatrix} \underline{u} \; ; \; \underline{x} = C\underline{z} + D\underline{u}$$

which is diagonally in blocks, the small dimension of the auxiliary state vector $\underline{z}$ (dim A_i in most cases (2,2) until (4,4)), and the decoupling with reference to inputs $\underline{u}$ results in a very effective model type and handling. Especially, this is seconded by computing the hydraulic heads as physical state variables $\underline{x}$ outside of the differential equation system by simple superposition of the auxiliary state variables and the inputs by assistance of C and D.

State Estimation

Progressive methods of systems analysis for state estimation /4,5/ permit by use of the reduced model to reconstruct the actual and complete state of the aquifer by weighted correction of the computed results based upon relative few measurements (only some few measurement locations and brief periods). By applying this methodology the general problem of determining initial states for forecasting computation has been solved and new criteria for lapses of measurements have been deduced.

Conclusions

The algorithms based upon the presented procedures of model formation are parts of a large software package for computing dynamics of the aquifer of the catchment area of water works /6/. So a powerful instrument has been created for seconding the regular regime of water works.

/1/ Busch, K.F., L. Luckner:
Geohydraulik
VEB Deutscher Verlag für Grundstoffindustrie
Leipzig 1972

/2/ Luckner, L., W. M. Schestakow:
Simulation der Geofiltration
VEB Deutscher Verlag f. Grundstoffindustrie
Leipzig 1975

/3/ Otte, G., C. Clauß;
Reduktion von Modellen linearer dynamischer Prozesse anhand verallgemeinerter Momente beschränkter Operatoren
27. IWK der TH Ilmenau, 25.-29.10.1982; A1.1-2

/4/ McLaughlin, D.:
An Assessment of the Relevance of Estimation Theory to Practical Water Resources Problems
Joint Automatic Control Conference 1980 Paper FA6-E

/5/ Blasberg, G., W. Vermeiren
Verwendung vereinfachter Modelle zur Steuerung des Grundwasserregimes eines Wasserwerkes
12. Jahrestagung "Grundlagen der Modellierung und Simulationstechnik", 7.-9.12.1983, Rostock, Vortrag 42

/6/ Blasberg, G.
Bewirtschaftung von Grundwasserwerken mit Kontroll- und Steuerprogrammen
KdT-Fachtagung "Wassergewinnung", 3.-4.12.84, Dresden

ANALYSIS OF REGIONAL WATER POLICIES IN LIGNITE MINING AREAS

Stefan Kaden[1)]

Summary. Regions with open-pit lignite mining frequently are characterized by conflicting interests of the mining industry, different water users, and of environmental protection authorities. There is an apparent need for the analysis of long-term policies to reconcile those conflicting interests. Such an analysis should be based on a policy-oriented decision support model system. The development of such a system for a test region in the German Democratic Republic (GDR) is part of the research work in the Regional Water Policies project carried out at the International Institute for Applied Systems Analysis in collaboration with the Institute for Water Management, the Dresden University of Technology and the VEB Braunkohlenkombinat Senftenberg, GDR.

The paper gives an overview on the methodological approach for the decision support model system under development and demonstrates its application for the test area.

Keywords. Water policies, open-pit mining, multicriteria analysis

1. Introduction

Due to the complexity of the socio-economic environmental processes in open-pit lignite mining areas, the design of water management strategies and water use technologies as well as mine drainage can only be done properly based on mathematical models. For short-term control and medium-term water management as well as the design of drainage systems (local problem) qualified methods and models exist [2]. Thereby, the complex interdependencies of the system are partly neglected. However, there is an apparent need for the development of methods and models supporting the *analysis* and *implementation* of long-term regional water policies, to reconcile the conflicting interests in open-pit lignite mining areas.

Two principle problems have to be solved:

1. *To find "good" long-term strategies oriented towards achieving a proper balance between both national and regional economic needs, regional social needs and the regional preservation of the environment.*

2. *To find and realize controlling policies in order to direct the regional development according to the estimated "good" long-term strategies.*

The fundamental tool for such analysis is an appropriate decision support model system (DSMS) suitable for analysing long-term regional water policies. For details on the study see [3].

2. Methodological Approach

The process of lignite mining and its impact on the water resources system is highly dynamic. Such problems have to be approached by time discrete dynamic systems models. The step size depends on the variability in time of the processes to be considered, on the required criteria and their reliability, and on the frequency of decisions affecting the systems development. Taking into account the policy making reality related to long-term regional water management two different step sizes discretizing the *planning*

1) International Institute for Applied Systems Analysis, A-2361 Laxenburg, Austria

horizon T (of about 50 years) are of major interest:

- the *planning periods* as the time step for management/technological decisions, (e.g. water allocation, water treatment, drainage technology)
- the *management periods* of one month for management decisions within the year related to short-term criteria as the satisfaction of monthly water demand (the classical criteria for long-term water resources management).

The discretization of the planning horizon into a restricted number of planning periods enables principally to apply optimization techniques for multicriteria analysis. Small time steps (for instance, $\Delta T_j = 1$ year) for the planning periods are favourable from the point of view of the evidence and accuracy of model results. Otherwise the number of planning periods should be minimized with respect to the available methods for multicriteria analysis, and computational facilities. As a compromise variable planning periods are used starting with one year and increasing with time. Taking into account the uncertainties of long-term predictions of model inputs (water demand, decisions on investment, etc.) and the required accuracy, decreasing with time, this approach is quite reasonable.

For monthly time steps the application of any optimization technique becomes unrealistic. To study monthly systems behaviour systems simulation is the only applicable tool. Furthermore, this simulation opens an easy way to consider stochastic inputs (hydrological data, water demand, etc.). Based on these assumptions a heuristic two-level model system is under development consisting of

- *planning model* for dynamic multicriteria analysis for all planning periods in the planning horizon
- *management model* for the stochastic simulation of monthly systems behaviour in the planning horizon.

For the planning model a nonlinear multicriteria programming system has been developed, using the reference point approach [6]. The method is based on the idea of "satisficing". Starting from aspiration levels of decision makers (reference points), efficient responses are generated (Pareto points "closest" to the aspiration levels). The best solution considering the preferences of the decision maker can be obtained by correcting the aspiration levels in an interactive procedure. The principle use of the method is illustrated in Figure 1 for two objectives.

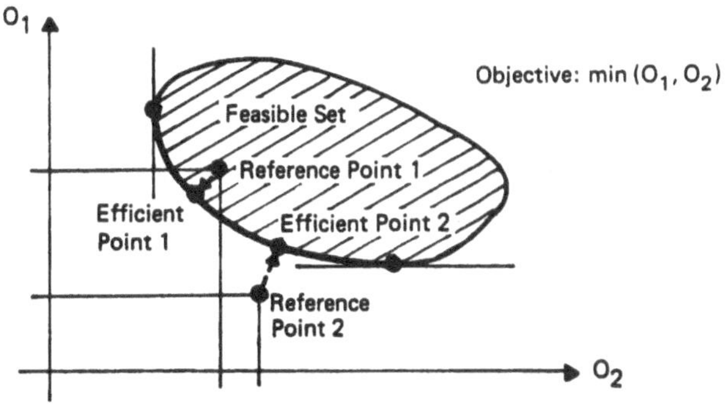

Fig. 1: Illustration of the reference point approach

In the case of many criteria the reference point procedure and the comparability of solutions might become complicated for decision making. For this reason the DSMS renders the interactive determination of criteria to be minimized, for the remaining criteria their bounds are considered.

The planning model is applied first, resulting in an efficient solution for planning periods. The determined control variables are used to estimate the parameters of a deterministic rule for the management model. Based on that, the *management model* serves as a stochastic simulation model simulating monthly systems behaviour. The Monte-Carlo Method is used to generate random inputs. From this simulation we obtain empirical distribution functions or frequency distributions for the monthly systems behaviour.

The most favourable case in running both models would be if the deviations between the results are negligible and the decision maker is satisfied with the results. Otherwise the planning model has to be used again with changed aspiration levels, or, the deterministic management rule has to be corrected.

The submodels for environmental and socio-economic processes to be studied have to be characterized by two major features. On the one hand, they should be simple enough mathematically to be integrated in a complex model system suitable for an interactive use. On the other hand, they have to reflect the important processes with an accuracy being necessary for appropriate decisions. Obviously, these features may be contradictory and a compromise has to be found. Depending on the state-of-the-art of modeling of a given subprocess, the availability of comprehensive models and data, different methods for the development of submodels have to be used [3], [4], [5].

In the past years the revolutionary development in electronic data processing has opened completely new possibilities for model applications in the decision making for large-scale, long-term planning. It is well-known that models for such purposes in the past did not find a wide application and impact in real policy analysis. As the main causes the following points might be seen:

- Modeler tried to *solve* long-term planning problems, anticipating decisions of the decision makers, neglecting subjective criteria in the decision making process.
- Generally, the developed models had to be used by specialists, the decision makers did interact with the model only through those specialists.
- The models frequently did not answer questions asked directly by the decision makers.

"The question, thus, is not whether to model, but how, and, most importantly, how to interface models with our more traditional ways of planning and decision making" [5]. To be accepted and used by the decision makers a Decision Support Model System must fit in the decision making reality (compatibility with common planning and decision making practice), and it has to be user-friendly, reliable, robust and credible.

The ongoing study is oriented towards those goals. With the methodological approach described above the policy making reality is reflected sufficiently, as we believe. The model system focuses on the necessary decisions and common criteria for long-term water management. The underlying time discretization corresponds to the common planning practice.

Based on the reference point approach for multicriteria analysis coupled with a stochastic simulation the model system is prepared for an interactive use. In addition the model handling and data management has to be designed interactively and user-friendly. The following aspects are considered:

- hierarchical data base (input and output data) with a robust screen oriented data display and editing system
- style and language of model use according to the planning and decision making reality
- use of computer colour graphics for visual display of model results.

The use of the hierarchical data base is menu-driven. Each data base level characterizes a menu and the user can either move downwards according to the menu or upwards to the previous level, or return to one of the models. For the data editing simple screen editor has been developed. Data checks realize the graceful recovery from failures. For the menu description as far as possible linguistic elements are used according to the practical language.

For the visual display of model results a flow chart representation of the area studied is used on a colour monitor. The flow chart is similar to Figure 2. The water quantity (flow) is characterized by the thickness of lines, and the water quality by the colour. To compare the criteria of different scenarios bar charts may be used (Figure 3).

3. The GDR Test Area

The test area is located in the Lusatian Lignite District in the lowlands of the south-eastern part of the GDR. Its area amounts to approximately 500 km^2. A scheme of the test region is given in Figure 2.

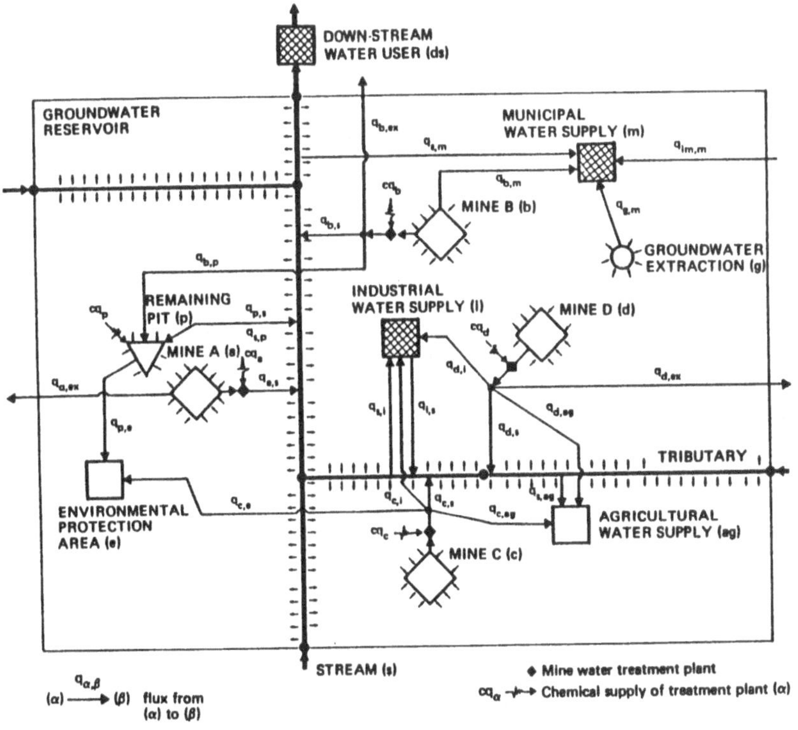

Fig. 2: Scheme of the test region

In Figure 2, the water users under consideration as well as the important decisions on the systems development (water allocation and water treatment) are depicted. A detailed description of the Test Area and of the mathematical models is given in [3]. In Figure 3, a few computational results of the planning model are demonstrated as a bar chart. The charts compare the values of some indicators of systems development depending on different aspiration levels.

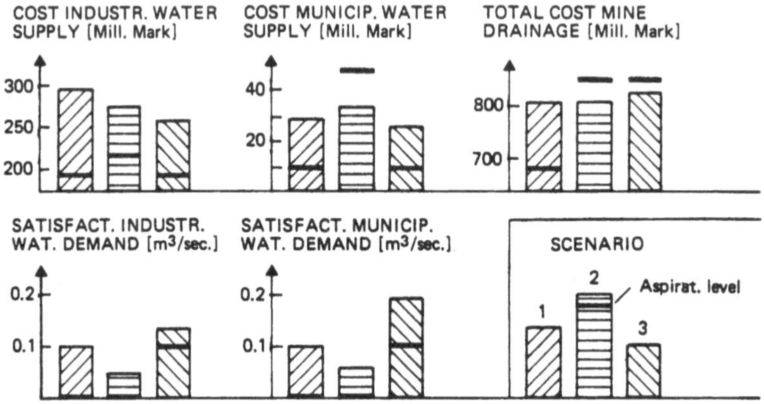

Fig. 3: Results of the planning model for the Test Area

4. References

[1] Fedra, K. and D.P. Loucks: Interactive Comparative Technology for Policy Modeling. Water Resources Research (1985), forthcoming.

[2] Kaden, St. and L. Luckner: Groundwater management in open-pit lignite mining areas. International Symposium on Groundwater Resources Utilization, Montreal, Canada, May 21-23, 1984, Proceedings I, pp. 69-78.

[3] Kaden, St. et al.: Water Policies: Regions with Open-Pit Lignite Mining. Working Paper, WP-04-85, International Institute for Applied Systems Analysis, (1985).

[4] Luckner, L. et al.: Methodology for the development of simplified regional models of water quality processes. Collaborative Paper, International Institute for Applied Systems Analysis, (1985), forthcoming.

[5] Tiemer, K. et al.: Methodology for the development of simplified regional water models of water flow processes. Collaborative Paper, International Institute for Applied Systems Analysis, (1985), forthcoming.

[6] Wierzbicki, A.P.: A Mathematical Basis for Satisficing Decision Making. Mathematical Modeling 3 (1982), pp. 391-405.

ANALYSIS OF WATER POLICIES IN REGIONS WITH INTENSE AGRICULTURE

P.E.V. van Walsum
S.A. Orlovski
International Institute for Applied Systems Analysis
Schlossplatz 1, A-2361 Laxenburg, Austria

ABSTRACT. In many rural areas, agricultural activities (including factory farming) conflict with the interests of other users of the environment, and also cause a deterioration of the biological quality of the environment itself. The study to which this paper refers is focused on the analysis of regional regulation policies providing for a satisfactory balance between socio-economic development and evaluation of the environment in the long run. Methodologically, it is based on the use of a two-stage decomposition approach, which includes scenario analysis in the first stage and policy analysis in the second. In this paper a brief outline is given of the scenario analysis module and the way it can be used as part of an interactive decision support model system.

INTRODUCTION

Intense socio-economic development in many regions of the world puts an increasing pressure on the environment both by consuming natural resources and by discharging pollutants that are hazardous to the population and to natural ecosystems. A substantial part of these impacts takes place through regional natural water systems. Here we consider regions where agriculture is the dominating activity.

Clearly, other water users in an agricultural region, like industry, the population, etc., should also be considered. But rather than going into their structural details, it suffices to consider them only in terms of demands that they make on the quantity and quality of water resources.

CONCEPTUAL AND METHODOLOGICAL FRAMEWORK

Hierarchical Institutional Structure of a Regional Socio-Economic Subsystem

We view a regional system under study as consisting of two major parts: the environmental subsystem and the socio-economic subsystem. Typically, regional socio-economic subsystems are of a complex hierarchical structure. The lower level elements (users of the environment) of this subsystem are those directly interacting with the environment. These interactions depend upon the production technologies implemented by the users, and they use these technologies according to their preferences. They are therefore often focused on local goals and not coordinated with each other. On the other hand, the upper level elements of the socio-economic subsystem (governmental, regional agencies, etc.) have preferences more closely reflecting the regional perspectives.

No formal description can encompass all the aspects of a real socio-economic subsystem. In this study, we use a simplified two-level representation of the socio-economic subsystem of the form shown in Figure 1.

Decomposition Analytical Approach

The "first stage" of our analysis is directed towards generating trajectories of the potentially rational development of the system under study. No behavioural aspects of the lower level elements are considered explicitly at this stage, and the analysis results in generating in some sense a reference trajectory of regional development. After having determined a reference scenario, the second stage of analysis is concerned with

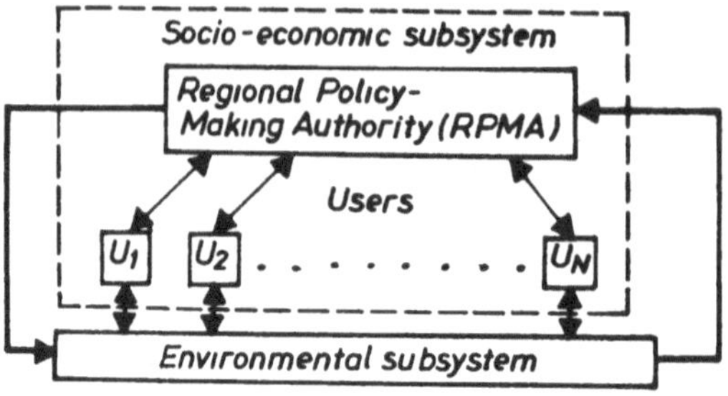

Fig. 1: Regional hierarchical structure

the search for those feasible regulation policies that influence the behaviour of the users and by doing that can direct the development of the whole system along the lines specified by the reference scenario obtained at the first stage. Schematically, this decomposition analytical procedure is illustrated in Figure 2.

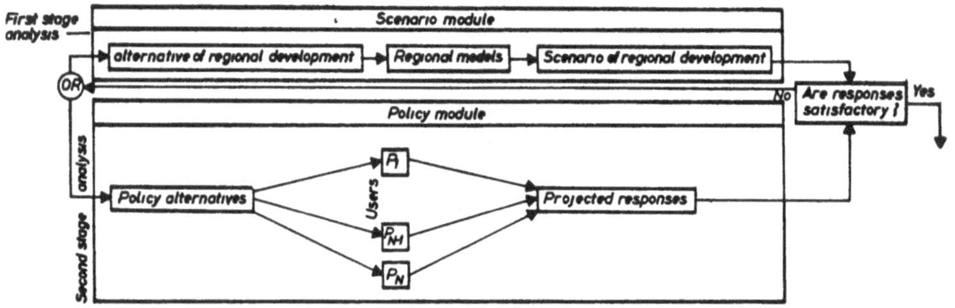

Fig. 2: Schematic of decomposition approach

The scenario module for performing the first stage analysis should consist of an integrated system of "simple" submodels with which screening analyses can be made. Simple submodels should be supported by more comprehensive models for the more accurate estimation of "promising" scenarios. After having given a short description of the region that this study focuses on as a practical example, we give an outline of the concrete realization of an integrated system of submodels to be used in the scenario module.

THE SOUTHERN PEEL REGION

Environmental Setting

The Southern Peel is an undulating area of about 30,000 ha in the south of the Netherlands. The lie of the land varies in altitude between 17 and 35 m above sea level. A major feature of the hydrogeology is the presence of a fault that divides the area into a Western part - the "Slenk" - which has a deep hydrological basis at 300-500 m below ground level, and an Eastern part - the "Horst" - which has a shallow hydrological basis at 8-36 m.

A large part of the area used to be covered by a layer of peat that grew as a consequence of extremely high groundwater levels. Most of the peat has been delved and used as fuel for heating The remaining peat areas are now protected from exploitation.

because of their value as recreation or nature areas. The nature areas can only keep their value if high enough groundwater levels are maintained.

Human Activities and Their Impacts

Roughly half of the agricultural land is used as pasture for dairy cattle; the remaining area is used for growing a variety of crops, of which maize is the most important one, followed by sugarbeets, potatoes and cereals. Farmers try to reduce moisture deficits by water conservation, subirrigation, and sprinkler irrigation.

As is characteristic of all regions with intense agriculture, farmers in the Southern Peel attempt to optimize the nutrient supply conditions of their crops. Both chemical fertilizers and animal slurries are used for this purpose. The soils in the Southern Peel are sandy and therefore have poor purification and fixation capabilities. So the excess nitrate is easily leached, thus increasing the nitrogen load on groundwater.

Most of the surface water pollution by agriculture is through surface runoff that has high concentrations of nutrients. Water for the pollution, industry, and factory farming is extracted from the aquifers in the Slenk area by public water supply companies. The resulting lowering of groundwater tables decreases the productivity of agriculture and creates deteriorating conditions in nature areas.

The quality of the extracted groundwater is still excellent, and nitrate levels in wells are hardly increasing yet. But measurements in phreatic aquifers under agricultural lands indicate that the concentrations in water "that is on its way to the wells" are alarming. A schematic diagram of the main impacts of human activities in the Southern Peel is given in Figure 3.

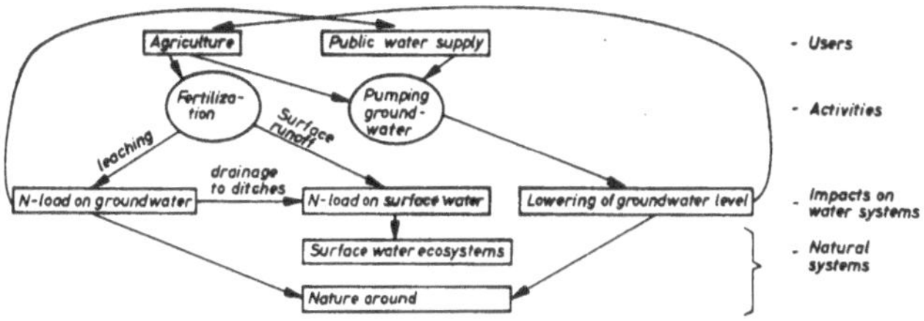

Fig. 3: Main impacts of human activities

SCENARIO MODULE

Water Quantity Processes

In the present conception the dynamics of groundwater processes is described by linear equations that have a (stochastic) basic component related to the unperturbed state of the system and components for the influence of control variables (e.g. the groundwater extractions for public water supply).

As a selection principle for "discarding" scenarios that are not feasible owing to the limited availability of moisture in the rootzone, the requirement is made that the average total moisture content of the rootzone in a subregion must not be lower than the total moisture contents required by the separate subtechnologies.

Nitrogen Processes in the Soil

Each technology that uses land has a specified level of the amount of nitrogen that is required for crop growth. This nitrogen can come from different sources, i.e. chemical fertilizer and various types of animal slurries. A linear function is used for balancing supply and demand of nitrogen available for crop uptake. Nitrate leaching to the phreatic groundwater is also described by a linear function of chemical fertilizer and animal slurry applications. A denitrification function (linear in the first level model, nonlinear in the second level one) is used for describing the influence of the depth to the (shallow) groundwater table.

Water Quality Processes

Groundwater quality

The "simple" model for (steady state) groundwater quality processes is based on the following assumptions:

1. All deep aquifers over the whole (Slenk) region can be regarded as one mixing cell. The phreatic aquifers are separate mixing cells that overlie the deep aquifers.
2. Decomposition of nitrate in the deep aquifers can be taken into account by a factor α (that depends on the organic matter content of the subsoil).
3. Adsorption and dispersion can be neglected.

Surface water quality

Regression formulas are used for describing the concentration of nitrogen in surface runoff during winter as a function of chemical fertilizer and animal slurry applications. The concentration of nitrogen in phreatic groundwater that drains to surface water during winter follows from the groundwater quality model.

Public Water Supply

If the demands of public water supply are given, then the total of the extractions in the subregions must satisfy respectively for the winter and summer period these demands plus the amount required for factory farming.

Natural Ecosystems

Ecological processes in nature areas are influenced in a complex way by the groundwater regime. We do not attempt to describe the dynamics of these processes. Instead, we assume that the conditions in "critical" years, e.g. years with "dryness" that on average occurs only once per decade, are good indicators for the level of satisfaction of water demand of nature areas.

COMPUTATIONAL IMPLEMENTATION

The integrated set of first level models are implemented within a multiobjective LP-framework: N-1 of the N objective functions are included in the form of constraints for which the user can supply the right-hand sides. The Nth objective function is then minimized by the LP-algorithm. This function has been taken as the sum of the investments (minus liquidations) required to reach the "target state" from the "current state" of the regional system.

Stochastic phenomena (i.e. weather conditions) are treated in a simple probabilistic manner.

For easy interpretation of the results an interactive colour graphics display system has been developed. Coupling of the first level models of the scenario module with more comprehensive models is being undertaken.

REFERENCES

Germeyer, Yu. (1976). Games with nonantagonistic interests, "Nauka", Moscow (in Russian).

Orlovski, S.A. and P.E.V. van Walsum. (1984). Water Policies: Regions with intense agriculture. Working Paper WP-84-40, International Institute for Applied Systems Analysis, Laxenburg, Austria.

Vatel, I.A. and F.I. Ereshko. (1973). Mathematics of Conflict and Cooperation. "Znanie", Moscow (in Russian).

Verdonschot, M.G. (1981). Invloed van bemesting en grondwaterstand op de processen in bodem en water. Nota 1250, ICW, Wageningen, Netherlands.

CONRTOL AND SIMULATION OF MULTIQUALITY WATER SUPPLY NETWORKS

Martin Reike
Lehrstuhl für Meß- und Regelungstechnik
Ruhr-Universität Bochum
F.R. Germany

SUMMARY

The present paper deals with the control and simulation of special water distribution networks. These networks operate with different sources of water. The quality can be adapted to the consumers demands by dilution and mixing. For the control of the system a hierarchical two level strategy is suggested. The first level determines the optimal flow rates in the net, the second deals with the dynamic control of the system. A digital simulation model was elaborated for a special case study. With the help of this model some problems of the control and simulation will be discussed. A typical simulation run will also be presented.

KEYWORDS: irrigation, water quality control, digital simulation, multiquality networks

1 INTRODUCTION

The future development of the industrial and agricultural production in the arid and semiarid areas of the world depends decisively on the availability of sufficient amounts of suitable water. Therefore it is necessary to pay more attention to the use of poor quality water, as brackish, salty or sewage water. An expensive purification of these low quality waters is economically not rational. The only way to use these types of waters is to dilute them with high quality waters, so that the quality can be adapted to the demand of industrial and agricultural consumers.

These considerations led to the development of large water networks that manage the distribution of the water as well as its dilution. Planning, design and control of these so called distrilution-networks+) brings up numerous problems which have not yet been examined. The research of complex water distribution networks was mainly restricted on networks in the industrial nations where plenty of water is available and where there is only need for single quality water supply networks. These networks deliver only drinking water without regarding the demands of consumers. This strategy is useful to industrial nations with plenty of water, although the water becomes more and more expensive due to ecological damage. Therefore even here it might sometimes be advantageous to build up multiquality or at least dual water supply systems.

The research results that have been found in connection with single or dual water supply systems can not generally be applied to distrilution-nets. On one hand, the new problems of the distrilution-nets arise from the numerous different water qualities in the net. On the other hand, there are many dislocated sources and consumers which make it more difficult to divide the system into smaller and easier to handle subsystems than is the case when ordinary single quality water networks are concerned. Therefore the research of distrilution-networks has to deal with relatively large and complex networks.

In the development of distrilution-networks three components have to be considered: Planning, design and control. Especially the control of distrilution-nets has hardly been examined. The present paper deals with the control of distrilution-networks. First of all one has to define the task of the control. Because of the marginal experience with building and operating of distrilution-nets it is often difficult to distinguish between the duty of the control-engineer and the planning- and design-engineer. The only clearly defined issue of the control engineer is a set of sources and consumers within an existing network. This distrilution-

+) Distrilution = Distribution + Dilution

net contains all pumps, valves and sensors necessary for the control. In this case the control system has to take care of a postulated flow distribution in the network in order to meet the quality and quantity demands of the consumers.

2 HIERARCHICAL CONTROL

For solving the control problem a two level control structure seems to be useful. The first (upper) level determines a stationary distribution of flowrates in the net which optimally fulfills a cost function. The second (inferior) level is responsible for the dynamic behaviour of the system, which means that it has to achieve the stationary flow rates by controlling pumps, valves etc.. The two level control subdivides the problem in one static and one dynamic problem. This makes it possible to clearly define the function of both levels. The optimal flow rates of the first level of control should be calculated by a centralized and superior computer. The dynamic control of the flowrates can be managed by centralized as well as by decentralized micro-computers or analog controllers. Fig. 1 shows the structure of the two level control.

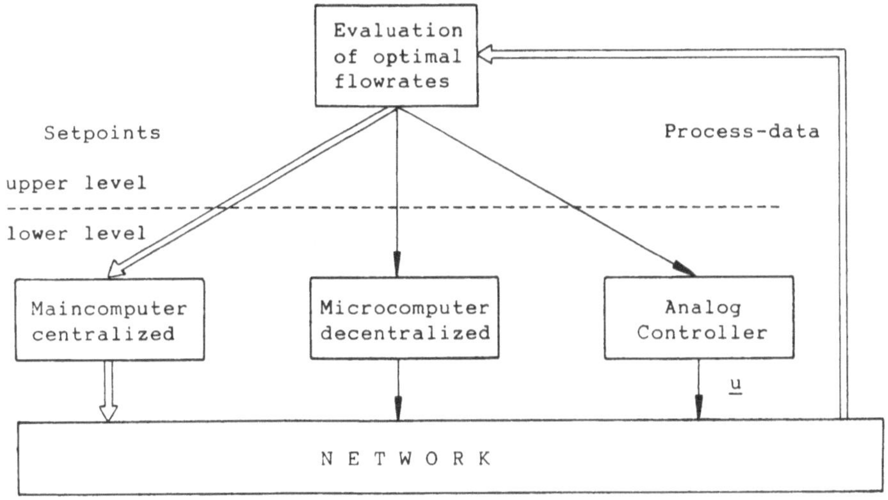

Fig. 1: Structure of Two Level Control

For the determination of optimal flow rates the net is described by a special static model. By means of the graph theory all the connections between every source and every consumer are determined. When minimizing an objective function the flow rates on these paths are obtained. On one hand, the objective function regards the cost for water and transport, i.e. mainly the cost of pumping. On the other hand, the effect on the crop yield of the irrigated areas caused by deviations between setpoint and actual value of quality and discharge is evaluated. Superposing the flowrates on the paths between sources and sinks one gets the optimal flow distribution of the network. This distribution has to be transformed into the setpoints for the lower level of the hierarchical control, i.e. setpoints for pumps and valves.

3 CASE-STUDY OF AN AGRICULTURAL DISTRILUTION-NETWORK

A typical agricultural distrilution-network for a great farm complex in Israel was chosen as a case study. The system can be looked upon as representative for irrigation networks in arid regions where the quantity as well as the quality of the water plays a major role.

Fig. 2 shows a scheme of the system. The network contains 12 dilution junctions as well as numerous valves, pumps and storage ponds from which only the most important are shown. By means of simulation it was to be determined how useful they are and as a result of this how costly they should be equipped.

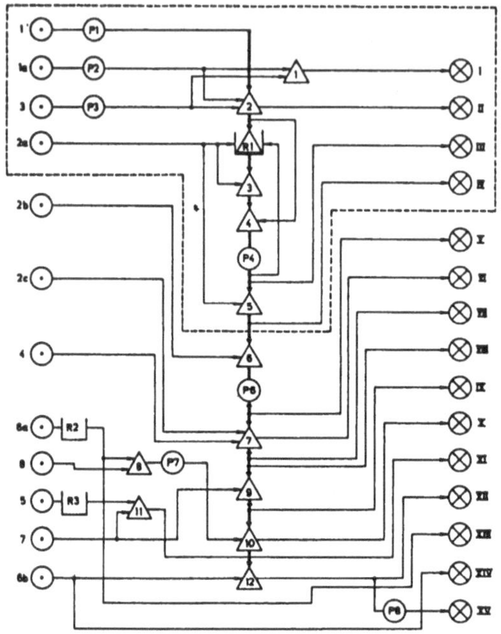

Fig.2: Case study of an agricultural distrilution-network

The network is fed by eight different sources of water.

1.) salty river water 600-1400 mg Cl/l
2.) water of a reservoir 250-300 mg Cl/l
3.) stored river water 500-1200 mg Cl/l
4.) salty springs 1200 mg Cl/l
5.) industrial sewage water 1250 mg Cl/l
6.) ground water 60 mg Cl/l max. 5000000 m^3/year
7.) fish pond return water 350-450 mg Cl/l
8.) rain fall 50 mg Cl/l Because of sediments no drinking water.

The consumers demand for water can be subdivided into four categories of different water quality.

1.) drinking water (XIV): Only water of source 6
2.) water for industry (XII) 50-400 mg Cl/l
3.) water for irrigation (I,III,IV,V,VIII,IX,X,XI,XIII,XV). The quality demand depends on the irrigated crops and the soil. 60-1000 mg Cl/l
4.) fish ponds (VI,VII) 300-500 mg Cl/l.

The demand of the consumers can change very slowly concerning the water quality, regarding the quantity it can change within hours time. Until today the control of the existing dilution junctions, valves and pumps is done manually. By means of a detailed simulation the efficiency of complex control structures and the installation resp. improvement of the dilution junctions should be determined.

4 • SIMULATION AND CONTROL

For the framed part of the system (fig.2) a simulation-programm was developed by the simulation language SIMUL 2. It is a blockoriented simulation language. The switching diagram is similiar to those that are used in case of analog simulation.

For determination of the pressure distribution in hydraulic networks commonly the Newton-Raphson iteration is used. In connection with SIMUL 2 a special iterative procedure was developed which can be seen as equivalent to the Newton-Raphson procedure concerning speed of iteration and accuracy. One problem of the simulation and of the control originates from the dominant and time varying dead times. For simulation of these dead times special algorithms had to be developed.

First of all the optimal flow rates in the network are determined (first level of control). To achieve these flow rates (second level) four pumps and seven valves are available. For the optimal control it is not always sufficient to control only discharge. Moreover, the postulated demand of consumers has to be satisfied. The optimal flow rates are only a guide line which comes from an open loop calculation. Deviations from the optimal flow rates are caused by parameter changes within the system. These are generally due to the differing quality changes of the sources. Although the optimization of the flow rates is repeated from time to time with updated parameters, smaller deviations should be compensated in the lower level of control. Therefore it is often necessary to build quality control loops, too. These introduce the problems of great deadtimes into the controlled system. To overcome these problems a predictive control was installed.

Another great problem for the control are the nonlinear limitations of the valves and pumps. Because the valve-settings and the valve lifting velocity are limited, the integrating part of the controller starts to run away from the actual value of valve setting and this causes severe disturbances in the system response. To overcome these problems special limited integrators had to be implemented in the controllers. Fig. 3 shows the improvement which can be achieved by using limited integrators compared to ordinary PI-controllers.

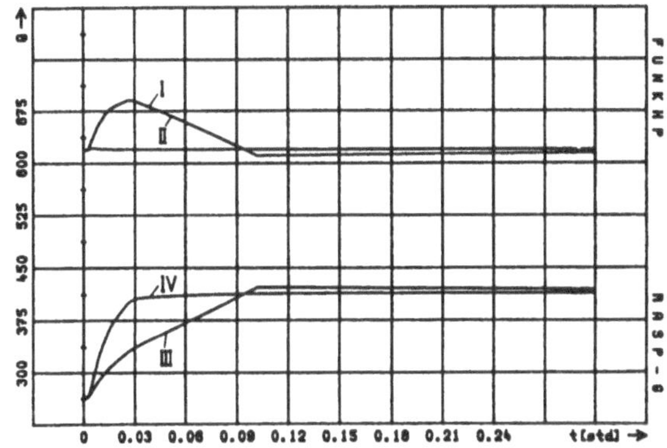

Fig. 3: Set-point change of flow-rate im pipe between dilution-junction 2 and pond R1.
I : Flow-rate of consumer II - ordinary PI-control.
II : Same as I, but limited integrator.
III: Flow-rate into pond R1 - ordinary PI-control.
IV : Same as III, but limited integrator in controllers.

/1/ Fasol, K.H.; Reike, M.: Simulation von landwirtschaftlichen Wasserverteil- und Vermischungsnetzen. ASIM 84, Wien, 1984

/2/ Pessen, D.; Reike, M.; Sinai, G.: Design and Simulation of Water Mixing Junctions in Irrigation Systems. Proc. Symp. Applied Control and Identification, IASTED, Kopenhagen 1983

This paper had been supported by the "Deutsche Forschungsgemeinschaft" (Fa 123/2-2).

MODEL-ASSISTED DECISION MAKING PROCEDURE FOR WATER QUALITY CONTROL OF LAKES AND RESERVOIRS

F. Recknagel, J. Benndorf, R. Kruspe [1] and K. Pütz [2]

The decision making procedure is based on the predictive dynamic model SAIMO which permits to simulate the dominant state variables of lake and reservoir ecosystems. The procedure consists of the following methods: expert's estimates, scenario analysis, stability analysis, dynamic optimization and risk analysis. In a case study it is applied to the eutrophication control of the Eibenstock reservoir in the GDR.

Key words: decision making, eutrophication control, predictive model, scenario analysis, structural stability, cost optimization, case study.

1. Introduction

Water quality control of lakes and reservoirs represents a complicated multi-objective decision making process which is characterized as follows:

1. It is aspired to meet the two primary goal functions: minimization of costs and maximization of the water quality.
2. Subjects of control are aquatic ecosystems. Consequently, high complexity, heterogenity, nonlinearity and dynamics must be taken into consideration.

In this situation predictive models can serve as objective and powerful tools for decision makers.

An integrated decision making procedure was developed for the purpose of eutrophication control of standing waters /6/. It is based on the dynamic model SAIMO which has proved to be valid and predictive for a broad class of lakes and reservoirs with different trophic states and morphometrics /1/, /5/. In a deterministic manner and taking into consideration the internal control mechanisms, the model SAIMO includes the following state variables: dissolved orthophosphate, dissolved inorganic nitrogen, three functional groups of phytoplankton, zooplankton, detritus and dissolved oxygen.

2. Integrated decision making procedure

The general course of the decision making procedure is shown in fig. 1. It is including expert's estimates, scenario analysis, stability analysis, cost optimization and risk estimates. Expert's estimates serve for finding the nominal control set which should consist of feasible measures refered to the specific conditions of the lake or reservoir under consideration. With the help of scenario analysis it is possible to identify alternative control policies on the base of the nominal control set. By the investigation of the structural stability of the system behaviour it is aspired to obtain information for a stable con-

1) Dresden University of Technology, Department of Water Science
2) Water Management Authority "Obere Elbe-Neisse", Dresden

trol of alternative policies. The next step is directed to derive the optimal control trajectories for the prefered policy. For this purpose a dynamic cost optimization with state-space constraints is used. Finally it should be possible to estimate the risk of the prefered control policy. This can be done by means of Monte-Carlo-simulations for calculating the probability distributions for relevant risk factors.

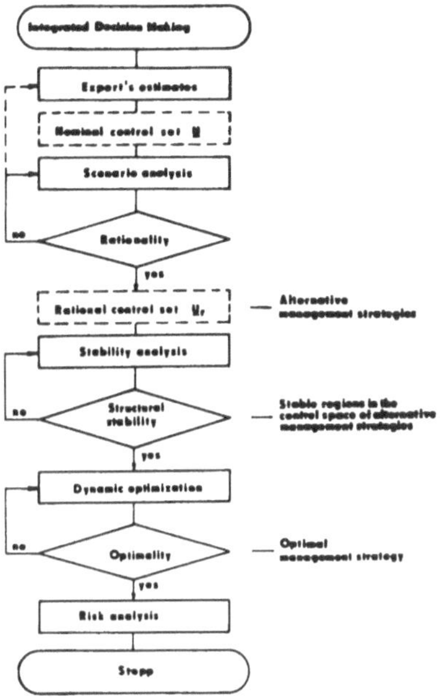

Fig. 1: Flow-chart of the integrated decision making procedure

3. Case study

The decision making procedure was applied to the eutrophication control of the Eibenstock reservoir in the GDR where risk estimates are not included. This reservoir is under construction and will supply several towns and communities with drinking water. Its catchment area is polluted by urban, industrial, natural and agricultural sources. To reduce the nutrient load to a level which should guarantee mesotrophic conditions in the reservoir, an extensive feedforward control policy was induced under consideration of pre-dams, sewage treatment plants and reconstruction measures /2/. With the help of the decision making procedure it has to be checked the effectivity of the following alternative eutrophication control policies:

1. External feedback control by a phosphorus elimination plant /3/.
2. Internal feedback control by biomanipulation of the food-web (see /8/) in combination with the artificial destratification of the water body during summer.

According to these policies on a maximal control level are defined scenario 1 and scenario 2 in the frame of the scenario analysis to provide point estimates of their effects. In fig. 2 the results of the scenario analysis by the simulated trajectories of the dissolved orthophosphate, phytoplankton and zooplankton are represented for the reference run and the specified scenarios. The evaluation of the scenarios is facilitated by the hand of the histograms for the integrals of the trajectories in the vegetation period of the year. Fig. 2 offers

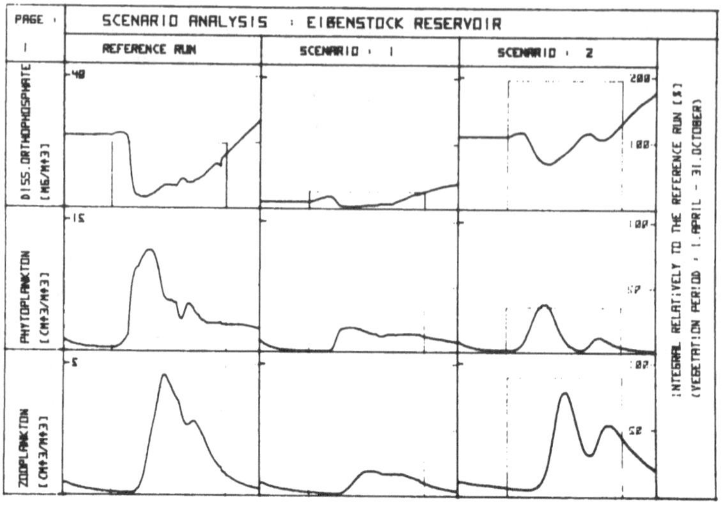

Fig. 2: Scenario analysis for the Eibenstock reservoir.
Scenario 1: Phosphorus elimination in the water inflow by 90%.
Scenario 2: Combination of biomanipulation and artificial destratification of the water body.

that both scenarios would be appropriate to reduce the phytoplankton biomass to about 35% relatively to the reference run by which the oligotrophication of the reservoir would be achieved. But there is a difference regarding the internal orthophosphate concentrations: while in scenario 1 the orthophosphate amounts to about 25%, in scenario 2 it amounts to about 200% relatively to the reference run. The internal nutrient accumulation in scenario 2 bears the potential danger of exceeding the desired trophic state (e.g. under extremely meteorological or toxicological influences).
In the next step the structural stability of the ecosystem behaviour was investigated assuming the dynamic control of the artificial destratification of the water body during the summer period. For this purpose the catastrophe theoretical approach as proposed in /7/ was applied. Fig. 3 shows the isocline surface of the zooplankton biomass. The control space is defined by the change of the external phosphorus load in the range of 10% and 200% relatively to the reference load (Z-axis) and the decade by decade extension of the artificial destratification (X-axis). Inspite of the fact that zooplankton acts as an effective internal control factor within the food-web of aquatic ecosystems, it is desirable to sustain the zooplankton behaviour in a stable manner. As to see in fig. 3 the stable region of the zooplankton behaviour is restricted to a control time of the artificial destratification between 10 and 16 decades. Because of this result the phosphorus elimination in the water inflow of the reservoir as considered in scenario 1 (see fig. 2) seems to be the prefered control

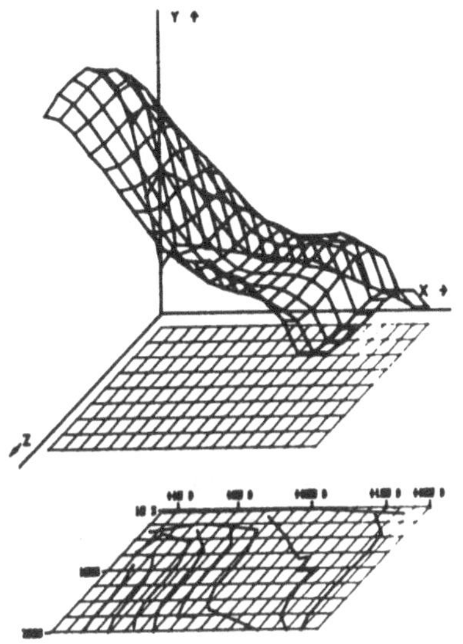

Fig. 3: Isocline surface of the zooplankton biomass (Y-axis) refered to the change of the external phosphorus load (Z-axis) and the dynamic control of the artificial destratification (X-axis). Isoline projections of the isocline surface (below).

policy for the purpose of a dynamic cost optimization. It was realized in the manner of a dynamic scalar optimization with state-space constraints (see /9/, /4/) where a direct coupling of a regularized NEWTON-method for minimization without derivatives with the model SAIMO was undertaken (see /6/). Fig. 4 reveals the phytoplankton trajectories, the control trajectories and the relative levels of operating costs for the uncontrolled reference run, for the permanent maximal phosphorus elimination by 90% and for the dynamic optimal phosphorus

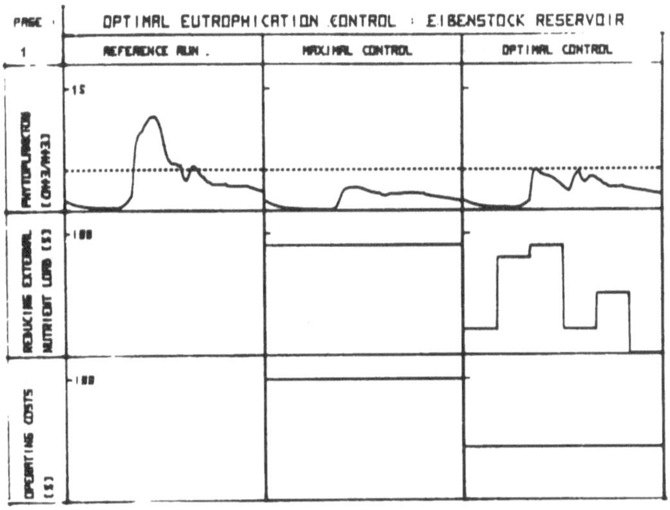

Fig. 4: Optimal eutrephication control of the Eibenstock reservoir by a dynamic phosphorus elimination in the water inflow

elimination. The control time of one year was divided into six constant control periods with a length of two months. The state-space constraint was defined by the phytoplankton limit value for mesotrophic conditions (marked by the dotted line in fig. 4) according to the water quality standard of the GDR. Fig. 4 offers that the determined control trajectory guarantees mesotrophic conditions in the reservoir and reduces the operating costs to about 40%.

4. Conclusions

The integrated decision making using the model SAIMO has proved to be an useful tool for decision makers in eutrophication control of lakes and reservoirs. In the next time the efforts will be concentrated to qualify the optimization procedure for considering more than one water quality criterion and different control variables in the sense of polyoptimization. On this way it should be possible to apply the procedure as a tool in the regional planning and management.

5. References

/1/ Benndorf, J. and F. Recknagel: Problems of application of the ecological model SAIMO to lakes and reservoirs having various trophic states. Ecol. Modelling 17 (1982), 129-145.

/2/ Benndorf, J., D. Uhlmann and K. Pütz: Strategies for water quality management in reservoirs in the German Democratic Republic. Water Quality Bulletin 6 (1981), 68-73.

/3/ Bernhardt, H. and J. Clasen: Oligotrophication of the Wahnbach Reservoir. Water Quality Bulletin 6 (1981), 74-78.

/4/ Kalceva, R., J. Outrata, Z. Schindler and M. Straskraba: An optimization model for economic control of reservoir eutrophication. Ecol. Modelling 17 (1982), 121-128.

/5/ Recknagel, F. and J. Benndorf: Validation of the ecological simulation model SAIMO. Int. Revue Ges. Hydrobiol. 67 (1982), 113-125.

/6/ Recknagel, F.: Angewandte Systemanalyse in der Wassergütebewirtschaftung von Standgewässern. Dissertation B. University of Technology, Dresden 1984, 1-169.

/7/ Recknagel, F.: Analysis of structural stability of aquatic ecosystems as an aid for ecosystem control. Ecol. Modelling (in press).

/8/ Shapiro, J., V. Lamarra and M. Lynch: Biomanipulation. An ecosystem approach to lake restoration. Contribution 143 from the Limnol. Res. Centre, Univ. of Minnesota, 1-32.

/9/ Wierzbicki, A.P. and S. Kurcyusz: Projection on a cone penalty functionale and duality theory for problems with inequality constraints in Hilbert space. SIAM J. Contr. Optimiz. 15 (1977), 25-56.

TWO ALGORITHMS OF OPTIMAL CONTROL OF THE WATER TREATMENT PLANT

Janina SZEBESZCZYK [1)]

The paper presents algorithms used to the calculation of the optimal production schedules for the water treatment plant. Optimized performance index is a pumping water daily cost. Optimization proceeds in two phases. In the first phase the approximated performance index is minimized. The optimal solution of the first phase is a starting point for the optimization realized in the second phase. In the second phase the optimized performance index contains the real pumping cost characteristics. In this phase two algorithms of optimization may be used: the separable programming algorithm and the multistage programming algorithm. The simulation results showed, that the application of proposed algorithms can significantly decrease the plant's energy costs.

Keywords : energy consumption cost, optimal schedule, nonlinear programming

1. Introduction

The stabilization of flows and water levels is used for the most popular strategies of the water treatment plant's control. Flow changes at the plant output and backwashing of filters are regarded as the main interferences [1] . The station control based on the daily schedules determined for predicted consumers demands, so as to minimize the daily electric energy costs has been proposed in this paper. The optimal schedule should satisfy all technological restrictions and coordination of flows between the pumping stations and purification objects.

2. The problem formulation

The typical water treatment plant consists of the raw water pumping station, the raw water reservoirs, filters, the pure water reservoirs, the pure water pumping station, the backwashing water pumping station. The main part of the plant energy cost is the cost of energy used for pumping of raw, pure and backwashing water. Raw and pure water pumping stations outputs can be regulated :

- by changing the configuration or number of working pumps and by the rotation speed control,
- by changing configuration or number of working pumps.

In the first case the power characteristic curves are discontinuous, nonlinear functions of their outputs. In the second case these characteristics are discrete. The output of the backwashing pumping station is usually regulated by pumps work time changing. In this case pumping station energy consumption is proportional to its output. Dividing 24-hour period into 1 equal parts of time Δt, the opti-

1) Silesian Technical University Gliwice Pstrowskiego 16, POLAND

mization problem can be formulated as : minimize

$$J = \sum_{k=1}^{l} C_k \sum_{j=1}^{3} E_{jk}(Q_{jk}) = \sum_{k=1}^{l} C_k \sum_{j=1}^{3} P_{jk} (\dot{Q}_{jk}) \ \Delta t \tag{1}$$

with respect to the control variables Q_{jk} for $k=1, \ldots, l$ $j=1,2,3$, subject to :

$$V_{1(k+1)} = V_{1k} + Q_{1k} - Q_{2k} - Q_{3k} \tag{2}$$

$$V_{2(k+1)} = V_{2k} + Q_{2k} - \sum_{n=1}^{s} Q^{o}_{nk} \tag{3}$$

$$V_z^{min} \leq V_{zk} \leq V_z^{max} \qquad \text{for } z = 1,2 \tag{4}$$

$$\dot{Q}_j^{min} \leq \dot{Q}_{jk} \leq \dot{Q}_j^{max} \qquad \text{for } j = 1,2,3 \tag{5}$$

$$\sum_{k=1}^{l_i} Q_{3k} \geq i \cdot V_p \qquad \text{for } i = 1, \ldots , p-1 \tag{6}$$

$$\sum_{k=1}^{l} Q_{3k} = p \cdot V_p \tag{7}$$

where: V_{1k} , V_{2k} - the volumes of water in the pure water reservoir and in the external reservoir at the begining of k-th interval, Q_{1k}, Q_{2k}, Q_{3k} - the volumes of water pumped by raw, pure and backwashing pumping stations in k-th interval, p - number of filters which must be backwashed in 24 hour period, V_p - volume of water necessary to backwash one filter, $\sum_{k=1}^{l} Q^{o}_{nk}$ - n-th customer water consumption.

3. The optimization algorithms

To eliminate many local optima and to decrease the dimension, the optimization problem has been solved in two phases. The approximated, convex performance index is used in the first phase. The following approximations of the pumping stations power curves have been used: linear approximation for the raw water pumping station and segmentary-linear, convex approximation for the pure water pumping station. Because of above assumptions the solution of approximated optimization problem can be solved by the separable programming method [2] . To every node of the segmentary approximation of function $E_{jk}(Q_{jk})$ the variable λ_{jki} $(i = 1, \ldots, r)$ is assigned. Then the problem may be considered as the linear programming task, which is solved basing on simplex algorithm at modified procedure of the selection of the variables to the base. The modification of conventional simplex algorithm reduces to the fact, that for every pair of indices (jk), in the base no more than two adjacent variables can exist. The linearisation used

in the separable programming increases the amount of decision variables but it enables solving the problem using very effective simplex algorithm. In the second phase the optimized performance index J contains the real pumping stations power characteristics. The ranges of raw and pure water pumping stations outputs are limited to the area around the first phase solution.
If the rotation speed control of pumps is possible (i.e. the continuous solutions of optimization problem are acceptable), the separable programming algorithm is useful.

If the changes of raw and pure water pumping stations outputs can be realized only by changing the number (or configuration) of working pumps, the multistage programming algorithm may be used. Algorithm is based on searching for optimum by the review of pumping stations outputs around the first phase solution. The vector, which co-ordinates are upper or lower values of pumping stations outputs in all scheduling periods is called the stage. The search begins from the stage, which state contains only lower values of outputs. In every next stage the states generated from the states of the previous stage by replacing one of the lower values by the upper value are analysed. The decision concerning the states, which may take part in the further generation is taken according to the rules :

1°. If in the given stage neither state performs the restrictions, then all states from this stage take part in further generation.

2°. If in the given stage is the "momentarily optimal" state (i.e. the state which has the lowest value of performance index from all the states of the stage and which performs the restrictions), then only these states take part in the further generation, which have the performance index values lower than the "momentarily optimal" state. Only for these states it is possible to generate (in the next stages) the states performing restrictions and with performance index value lower than "momentarily optimal" state.

Proposed algorithm makes possible to realize the review with significant savings of computer memory and time of calculations.

4. The simulation results

The described methods were used to calculate the daily schedules for existing water treatment plants and for typical daily distribution demands, i.e. the uniform distribution, the distribution for communal consumers, the distribution for industrial consumers. Decreaisng of the daily electric energy cost in comparision with this cost for the traditional control method is :

- for the schedules calculated by using the separable programming algorithm : 15 - 30 % [3] ,
- for the schedules calculated by using the multistage programming algorithm : 8 - 20 % .

The proposed methods make possible to take into considerations the real characteristics of pumps and pipelines and to satisfy all technological and operational requirements concerning among others : backwashing of filters, accumulation of desired amount of water in the reservoirs, frequency of turning on and off the pumps.

5. Literature

[1] KLINCK M. : "Simulation Aided Modeling of the Dynamic Behaviour for Some Elements of a Surface Water Treatment Plant". Modeling and Simulation of Water Res.Syst. North Holland Publ.Comp.,1975

[2] MILLER C.E. : "The Simplex Method for Local Separable Programming" Recent Advances in Math.Programming. Mc Graw-Hill, New York 1963

[3] SZEBESZCZYK J. : "Optimization Programming-Nonlinear Algorithm" Applied Informatics IASTED Symposium, Lille 1983.

DISPECER-H - A DECISION SUPPORTING SYSTEM IN WATER RESOURCES DISPATCHING

Alexandru-Dan Donciulescu, Florin-Gheorghe Filip [1)]

Abstract. DISPECER-H is a software package designed to assist the operational control of a water system (WS). Based on a model of WS dynamics and also on optimal water resources distribution proposed by the DISPECER-H, the human dispatcher can decide quickly and correctly on water distribution. The paper briefly approaches a WS within a decision support system. The model and the decision-making problem are presented The functions and performances of the system are also described.

Keywords: water systems; dispatcher systems; decision support systems; hierarchical optimization; operational control.

1. Introduction

A water system (WS) consists of a number of water accumulation elements (lakes) naturally inflowed (by rivers, torents) from which water is supplied to a number of consumers (industrial, agricultural and households). The task of the dispatcher (manager of water in the WS) is to control the moving waters in the next time interval, so that the consumers' demand should be satisfied, keeping a certain water level in lakes (to prevent flood or draught). The controls established by the dispatcher are implemented in installations which make waters move among lakes or outside the WS.

Solving the dispatcher's problem implies a representation of the WS as an interconnection of installations and lakes. Each installation is characterized by a control variable representing the water flow for the next time interval.

The control problem is to determine, over the next time horizon the "operation schedule "consisting in the sequences of the working rates of the installations. This problem can be formulated (model, constraints, criterion) as an optimization problem, whose solution is the optimal schedule. That schedule can be accepted by the dispatcher and implemented as such, or it can be the starting alternative for the final decisionmaking by the human, analysed and changed in accordance with his skills and current conditions.

2. Problem formulation

Consider the representation of the WS in Fig.1. The model of the WS represents the dynamics of water resources in each lake $i = 1,\dots n_x$:

$$x_i(k+1) = x_i(k)+\sum_{j=o}^{\theta T} B\underline{m}(k-j)+\sum_{\ell=1}^{n^i_{wi}} \sum_{j=o}^{\theta M} wi^i_\ell(k-j)+\sum_{\ell=1}^{n^i_{we}} we^i_\ell(k)+\xi_i; \quad k=1,2,\dots,KF \qquad (1)$$

1) Central Institute for Management and Informatics
Bd.Miciurin 8-10 71316 Bucharest, Romania

where: $x_i(k)$ = water volume in lake i at the beginning of kth time interval; $\underline{m}(k)$ = the control vector of n_m dimension, of water volume moved during kth interval through the installations; $\underline{wi}^i$ - vector of the n_{wi}^i water inflows in ith lake (which can be measured or forecast); $\underline{we}^i$ vector of n_{we} water outflows (water demands which can be negotiable or fixed); ξ_i = water losses; B = a $n_x \times n_m$ matrix; θT, θM - maximal delays of controls, respectively water inflows affecting lake's volume. Note also: ΔT - the time horizon on which the schedule is made; KF - the number of time intervals of ΔT.

Remarks: 1. Some of variables m can be substituted for variables we in case the demand of the consumers is negotiable.

2. Eq. (1) represents a linear material balance model which is convenient for computation. Actually the user introduces and obtains data about the lake levels (not volumes).

The functioning of the WS can be constrained within upper and lower limits for the water volumes x and transfer rates m :

$$\underline{x}_m \leqslant \underline{x}(k) \leqslant \underline{x}_M \; ; \; \underline{m}_m \leqslant \underline{m}(k) \leqslant \underline{m}_M \qquad (2)$$

The decision problem is to determine the control sequences $\{\underline{\overset{*}{m}}_{[1;KF]} \triangleq \underline{\overset{*}{m}}(k), k=1,\dots,KF\}$ which satisfy to the greatest extent the consumer demands and ensure the levels in lakes near to recommended values, $\underline{xd}$. The values $\underline{m}$ have to evolve around some technological and/or economic values $\underline{md}$. Therefore the criterion, which is to be minimized, is:

$$J = \sum_{k=1}^{KF} \left[\|\underline{m}(k) - \underline{md}(k)\|_R^2 + \|\underline{x}(k+1) - \underline{xd}(k+1)\|_Q^2 \right] \qquad (3)$$

where $\|\underline{v}\|_P^2 = \underline{v}^T.P.\underline{v}$ and Q, R are weighing matrices.

The problem of computing the controls which minimize (3) w.r.t. constraints (1) and (2) is solved using a hierarchical optimization method [2] based on Tamura's algorithm [7], in which improvements have been made in convergence speeed.

3. DISPECER-H Description

DISPECER-H is an interactive software system providing all the functions required to assist the dispatcher in operational control of the WS. The main functions are: a) the collection of data about the actual state and available resources estimation: actual level in lakes; actual production rates of the installations; actual water input of the affluents; forecast of water inputs $\underline{wi}_{[0;KF]}$, which are computed by a Box-Jenkins time series analysis; collecting the water demands of the consumers $\underline{we}_{[0;KF]}$; b) computing the optimal schedule of the WS by solving the optimization problem; c) simulation of the evolution of water levels in lakes; d) analysing (at the dispatcher's choice) other variants of the schedule: introducing a new schedule; modifying the schedule; simulating schedules; diagnosing/correcting the schedules feasibility; e) reports delivery on request (including graphic ferm) about the former states of the WS and its future evolution; f) managing the data base of the representation of WS problem (model, criterion, constraints) and also the user interface.

The system offers facilities for automatic generation of the mathematical representation of the problem, based on data known by the dispatcher. There are also some functions easy to operate which update the data according to the changes in WS: changing the time horizon and number of intervals; updating θM and θT values; considering measurements from automatic devices; updating level-volume and level-flow characteristics, including new affluents and consumers etc.

DISPECER-H is designed for SM-4 or PDP-11-like machines provided with RSX-11M compatible operating system. The operation is wholly interactive, computer guided, on teletype or display (possible graphic) terminals. The dialogue efficiently uses the facilities of a particular terminal. All functions except optimization meet the operating speed of the human. The optimization time is performed in less than 1.5 minutes even for large systems.

The normal operation of the DISPECER-H system implies: a) input the current level in lakes (which is automatically transformed into volumes), the flows of the affluents and the water demand from the consumers for the next time period; b) optimal schedule computing; c) on request analysing schedule variants, diagnosing their feasibility and making corrections, all computer assisted; d) delivering reports about historical data or the scheduled time horizon.

Running a particular function of the DISPECER-H System is totally on the dispatcher's initiative. Implementation and operation are simple and do not require a special qualification or training of the operator.

4. Conclusions

DISPECER-H, is a new product. It belongs to DICOTR (dispatching and operational control) family and represents a specific version of DISPECER system [4] which can be used in a broader class of applications in the environment of continuous systems interconnected via reservoires It has been tested in an actual WS (5 lakes with hydropower stations and 6 big consumers) with very tight operation constraints. The optimization takes less than 15 sec. for a horizon of 12 hours. There are very good results for the time being: a better insight and evaluation of the implications in real life of a certain decision of the human dispatcher. Current researches are carried-out to design a hierarchical structure to implement the on-line dynamic coordination method [1,3].

REFERENCES

1. Filip F.G., D.A.Donciulescu: On an on-line dynamic direct coordination method in process industry. Automatica. 19 (1983) 3, 317-320.
2. Filip F.G., D.A. Donciulescu, R. Gaşpar, M. Muratcea, L. Orăşanu, Multilevel optimization algorithms in computer aided control in process industry. Computers in Industry 5 (1984) 4.

3. Findeisen W. A view on decentralised and hierarchical control. Syst. Anal. Model. & Simul. 1 (1984) 2, 63-99.
4. Guran M., F.G. Filip, D.A. Donciulescu, L. Orăşanu. Hierarchical optimization in computer aided dispatcher systems in process industry. Large Scale Systems (to appear).
5. Hopfgarten S., H. Puta, K. Reinisch, C. Thümler. Applications of hierachical methods to the structural expansion and control of a regional water distribution network. In A. Straszak (Ed.) Preprints 3-rd IFAC/IFORS Symp. LSSTA, Warsaw 1983, 209-215.
6. Malinowski K., K. Salewicz, T. Terlikowski. Hierarchical dispatching control structure for a multireservoir system. In A. Straszak (Ed.) Preprints 3-rd IFAC/IFORS Symp LSSTA, Warsaw 1983, 318-327.
7. Singh M.G., A.Titli, Systems: Decomposition, optimization and control, Pergamon Press. Oxford 1978.

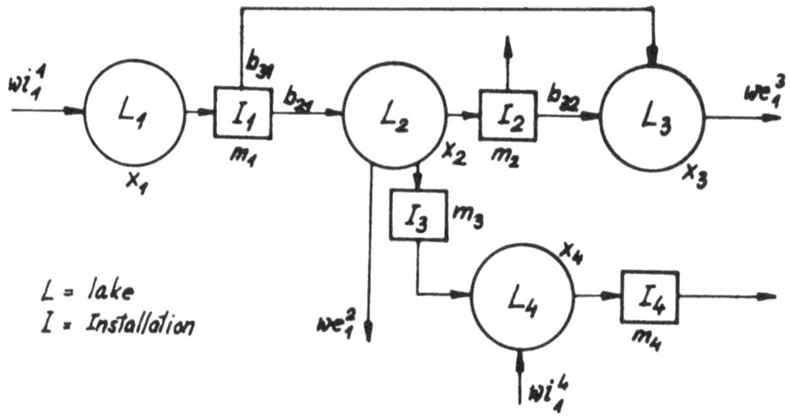

Fig.1. Representation of a part of WS

UTILISATION OF A MODEL OF WATER SUPPLYING NET FOR DESIGNING A MEASURING SYSTEM

Urszula POCIASK[1)]

In the paper the method is shown of application of the model of a water supply network for determ nation of the flow intensity values in the net branches and pressure values in knots. Using the most credible estimator under the assumption of the square characteristic of the pipe and stationary, Gaussian errors the formulae for the value of limit errors and measurements number were shown. By application of this method it is possible to reduce the number of measuring instruments in control of the network state.

Keywords : the model of the network, nonlinear estimator, limit error

1. Introduction.

The controlling of the water-distribution in a water-supply net requires a permanent control of two parameters: of the pressure in the knots and of the flow-intensity in the net branches. The ascertainment of the instantenous valies of the current intensity "q" and of the pressure "H" can be determined :

- by direct measuring of both parameters,
- with the methods utilizing the results of measuring of one parameter only and based on the model of water supplying net.

The first of above mentioned method requires an application of as many pressure gauges as there are knots in the net and as many flowmeters as branches there are in the net. The second of above mentioned ways allows a partly reduction of the number of measuring instruments by utilizing the physical net model. In the following part there will be presented the method how to de ermine the estimators of pressure and of flowintensity and the number of measuring series in a given measuring knot in a net including "r" knots and "m" branches at the assumtion, that the physical net-model is known.

2. The determination of the estimator of instantenous pressure and flow-intensity values.

Let us assume, that the dependence between pressure-drop and flowintensity in a branche is a quadratic one[2)]:

$$h = \nu q^2 \quad \text{or} \quad q = c h^{1/2} \tag{1}$$

where : ν , c - constant values.

Let υ to be vector of the additives observation errors. The estimators q and h with the biggest value of credibility for $\underline{z}$ with normal distribution N (0,R) will take the shape[3)]:

- at the direct measuring of pressure :

---

1) Silesian Technical University
Pstrowskiego 16 Gliwice POLAND

2) T.Gabryszewski - Wodociągi. PWN Warszawa 1975

3) Fred C.Schweppe - Układy dynamiczne w warunkach losowych . WNT Warszawa 1978

$$\hat{q} = \left(\frac{\sum_{i=1}^{n} h_i}{n \nu} \right)^{1/2} \tag{2}$$

where : $h_i = H_{ki} - H_{pi}$; H_{pi}, H_{ki} - values of pressure at the beginning and at the end of a branch in the i-measurements, n - number of measurements.

- at the direct flowintensity-measurements :

$$h = \left(\frac{\sum_{i=1}^{n} q_i}{n} \right)^2 \tag{3}$$

where q_i - value of flowintensity in the i-measurement.

The corresponding average value and the variation of the estimation error a the assumption that the vector "$\underline{z}$" is a stationary one will be :

$$E(\Delta \hat{q}) = \frac{\sigma_z^2}{8 n \nu^2 n^3} \qquad \sigma_{\Delta \hat{q}} = \frac{\sigma_z}{2 \nu q \sqrt{n}} \tag{4}$$

$$E(\Delta \hat{h}) = \frac{-\sigma_z^2}{n c^2} \qquad \sigma_{\Delta \hat{h}} = 2\sqrt{n} \; \frac{\sigma_z}{c\sqrt{n}} \tag{5}$$

$$R = \begin{bmatrix} \sigma_z^2 & & 0 \\ & \ddots & \\ 0 & & \sigma_z^2 \end{bmatrix} \qquad R = \begin{bmatrix} \sigma_z^2 & & 0 \\ & \ddots & \\ 0 & & \sigma_z^2 \end{bmatrix}_{n \times n}$$

It can be shown that for corresponding great "n" and small " σ_z^4 " the error distribution " Δ q" and " Δ h" is a Gauss distribution.

3. The determination of measuring serie-number

3.1. The direct measurement of pressure

Let us to mark by : H_{pj} - the pressure value in the first knot, of j-branch ; H_{kj} - the pressure value in the last knot of j-branch ; n_{pj} - measurements number of H_{pj} ; n_{kj} - measurements number of H_{kj}.

The value of the pressure drop estimator in the j-branch minimilizing the matrix of covariation of estimator error is

$$\hat{h}_i = \frac{\sum_{i=1}^{n} H_{kji}}{n_{kj}} - \frac{\sum_{i=1}^{n} H_{pji}}{n_{pj}} \tag{6}$$

The variation $\sigma_{\Delta \hat{h}}$ of estimator error $\hat{h}_j$ at the assumption of stationary and Gauss-type distributions at measurements of H_{kj} and H_{pj} is :

$$\sigma_{\Delta \hat{h}_j} = \sqrt{\frac{\sigma_{zp}^2}{n_{pj}} + \frac{\sigma_{zk}^2}{n_{kj}}} \tag{7}$$

where : σ^2_{zp} , σ^2_{zk} - variations of additives measuring errors of H_{pj} and H_{kj} .

The determination error of q_j calculated on the base of formulas (4) for the credibility horizon α is in the section :

$$\frac{\sigma^2_{\Delta\hat{h}_j}}{8v_j^2 q_j^3} - t_\alpha \frac{\sigma^2_{\Delta\hat{h}_j}}{2v_j q_j} \leqslant \Delta\hat{q}_j \leqslant \frac{\sigma^2_{\Delta\hat{h}_j}}{8v_j^2 q_j^3} + t_\alpha \frac{\sigma^2_{\Delta\hat{h}_j}}{2v_j q_j} \qquad (8)$$

As the values $\sigma_{\Delta\hat{h}_j}$, v_j, q_j are positiv ones, the maximum value of error $\hat{q}_j$ in the sphere $q_{j\,min} - q_{j\,max}$ amounts to

$$\Delta\hat{q}_{jmax} = \frac{\sigma^2_{\Delta\hat{h}_j}}{8v_j^2 q_{jmin}^3} + t_\alpha \frac{\sigma^2_{\Delta\hat{h}_j}}{2v_j q_{jmin}} \qquad (9)$$

In order not to exceed the measuring error in the given knot and branch the settled value Δq and ΔH at the probability α, there must be fullfiled following conditions .

$$\Delta H \geqslant t_\alpha \frac{\sigma^2_{zHl}}{\sqrt{n_{Hl}}} \quad l = l \dots n \qquad \Delta q \geqslant \Delta\hat{q}_{jmax} \quad j = l \dots m \qquad (10)$$

where: σ_{zHl} - variation of additive error of pressure measurements at the l-knot, n_{Hl} - number of pressure measurements in the l-knot.
At the result of solving the inequalities (10) we get the values n_{Hl}, which determine the number of series in the l-knot. Moreover there will be two different quantities. This nonconformity follows from the fact, that in the same knot there may start or end several branches. The number of series will be determine by the greatest among the obtained values.

3.2. The direct measurements of the flowintensity

The estimator value of the flowintensity in the j-branch (which minimilizes the matrix of covariation of the estimation error) and the value of its variation amounts to :

$$\hat{q}_j = \frac{\sum_{i=1}^{n_{qj}} q_{ji}}{n_{qj}} \qquad \sigma_{\Delta q} = \frac{\sigma_{zq}}{\sqrt{n_{qj}}} \qquad (11)$$

where : n - measurements number g_j , q_{ji} - the result of the i-measurement, σ_{zq} - variation of the additive measuring error q_j.
On the basis of formula (3) the value of pressure drop estimator in the j-branch amounts to :

$$\hat{h}_j = \left(\frac{\sum_{i=1}^{n_{qj}} q_{ji}}{n_{qj}\, c_j} \right)^2 \qquad (12)$$

Assuming a normal distribution of estimation error Δh_j, whot is correct for corresponding big v alues h_{qj} - the maximum value Δh_{jmax} on the base of formulas (5) amounts to :

$$|\Delta\hat{h}_{jmax}| = \frac{\sigma^2_{zqj}}{n_{qj}\, c_j^2} + t_\alpha \frac{\sigma_{zqj}}{c_j} \sqrt{\frac{h_{jmax}}{n_{qj}}} \qquad (13)$$

where : h_{jmax} - maximum value of pressure drop in the j-branch.
The absolute value of pressure in the l-knot defines the dependence :

$$\hat{H}_l = \hat{H}_0 + \sum_{j=1}^{\beta} \hat{h}_j \qquad (14)$$

where: $\hat{H}_0$ - the value of pressure estimator in the corresponding knot, β number of branches connecting the l-knot with the corresponding knot.

The selection of the corresponding knot and branches connecting it with the 1-knot has to be done on the criteria-basis for minimizing the determination error $\hat{H}_1$. In order not to exceed the measuring error in the given branch and knot, the given value ΔH and Δq with a probability α there has to be fullfilled following inequalities :

$$\Delta q \geq t_\alpha \frac{\sigma_{zq_j}}{\sqrt{n_{q_j}}} \qquad j = 1 \ldots m \tag{15}$$

$$\Delta H \geq t_\alpha \frac{\sigma_{zH_0}}{\sqrt{n_{H_0}}} \qquad \Delta H \geq \sqrt{\left(t_\alpha \frac{\sigma_{zH_0}}{\sqrt{n_{H_0}}}\right)^2 + \sum_{j=1}^{\beta} \left(\Delta \hat{h}_{jmax}\right)^2} \tag{16}$$

where: σ_{zH_0} - variation of measuring errors $\bar{H}_0$, n_{H_0} - number of measurements.

The serie number n_{H_0}, n_{qj} will be defined (similar as in the case of direct measurements of pressure) as the greatest value among the obtained values after solving the inequalities (15) and (16).

4. The selection of measuring points

It is possible to distinguish three structures of measuring system, which is utilizing the physical model for determination of flowintensity estimators and pressure estimators : - the system including pressure gauges only, - the system including flowmeter only, the system including manometers and flowmeters also.

Systems including manometers only or flowmeters only will be systems with the much lower number of instruments, compared with direct measuring of both parameters. And so for the analysed net it enables to economise m flowmeters or n manometers. The utilization presented in the paragraphs 2 and 3 including the method of calculating the estimators and number of measuring series, demands : - the knowledge of the net model, - the performance of simulation research (for determining of q_{jmin} or h_{jmax}).

The net model and the results of simulation researches as well are already prepared at the time of designing of water supplying net and in connection with this, they are usually easy available. The necessity of application a mixed structure may happen in the case when :

- the technical conditions do not allow the realization of one parameter system in regard of measured quantity, - the number of measurements which secures no exceeding of error values is great in the given knot or branch.

The measuring system, which utilizes the physical model for calculating the wstimator values has to be supplied with technical equipment, which enables the determination of estimator values in the real time. Therefore the application of the above presented method is particulary useful for big systems in which it is possible to utilise the equipment necessary for collecting of informations.

THE CLOSED-LOOP CONTROL FOR THE WATER SUPPLYING SYSTEM

Janusz ZELEZIK[1)]

A problem of optimisation of water supply system has previously been formulated and the method of determination the optimum open-loop trajectories has been developed. In this paper a problem of finding the closed-loop control policy is considered. Making certain assumptions this task is reduced to solving the conventional linear regulator problem. Results of calculations for simple water supply system are presented.

Key words: Water supply, Optimal control.

1. The Optimal Open-Loop Trajectories

The water supplying system consist of : pump stations, network of pipes, reservoirs, consumers of water. In [1] , [2] the algorithm of determining of optimal (considering the cost of pumping water) open-loop control for the system is elaborated. The approximate dinamic model of the system is utilized :

$$x(k+1) = A.x(k) + B.u(k) - c(k) \tag{1}$$

where: x(k)- N-dimensional state vector in moment k ; corresponding to water volumes in reservoirs, u(k)- M-dimensional vector of control variables (flows from pump stations), c(k)- vector corresponding to agregated consumer demands (they changes in day cycle), A,B - coefficient matrices, k - stage variable, $0 \leq k \leq K$ (K - control horizon). The performance index (consist mainly of pumping cost):

$$J = \sum_{k=0}^{K-1} \left[u(k)-u^A(k)\right]^T \cdot R(k) \cdot \left[u(k)-u^A(k)\right] + \sum_{k=1}^{K} \left[x(k)-x^A(k)\right]^T \cdot Q(k) \cdot \left[x(k)-x^A(k)\right] \tag{2}$$

is minimized, where $R(k) = \text{diag}\left\{r_i(k)\right\} = \text{diag}\left\{r_i \cdot T(k)\right\}$,
$Q(k) = \text{diag}\left\{q_i(k)\right\}$ are matrices of weigting factors,
$u^A(k)$, $x^A(k)$ - vectors of parameters,
T(k)- electricity unit tariff values.
Also the following technological constraines should be fulfiled:

$$x^m \leq x(k) \leq x^M \ ; \quad u^m(k) \leq u(k) \leq u^M(k) \tag{3}$$

and with regard for periodicity of water demand the state of system after one day should be the same :

$$x(K) = x(0). \tag{4}$$

Determination of optimal open-loop control strategies is the deterministic problem. It is asumed that water demands in the whole control horizon are known (usually one day). Using the effective Tamura algorithm the optimal trajectories $u^o(k)$, $x^o(k)$- k = 0,1,2 ... (K-1) are determined. The open-loop control executes by cyclic with a 24 h period realisation calculated nominal control strategy $u^o(k)$. Must be

1) Institute of Automation
Silesian Technical University, Gliwice (POLAND)

mentioned, that a certain set of points S_1 of optimal trajectories exists in which $u^o(k)$, $x^o(k)$ on the upper or lower bounds (3) lies.

2. The Closed-Loop Control

The next step on the control procedure is finding the closed-loop control law. The real state trajectories $x(k)$ differ from calculated optimal open-loop trajectories $x^o(k)$, because there are fluctuations of water demands and the control strategies are not exactly realised.

One method of feedback introducing can be using of shifting optimisation horizon. Having the measurements of state vector $x(k)$ and assuming these values as the initial conditions, the new control trajectory is calculated. For example such procedure in one control step (k=0) will be analised (this procedure can be directly generalised for another steps, when we note the current moment as k=0 and following moments 1,2, ...).

At the begining of control step the state variables are mesured :

$$x(0) = \hat{x} \quad \text{(given)} \tag{5}$$

Mostly $x(0) \neq x^o(0)$, then the new control strategy $u^*(k)$ is calculated. The performance index (2) is minimised, but the constrains (1), (3), (5) and

$$x(K) = x^o(K) \tag{6}$$

must be fulfiled. The constrain (6) means, that the state deviation from nominal trajectory after one day must be liquidated. From the new control strategy only $u(o) = u^*(0)$ is realised and after it the measurements and minimization are repeated. The fault of this method is that the complicated calculations must be done on-line.

The practical calculations were performed. For small deviation $\Delta x(0) = x(0) - x^o(0)$ it was noticed, that new determined trajectories $u^*(k)$, $x^*(k)$ are monstly equal to the upper or lower bounds (3) in the same set of points S_1, what was for $u^o(k)$ and $x^o(k)$.

For the next considerations it is assume that for small deviations $\Delta x(o)$ these trajectories of control and state lies on the bounds always in the same set of points S_1. The set S_1 is known from $u^o(k)$ and $x^o(k)$. Then the on-line realised optimization problem can be differently formulated. In every control step it is sufficient the modificated performance index J^* to minimise.

$$J^* = \sum_{k=0}^{K-1} \left[u(k) - u^B(k)\right]^T \cdot R^*(k) \cdot \left[u(k) - u^B(k)\right] + \sum_{k=1}^{K} \left[x(k) - x^B(k)\right]^T \cdot Q^*(k) \cdot \left[x(k) - x^B(k)\right] \tag{7}$$

The minimisation is done to the system (1) with initial condition (5), but without the constrains (3). The matrices $R^*(k)$, $Q^*(k)$ and parameter vectors $u^B(k)$, $x^B(k)$ are received from $R(k)$, $Q(k)$, $u^A(k)$,

$x^A(k)$ after introducing the following changes :

$$q_i^*(K) = G \; ; \; x_i^B(K) = x_i^o(K) \qquad i = 1,2,\dots N \qquad (8)$$

$$\text{if } \left[u_j^o(k) = u_j^m(k) \text{ or } u_j^o(k) = u_j^M(k)\right]$$

$$\text{then } \left[r_j^*(k) = G \; ; \; u_j^B(k) = u_j^o(k)\right] \quad j = 1,2 \dots M; \; k=0,1\dots(K-1) \qquad (9)$$

$$\text{if } \left[x_i^o(k) = x_i^m \text{ or } x_i^o(k) = x_i^M\right] \text{ then } \left[q_i^*(k) = G \; ; \; x_i^B(k) = x_i^o(k)\right]$$

$$i = 1,2\dots N \; ; \; k = 1,2\dots K \qquad (10)$$

where : G – the number much greather than another weighting factors q_i , r_i .

The performance index modification is done in order to assign the greatest loses to the deviations of trajectories u(k), x(k) in the points, in which the values these trajectories are known. The recently formulated optimization task is the conventional linear regulator problem for system (1) and performance index (7). The control strategy is according to known rule:

$$u(k) - u^o(k) = S^*(k)\left[x(k) - x^o(k)\right] \qquad (11)$$

where the feedback gain matrix $S^*(k)$ is calculated backward in time for $k = (K-1),(K-2) \dots 0$ from equations :

$$S^*(k) = -\left[B^T W(k+1) \cdot B + R^*(k)\right]^{-1} \cdot B^T W(k+1) \cdot A \qquad (12)$$

$$W(k) = A^T \cdot W(k+1) \cdot A + A^T W(k+1) \cdot B \cdot S^*(k) + Q^*(k) \qquad (13)$$

where $W(K) = Q^*(K)$.

From the calculated sequence matrices $S^*(k)$ for control (in step k=0) only finaly matrix $S(0) = S^*(0)$ will be used. The feedback gain matrix S(1) for k=1 is similary determined (calculation according to Eq (12), (13) sequently for k=0, (K-1), (K-2), ... 1 and substituting $S(1) = S^*(1)$). After off-line calculations the whole sequence of matrices S(k), for k = 0,1,...(K-1), the closed-loop control is simply realised :

$$u(k) = S(k) \cdot \left[x(k) - x^o(k)\right] + u^o(k) \qquad (14)$$

and complicated on-line calculations are not demand.

3. Example

The results of optimization for the simple water supply system (consisting of two pumping stations supplying one reservoir, from which consumers are taking water) are presented in Fig.1. For 24 h cycle are shown : consumer water demand c(k), electricity unit tariff values T(k), determinated the open-loop control strategies $u_1^o(k)$, $u_2^o(k)$ and trajectory of state $x^o(k)$, values of feedback gains calculated for the closed-loop control. The range of state deviations $\Delta x(k)$ arround the optimal open-loop trajectory is marked. In this range the proposed closed-loop control is optimal. The simulation investigation confirmed, that for greather deviations $\Delta x(k)$ this closed-loop control procedure is also satisfactory.

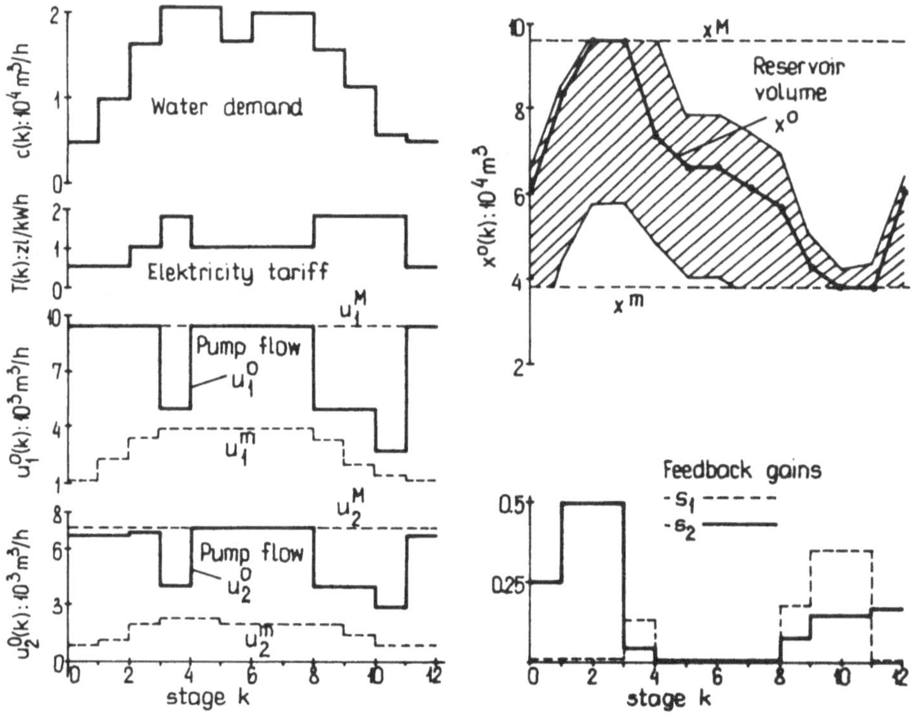

Fig.1

4. References

[1] COULBECK B., STERLING M.J.H. - Optimised control of water distributions systems. Proc.IEE, Vol.125, 1978, N° 9.

[2] ZELEZIK J. - Optimal scheduling in water supply system. Z.N.Politechniki Śląskiej, s.Automatyka, N° 69 (in Polish)

SIMULATION OF LARGE SOCIO-ECONOMIC SYSTEMS

D. Schmolke, H. Starke, F. Stuchlik[1)]

The modelling and simulation of large systems with time dependent elements and a very high complexity is regarded. The model classes used in the modelling and simulation system and the services of the software system are described.

1. Introduction

First, some introductory remarks on the topic of this paper. A class of models is regarded. These models serve for the simulation of socio-economic processes with high complexity. In this sense a model is a mapping of objects, relations and manipulable quantities of a system. Objects and relations could be represented by graphs. These graphs are a basic for the simulation of the systems behaviour. For this reason each object is described by a state variable. Values are assigned to all state variables over equidistant values of the growing time variable (time series). These values are partially known only for the interval of past time (time interval of analysis) and they are unknown for the period of future.

The aim of simulation is to determine the unknown values of the state variables. The relations between the objects describe the structure of the system. They could depend on time also. The objects are represented by a set of nodes $D = (x_1, x_2, \dots, x_n)$ and the relations by a set of edges $R = (y_1, y_2, \dots, y_m)$ both belong to one directed graph. The dynamic system structure and the set of edges attributes $A = (a_{ij})$, $i=1,2,\dots,6$: $j=1,2,\dots,m$ characterize the course of the process, i.e. the sequence of system states. The triple $G = (D,R,A)$ completely describes the graph. The edges attributes consist of parameters selected from the set of a_i, $i=1,2,\dots,6$.

The parameter a_1 is a factor of the product with the content of the first node x_i of the edge $y_r = (x_i, x_j)$, $i,j=1,2,\dots,n$ Parameter a_2 is a duration given in discrete time units and i.e. the relations with respect to time of node x_i and x_j, $i,j=1,2,\dots,n$ to edge $y_r = (x_i, x_j)$, $r=1,2,\dots,m$. The parameter a_3 is a period given in discrete time units and i.e. the cyclic validity of edge $y_r = (x_i, x_j)$, $i,j=1,2,\dots,n$; $r=1,2,\dots,m$. The parameter a_4 is a logical attribute. It means the edge $y_r = (x_i, x_j)$ represents a split. The parameter a_5 denotes one of the base operations +, –, *, /. Each operation concerns the kind of

---

1) College of Advanced Technology Otto von Guericke
Section Informatics
3010 Magdeburg, German Democratic Republic

mapping of the first node's content into the second node's one. The operations * and / require two connected edges each with two first nodes and a common second node. The parameter a_6 characterizes the kind of mapping from the contents of $x_1, x_2, \ldots, x_q$ into x_p. $x_1, x_2, \ldots x_p$ are first nodes of edges an x_p the common second node of all edges (x_1, x_p), (x_2, x_p), $\ldots, (x_q, x_p)$.

2. Model classes

The attributes set A permits several classes of socio-economic systems to be modeled and therefore determined model classes to be verified.
In this connection the parameter a_4 shortly named "split parameter" describes one essential system quality. This quality concerns a so called "split net", i.e. a subgraph. A subgraph of the graph G is said to be a split net if the following two conditions are fullfilled:

a) Let $x_{f1}, \ldots, x_{fq}$ and $x_{s1}, \ldots, x_{sp}$ be the set of first nodes and the adjacent set of second nodes, respectively. Then for each x_{fi}, $i=1,2,\ldots,q$ exists at least one x_{sj}, $j=1,2,\ldots,p$ which is a second node for an x_{fk} with $k \neq i$ (net condition).
b) $SUM(<x_{fi}>, i=1,2,\ldots,q) = SUM(<x_{sj}>, j=1,2,\ldots,p)$ and $<x>$ denotes the content assigned to the node x (split condition).

For reasons of the system charakterizations (structure, course of the process) the models of socio-economic systems could be classified in the following classes:

1. recursive and nonrecursive models

In this connection recursion means that a value of the object x_i $i=1,2,\ldots,n$ in the moment $t+t_1$ is a function of the object value in the point of time t: $x_i(t+t_1) = f(x_i(t),\ldots)$.

2. models with and without split nets
3. connected and decomposable models

The "decomposition of the graph G" denotes the generation of subgraphs $G_1, \ldots, G_k$ of G and the edges of the subgraphs are disjunct.

4. models with fixed horizonts of start and stop (closed models)

In this case selected objects of the model have got fixed values for the moments of begin and end of simulation.

3. Model description

The graphs are represented by records of the objects and records of the strukturen. A record of a object consists of the object name and 6 descriptors 20 Byte long each used to describe the data. All data are stored in a special organized data base of the simulation system. The access to the data of each object is realized by these 6 descriptors.

The choice of the descriptors requires great attention. It is necessary to check existing code systems to find out which of them are suitable for an integration in the descriptors list in order to use available data bases.
Each directed edge of the graph is to be assigned to a structure record. This record contains the names of the two associated objects and information about the attributes of edges.

4. Software components of the modelling system

4.1. Generation and manipulation of the model description

The numbers of the object and the structure records are very great. The model descriptions are developed continuously. Thus both have to be easy to correct and to extend. For this reason elements of a language are implemented which are useful for the easy generation of a lot of new records. Tables externally stored can give assistance to this process. The manipulation of all models is supported by an editor which has an access to all files with a fixed record structure.

4.2. Interpretation and simulation

The socio-economic processes under consideration are very extensive and the corresponding models too. In a first step thus the model descriptions are syntactically and semantically checked by the computer program. A sequence of computation of the data flow in the graph is determined at the same time. The aim is to minimize the requirements of the simulation system for main storage capacity and CPU-time. The descriptions interpreted by the program are managed by a data bank operating system.

4.3. Data management

The data management of the modelling and simulation system is supported by a relational data bank system. This system manages model descriptions, collected statistical data from time interval analysis, data about aiming points and limitations (i.e. lower and upper bounds) and last not least the values of objects determined by simulation.

4.4. Presentation of results

The modelling and simulation system gives a extensive assistance to the preparation, presentation and interpretation of the results. This is made by a list generator, a program for graphic generation and a dialog inquiry-program.

4.5. Applications of the modelling and simulation system

The modelling and simulation system is used for the prognosis of

-the demographic developments
- the development of parts of the higher education system
- the staff requirements of the national economy and so on.

A decomposition of very complex socio-economic processes into partial processes is useful. This decomposition gives assistance to both verification and validation of partial processes and the whole process. The coupling of partial processes is made by a transfer of the simulation data (supported by the DBMS).

5. References

[1] Boehm, F.; Weise, G.: Graphen in der Datenverarbeitung. VEB Fachbuchverlag, Leipzig 1981

[2] Cvetkovic, D.M.; Doob, M.; Sachs, H.: Spectra of graphs, theory and applications. Deutscher Verlag der Wissenschaften, Berlin 1980

[3] Heike, H.D.: Modellierungssoftware. Walter de Gruyter, Berlin 1981

[4] Lockemann, P.C.; Mayr, H.C.: Rechnergestuetzte Informationssysteme. Springer-Verlag, Berlin - Heidelberg - New York 1982

[5] Schwarze, G.: Simulations-, Datenbank-, Modellbanksysteme. edv-aspekte 1 (1982) 4, 18-22

Experiences in Modelling Complex Demographic Processes

Thomas Büttner[1)]

Multistate mathematical demography, Population analysis and forcast, Marital status of Population.

The Paper deals with a special aspect of the family-dimension of population dynamics: the population disaggregated by age, sex and marital status and its development. First, the methodological approach of the multistate analysis is outlined and second some results of an application are given.

The connection between population development and other social processes needs an intensivied scientific treatment under the condition of an intensively extended economy, of socialist ways of living and of long term changes in the patterns of demographic reproduction. Thus, the so called process aging population, involving significant changes in the age structure of the GDR's population, describes a complex and complicated phenomenon, having consequences to many spheres of life. Social sciences have to recognize and, finally, quantify this changes and its implications. Therefor modelling of essential demographic processes are an important precondition.

Methodology: The central concept in demography is the life table, describing the evolution of a (hypothetical) cohort (a generation, let say) of individuals born at the same time and exposed to an unchanging agespecific schedule of vital rates. The starting point for constructing a life table is the differential equation that defines l(x), the probability of surviving to a certain age x (survivorship function):

$$\frac{d}{dx} \; l(x) = -\mu(x) \; l(x) \quad ^{2)} \qquad (1)$$

The classic life table approach was only able to deal with the transition between two states of existence, mostly from being alive to being dead. To tackle more complex problems a generalisation of classic life tables to increment-decrement life tables (multistate life tables) were developed in the last decade. (1), (4).

This approach is also founded on equation (1), but with matrices replacing scalars. The procedure of estimating an age-specific matrix of transition probabilities P(x) between several states (including death) from observed data is described in the literature.(2) (3).

1) Institute for Sociology and Social Policy, Academy of Sciences, GDR, 1080 Berlin, Otto-Nuschke-Str. 10/11

2) Note, that the chance/risk of dying between age α and $\alpha + d\alpha$ for an α- year-old person ist $\mu(\alpha)\, d\alpha$; in (1) x stands for exact age

The survivorship function is then given by

$$L(x+h) = L(x)\,P(x) \qquad (2)$$

In the multistate perspective of population analysis there ist a set of several life table functions (statistics), all of them originating from the transition probabilities P(x).
A discret age-time model of multistate population development expresses the population projection process by means of matrix operations. Given the multistate population as a vector K^t, disaggregated by age and states at t=0, and transforming the transition probabilities P(x) into the survivorship proportions S(x), that refer to individuals who are not at exact age x but in age groups x to x+h, by

$$S(x) = [I + P(x+h)] \cdot P(x) \cdot \{I + P(x)\}^{-1} \qquad (3)$$

(note, that I is the identity matrix), then the basic projection equation may be expressed as follows:

$$K^{t+1}(x+h) = S(x)\,K^t(x)$$

(For computing $K^{t+1}(0)$, d.e. the surviving babies born in the unit time interval, see (3))
Application: The flows between marital states, recorded in the official statistic, are given as follows[1)]

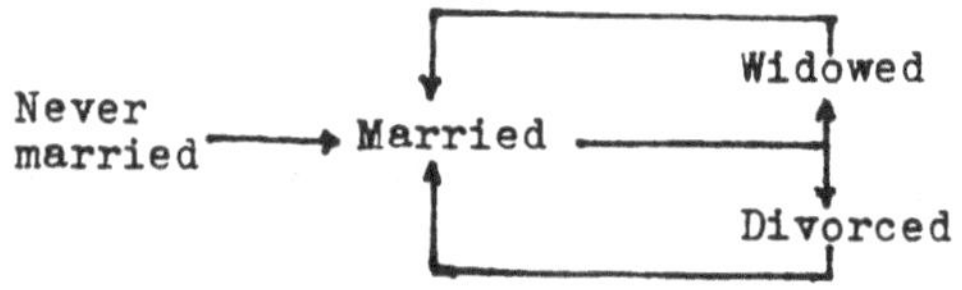

For analysing the patterns of marital status behaviour, after estimating P(x), the life table statistic 'Life expectancy by state of existence at age x' was computed. Table 1 gives an impressive comparison between these statistics for 1975 and 1982. Reducing a large number of observed data to condensed figures it is visible that the remaining married life time decreases but the years to be divorced increased for an 2o-year-old female between 1975 and 1982 significantly; the same for males. An aspect of this phenomenon is the increase of "paperless" marriages ('cohabitions' within younger age groups.

---

1) Note, that from each state (and each age, consequently) deaths are possible.

Table 1. Expectancies of remaining life time in each marital state of 2o - year - old females

Status at age 2o	Remaining life expectancy (in years) Never married	Married	Widowed	Divorced	Total
			1975		
Never married	5.3o	36.31	8.94	5.o8	55.63
Married	0	4o.63	9.31	5.86	55.8o
Widowed	0	29.53	2o.45	3.89	53.87
Divorced	0	36.15	9.1o	1o.31	55.56
			1982		
Never married	9.87	32.14	8.oo	6.o8	56.o9
Married	0	39.36	8.93	8.16	56.45
Widowed	0	29.48	2o.o8	5.28	54.84
Divorced	0	32.72	8.46	14.99	56.17

Population projections are frequently used for making present and possible tendencies clearly visible. More than that they are important sources of data for policy making.

The changes in marital behaviour refered above in connection with the given age structure of the population may effect the percentage distribution of all marital states and both sexes.

There is a trend that the pórtion of married persons will decrease and that the unmarried will increase. In general, the future population development will bring society face to face with a host of far-reaching problems; especially in terms of care to be provided for persons of advanced age - most of them will be unmarried women.

References:

(1) Land, K.C., Rogers, A.: Multidimensional Mathematical Demography: An overview, International Institute for Applied Systems Analysis (IIASA), Laxenburg 1982, RR-82-35.

(2) Ledent, J.: Some methodological and empirical considerations in the construction of increment-decrement life tables, IIASA, Laxenburg 1978, RM-78-25.

(3) Rogers, A., Willekens, F.: Spatial Population Analysis: Methods and Computer Programs, IIASA, Laxenburg 1978, RR-78-18.

(4) Rogers, A. (Ed.): Essays in Multistate Mathematical Demography, IIASA, Laxenburg 198o, RR-8o-1o.

(5) Tables of Working Life: The Increment-Decrement Model, U.S. Department of Labor, Bureau of Labor Statistics, Washington 1982, Bulletin 2135.

MODELLING AND SIMULATION OF FLEXIBLE MANUFACTURING SYSTEMS (FMS)

M. Frank [1)]

Abstract: Designs of flexible manufacturing systems (FMS) must be inspected using model experiments based on appropriate simulation software (SS). Problems connected with the development of SS will be discussed. Moreover, a special modelling and simulation system suitable for FMS will be presented.

Key words: Discrete simulation, flexible manufacturing system

1. Problems of behaviour analysis of FMS

In order to design efficiently working FMS the designer must analyse its behaviour with the help of simulation tools before the system will be implemented. This demand is objectively conditioned and is connected with the problem of including all relevant aspects within only one model.

FMS are very intricate. This fact can only partially be recognized in fig. 1 . Referring to this simplificated representation of main and help functions of FMS - derived from /3/ - one has to concern:
1) Each type of workstations outlined in the midle of fig.1 represents a class of such stations, 2) all edges represent connections that must be realised by an integrated transport system, 3) each represented subprocess is subordinated to a control subprocess, the performance of which effects the efficiency of the basis process substantially.

The theoretical fundamentals and technological means of modelling and simulation presently available demand problem decomposition. Fortunately, decomposition seems possible without substantial loss of adequacy of results. If necessary decomposition enables the complex analysis to be replaced by a sequence of simpler investigations.

2. Hypothesis for designing of a FMS simulation system

Because of the lack of a founded methodology for decomposition of FMS concerning behaviour analysis designers of simulation tools are forced to work hypothetically. The extension of an already existing simulation system called TOMAS referring to FMS is based on the following hypothesis (see fig.2):
1) The layer of process control has to be considered in the model serving the simulative behaviour analysis. But preconditions that must be provided by process planning and factory planning can be assumed to be fulfilled.

1) Ingenieurhochschule Dresden, DDR-8019 Dresden, Hans-Grundig-Str.25

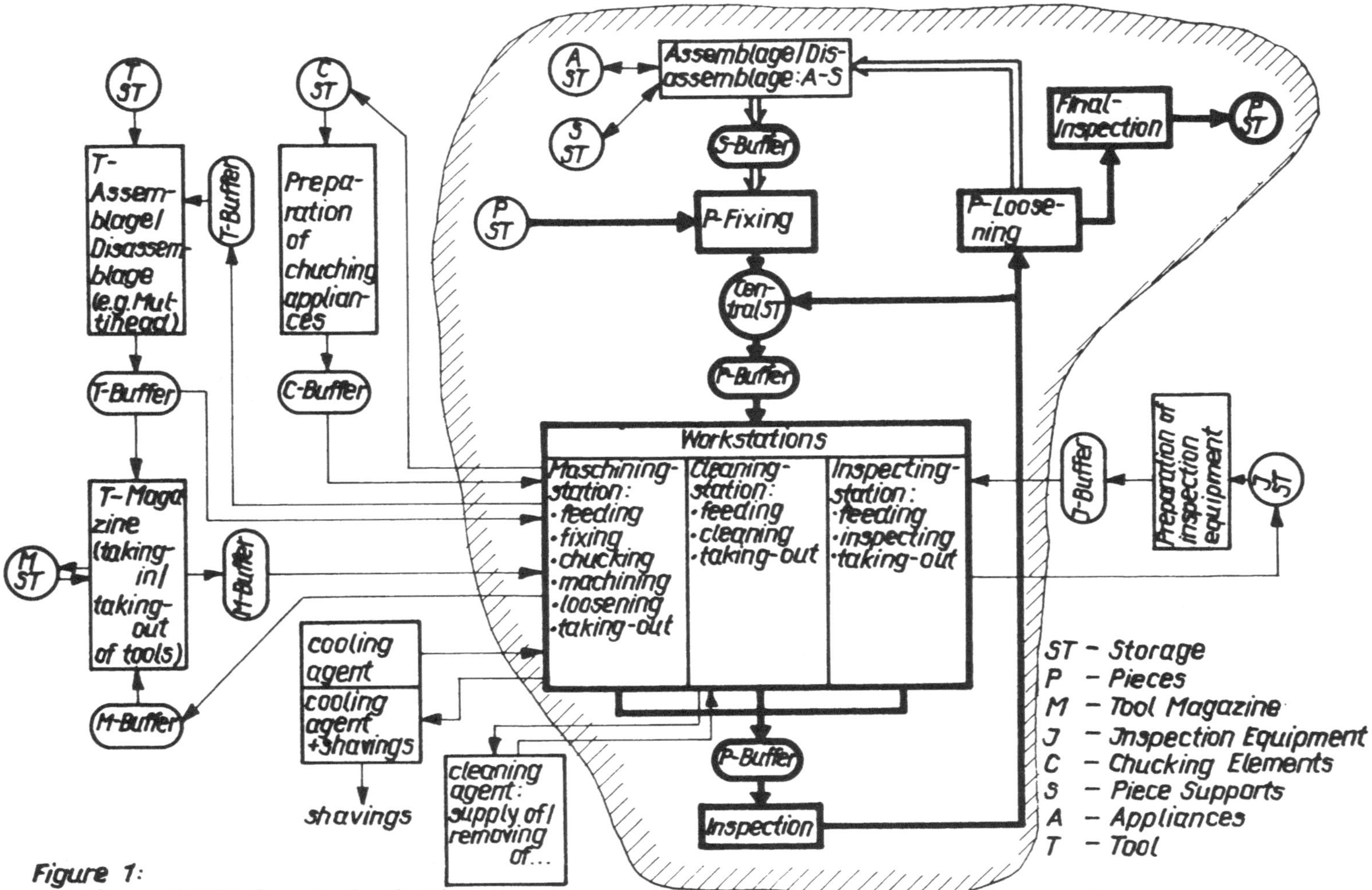

Figure 1:
Functions of FMS (rough structure)

2) It is unnecessary to model device control explicitly. Its effects on machining equipment can be integrated into the behaviour description of the machining units etc.

	Technological preparation, control, execution	Organizational preparation, control, execution
Process planning	Technological ~ (machining sequence, NC-programming etc.)	Organizational ~ (timing, machining programme-composition etc.)
Process control	Technological ~ NC-programme-management, NC-data distribution etc.	Organizational ~ (task-, transport-, tool-dispatching etc.
Device control	Workstation ~ (processing of NC-data)	Means of transport ~ (processing of transport control data)

Figure 2: Layers of FMS Control System

3)Analysis of technological behaviour (effectivity, quality) can be seperated from analysing organizational behaviour (quantity, time). That means one is allowed to assume effects from one of these behaviour components on the other one which will be analysed as given.

4) Analysis of quantity-time-behaviour can initially be reduced to the main process, the material flow. That means effects of toel flew etc. can be included in the model of the main process in an aggregated way by distribution functions, e.g..

According to these hypothesis the extended simulation system TOMAS/ES-2 has been designed in a way that 2-layer models of FMS can be built which represent the basis main process and the assigned process control concerning the quantity-time-behaviour.

3. Characteristic of TOMAS/ES-2

TOMAS/ES-2 has been derived by extension from TOMAS/ES-1 /5/ which has been designed and implemented for modelling and simulation of technological processes with a directed flow of pieces where an explicit modelling of control processes can be more or less neglected. Such production processes are typical for industrial branches with mass or large-lot production. Essentially, extension means insertion of 2 modelling elements into the modular system and functional completion of modules available. The new elements model

- central dispatching operations (scheduling of machining and transport)
- centralized transport control.

The other features of TOMAS/ES-2 are in accordance with those of TOMAS/ES-1: That means:

1. It is matched to non-programming users by making use of computer controlled model building and manipulation based on a combined menu-form-technique.

The user's modelling activities are aided by a system owned structuring concept /4/ characterized by the terms operand, operator, operation rules. That is the process to be simulated must be structured by the user as a graph where the start nodes are sources of operands, the inner nodes are different operators and the aim nodes are hollows for operands. The graph edges represent potential flows of operands. Sources and operators must be matched to the concrete process conditions by operation rule identifiers and parameters.

2. Starting from the user's inputs the system generates a processable simulation model automatically based on a system owned formalization concept derived from the automata theory.

The finally generated internal model represents a network consisting of stochastic asynchronous automata /1/, /2/.

3. The system supplies a standard output table of result data and allows printing of histograms and time sequence diagrams of chosen variables.

These features are the major ones.

References

/1/ Frank, M.: Ein automatentheoretisches Modellkonzept für ein fachgebietsbezogenes Simulationssystem zur Untersuchung diskreter technologischer Prozesse.
Dissertation B, Technische Hochschule Ilmenau 1979

/2/ Frank, M.: A Process-Oriented Modelling Concept for Special Simulation Systems. in: Discrete Simulation and Related Fields (edited by A.Javor)
North-Holland Publishing Company, Amsterdam.New York.Oxford 1982

/3/ PDV-Berichte, Software für flexible Fertigungssysteme.
Kernforschungszentrum Karlsruhe, Januar 1980

/4/ Schmidt, B.: Systemanalyse und Modellbildung. Informatik-Fachberichte 56 (Proceedings 1. Symposium Simulationstechnik Erlangen)
Springer Verlag Berlin.Heidelberg.New York 1982

/5/ TOMAS - Fachgebietsorientiertes System zur Modellierung und Simulation technologischer Prozesse
Anwenderdokumentation, Ingenieurhochschule Dresden, 1984

SYSTEM ANALYSIS OF INNOVATION PROCESSES IN THE FIELD OF FLEXIBLE AUTOMATION

Prof. Dr. habil. H.-G. Lauenroth [1)]

The aim of the system analysis of the innovation process of flexible automation is the determination of the potential of automation, the formation and the variation of automation strategies, and the evaluation of strategy variants. The concept, the algorithmic system, a scenario, and an interactive program of the computer aided decision system REMISA are described.

Keywords: Algorithmic system, flexible automation, innovation, strategy, system analysis

1. Innovations and Innovation Processes

The realization of the economic strategy to develope the national economy of the G.D.R. is depending on the accelerated introduction of the results of the scientific and technological progress. Innovation processes are fundamental elements of this progress. They are processes of the creation, development, use, and diffusion of new products with improved parameters (product innovations) and of new technologies with higher efficiency (process innovations).
Innovations I in the field of flexible automation

$$I = f(D, S, R) \qquad (1)$$

are influenced by 3 factors:

- existing or latently subsisting demand D for rationalization of production processes or of information processes,
- usable scientific or technological problem solutions S in the area of automatic devices and equipments,
- given or possible conditions of realization R regarding hardware, software, material, financial, and other resources.

The consequences of the innovation processes are long term consequences in technical, economical, and social view. Therefore strategies are needed for the planning and for the controlling of the process of flexible automation in the next twenty years to promote the product innovations and the process innovations.

---

1) Academy of Science of the G.D.R.
Institute of Theory, History, and Organization of Science,
G.D.R. 1100 Berlin, Prenzlauer Promenade 149 - 152

Innovation processes are characterized by

- complexity and complicatedness
- influence of stochastic factors
- ability of adaptation and of development.

These characteristics are main characteristics of large scale systems and therefore can be used the methodology and the models of applied system analysis to elaborate innovation strategies.

In this context, three problems have to solved:

- analysis of the innovation situation (fields of demand, fields of solution, typs of innovations, potentials)
- synthesis of innovation processes (factors of influence, phases, basic strategies and strategy variants)
- evaluation of innovation strategies (qualitative results, quantitative variants of efficiency).

The methodological way for elaborating innovation strategies is formalized by a heuristic algorithm [1].

2. Innovation Field of Flexible Automation

The process of flexible automation is a fundamental innovation process, stimulated and promoted by development of the microelectronics. The essential elements of the innovation field of flexible automation are the field of demand:

- production processes to be automatized (pre-manufacturing,assembling, measuring, maintenance, transport, storage)
- information processes to be automatized (computer aided planning/CAP, computer aided design/CAD, computer aided manufacturing/CAM, computer aided accounting/CAA),

the field of solution:

- automatic production technologies, devices, and equipments (NC/CNC machines, industrial robots, integrated flexible manufacturing systems)
- automatic information technologies, devices, and equipments (microprocessors, microcomputers, data processing systems, telecommunication systems),

and the field of realization:

- Material resources (equipments, material, buildings)
- soft/orgware (projects, algorithms, computer programs, operating systems)
- financial resources (costs for equipments, periphery, organization)
- manpower resources (specialists, qualification potential).

Founded on the basic types of innovation strategies two main directions in strategy formation of flexible automation can be identified[2]:

I. Extension of existing elements of the flexible automation, especially by opening new fields of using (assembling processes), or accelerated introduction of industrial robots of the second generation,

II. Design and realization of complex large scale projects of hierarchical production-, control-, management-, and information systems with integrated CAP/CAD/CAM/CAA-solutions and manless production units.

Economical effects of flexible automation must be regarded in complex relations. There are technological, sociological, and innovational factors which influence the economical results of flexible automation:performance data of the equipment, innovation level, fixed and variable costs, growth of productivity and profit, saving manpower, equipment working time, material, and energy, improvement of product quality, stabilization of industrial production processes, social effects regarding content and conditions of labour, and dynamic innovational efficiency[3].

3. Scenarios of the Computer Aided Decision System REMISA

The development, variation, and evaluation of automation strategies is connected with a large expense of data processing and computing time. Therefore a computer aided decision system REMISA has been created, consisting of 5 elements: the user in man-machine dialogue, the communication and control element, the model of the innovation process, the model and method bank, and the data bank. The function of the system is characterized by flexibility regarding targets, restrictions, parameters, and methods of simulation and evaluation.

This function is realized step by step by an algorithmic system

$$S_{\Pi} = \left[\left\{ \Pi_P, \Pi_{St}, \Pi_E \right\}, \left\{ R_1, R_2, \ldots, R_6 \right\} \right] \qquad (2)$$

with algorithms of the determination of the potential of automation (Π_P), for formation and variation of automation strategies(Π_{St}), and for the evaluation of strategy variants (Π_E); these algorithms are connected by a set of relations R_i (Fig. 1).

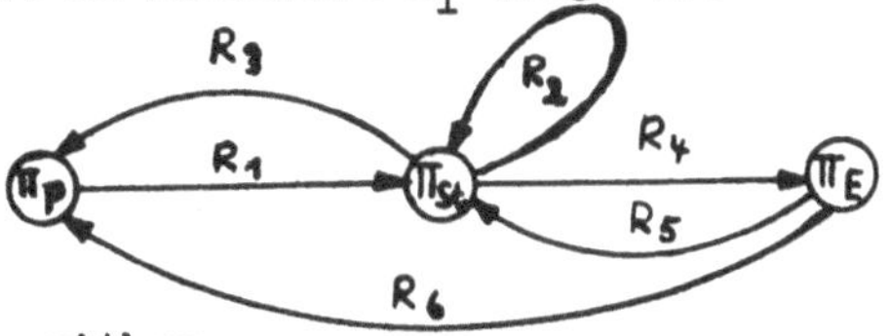

Fig. 1: System of algorithms

A scenario and an interactive computer program FA TREND has been developed for the computing of variants of the industrial potential of those processes which are worth for automation. Depending on the possible degree of automation, on productivity increasing, intensity of using robots, and the expense for the devices, output data can be computed about needed automatized systems and industrial robots, number of manpower and their productivity, costs and efficiency.

References

[1] Lauenroth, H.-G.; Weber, M.: Inhalt, Prinzipien und Methoden der Systemanalyse von Innovationsprozessen. messen-steuern- regeln 26 (1983) 10/11, 542-545, 630-636 .

[2] Haustein, H.-D.: Zur Strategie der flexiblen Automatisierung: Die rechnergestützte Entwicklung und ihre Produktivitätswirkung. Wirtschaftswissenschaft 31 (1983) 7, 1002-1018

[3] Maier, H.: Wissenschaftlich- technische Neuerungsprozesse, Effektivität und Strategienbildung. Wirtschaftswissenschaft 29 (1981) 11, 1313-1329

ADAPTIVE CONTROL OF FLEXIBLE MANUFACTURING SYSTEMS

Peter Bachmann[1] & Knut Richter[2]

The main components of a so-called Flexible Manufacturing system are described. An adaptive control strategy for this system is outlined. The strategy includes recursions describing the change of the state of the system and local control problems appearing as boolean linear optimization programs. Numerical results are cited.

Key words: Flexible Manufacturing System, adaptive control, discrete optimization, approximation algorithms, NP-completeness

Flexible Manufacturing Systems have been designed for several years to introduce automation into small- and medium-scale manufacturing. A Flexible Manufacturing System (FMS) may consist of a number of machines, of one Main Store (MS) and of one Transportation Device (TD) and such a system is being controlled by a uniform strategy. Any transportation operation (TO) between the MS and the machines is performed by the TD, that's why controlling the system means actually controlling the TD.

Theoretically, mathematical models of sequencing type [3] or dynamic models of discrete optimization [4] can be designed for FMSs. There are, however, at least two reasons not to develop such models:

(i) The models will be NP-hard.

(ii) An optimal solution will not be stable in the sense that small changes in the data will make the solution useless.

As a consequence, an adaptive model has been designed and an outline of the model and of the corresponding control strategy will be given in this paper.

1. THE FLEXIBLE MANUFACTURING SYSTEM

Set I of jobs is at fixed places in the MS and any job will be there if it is not being moved by the TD or being processed on a machine. After having processed some job on a machine this job is to be transported back to the MS, since immediate transportation from one machine to another one is not admissible. Jobs cannot be divided and no more than one job may be processed on one machine at a time.

Set S of machines is used to process the jobs and every job can be processed on at most one machine at a time. All jobs are associated with a fixed sequence of machines they have to pass. Each machine is provided with one entry store place and with one exit store place for

1) TH Karl-Marx-Stadt, Sektion Mathematik

2) TH Karl-Marx-Stadt, Sektion Wirtschaftswissenschaften

one job to be processed next and for one processed job waiting for transportation to the MS, respectively.

The Transportation Device is being used to move the jobs from the MS to the entry store places and back from the exit store places to the MS. Only one job can be transported at a time, sometimes the TD will perform other TOs like providing the MS with new jobs or removing completed jobs from the MS.

The control problem can be now formulated as follows: A sequence of TOs is to be determined such that all jobs are transported according to their fixed sequences of passing the machines, no job can be moved to a machine with full entry store place, the TD must not perform unnecessary TOs, the induced processing sequence will secure the jobs to pass the system as fast as possible on an average.

A mathematical model based on this situation and consisting of a finite number of recursions has been developed [1,2] . The state of the system at time t+T is generated by the recursions using the state at time t and the TOs performed within the time interval (t,t+T). Thus a planning horizon is given by T. The position of the TD, the state of the MS, i.e. the occupied and empty store places, the state of the machines and of the associated entry/exit store places, i.e. which jobs are being processed or being stored respectively, the state of readiness, i.e. the information how many machines the jobs have already passed, and other data form the state of the system at a given time.

Using such a model no need arises in restricting the general time horizon and the number of jobs respectively. On the contrary, the subsequent explanation will show that new jobs may enter the system, ready jobs may leave it and that the model can be used as long as no changes associated with the MS, the TD and with the machines occur. The mentioned recursions, however, will not be described here.

2. THE CONTROL STRATEGY

The complexity and the size of FMSs do not make much sense in determining sequences of TOs over a large time horizon because of the mentioned above reasons. Therefore the planning horizon T will be chosen sufficiently small and it must be guaranteed that the necessary information for the deterministic planning within (t,t+T) is provided.

While implementing a generated sequence of TOs for a given horizon a Control Procedure is shifting the horizon, is completing the information and it will find another sequence based on the real state reached by the system. In other words, the Control Procedure is accompanying the real system continuously and it will uptodate the sequence of TOs if the real state differs significantly from the calculated one and if the previously determined planning horizon is reached.

Formally, the Control Procedure consists of two subroutines:
(i) Supervision Procedure: The state of all components of the system is being continuously supervised and being stored. The procedure is checking whether new TOs must be performed and whether a new sequence of TOs is to be determined. In the latter case the planning horizon T will be found and the Choice Procedure will be called up.
(ii) Choice Procedure: The admissible TOs are to be determined which can be implemented within the planning period $(t,t+T)$ and a sequence of these TOs will be found by solving the so-called Local Control Problem.

The use of this strategy has the following advantages. The Local Control Problems arising in the Choice Procedure can be solved efficiently what cannot be said about problems covering a significantly larger planning horizon. Flexible reactions based on determining new sequences of TOs can be reached in the case of any disturbance of the system. Since the model is being used in real time the real state of the system will provide the basis for determining and solving the arising mathematical problems. That the "thoretical" optimum of the process -as ever it may be defined - cannot be reached by this strategy is the general disadvantage of this approach.

3. THE LOCAL CONTROL PROBLEM

The state of the system can be evaluated at any time t by the function

$$F(t) = \sum_{i\in S(t)} F_i(t), \qquad (1)$$

where

$$F_i(t) = (2|S_i| - N_i(t))(t - \bar{t}_i)/T_i, \quad i\in S(t),$$

and the number of machines ith job has to pass is denoted by $|S_i|$, the number of TOs performed for ith job until time t by $N_i(t)$, the time at which ith job entered the FMS by $\bar{t}_i$, the total processing time of ith job by T_i, the set of transportable jobs by $S(t)$.
It is clear that value $F_i(t)$ will rise if t is increasing and ith job is excluded from processing. Therefore, if the general aim consistsin minimizing $F(t)$ every job must be processed automatically after some time. Function (1) has been used to construct Local Control Problems which - after several simplifications - appear as boolean linear optimization problems of the type

$$\sum_{j=1}^{n} a_j x_j \longrightarrow \max, \quad \sum_{j=1}^{k} x_j \le T_k, \ k=1,2,\dots,n,$$
$$\sum_{j\in E_s} x_j \le 1 \text{ or } 2 \text{ or } x_s, \ s\in S, \quad x_j \in \{0,1\} \ j=1,2,\dots,n \qquad (2)$$

which chooses the feasible TOs from $\{1,2,\dots,n\}$ according to technological conditions and to objective (1). Problem (2) can be proved to be a special case of a problem of maximizing a linear function over an independence system with rank quotient = 1/2. A greedy algorithm based on this result and having the worst-case-estimation $2f_G \geq f_o$ can be constructed where the value of the greedy solution is denoted by f_G and the maximal value by f_o. The use of such a greedy algorithm is not only necessary because of the real time control. The mentioned maximization problem over the independence system has the property that there is no polynomial oracle algorithm to solve the problem. That's also why the use of a fast approximation algorithm like the greedy algorithm is recommended.

4. NUMERICAL RESULTS

The described recursions and algorithms have been implemented in PL/I on the computer EC 1040 and 100 instances with randomly generated data have been solved. In these instances 20 jobs have to be processed on 5 machines. Every job has to pass from 2 to 10 machines, i.e. there are jobs passing some machines several times. The entry times of the jobs into the system were also randomly generated. The total time to solve the problems was 17.2 sec. on an average, i. e. with respect to the time the model can be used for real time control. According to the numerical tests the average time an ith job has spent in the FMS was equal $2T_i$.

References:

[1] Bachmann P., Forschungsberichte zur Aufgabe IGFA-Teilefertigung, Teil 1 - 6, TH Karl-Marx-Stadt, Sektion Mathematik, 1981-84

[2] Bachmann P., Mathematische Analyse und näherungsweise Lösung eines verallgemeinerten Matroidproblems, Diss. TH Karl-Marx-Stadt 1984

[3] Brucker P., Scheduling, Studientexte Informatik, Akademische Verlagsgesellschaft Wiesbaden 1981

[4] Richter K., Dynamische Aufgaben der diskreten Optimierung, Akademie-Verlag 1982

ANALYSIS, MODELLING AND CONTROL OF COMPLEX MANUFACTURING SYSTEMS

G. VOIGT, J. LIPPOLD [1)]

An universal system of computer-aided design, computer-aided-manufacturing, computer-aided planning, computer-aided control and computer aided-accounting of a flexible fabrication is introduced. The planning and control are carried out by aid of man-computer-dialogue and simulation.
Keywords: Flexible manufacturing, computer-aided design, computer-aided manufacturing, computer-aided planning, computer-aided control, man-computer-dialogue, simulation.

Complex manufacturing systems are typical examples of large-scale social, economical and technical systems. The increasing demand for flexibility and effectivity of fabrication require the building up of universal systems of computer-aided preparation, course and accounting of manufacturing [2]. Such a system roughly designed and partially realized by us is shown in fig. 1. The structure and function of the original system are reflected by a hierarchically structured system of discontinuous discrete models [4], [5]. The system of models is structured as a multi-goal-multi-level-system and/or as a multi-layer-system [3].

By designing the system we mainly relied on the principle that human actions and decisions are effectively to be connected with the operations of a computer network. This is realized with the following strategy. The elements and levels of the system act and decide responsibly in given boundaries. This is made possible by fuzzy couplings, fuzzy goals and fuzzy instructions and by using man-computer-dialogue-systems. The dialogue is worked out in computer-guided and/or operator-guided manner with the aid of fuzzy programs (this depends on the problem to be solved). In fuzzy programs algorithmic sections are connected with sections in which the operator answers questions asked by the computer (computer-guided part of the dialogue). The decisions are found by information about the situation. The set of these information is called situation indicator. Furthermore the decisions are based on a set of fuzzy advices given by the computer. The operator is also able to ask additional information, made available and calculated by the computer (operator-guided part of the dialogue). The set of fuzzy advices is developed automatically by a learning process. Based on theoretical considerations the set of advices is developed step by step by collecting, storing and calculating experience. For instance by sliding scheduling of manufacturing programs extending of five years there is worked out a set of variants by means of a fuzzy program. The foundations of the manufacturing programs are the forecasted demand of each position of material, the forecasted demand of processing time at each working place, the forecasted manufacturing costs, including the costs of circulating

1) Prof. Dr. Gerhard Voigt, Dr. Joachim Lippold
University of Engineering at Mittweida
9250 Mittweida GDR

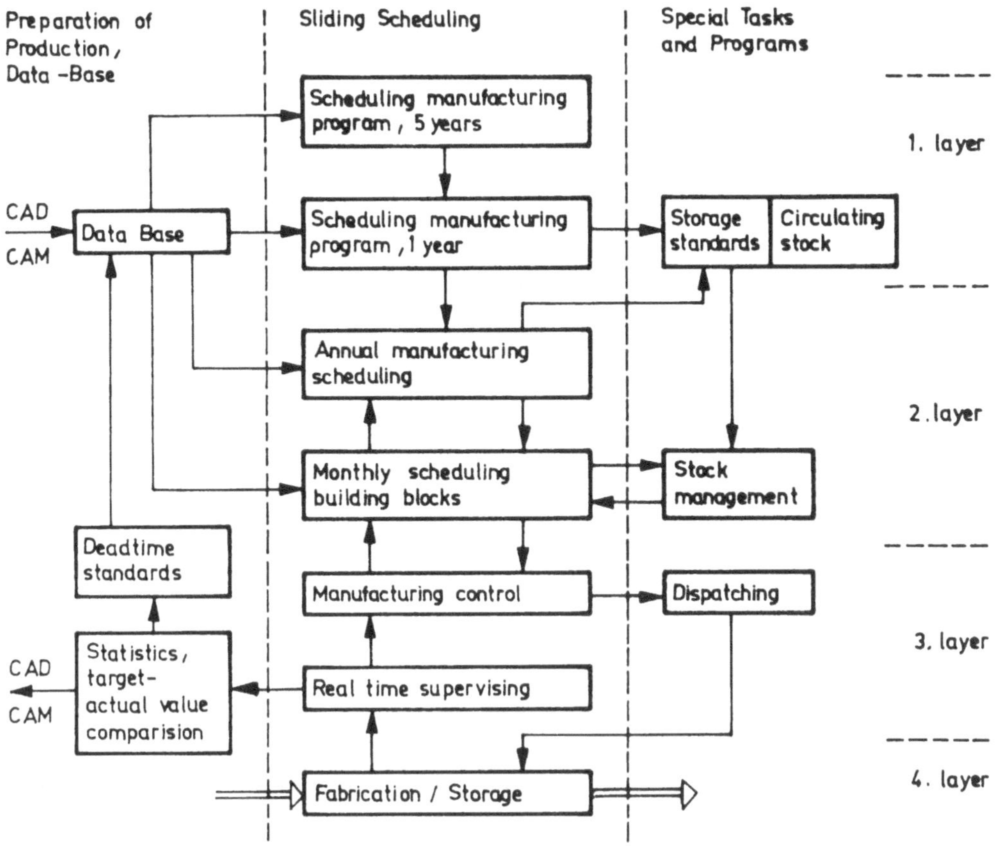

CAD Computer-aided design
CAM Computer-aided manufacturing

Fig. 1 A general system of computer-aided preparation, course and accounting of manufacturing

stock and forecasted gain. The forecasted demands are balanced with the stock probably being available. Capacities are equalized with demanda. Sets of manufacturing programs and the equalization of capacities with demands are worked out in man-computer-dialogue because in general it is not possible or appropriate to establish algorithms for each necessary decision.

In the universal system of computer-aided preparation, course and accounting of manufacturing there is used a principle of step by step making up to date and precizing of schedules, programs, instructions and advices given. This leads for instance to the principle of sliding scheduling. This principle is shown in figure 2 for a manufacturing program extending over five years. In each year n at the scheduling point r (time of the computer run) there are worked out plans for the years n+1 to n+5. The precision of the plans rises step by step from n+5 to n+1 with growing knowledge about customerš requests, about technical and technological parameters and so on. The same principle is used by annual.

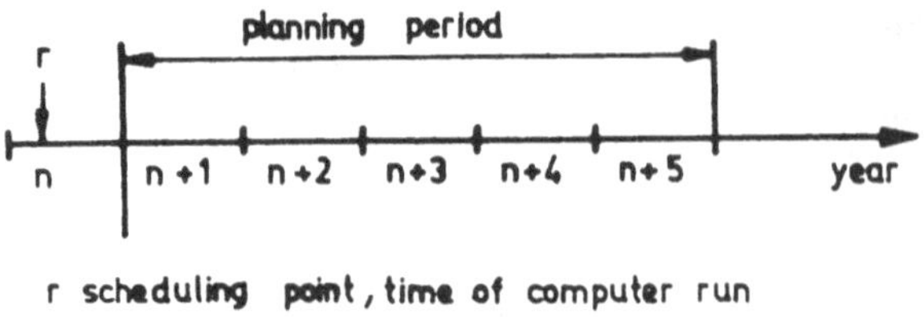

Fig. 2 Principle of sliding scheduling

manufacturing scheduling, by monthly scheduling of fabrication building blocks and by the real time control of the manufacturing process. In scheduling manufacturing programs extending over five years or one year deterministic models and expectation values are used. Demands and capacities are regarded as even distributed all over a year. In annual manufacturing scheduling and monthly scheduling of building blocks fabrication plans for monthly periods are worked out taking into consideration the monthly fluctuations of demands and capacities. Here deterministic models and expectation values are also used. By the manufacturing control the deadlines for the beginning and finishing of the charges and the treating times per charge are forecasted and fixed. The manufacturing process is controled based on real time supervision. Manufacturing control is performed by the aid of stochastic models. Before making any decision the probable behavior of the system is forecasted by simulation.

As a result the following statement shall be made. The main problem in analysis, modelling and controlling of complex manufacturing systems is the concerted action of a lot of different findings and methods of different scientific disciplines such as technology, mathematics, information-processing, cybernetics and so on. The development of new methods and the use of highly specialized methods is in second order interest.

References

[1] Klabuhn, H.-D.; Matern, M.: Mensch-Maschine-Kommunikation im Schnittpunkt von Automatisierung und Arbeitstätigkeit, msr, 27 (1984) 11

[2] Lauenroth, H.-G.; Weber, M.: Inhalt, Prinzipien und Methoden der Analyse von Innovationsprozessen, msr, 26 (1983) 10,11

[3] Mesarovič, U.D.; Macko, V.D.; Takahara, Y.: Theory of Hierarchical Multilevel Systems, New York, London 1970

[4] **Voigt, G.; Cramer, S.: Diskontinuierliche Fertigungsprozesse, Analyse, Modellierung, Steuerung, Grundlagen und Anwendungen, Akademie-Verlag Berlin, in Vorbereitung**

[5] Voigt, G.: Entwicklungstendenzen bei der Analyse, Planung und Steuerung technologischer Prozesse im Gerätebau, Feingerätetechnik, Berlin 32 (1983) 1

SYSTEM MODELLING OF INTEGRATED CIRCUITS

D.J.Zanevitchius, A.I.Bashkys, L.L.Sakalauskas[1)]

A non-linear transistor model on transconds[2)] is proposed in this paper. The problems of the system modelling of integrated circuits (IC) on a micro-computer by the help of the above model are considered. The description of the sofware DIALOG is given here as well. The results of the concrete IC analysis when using the above software are introduced.
Keywords: transistor model, transcond, system modelling, micro-computer, software DIALOG, abstraction.

Large IC and, especially, very large IÇ are the most complicated objects ever done by a human being. The analysis and synthesis difficulties of such objects are complicated because of the absence of the decomposition approach for IC, what means that IC are to be considered as a system.

On the other hand, as a non-linear system of the large scale production, IC may be considered as a stochastic system as well. The investigation of such a system requires the mathematical modelling together with the usage of non-linear methods and statistical modelling. The modelling is required to be performed on a micro-computer. When modelling such systems the computation time is the determinanting factor which is mainly defined by the type of the function, the model of the non-linear IC element - a transistor is described by.

The initial differential equations have been assumed as the non-linear ones when deriving the adoptable classical non-linear transistor models (e.g. the Ebers-Moll model). Hence the non-linear transistor characteristics are represented in the form of exponential functions exp or functional exp, e.g. $erfc$.

The IC model, designed on the base of the classical non-linear transistor models, is the system of non-linear differential or transcendential equations, the solution of which is possible by the iterative methods only, what requires much computation time.

If we consider that the parameters (the diffusion coefficient, the average life-time. etc.) in the initial differential equations describing the transfer processes of the charge carriers in the transistor base region are not the constant ones (what has not been taken into account in the classical models) then, in most cases, these equations are solved in the

1) Academy of Sciences of the Lithuanian SSR, 232600 Vilnius, K. Poželos 52, USSR.

2) The term "transcond" is referred to as a non-linear, controlled conductivity, in the form of which the equivalent circuit model is represented.

form of the second rank functions spx, $sapx$, cpx [1]. On contrary to the transcendential functions the employment of the second rank functions leads to the IC representation in the form of the differential or algebraic equations which, in most cases, may be solved in the explicit form. Besides, the second rank functions poseess the finiteness, what allows us to obtain the analogue-digital models, what is very important when modelling digital circuits.

If the diffusion coefficient and the average life-time of the minority charge carriers are assumed as density-dependent n by the law n^{-1} then the transistor model is represented as the second rank function.

$$
\begin{aligned}
i_b &= (1-\alpha) J_{eo}(1+\tau_\beta s)\, sp\gamma_{ue} + (1-\alpha_J) J_{KO}(1+\tau_{\beta J} s)\, sp\gamma_{uK},\\
i_K &= -\alpha J_{eo}\, sp\gamma_{ue} + (1+\tau_{\alpha J} s) J_{KO}\, sp\gamma_{ue},
\end{aligned}
\tag{1}
$$

where $\gamma_{ue} = \frac{U_{eb}}{\psi_e}$, $\gamma_{uK} = \frac{U_{Kb}}{\psi_K}$, ψ_e, ψ_K, J_{KO}, J_{eo}, α, α_J, τ_β, $\tau_{\beta J}$, $\tau_{\alpha J}$ – are the transistor model parameters; $s = \frac{d}{dt}$; i_b, i_K – the base and collector currents; U_{eb}, U_{Kb} – the voltages of the emitter-base and the collector-base. The model obtained is referred to as the non-linear transistor model on transconds.

The IC mathematical model is desired to be the systematic one and to cover both the IC circuit technology and the main technological process models of the manufacturing, e.g. that of the diffusion. If the diffusion coefficient of the atomic impurities is assumed as concentration-dependent N then the distribution profile of them is determined by the $sap^{-1}x^2$ function, what enables us to obtain the explicit dependences for both the main transistor parameters and IC on the design and technological parameters of the diffusion layers.

The usage of the IC system model representes in the form of the explicit analytical expressions enables us to deny the iterative solution methods and to shorten the calculation time from 10 till 100.

In the Academy of Sciences of the Lithuanian SSR the software DIALOG has been worked out for the statistical modelling of the digital IC designed on the ECL valve base. In the above software the transistor is represented in the form of the model on transconds. The data abstraction has been used to shorten the volume of the required memory. When abstracting a number of parameters M included into the IC mathematical model are devided into separate groups of the same type. Besides, the correlation matrixes may be devided into separate cells, what enables us to reduce the number of the operations when processing the correlation matrixes from $O(M^3)$ to $O(M)$.

This software has been worked out on one of the algorithmic languages FORTRAN-IV and consists of the following subprogram modules: the statistic IC modelling and the yield, the sensitivity coefficient calculation, the calculation of the parameter distribution curves, etc. This software may be applied as the optimization means when working under dialogue regime.

Fig. 1 represents some of the results of the concrete IC optimization by the application of the software DIALOG when the variables are both - the specific resistance of the IC base layer R_s and the width of the IC diffusion resistors d, while the objective function is the yield.

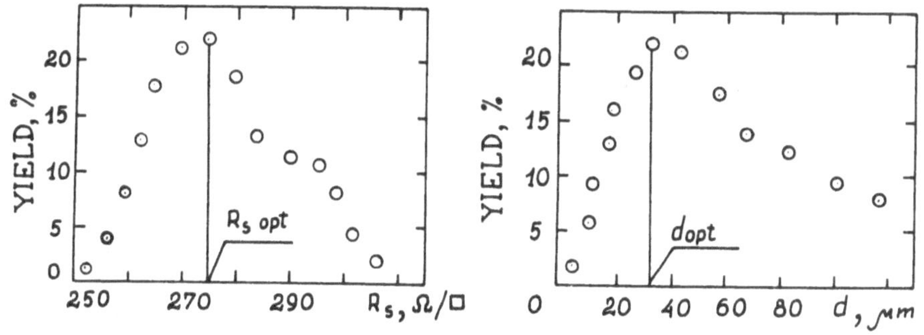

Fig. 1. The concrete IC yield dependences on R_s, d.

The fact that the transistor model on transconds allows the IC parameters to be obtained in the explicit form, gives us the opportunity to apply if for the symbolic IS modelling. For this purpose a special algorithm has been worked out which formalize the obtaining of the IC parameter analytical expressions. The realization of such an algorithm allows us to obtain the analytical expressions of the investigated IC parameters when applying a computer.

Literature

[1] Zanevichius, D.J.: Nonlinear Simulation of Integrated Circuits. Matematicheskoe i mashinnoe modelirovanie v mikroelektronike. Academy of Sciences of Lithuanian SSR, Vilnius, 1984, 3-17.

USE OF WAVE DIGITAL NETWORKS FOR TIME DOMAIN SIMULATION OF MIXED LUMPED AND DISTRIBUTED CIRCUITS

Authors: K. Cséfalvay[1/]., V.Krizs[2/], L. Petróczki[3/]

Abstract

The paper deals with the wave digital simulation of circuits. A simulation program based on this method is presented. The basic idea of wave digital representation, the network modelling procedures are described and an example is also shown. The program was developed as a kernel of a future CAD system.

Keywords: Simulation, wave digital model, wave parameters, distributed and nonlinear networks.

1. Introduction

Wave digital modelling of networks was introduced by Fettweis in the second half of the 70's [2] . Originally, the method was applied to digital wave filters. (Fettweis [2], Meerkötter[3])
We have extended the field of application [4] . In our models linear distributed parameter components and nonlinear resistive single ports are also allowed.
In our method, at first the continuous time wave parameter model of the simulated network is formed. The time domain excitation functions of the original network are sampled, consequently the excitations and responses of the wave parameter representation will also be sampled, producing the wave digital model of the original network.
Since both the excitation functions and the response functions are approximated by sampled values, we call our model the wave digital model approximating the continuous time wave parameter model of the network.
The wave digital model provides fast and efficient means to monitor the time domain responses (voltages and currents) at any node of the simulated network, at any discrete instant of time. These responses may be subjected to further processing, e. g. DFT.
Sampling offers to use as excitation any function of time, including random, pseudo-random and tabled functions. Obviously, if the input to a network component can be tabled, so is the output as well, therefore nonlinear components can be characterized by their output tables, or by their sampled output functions. The sampled characteristics enables to define a wave digital circuit corresponding to the nonlinear component.

1/ Assistant lecturer. Technical University of Budapest. Faculty of Electrical Engineering.

2/ R and D advisor. Works of Electronic Measuring Instruments (EMG). Budapest

3/ Development engineer. Works of Electronic Measuring Instruments (EMG). Budapest

2. Wave parameter representation of networks

2.1. Converting the conventional U-I representation into A-B wave parameter representation.

Relationship between the conventional U-I parameters and the A-B incident wave and reflected wave parameters is as follows:

$$A = U + Z_o I$$
$$B = U - Z_o I \qquad /2.1/$$

Where Z_o is the arbitrarily selected normalizing impedance.
In the following the wave parameter characterization of a few, more frequently encountered components are given.

2.1.1. Impedances

In the U-I representation the impedance is characterized by Z impedance /or Y admittance./. In the wave parameter representation the S wave characteristics determine the relationship between the incident wave A, and the reflected wave B. B =SA and S = B/A, consequenly we get from /2.1/:

$$S = \frac{Z - Z_o}{Z + Z_o} \quad \text{or} \quad S = \frac{Y_o - Y}{Y + Y_o} \qquad /2.2/$$

For nonlinear ports /2.1/ is replaced by the U = U/I/ or by the I= I/U/ characteristics, respectively. These characteristics may either be given in analytical form or as function tables. If the characteristics is available in analytical form the nonlinear B = F/A/ wave characteristics can be computed in a similar way as shown above.
If with the analytical solution mathematical difficulties arise, then the corresponding A-B pairs should be formed, and the results are to be presented as a table.

2.1.2. Generators

The characteristics of a voltage source is $U = Z*I + E$
The wave parameter characteristics /using 2.1/ will be: $B = SA + S_E E$

$$\text{Where} \quad S = \frac{Z - Z_o}{Z + Z_o}, \quad S_E = \frac{2*Z_o}{Z+Z_o} \qquad /2.2/$$

2.1.3. Twoports

Arbitrary values are selected for Z_{o1} and Z_{o2} normalizing impedances and the /2.1/ system of simultaneous equations is applied to both ports, and using algebraic methods, the wave characteristics can be obtained from any other one /Z,Y,H,K,A,B/ as: **B** = S **A**.

2.1.4. Distributed networks

The component of the distributed network is considered as a twoport, and it is characterized either by its Z,Y or its A characteristics.

If the distributed network is a transmission line having a Z_c characteristic impedance and a $p = \alpha + j\beta$ propagation coefficient, and selecting the normalizing impedances as $Z_{o1} = Z_{o2} = Z_c$ the wave parameter representation is obtained very simply as: $B = S*A$

$$S_{11} = S_{22} = 0 \qquad S_{12} = S_{21} = e^{-ph} \qquad /2.3/$$

where h is the line length.

2.1.5. Coupled transmission lines.

Coupling between transmission lines can be accounted for by introducing a 2n-port component representing the coupling between n transmission lines. The wave parameter characterization will be simple, if the normalizing impedance matrix is selected so as to be equal to the characteristic impedance matrix.

The wave characteristics will be: $B = S*A$

$$\underline{\underline{S}}_{11} = \underline{\underline{S}}_{22} = \underline{\underline{0}} \qquad \underline{\underline{S}}_{12} = \underline{\underline{S}}_{21} = e^{-\underline{\underline{P}}h} \qquad /2.4/$$

Since matrix $\underline{\underline{P}}$ is symmetric, therefore

$$\underline{\underline{S}}_{12} = \underline{\underline{S}}_{21} = \underline{\underline{Q}} \cdot e^{-\underline{\underline{p}}h} \cdot \underline{\underline{Q}}^{-1} \qquad /2.5/$$

2.2. Connecting component to make a wave parameter network.

The conventional networks are built of n-port components, and ports are connected either in parallel or in series or they are cascaded. These connections are modelled in wave parameter representation by serial or parallel adaptors. The characteristics of these adaptors are obtained by replacing the currents and voltages by the respective wave parameters using /2.1/.

The topology of the modelled network will differ from that of the original one.

The topology of the cascade connection is the same in both models.

If the normalizing impedances of two connected ports differ, then a matching adaptor should be inserted between the two components.

2.3. Discrete approximation. Wave digital network.

First the wave digital model of the conventional network is found in the complex frequency domain, then the bilinear transform is applied to the wave parameter characteristics. The bilinear transform gives the relationship between the input and output variables in the form of a difference equation. The outputs of the delay elements are the state variables of the model. The simultaneous differential equations of the model are composed from the difference equations of the components and from equations considering the connection topology of the model.

The variables of the simultaneous differential equations are the state variables and the node variables.

Solving the difference equations in a prescribed order, the state variables and node variables can be computed in consecutive time instans. The prescribed order of the computation is equivalent to following the path of the signal propagation.
The responses of the model are computed from the waves passing through the selected nodes, using the definition formulas /2.1./.

3. The "Time" program system.

The program is written in BASIC, it is running on EMG 777, a Hungarian made graphic professional computer.
The wave parameter model of the conventional network should be formed first by the user before simulation.
For the time being, the component set consists of the following:

- voltage sources
- R,L,C elements
- linear, distributed transmission lines
- linear, distributed, lossy transmission lines
- monotone, nonlinear resistive ports
- digital filter components. /canonic realization/

The input for the program is the type, the serial number of the component and the nodes connecting to it. The program automatically establishes the wave digital model, computes the parameters of the adaptors and determines the path of the signal propogation. The solving of the network difference equation is done along this path, starting from the excitations and state variables. There is no need to store the matrix of the coefficients of the difference equations, therefore the program is capable of handling models of great networks.
The main features of the program are:

- very "friendly" and interactive.
- any time function can be used as excitation input.
- nonlinear elements are represented by their U-I characteristics, and are stored in the element library.
- the current or voltage functions at any arbitrarily selected node or nodes are provided for.

The method /and the program/ is expected to find its applications in fields where its features are the most advantageous such as:

- theoretical examinations of lossy and loss-free transmission lines
- overvoltage protection of transmission lines with nonlinear single port limiters.
- analysis of interconnecting lines among high speed logic circuits

4. **Example**

Let us examine the switch-on transients of a lossy transmission line terminated by a capacitor at one end and excited by a step voltage at the other. The circuit and its wave digital model are shown on fig.4.1. Fig. 4.2 is the hardcopy output of the "Time" program. As a response the voltage across the capacitor is monitored.

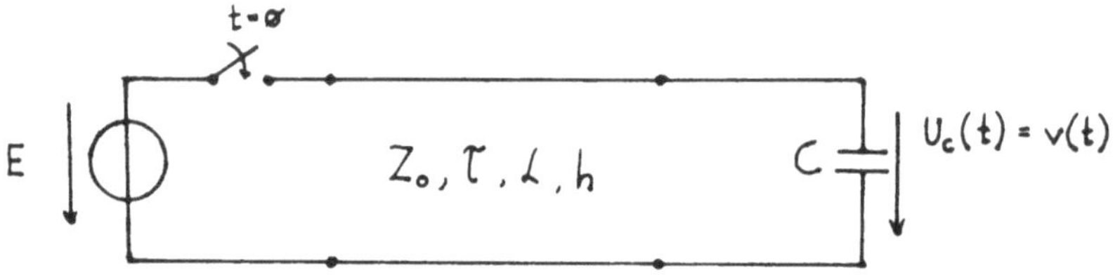

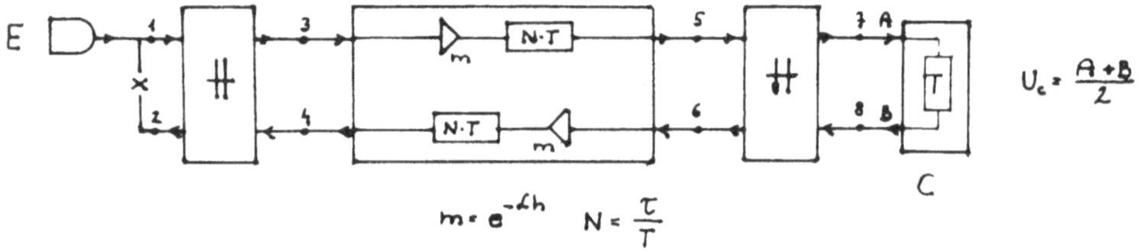

Fig. 4.1.

Lossy transmission line and its wave digital model.

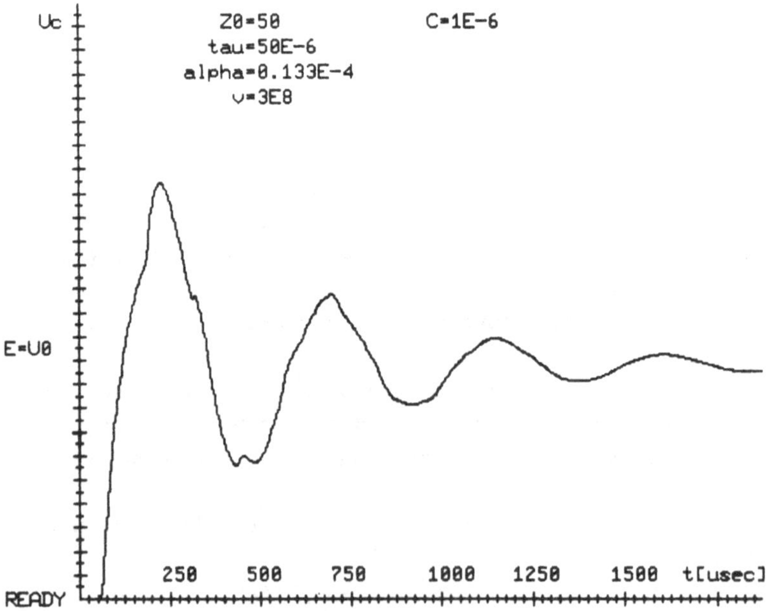

Fig. 4.2.

Response of the circuit of fig. 4.1. on step voltage.

References

[1] Simonyi Ernő: The basics of digital signal processing (In Hungarian) Budapest 1984

[2] Fettweis: Canonic realization of ladder wave digital filters. Int.J.Circuit Theory Appl. Dec.1975.

[3] K. Meerkötter: Canonic realization of wave digital filters involving unit elements. Int J.Circuit Theory Appl. Oct.1983.

[4] Cséfalvay K.: Discrete approximation and wave parameter simulation of mixed component networks. Doctoral thesis. (Manuscript in Hungarian) Budapest 1984.

FINITE ELEMENT METHOD TO 3-DIMENSIONAL SIMULATION OF SEMICONDUCTORS

Reschke, D. and Schönefeld, R. [1)]

Summary: Different semiconductor models are presented. The mainly used numerical methods to simulate these models are shortly compared to favour the FEM. Some considerations to the choice of the elements, the form functions, and the boundary conditions are made, where the case of boundary conditions of the 3.kind requires a special projection to optimize accuracy. The executable programs BIFIN and TRIFIN are characterized shortly.

1. Introduction

The two main directions in managing the whole circuit design problem are system design and technological design. System design is subdivided in system specification, partitioning, logic design, network design, up to layout. The technological design consists of the process design in dialog with the process simulation and the component design in dialog with the component simulation. Component simulation informs about physical processes in the component and about the overall behavior of the single component. The importance of such simulations increased strongly with the subminiaturization of the components for the following reasons:

- analytical solutions are non existent
- experimental methods are failing because of deficiency of suitable measuring methods
- some semiconductor effects demand a 3-dimensional computation

2. Alternative semiconductor models

Under the assumption rot $\vec{E} = 0$ (vortex free field) and in case of completly ionized defects the semiconductor model equations are derivable from the MAXWELL equations comprising the POISSON equation (1) and the continuity equations (2), (3)

$$\Delta \Psi = -\frac{q}{\varepsilon} (p - n + N_D - N_A) \qquad (1)$$

$$\frac{\partial n}{\partial t} = \frac{1}{q} \operatorname{div} \vec{j}_n - R(n,p) \qquad (2)$$

$$\frac{\partial p}{\partial t} = -\frac{1}{q} \operatorname{div} \vec{j}_p - R(n,p) \qquad (3)$$

where

Ψ	electrostatical potential	N_D, N_A	ionized defects
q	electric unit charge	$\vec{j}_n$, $\vec{j}_p$	current density
ε	dielectric constant	$R(n, p)$	generation-recombination rate (nonlinear)
p, n	carrier density		

---

1 Institute of Technology Ilmenau, DDR - 6300 Ilmenau, PSF 327

To govern some problems in a better way there were introduced new variables which led to other formulations of the semiconductor equations. A new quantity is the FERMI level corresponding to the carrier densities. The alternative model is described by

$$\Delta \psi = \frac{q}{\varepsilon}\left[e^{\psi - \varphi_n} - e^{\varphi_p - \psi} - N \right] \qquad (4)$$

$$\frac{\partial n}{\partial t} = \frac{1}{q} \nabla (q \cdot \mu_n \cdot n \nabla \varphi_n) - R(n,p) \qquad (5)$$

$$\frac{\partial p}{\partial t} = -\frac{1}{q} \nabla (q \cdot \mu_p \cdot p \nabla \varphi_p) - R(n,p) \qquad (6)$$

where

$N = N_A - N_D$ $\qquad$ μ_n, μ_p carrier mobility

φ_n, φ_p FERMI potential

By introduction of the new variables $u = e^{-\varphi_n}$ and $v = e^{\varphi_p}$ from (4) to (6) results a further model with the advantage that the new variables u and v are nearly constant over a large scope. This allows to reduce the number of mesh points in a numerical solution. Additional models can be found by neglecting of R (n, p) and introducing a vector potential as well as substituting the electric field strength into the continuity equations.

3. Numerical methods

Appropriate numerical methods for simulation of the in 2. given models are

i) finite difference method (FDM)
- single mesh and
- multi mesh methods

ii) finite box-difference method (FBDM)

iii) finite element method (FEM)

The FDM has the essential disadvantage that mesh points or nodes are located on straight lines parallel to the axis and a mesh refinement for an increasing of the solution accuracy in a certain area also creates additional nodes in "noninteresting" areas of the solution. A further disadvantage of the FDM is the complicated incorporation of boundary conditions in case of curved boundaries, especially with recard to an automatic generation of the mesh on a computer. The FBDM allows to end a mesh line in the solution area on a so called finite box. This reduces the number of nodes. But the FBDM has nevertheless three disadvantages, which are first the restrictions for the ending of mesh lines, second the additional conditions for the mesh size ratio, and third the relative complicated handling of boundary conditions.

The fundamental statements of the FEM are either variational methods (RITZ) or residual methods (GALERKIN). The FEM allows an optimal subdivision of the solution area and a relatively simple managing of the boundary. FEM is good suited for an automatic preparation on computers.

4. Choice of elements, form functions, boundary conditions

Because of the mentioned benefits the FEM was chosen. With the choice of elements some aspects are to be considered. The triangle is the simplest closed 2-dimensional element with straight edges. With triangular elements curved lined solution areas can be area-covering filled and moreover they allow a good approximation of the boundary similar as a polygon. A quadratic approximation of the form functions results from a tradeoff between effort and accuracy. For this reasons 6-node triangular elements have been chosen in the 2-dimensional case. An analog to the triangle is the tetrahedron in the 3-dimensional space. To meet program compatibility it is beneficial to use tetrahedron elements in the 3-dimensional case, because they have 6-node triangular elements as boundary areas. Fig. 1 shows the used elements and the numbering of the nodes in the implemented programs. To realize quadratic form functions 10-node tetrahedrons have been applied for three dimensions.

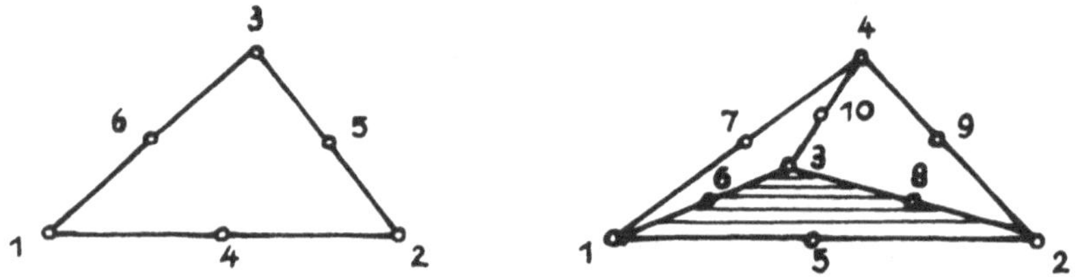

Fig. 1

After introduction of triangular as well as tetrahedron coordinates, which are beneficial for mathematical integration, one obtains 6 and 10 form functions and element matrices of 6 x 6 and 10 x 10 respectivily. For realization of boundary conditions of the 3.kind in the 3-dimensional case, how they occur on the interface Si-SiO_2, the following consideration is necessary. The case of an arbitrarily in the space located boundary area is the most common. The relation of a surface integral in space and an area integral by a parameter representation of the area is given by (7)

$$\iint_F f(x,y,z)\,dF = \iint_D f(\alpha(u,v), \beta(u,v), \gamma(u,v))\sqrt{EG-F^2}\,du\,dv \qquad (7)$$

In (7) is $x = \alpha(u,v)$; $y = \beta(u,v)$; $z = \gamma(u,v)$ the parameter representation of the area F. u and v hold for the area D in the u, v - plane. In case of a projection of the area in the xy-plane one obtains after calculation of the quantities E, G and F the formula

$$\iint_F f(x,y,z)\,dF = \iint_D f(x,y,\gamma(x,y))\sqrt{1+P^2+Q^2}\,dx\,dy \qquad (8)$$

where $P = \frac{\partial \gamma}{\partial x}$; $Q = \frac{\partial \gamma}{\partial y}$

Corresponding formulas can be found for both the other two projections. For the computation error it is not arbitrarily on which of the three coordinate planes the area is projected. In general it is promising to choose that projection plane in which the area appears as the largest, because otherwise in the most unfavourable case infinitly smal areas are multiplied with infinitly large quantities $\sqrt{1 + P^2 + Q^2}$. It should be approached to project in such a plane in which the root expression ist the smallest. In the program package there is a program PROJEK minimizing the root expression to define the optimal projection plane.

5. Program systems

Two PASCAL-program systems have been developed. BIFIN for simulation of 2-dimensional problems and TRIFIN for such of three dimensions. Each of them handles linear and nonlinear elliptic and parabolic partial differential equations with boundary conditions of the 1., 2., and 3.kind. Furthermore the continuity equations can be computed. For parabolic typs there is a fixed or an automatic timestep. BIFIN and TRIFIN are structured by a lot of subroutines, which by itself are subdivided in programs for input/output and processing. For each solution case exists a main subroutine.

6. References

[1] Reschke, D. : Untersuchungen zum Aufbau von Programmbausteinen für die Simulation von Halbleiterbauelementen mit zwei- und dreidimensionalen Modellen unter Verwendung der Methode der finiten Elemente. Diss. B, TH Ilmenau, 1985

SIMULATION OF RADAR CLUTTER PROCESSING

Jarmila Malá, Jaroslav Vacík [1)]

Summary

This paper deals with the problem of filtering a useful radar signal from the clutter. In the first part a short description of the problem is under the study, further the methods and technics applied below are stated. The unwanted signal, so called clutter, which is formed by radar echoes from ground objects, meteorological objects (cloud, rain, snowfall) is treated as a random signal with Gaussian distribution function and known spectral density (or autocovariance function). The effects of the signal sampling and discretisation are studied too.

Keywords : Signal filtering, radar clutter, hybrid computer simulation, random signal generation

Introduction

The radar echo is composed of the useful signal (target) and the unwanted one (clutter, receiver noise). In the normal environement for air-traffic-radar control in the local airport radar station the power of useful echoes is satifactorilly beyond the receiver noise and this component of the signal is not interesting for the purposes stated above. But the clutter component is very important for the safety of air traffic. Radar echoes returning from stationary targets can be eliminated by means of moving target indicator (MTI). So the main interest is devoted to the target of meteorological character (clouds, heavy rain, snowfall, hail) and other moving objects (flocks of migrating birds, etc.). First of all some published data about spectral densities of such unwanted targets were used. On basis of these data computer models of clutter were constructed.

Time and Amplitude Discretization

For the generation of a random signal with Gaussian probability density and prescribed power spectral density (or autocovariance function or matrix of covariance coefficients-Toeplitz matrix) the folowing procedure was used: a random signal generator was used as the input to a shaping filter. Its transfer function was chosen so that the power spectral density (PSD) of the output signal was as close as possible to the predescribed one . (In some cases a linear shaping filter could not be found so that the output had not the prescribed PSD; then some approximation had to be used).

---

1) Institute of Information Theory and Automation, Academy of Sciences Pod vodárenskou věží 4, 182 08 Praha 8, Czechoslovakia

The influence of time and amplitude discretization was studied in the following way: Some criteria for sampling frequency and number of discretisation levels were found (in fact equal to the length of the computer word) with regard to the accuracy of the computed PSD or covariance matrix coeficients (Toeplitz matrix). When the suitable limits for the sampling frequency and the number of discretisation levels were found the next problem was to design the algorithm for the adaptive filtering removing the clutter of different origin from the received radar data.

Model of Radar Clutter

In the following part of our work a mathematical model of radar clutter was used according to / 2 /.

Let us consider the clutter target as a cloud of dipole scatterers with random rotational and drift velocities where σ_r and σ_d are the standard deviations of the rotational and radial drift velocities, respectively. As the transmitted wave reaches the scatterers, echoes are received at a rate depending on the local density of the cloud. In / 6 / a clutter model was derived by treating the scatterers as dipoles. The functional dependence of the radar echo from a dipole scatterer located at azimuth Θ and a range cell k can be expressed as follows:

$$s_{ik}(t_k, \nu_r) = s'_{ik}\left\{C + U\exp\left[j2(2\pi\nu_r t_k + \alpha)\right] + L\exp\left[-j2(2\pi\nu_r t_k + \alpha)\right]\right\} \tag{1}$$

where

s_{ik} is the broadside echo voltage of the dipole with the electric vector parallel to its axis

ν_r is the rotational frequency of the dipole

α is the initial angle between the dipole axis and the plane perpendicular to the direction of propagation

C, U, L are random variables depending onthe polarisation and orientation of the rotational axis of the dipole

The total clutter signal $I(p)$ at the p-th sampling instant is given by

$$I(p) = \sum_i \left\{\left[\sum_k s_{ik}(p, \nu_r)\, X_{SF}(\tau_k, \nu_d)\right] a_p(i)\right\} \exp\left(j2\pi\nu_d p T_R\right) \tag{2}$$

where

X_{SF} is the cross-ambiguity function between the envelope of the transmitted pulse and the weighting function of the receiver matching filter

ν_d is the Doppler shift given by the average radial velocity of the cloud

T_R is the interpulse period of radar transmitter

The derivation of the normalized autocovariance function of the clutter for this case can be found in / 2 / in the form:

$$R_I(q) = \frac{2}{3}\left[1 + \frac{1}{2}\cos(4\pi q \nu_{or}) \cdot exp(-8\pi^2 q^2 \sigma_r^2)\right] \cdot exp(-j2\pi q \nu_{od} - 2\pi^2 q^2 \sigma_d^2 - q^2/2T_o^2) \quad (3)$$

T_o is the one way 3-dB beamwidht of the antenna

$T_o, \sigma_d, \sigma_r, \nu_{od}, \nu_{or}$ are normalized to the interpulse period

This model was used for all further investigations because, by choosing suitable value of parameters, all shapes of PSD which are of interest can be simulated. A computer program was designed according to the formula (2) and clutter generated by it was recorded and further processed. First of all, the autocovariance function of the clutter signal was computed and displayed in graphical form for different parameters ν_{od}, ν_{or} , σ_r , σ_d and N (N is the number of scatterers in the main lobe of antenna pattern; it depends on the dynamic range of the receiver and the density of the cloud).

Application of the Model

Next step in our work was to utilize this model of the clutter and the results from the first part (i.e. the limits for time and amplitude discretisation). The clutter was added to simulated target echoes with different signal-to-noise ratio. Several methods were considered then for extracting signal from the clutter.

A) This problem is equivalent to the problem of extracting signal from correlated noise. In the case of Gaussian noise the optimum detection rule is given by the two step operation "prewhiten and match". If the noise correlation is a priori unknown, the whitening operation should be carried out adaptively. To transpose the clutter from moving objects to the zero Doppler frequency by means of phase locked loop, and then to process the signal by MTI like / 5 /. Processing of the signal is quite simple but the MTI must preceede this adaptive loop, as targets with zero Doppler frequency are transposed to a non-zero Doppler and the problem remains unsolved. Removing signal from targets with non-zero Doppler velocities may lead to a hazardeous situation as the flight control might guide the plane to a dangerous region of the space (into bird flocks, heavy clouds, clear air turbulence areas e.t.c.).

B) Processing of radar signal by means of computation of Toeplitz matrix, invert it and handle the signal by this inverse filter like / 1 /. This needs very fast blocks for computation of the autocovariance coefficients and matrix inversion in real time.

C) Process the radar signal by means of Fast Fourier Transform in order to obtain spectral components of signal, distinguish between wanted targets and the clutter, and then tune coefficients of the filter to this particular clutter. This method seems to be faster than B).

D) To use Burg's maximum entropy method / 3 / for computation of spectral components and proceed as in C). This method seems to be most effective in terms of speed and amount of computation.

Conclusion

Several methods of clutter filtering were presented above. The general results and discussion of effectivness of these methods given in the paper will be completed by numerical outcomes of our computations.

Literature

/1/ Buehring W., KlemmR.: An Adaptive Filter for Supression of Clutter. Frequenz, Sept. 1976, 238-243

/2/ Hawkes C.D., Haykin S.: Molelling of Clutter for Coherent Pulsed Radar. IEEE Trans. on IT, Nov. 1975, 703-707

/3/ Kesler S.B., Haykin S.: The Maximum Entropy Method Applied to the Spectral Analysis of Computer-simulated Radar Clutter. IEEE Trans. on IT, Vol. IT-24, No 2, March 1978, 269-272

/4/ Klemm R.: Design Considerations for a Digital Adaptive Least Squares Filter for Real-Time Processing. Radar 1977, 271-274

/5/ Short R.D.: An Adaptive MTI for Weather Clutter Suppression. IEEE Trans. on AES- 18, No 5, 552-562

/6/ Wong J.L., Reed I.S., Kaprelian Z.A.: A model for the Radar Echo from a Random Collection of Rotating Dipole Scatteres. IEEE Trans. on AES-3, No 2, 171-178

NUMERICAL SIMULATION OF THERMAL STATE OF SOME THERMAL SYSTEMS

Krookovsky P.G.[I)]

Principles of constructing the applied program for solving two-and three-dimensional problems of stationary and non-stationary heat conduction for non-uniform regions of solution of any step-wise form under any boundary conditions of heat transfer are considered. An example of solving a three-dimensional problem for heat exchange unit is given . On the basis of the method of solving reverse problems of heat conduction the analysis of the existing structures is given and more exact method of calculating heat transfer parameters of multi-layer structures is proposed.

Key words: numerical methods, technical systems, heat transfer, parameters identification.

One of the main methods of studying the thermal state of industrial and power systems which play an ever growing role is the method of mathematical modelling. The possibilities of this method of studying thermal systems (where individual elements exchange thermal energy) are determined by the correspondence of the mathematical model chosen to the real process in the system. The model adequacy is provided by an exact mathematical formulation of the problem and the data on the coefficient included into the formulation. The lack of information on the model form and its coefficients can be compensated by solving the problem of structural and parametric identification, respectively, by the given supplementary information about the state of the thermal system. Such problems are called reverse problems of heat conduction (RPHC) in the theory of heat transfer. In any case, while solving direct and reverse problems of heat transfer for thermal systems it is necessary to have methods, algorithms and programs to discover general mathematical models of thermal systems availability of which is the necessary condition of creating the methods and problems for RPHC solution.

The general enough mathematical model of energy exchange in thermal systems consists of the differential equation of heat conduction

$$\mathrm{div}\left[\lambda(T)\,\mathrm{grad}\,T\right] - c_v(T)\frac{\partial T}{\partial \tau} + q_v(T) = 0, \quad (I)$$

I) Institute of Thermal Thermophysics, USSR, Kiev

written for the coordinate system chosen and at the corresponding boundary and initial condition. In order to solve equation (I) at any given boundary and initial conditions, the author has chosen the finite difference method (explicit difference scheme) [I] which reduces the solution of equation (I) to that of the system of algebraic equations. The algorithm and the program having been developed make it possible to solve two- and three-dimensional non-stationary and stationary $(\partial T/\partial \tau = 0)$ problems of heat conduction in non-uniform regions of the arbitrary step-wise solution with the boundary conditions of heat transfer in the Cartesian cylindrical and spherical coordinate systems. The system of algebraic equations is solved by the iteration method of the upper relaxation. The programming language is FORTRAN/EC, the carrier is in the form of cards, the volume of FORTRAN/EC operators is 500.

As an example of the program application let us consider the thermal state of the physical model of the heat transfer apparatus (HTA) a portion of which is shown in Fig.I. HTA consists of the metallic housing with cavities for pumping the hot and the cold carriers divided by a partition. The HTA body is thermally isolated from outside, the isolation being in perfect thermal contact with the body. The heat exchange in HTA body cavities and isolation with the environment is described by Newton's law of convection with the given heat transfer coefficients α . To calculate the temperature field in the given thermal system of HTA (heat carriers-HTA body-isolation-environment) a difference three-dimensional network was chosen with the total nodes number 5.5 thousand. Fig.2 shows one of the calculation results for stationary thermal state of the HTA in the form of isothermal lines. The time of the problem solution is I5 min. for the computer EC I040. On the basis of the program of the objects temperature field calculation, the method and the program have been developed for identification (RPHC solution) of external heat transfer parameters, in particular α in the HTA. The method and the program are based on using the method of sensitivity functions allowing to substitute the initial mathematical model of heat conduction for the system of linear algebraic equations (SLAE) which relates directly the external heat transfer pa-

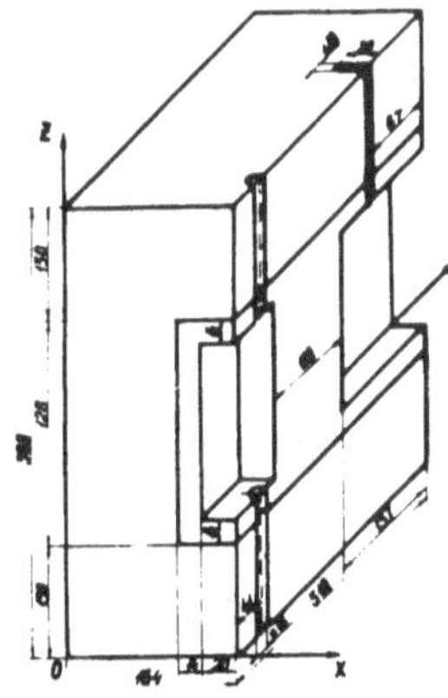

Fig.I. Scheme of heat transfer apparatus

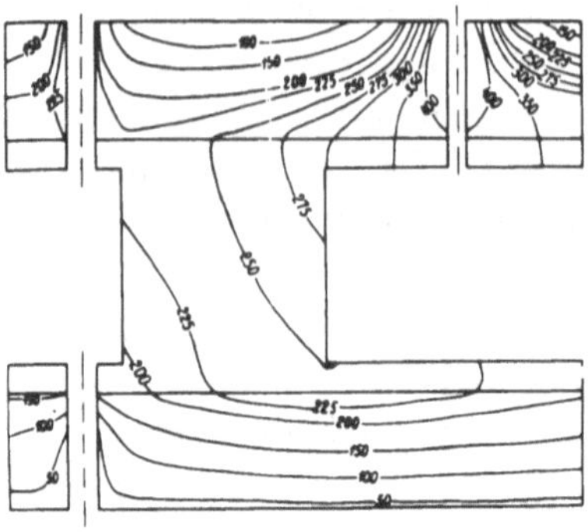

Fig.2. Temperature field in the middle section of HTA.

rameters to the given temperatures T_E inside the HTA. In the presence of measurement noise in T_E , the method of regularization by A.N.Tikhonov [2] is used in which the regularization parameter is selected from the condition of T_E proximity to T_M temperatures obtained by the problem (I) numerical solution.

It is interesting to mention the analysis of thermal state of multi-layer systems (structures) the main feature of which consists in a substantial dependence of the structure temperature field on the values and the law of distribution of contact thermal resistances between individual layers (metal, for example). The values of such thermal resistances can be assessed using the method of RPHC. In Fig.2a the structure is examplified which consists of 5 metal layers with the thickness $\delta_M = 2 \cdot 10^{-3}$m. Thermal conductivity λ_M and volumetric heat capacity C_{VM} are much higher than λ_A and C_{VA} . The mathematical model of thermal energy exchange in such a thermal system is described by equation (I) for one-dimensional non-stationary process of heat conduction with the boundary conditions of the first kind ($T(0,\tau) = 100°C$, $T(L = 27,08;\tau) = 0°C$). RPHC is formulated in the following extreme statement [3] . It is necessary to find the law $\lambda(x)$ for which the value

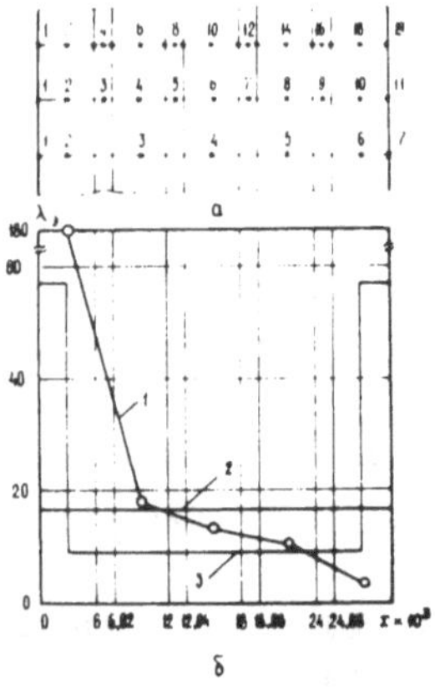

Fig.3. Scheme of multi-layer plate with nodes spatial location(a) and λ_E dependence on the coordinate x (δ).

$$\Phi = \sum_{J=1}^{K} \sum_{i=1}^{m} \left(T_{MJ,i} - T_{EJ,i} \right)^2 \qquad (2)$$

is minimum. In equation (2) m is the number of points with the given T_E, K is the number of time intervals in the difference network. In some works ([4] ,for instance) it is proposed to find such an effective λ = =const for which the value Φ would be minimum (m=I). The thermocouple

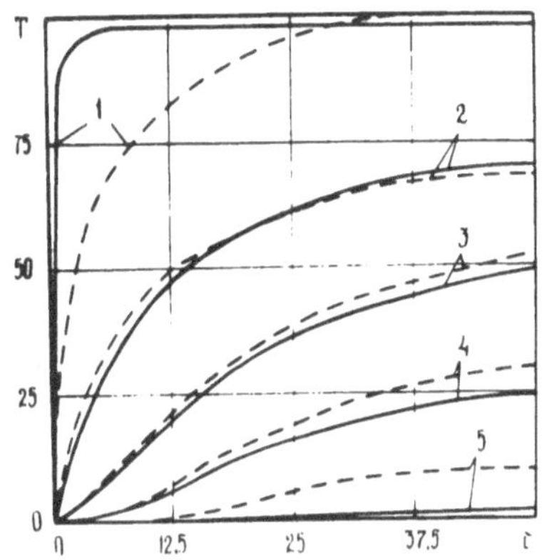

Fig.4. Dependence of temperature $T(^{o}C)$ on time for central points of metal layers: $I - = 3 \cdot 10^{-3}$m; 2-9·0I; 3-I5.03; 4-2I.05; 5-27.07; solid line $-T_M$; dashed line $- T_E$.

coordinate is $x_E = L$. It is shown in Fig.3b (curve I, circles) which values of λ_E would have been obtained depending of x_E. Hence at m=I the search for λ_E=const can yields scatter of two orders of magnitude. RPHC solution at m=5 gives λ_E =I6.8 W/(mK) (curve 2). Comparison of T_M with T_E at such λ_E (Fig.4) demonstrates large discrepancy in these temperature (Φ= I3370) which is inacceptable in practice. Close values of T_M and T_E can be obtained if dependence $\lambda_E(x)$ is assumed to be a three-step function (Fig.3b, curve 3) with a simultaneous search of these steps values. In this case at m=I (which is worse than m=5) and at any x_E, $\Phi \approx 0.5$. While searching for minimum Φ, Powell method is used [5]. The best results as to the proximity of T_M to T_E ($\Phi \approx 0$) facilitate the presentation and search of $\lambda(x)$ in the form of a nine-step function, i.e. search for λ_M and λ_E for the corresponding layers. From the point of view of calculation this is the search for minimum Φ in the function of λ_M and λ_E. Solution of such RPHC is resistant to errors in T at the value of m $\geqslant$ I.

Literature

I Самарский А.А. Теория разностных схем. -М.: Наука, I977.

2 Тихонов А.Н., Арсенин В.Я. Методы решения некорректных задач. -М.: Наука,: I979.

3 Коздоба Л.А., Круковский П.Г. Методы решения обратных задач теплопереноса. -Киев, Наукова думка, I982.

4 Schimmel W.P., Beck J.V. and Donaldson A.B.: Journal of Heat Transfer, Ser.C, 9, (I977), 3, 466-470.

5 Powell M.J.: Computer Journal, 9, (I965), 4, 303-307.

THE BEST EQUIVALENT CIRCUIT FOR THE HEAT CONDUCTION IN A COMPACT OBJECT

GUSTAV KÜNZEL [1)]

The paper brings a proof that the equivalent thermal circuit or the system of higher than the second order is not a suitable representation of a compact object. The most suitable order of the circuit is 2 (a two-capacitance thermal equivalent circuit) which agrees strikingly and obviously reasonably with practice in the control system analysis. Moreover it is possible to determine two unequal time constants of one optimum model. The equivalent circuits of n-order do not generally transform the heat conduction of a compact object, and therefore they are not correct. The used interpretation is fallacious.

1. Chain circuit models of heat conduction in the past

Electrical passive analogs for the solution of heat conduction dynamics through the walls were constructed in Europe and the United States since 1936 at least. L. C. Beuken is considered to be their author, and therefore they are called Beuken s models. They are described in many works [1, 2, 3, 4, 5, 6].

The problem is considered as a one dimensional heat conduction across a non-limited layer of some solid having a final thickness l. The heat flows only in the direction x, which is normal to the layer. This wall is considered to be divided into n equal layers possessing l/n thickness. To each of the sections a heat capacity C/n and a thermal resistance W/n are assigned. For a good model a big number [3] $n \geq 10$ of loop circuits is recommended (Fig. 1).

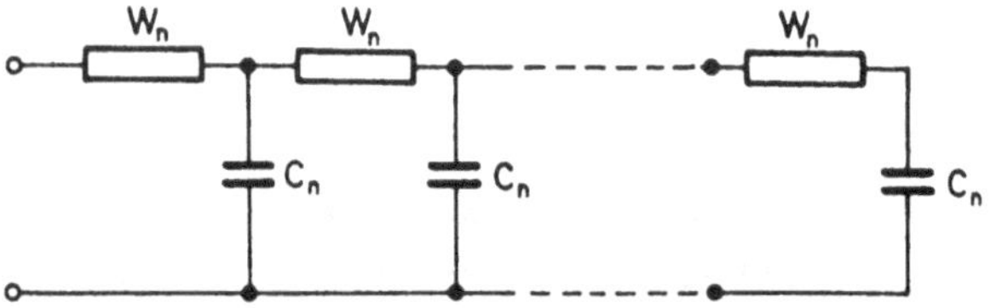

Fig. 1: The simplest variant of Beuken´s model circuit.

The ˥, Π or T forms are not substantial. The correctness of a model circuit is based on the formal identity of the equations

$$\frac{\partial^2 v}{\partial x^2} = \frac{1}{\varkappa}\frac{\partial v}{\partial t} \; ; \quad \frac{\partial^2 v}{\partial x^2} = W_1 C_1 \frac{\partial v}{\partial t} \; ; \quad W_1 = \frac{1}{KS} \; ; \quad C_1 = \frac{KS}{\varkappa} \qquad (1)/(2)$$

where W_1 and C_1 are the heat resistance and the heat capacity of a prism having the cross-section S and being one metre long. The time constant of one section of the layer is $T_n = W_n C_n = l^2/\varkappa n^2$. The transfer function of the circuit in Fig. 1, if it has n loop circuits, is

$$S(p) = \frac{1}{(1+T_n p)^n} = \frac{1}{\left(1+\frac{l^2}{\varkappa n^2}p\right)^n} \qquad (3)$$

1) Ing. Gustav Künzel, CSc., TESLA Lanškroun, 563 24 Lanškroun, ČSSR

The corresponding temperature step response is

$$\frac{v}{V} = L^{-1}\left\{\frac{1}{p(1+T_n p)^n}\right\} = 1 - \exp\left(-\frac{t}{T_n}\right)\sum_{\nu=0}^{n-1}\frac{t^\nu}{\nu!\,T_n^{\nu}} \tag{4}$$

for $\varkappa/l^2 = 1$. The frequency response in Bode´s form is

$$\left|S(j\omega)\right|_{dB} = -10\,n \log_{10}\left[1 + \omega^2\left(\frac{l^2}{\varkappa n^2}\right)^2\right] \tag{5}$$

2. The model circuits with heat capacitances are generally not correct

The graphs of the functions (4) and (5) in Fig. 2 and 3 are for n = 1, 2, 3, ... and they show that the responses do not converge to the solution F, of the Fourier equation (1) for the conditions

$$\begin{aligned} &v(x,t) = 0\ ;\ t = 0\ ; \\ &v(x,t) = V\mathbb{1}(t)\ ;\ x = l\ ; \quad \frac{\partial v}{\partial x} = 0\ ;\ x = 0\,. \end{aligned} \tag{6}$$

The transfer function corresponding with Fourier equation is

$$S(p) = \frac{\cosh\left(x\sqrt{\frac{p}{\varkappa}}\right)}{\cosh\left(l\sqrt{\frac{p}{\varkappa}}\right)}\ , \tag{7}$$

the step response is

$$\frac{v}{V} = 1 - \frac{4}{\pi}\sum_{n=0}^{\infty}\left\{\frac{(-1)^n}{2n+1}\exp\left[-\frac{\varkappa}{l^2}(2n+1)^2\frac{\pi}{4}t\right]\right\} \tag{8}$$

and the frequency response in Bode´s form is

$$\left|S(j\omega)\right|_{dB} = -20\log_{10}\frac{2}{\left|\cosh\sqrt{\frac{\omega l^2}{2\varkappa}}\right|} \tag{9}$$

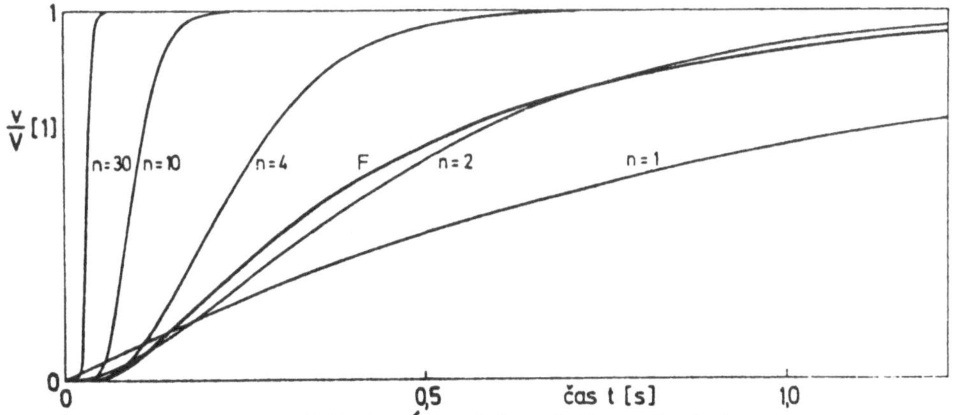

Fig. 2: Step responses of Beuken´s model and the actual F one.

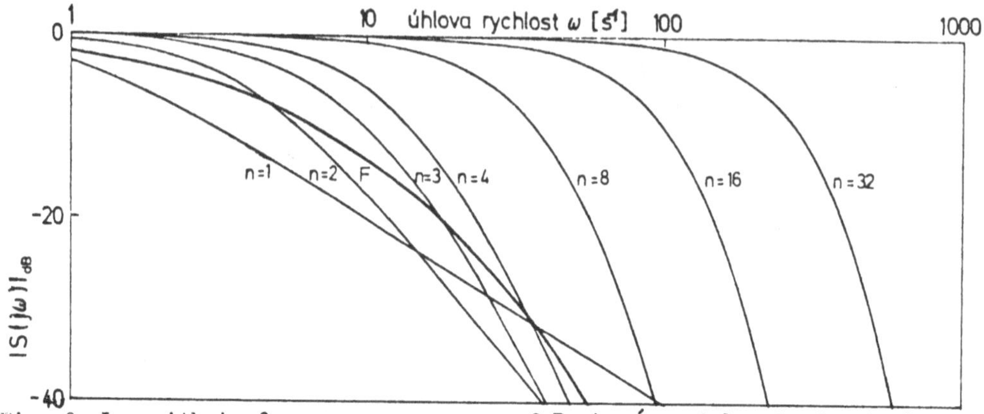

Fig. 3: Logarithmic frequency responses of Beuken´s model and of actual layers according to Fourier.

A. The expressions (4) and (8) as well as (5) and (9) are not equal neither in the case if $n \to \infty$. Therefore the model circuits according to Fig. 1 are not generally correct and they will not be more correct with increasing n.

B. The best approximation is represented by a two-capacitance WC circuit with n = 2. A mistake [1, 2, 3, 4, 5, 6] comes from a wrong interpretation of the derivation of the Fourier partial differential equation. The assignment of the circuit (Fig. 5) to the volume element is not justified.

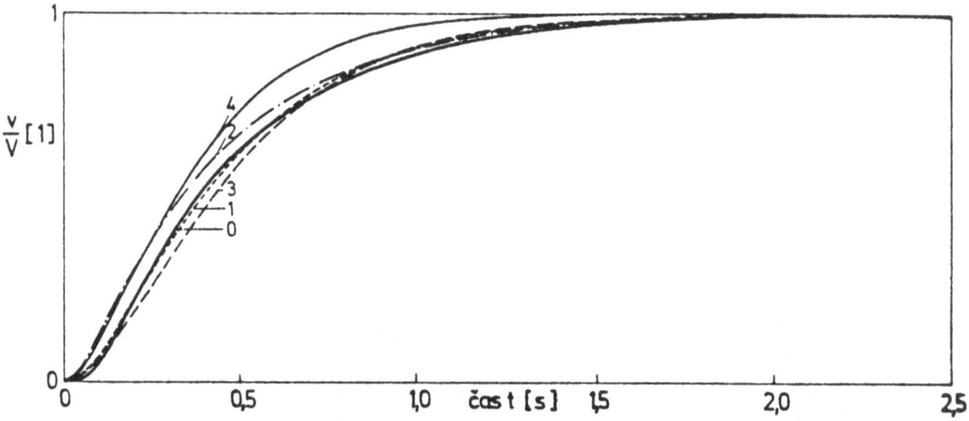

Fig. 4: Step responses: 0 - the actual one according to Fourier; 1 - the best symmetrical circuit; 2 - the best asymmetrical circuit; 3 - a circuit with n = 2.

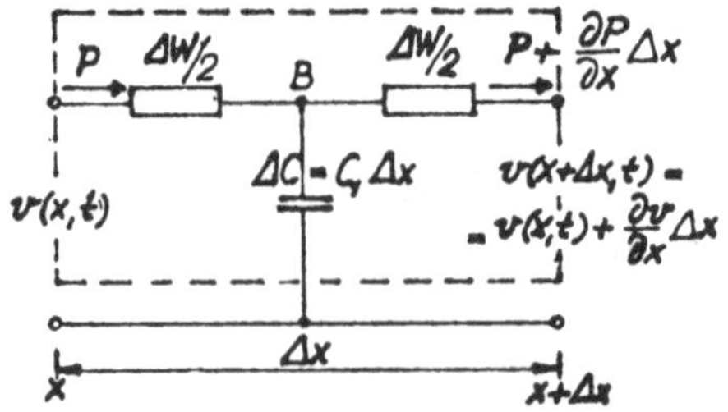

Fig. 5: The scheme by which a derivation of Fourier´s equation is uncorrectly interpreted.

The circuit with concentrated parametres can be described only by means of ordinary differential equations.

The parametres of the two-capacitance model circuit can be optimalized. The best asymmetrical circuit according to Fig. 6, Fig. 4 has both the heat capacitances equal C/2 and heat resistances are W_1 = 0,278 W, W_2 = 0,686 W, and W_3 = 0,036 W. $W_1 + W_2 + W_3 = W$. The best symmetrical model circuit has the same capacitances C/2, but $R_1 = R_3$ = 0,1 W and W_2 = 0,8 W.

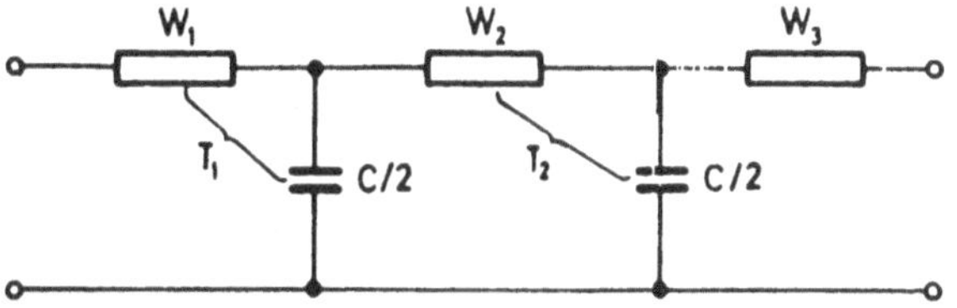

Fig. 6: The two-capacitance model circuit.

References

[1] Beuken, C. L.: Wärmeverluste bei periodisch betriebenen elektrischen Öfen. Dissertation Freiberg 1936. Triltsch + Huther Berlin 1936

[2] Beuken,C. L.: Wärmeverluste bei periodisch betriebenen elektrischen Öfen: eine neue Methode zur Vorausbestimmung nicht stationärer.. Comptes - Rendus du Conggres International des Applications Electrocalorifiques. Scheveningue, Nederland, 1936.

[3] Brokmeier, K. H.: Über ein Beukenmodell kleinster Abmessungen. Dissertation. TH Braunschweig 1950.

[4] Hegbom, T.: Das umgebaute Beukenmodell im Elektrowärme - Institut Essen. Elektrowärme 16 (1958) Nr. 2.

[5] Horn, B.: Untersuchung der Wärmeisolation einer zentralen elektrischen Blockspeicherheizanlage mit dem Beukenmodell. Elektrowärme international 27 (1969), Nr. 2.

[6] Eickhoff, E.: Das neue Beukenmodell des Elektrowärme - Instituts Essen e. V., sein Aufbau und seine Anwendungsmöglichkeiten. Internationale Z. f. Elektrowärme 25 (1967), Nr. 7.

[7] Kraus, A. D.: Cooling Electronic Equipment. Prentice - Hall, Inc. Englewood Chiffs, N. Y. 1970

[8] Künzel, G.: Mathematical models of heat transfer in constructions of electronics. ACADEMIA, Prague 1985 is printed.

[9] Venikov, V. A.: Teorija podobija i modelirovanija. Moskva. Vysšaja škola, 1976.

[10] Tretjačenko, G. N.: Modelirovanie pri izučenii pročnosti konstrukcij. Kiev, Naukovaja dumka 1979.

[11] Tetelbaum, I. M., Tetelbaum, Ja. I.: Modely prjamoj analogii. Moskva, Nauka, 1979.

[12] Doetsch, G.: Anleitung zum praktischen Gebrauch der Laplace-Tranformation und der Z - Transformation. R. Oldenbourg verlag München. Wien 1967.

MODELLING OF AQUIFER THERMAL ENERGY STORAGE

Elwira Małgorzata Wolska [1)]

This paper considers mathematical modelling of aquifer thermal energy storage. Water of the temperature higher then underground water temperature is stored in aquiferrous layers, which are surrounded by weak permeable layers. Heat flow between aquifer and weak permeable layer is consider in the presented example.

Keywords: modelling, aquiferous layer, weak-permeable layer, porous medium

1. Mathematical model

Consider an horizontal aquifer in charge, limited by two weak-permeable layers. Water of temperature higher then underground water is injected and pumped by a throughly penetrating well (fig. 1).

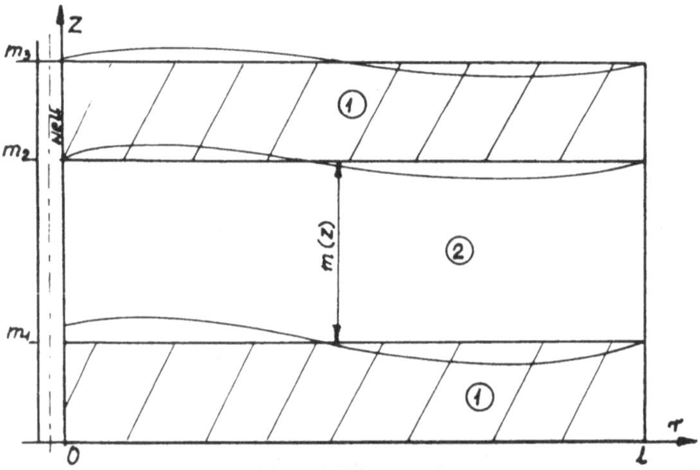

Fig. 1. Consider domain (1) weak-permeable layers (2) aquiferous layer

The following assumptions are consider

- aquiferous and weak-permeable layers are homogenous and isotropic
- water and porous medium are incompressible
- thermal equilibrium between water and solid is instateneous
- there are no flow instabilities
- in the consider domain Darcy law and continuity law are valid
- dependance of the water and porous medium density ρ and water viscosity μ on the temperature T is neglected
- the scheme is axial symetric

----

1) Institute of Biocybernetics and Biomedical Engineering, KRN 55, 00-818 Warsaw, Poland

- the water flux between aquifer and weak-permeable layers is neglected
- the conductive heat flow between layers is consider
- the initial temperature in porous medium is constant

The consider problem is described by following set of equations

- aquiferous layer

$$\operatorname{div} \vec{V} = 0 \tag{1}$$

$$\vec{V} = - k(\overrightarrow{\operatorname{grad}}\, p - \rho \vec{g}) \tag{2}$$

$$\operatorname{div}(\lambda^* \overrightarrow{\operatorname{grad}}\, T) - (\rho C)_f \vec{V}\, \overrightarrow{\operatorname{grad}}\, T = (\rho C)^* \frac{\partial T}{\partial t} \tag{3}$$

- weak-permeable layers

$$k = 0 \tag{4}$$

$$\operatorname{div}(\lambda_e \overrightarrow{\operatorname{grad}}\, T) = (\rho C)_e \frac{\partial T}{\partial t} \tag{5}$$

where:

$$\lambda^* = n \lambda_f - (1 - n) \lambda_s \tag{6}$$

$$(\rho C)^* = n \rho_f C_f + (1 - n) \rho_s C_s \tag{7}$$

V - velocity [m/s]
k - filtration coefficient [m/s]
p - pressure [atm]
ρ - density [kG/m^3]
g - gravitational acceleration [m/s^2]
λ^* - porous medium thermal conductivity [W/mK]
λ_f - water thermal conductivity [W/mK]
λ_s - solid thermal conductivity [W/mK]
λ_e - weak-permeable layer thermal conductivity [W/mK]
$(\rho C)^*$ - heat capacity of porous medium [J/m^3K]
$(\rho C)_e$ - heat capacity of weak-permeable layer [J/m^3K]
n - porosity
T - temperature [K]
T_0 - initial temperature [K]
T_1 - injection temperature [K]
H - height of the layer [m]
$_0$ - flow rate [m^3/s]
r, z - radial, axial coordinates [m]
t - time [s]

2. Resolution for a particular case

The porous medium is surrounded by two horizontal weak-permeable layers, water is injected at constant flow rate.
The following equations are to be solved
- aquiferous layer

$$(\rho c)^{*} \frac{\partial T}{\partial t} = \lambda^{*} \left(\frac{\partial^2 T}{\partial r^2} + \frac{1}{r} \frac{\partial T}{\partial r} + \frac{\partial^2 T}{\partial z^2} \right) - (\rho C)_f \frac{Q_0}{2\pi H} \frac{1}{r} \frac{\partial T}{\partial r} \qquad (8)$$

in the domain
$0 \leq r \leq 140$ m, $\quad 5$ m $\leq z \leq 10$ m

- weak-permeable layers

$$(\rho C)_e \frac{\partial T}{\partial t} = \lambda_e \left(\frac{\partial^2 T}{\partial r^2} + \frac{1}{r} \frac{\partial T}{\partial r} + \frac{\partial^2 T}{\partial z^2} \right) \qquad (9)$$

in the domain
$0 \leq r \leq 140$ m, $\quad 0 \leq z < 5$ m, $\quad 10$ m $< z \leq 15$ m
with the boundary conditions

$$\begin{aligned}
&T(r,z,0) = T_0 && \\
&T(r,z,t) = T_1 && z \in [m_1, m_2] \\
&\frac{\partial T}{\partial n}(r,z,t) = 0 && z \in [0, m_1) \wedge z \in (m_2, m_3] \\
&T(r,z,t) = T_0 && z = 0 \wedge z = m_3 \\
&T(r,z,t) = T_0 && r = 1
\end{aligned}$$

On the boundary between the layers the continuity law of heat flow

$$\lambda^{*} \frac{\partial T}{\partial z} = \lambda_e \frac{\partial T_e}{z}$$

and the equality of the temperatures

$$T = T_e$$

have to be valid.

3. Results

The case of constant thickness of aquiferous layer and constant filtration coefficient was solved. Duration of the water injection is 5 months. Duration of the pause (water storage) is 2 months. Filtration coefficient $k = 10^{-4}$ m/s porosity $n = 0.2$, flow rate $Q_0 = 7.4\ 10^{-6}$ m^3/s. In the consider example $T_1 = 30$ K, $T_0 = 8.5$ K. The results are presented on the figures.

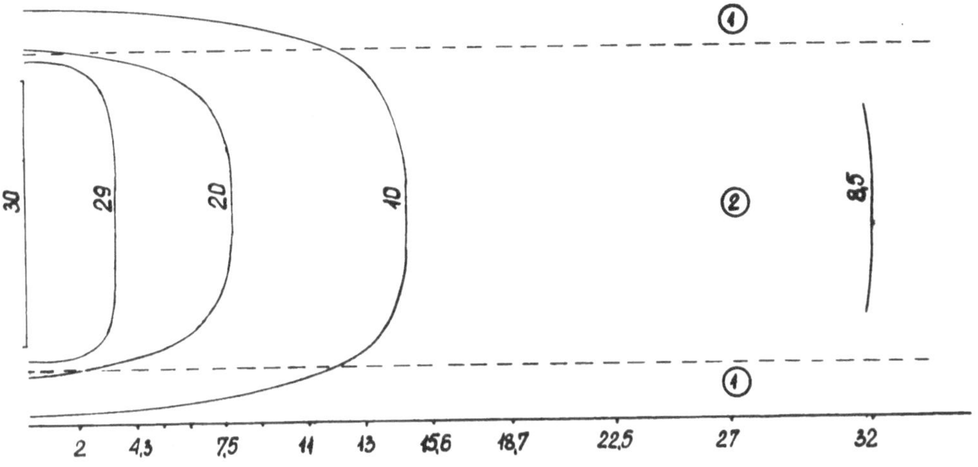

Fig. 2. Heat distribution after 5 months injection

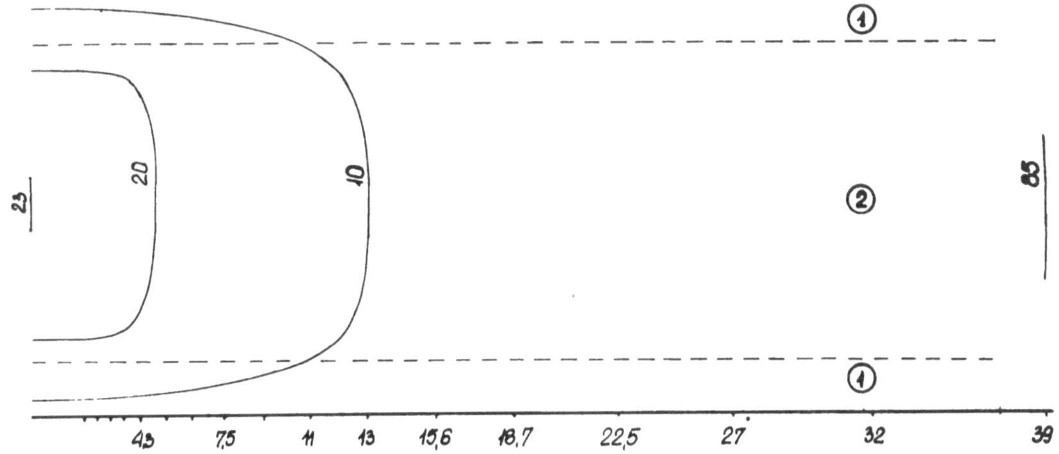

Fig. 3. Heat distribution after 2 months pause

4. References

1 Bear J. - Dynamic of fluids in porous media, Elsevier Scientific Publishing Company, Amsterdam, Oxford, New York, 1977.

2 Alexandre M. - Approche numerique desphenomenes de diffusion convection lors d'un stockage d'eau chande en aquifere, Thèse de Docteur - Ingenieur, Institut National Polytechnique de Toulouse, 1982.

AN ALGORITHM OF DETERMINING THE OPTIMAL CONTROL SIGNAL IN A CERTAIN COOLING PROCESS

Edward Kącki   Włodzimierz Korycki [1)]

In the paper, an application of the dinamic programming method for optimization of a cooling process described by a limit problem for the conductivity equation is presented. A square integral functional has been accepted as an evaluation criterion. In the paper the problem of optimization has been formulated and analysed as regards its correctnes, also the appropriate theorems have been formulated. Next, after a discrete approximation based on dynamic programming the algorithm of determining the optimal control signal has been elaborated and its properties have been examined.

Key words: algorithm, dynamic programming, distributed parametr, cooling process, optimal control.

1. Introduction

The process, considered in this paper is of the general character and includes not only a case of the boundary control but also distributed control inside a flat /two-dimensional/ plate $0 \leq x \leq a$.
The first case is described by the boundary data

$$\left.\begin{aligned} \left(\frac{\partial t}{\partial x}\right)_{x=0} - \gamma\left[t(0,\tau) - m(\tau)\right] = 0 \\ \left(\frac{\partial t}{\partial x}\right)_{x=a} + \gamma\left[t(a,\tau) - m(\tau)\right] = 0 \end{aligned}\right\} \qquad \text{for} \quad \tau \geq 0$$

where $t(x,\tau)$ for $x,\tau \in \langle 0,a\rangle \times \langle 0,\infty\rangle$ - temperature
$m(\tau)$ - commend signal /ambient temperature/
γ - surface film conductance
This problem can be reduced to the case of distributed control as follows:
we introduce a new variable $v(x,\tau)$:

$$v(x,\tau) = t(x,\tau) - m(\tau)$$

Now we can consider the increase of temperature $v(x,\tau)$ taking account of the distrubuted control case in the cooled plate.

2. Formulation of the problem

Function $u(x,\tau)$, describing the cooling process, satisfyes an equation

$$\frac{\partial^2 u}{\partial x^2} - \frac{\partial u}{\partial \tau} = g(x,\tau) \qquad \text{for} \quad x,\tau \in Q = (0,a)\times(0,T) \tag{1}$$

and conditions

$$u(x,0) = \varphi(x) \qquad \text{for} \quad x \in (0,a) \tag{2}$$

$$\left.\begin{aligned} \left(\frac{\partial u}{\partial x}\right)_{x=0} - \gamma\, u(0,\tau) = 0 \\ \left(\frac{\partial u}{\partial x}\right)_{x=a} + \gamma\, u(a,\tau) = 0 \end{aligned}\right\} \qquad \text{for} \quad \tau \in (0,T) \tag{3}$$

[1)] Technical University of Łódź, Poland

where: $a>0, \gamma>0, T>0$ and $q(x,\tau)$ is the wanting command signal minimizing functional

$$J[q]=\iint_Q [\lambda u^2(x,\tau)+(1-\lambda)\, q^2(x,\tau)]\, dx\, d\tau \qquad (4)$$

in a set of admissible controls Ω

$$\Omega=\{q:\ q\in L^2(Q),\ \|q\|_{L^2(Q)}\le 1\} \qquad (5)$$

$\lambda\in(0,1)$, $u\in L^2(0,T;H^1(0,a))$, $H^1(0,a)\subset L^2(0,T)$ is a Sobolev space.

We assume, that the function φ, describing the initial temperature decomposition belongs to $L^2(0,a)$.

3. Existence and uniqueness of the solution of the problem

Existence and uniqueness of the solution of the formulated boundary problem with the given function $q(x,\tau)$ and the accepted conditions directly follow from the well-known theorem /see [4], [6]/.

To show the existence and uniqueness of the optimal control it is necessary to consider the following facts:

1/ the set of admissible controls Ω is convex

2/ if the solutions $u_1, u_2 \in L^2(0,T;H^1(0,a))$ of the boundary problem correspond to the controls q_1, q_2 then for any $\alpha\in\langle 0,1\rangle$ the solution $\alpha u_1+(1-\alpha)u_2\in$ $\in L^2(0,T;H^1(0,a))$ corresponds to the control $\alpha q_1+(1-\alpha)q_2\in$

3/ functional (4) is convex on Ω

4/ Ω is weakly-compact, so is also weakly closed [5]

Theorem. For the quality coefficient $J[q]$ defined by (4) there exists exactly one optimal control in the set of admissible controls Ω.

Proof. Functional $J[q]$ is weakly lower-semicontinuous on the weakly closed set Ω, so according to the generalized Weierstrass theorem it achieves its minimum value.

From the form (4) of functional $J[q]$ it follows that there can't exist two different optimal controls $q_1^0 \ne q_2^0$ [3].

4. Optimal control algorithm

We divide the intervals $(0,a)$ and $(0,T)$ into M subintervals of the length h and N subintervals of the length l respectively.

As a result we obtain a rectongular net, plot on the two-dimensional domain Q. Then we introduce an average operators [3] P_{hl} and S_{hl} defined on the sets of functions u and q

$$P_{hl}\, u(x,\tau)=u_{hl}(x,\tau), \qquad S_{hl}\, q(x,\tau)=q_{hl}(x,\tau) \qquad (6)$$

After ascribing the appropriate averoging values U_{hl} and q_{hl} to the knots of the net, we can discretize the analyzed problem.

Theorem. Equation (1) with the conditions (2) and (3) withe the accepted assumptions can be approximate with the discrete vector equation

$$u_{k+1}=[I-\beta R+\beta S]u_k + l\, q_k \qquad \text{for } k=0,1,\dots,N-1 \quad (7)$$

where $u_k = [u_1(k), u_2(k), \dots u_{M-1}(k)]$, $q_k = [q_1(k), q_2(k), \dots q_{M-1}(k)]$

$$u_i(k) = u_{hl}(ih, kl) \quad \text{for } i = 1,2,\dots M-1 \quad k = 0,1,\dots N$$

$$q_i(k) = q_{hl}(ih, kl) \quad \text{for } i = 1,2,\dots M-1 \quad k = 0,1,\dots N-1$$

$$r_{ij} = \begin{cases} 2 & \text{for } i=j \\ -1 & \text{for } |i-j| = 1 \\ 0 & \text{for } |i-j| > 1 \end{cases} \qquad s_{ij} = \begin{cases} \frac{1}{1+\gamma h} & \text{for } i=j=1 \text{ and } i=j=M-1 \\ 0 & \text{for the remaining } i,j \end{cases}$$

Functional(4) after the discretization takes form

$$J^*[\{q_k\}] = \sum_{k=0}^{N-1} [\lambda (u_k, u_k) + (1-\lambda)(q_k, q_k)] \tag{8}$$

This functional is weakly compact on the weakly compact set Ω_τ so it achieves it's only minimum on this set. Ω_τ is a family of the sets approximating Ω [3].

The optimal control algorithm of the discrete problem we design using the Bellman's optimality principle. The Bellman's equation for the quality coefficient(3) takes form

$$B^k(u_k) = \min_{\{q_m\}} \sum_{n=k}^{N-1} [\lambda (u_n, u_n) + (1-\lambda)(q_n, q_n)] \tag{9}$$

Theorem. For k=0,1,...N-1 the Bellman's function is of the form

$$B^k(u_k) = (P_k u_k, u_k) \tag{10}$$

where

$$P_{N-1} = \lambda I$$

$$P_k = \lambda I + A P_{k+1} \{ I - l^2 [(1-\lambda) I + l^2 P_{k+1}]^{-1} P_{k+1} \} A \tag{11}$$

Matrixs P_k are symmetric, positive definite and independent from the state of the system.

The optimal control in the moment t=kl linearly depends on the state of the system in this moment. This property is described by a formula

$$q^o_{N-1} = 0$$

$$q^o_k = -l [(1-\lambda) I + l^2 P_{k+1}]^{-1} P_{k+1} A u_k \qquad \text{for } k=0,\dots N-1 \tag{12}$$

The algorithm, given by the formula (10) - (12) is numerically stable - this fact avises from the optimality principle.

It has been shown, that the sequence of optimal controls $\{q^o_k\}$ is, for k,l $\to$ 0, weakly convergent to the optimal control q^o in Ω (5). Hence, the sequence $\{u_k\}$, describing the state of the process is also weakly convergent to $u^o \in L^2(0,T; H^1(0,a))$. Analogically to the papers [2] [3] we can show that sequences $\{q^o_k\}$ and $\{u_k\}$ with the accepted assumptions strongly converge to q_o and u_o respectively (u_o and q_o are the solutions of the optimization problem, described by the equations (1) - (5)).

5. Conclusions

The algorithm presented in this paper is very convenient for programming and gives good conditions to calculate the optimal command signal and to analyse the optimal cooling process. It also enables to analyse the influence of the thermal parameters upon the run of the optimal process.

References

[1] Cea J.: Optimisation: theorie et algorithmes. Dunod, Paris 1971

[2] Just A.: O pewnym algorytmie optymalizacji procesu opisanego równaniem przewodnictwa opartym o zasadę programowania dynamicznego, /doctoral dissertation/. Łódź 1979

[3] Kącki E., Korycki W.: Optimization algorithm of a process described by parabolic type equation based on dinamic programming. Informatika, Nr 1, 1985 /in press/

[4] Lions J.L.: Controle optimal de systemes gouvernes par des equations aux derivees partilles. Dunod Gauthiev-Villars, Paris 1968

[5] Malanowski K.: Zagadnienia sterowania optymalnego dla układów opisywanych równaniami cząstkowymi. AGH, Kraków 1973

[6] Sokołowski J.: External approximation of a parametric optimization problem for parabolic equations, Archiwum Automatyki i Telemechaniki, Vol.5., Nr 4, 1976

SOLUTION OF OUTER INVERSE THERMAL CONDUCTION PROBLEMS BY HYBRID SIMULATION METHODS

Matsevity Yu.M., Tsakanyan O.S., Ivanov V.A.[1)]

The hybrid solution of the outer inverse thermal conduction problem with the use of parametric optimization method is considered. The influence of different kinds of errors on the quality of obtained solutions is estimated.

Keywords: inverse thermal conduction problem, parametric optimization method, hybrid computer, boundary conditions approximation error.

Determining the heat transfer conditions on the surface of an object from limited data about its temperature field is the subject of the outer inverse thermal conduction problem.

The mathematical model of the heat transfer phenomenon consists of the differential equation

$$\frac{\partial}{\partial x}\left[\lambda(T)\frac{\partial T}{\partial x}\right]+\cdots+\frac{\partial}{\partial z}\left[\lambda(T)\frac{\partial T}{\partial z}\right]=C_v(T)\frac{\partial T}{\partial \tau}+q(T), \qquad (1)$$

initial conditions

$$T=f(x,y,z),\ \tau=0, \qquad (2)$$

and boundary conditions

$$T_n=f(x,y,z,\tau), \qquad (3)$$

$$q_n=f(x,y,z,\tau,T_n) \qquad (4)$$

$$\alpha(x,y,z,\tau,T_n)(T_n-T_c)=-\lambda(T)\frac{\partial T}{\partial n}, \qquad (5)$$

where T is the temperature; x, y and z are spatial coordinates; τ is the time coordinate; $q(T)$ is the internal heat source; α is the heat exchange coefficient; $\lambda(T)$ and $C_v(T)$ are the thermal and physical characteristics of the object; subscripts: Π - surface, C - medium.

If the inverse problem is considered as an object control problem, then it implies the determination of such a control action which realizes the motion of an object over the specified trajectory during transition from the initial to the finite state. The input values of the object are the boundary conditions, and the output values are the temperatures at the object observation points.

The result of solving such problems is the determination of control actions (surface temperature, thermal flows, thermal exchange coefficients which are functions of spatial and time coordinates) ensuring minimal deviation of the simulated value (temperature T)

1) Insttitute for Problems in Machinery, Academy of Sciences, Ukr. SSR, Dmitry Pozharsky Str. 2/10, 310046, Kharkov-46, USSR.

from the specified value (T^*) in the points of observation.

The problem of determining control actions from a specified set

$$\left.\begin{array}{l} T_{n\,min} \le T_n(\tau) \le T_{n\,max} \\ q_{n\,min} \le q_n(\tau) \le q_{n\,max} \\ \alpha_{min} \le \alpha(\tau) \le \alpha_{max} \end{array}\right\} 0 \le \tau \le \tau_1$$

is essentially a variational one, and its solution consists in minimizing the error functional

$$F = \int_0^{\tau_1} [T(x^*, y^*, z^*, \tau) - T^*(x^*, y^*, z^*, \tau)]\, d\tau$$

where x^*, y^*, z^* are the coordinates of the observation points.

The solution of the inverse thermal conduction problem is a time consuming computational process requiring a considerable amount of computation time which increases with an increase of the number of determined control actions and dimensionality of equations comprising the mathematical model of the object.

The limited amount of data about the temperature field of the object leads to a search of the boundary conditions in the form of approximating parametric functions, and the solution of the inverse thermal conduction problem can be reduced to the determination of the parameters of these functions.

The mentioned parameters can be determined, for example, by parametric optimization [1]. Since this method requires a considerable amount of computation time, the main factor influencing the choice of the computer is its speed of operation. This factor, when solving field theory problems, is the highest for passive analog devices. Their speed of operation can be increased by automating data input and output of results through the linkage of an analog processor with a general-purpose electronic digital computer or a special digital computer.

The inverse thermal conduction problems are usually solved on passive analog models by including these models into a control loop with a closed feedback link [2] . The determination of local boundary conditions, defined as a piecewise-constant function along the object boundary, is carried out thereat with the help of n independent control loops (according to the number of boundary sections) each of which tracks its own point of observation. Such an approach is possible if independence of control loops is provided, i.e. fulfilling of conditions which exclude their mutual influence through a passive model (the model should be sufficiently inertial, and the points of observation should be spaced far away from one another). If these conditions are not met, obtaining of a solution is practically impossible since any change of the control action in any of the control loops disturbs the equilibrium of adjoining loops, and the solution shall oscillate.

To eliminate this undesirable phenomenon, it is necessary to use one multi-input control loop which tracks the changes in all the observation points of the model. The output of such a loop is switched over in a preprogrammed sequence to control parameters which determine the boundary conditions. Such an automatic control system may be efficiently implemented on a hybrid computer made up of a network analog processor and a special hybrid computer operating according to the parametric optimization algorithm.

The hybrid realization of this algorithm consists in successive searching of minimums of the error functional for each of the parameters (the remaining ones being fixed) until the obtained minimum equals a machine zero, or until subsequent changing of parameters shall not cause its variation.

When solving the non-stationary problem the described procedure has to be realized for each time step. To decrease the errors accumulated from step to step, we use the technique of multiple solution at the given time step for various initial values of parameters, as a result of which the final value of the boundary conditions for this step is determined as the mean arithmetic value of obtained solutions. The so defined boundary conditions are used for solving the direct problem at the given time step, and the obtained temperature field is the initial condition for solving at the next time step.

Qualitative analysis of results obtained with the help of the described method is carried out when solving a test inverse thermal conduction problem with boundary conditions of the first kind

$$(T_{n1} = T_{n3} = 0{,}4 + 3{,}6\,y;\ T_{n2} = 4(x-1)^2;\ T_{n4} = 0{,}4)$$

for a flat rectangular plate with the sides ratio 1:2.

Investigated were the influence of the error of specifying the initial data on the hybrid computer, its instrumental error, and the error of approximating the boundary conditions with piece linear functions on the results of solution to the inverse thermal conduction problem.

Boundary conditions of the first kind, defined as $T_n = 4(x-1)^2$, were approximated in the interval [0.1] by two, three and four segments. Thus obtained piece linear functions have the respective mean arithmetic deviations from the given function: 2.22%, 1.11%, 0.44%.

Fig. 1 shows the dependences of the error of solution of the inverse problem on the distribution of observation points with respect to the surfaces for which the boundary conditions are determined. Curves 1, 2 and 3 relate to different cases of approximation of boundary conditions (with two, three and four linear segments, respectively). Solid curves are obtained at the relative data error equal to 0.25%; dashed curves - for 0.025%.

The given results once more confirm the direct relationship between the error of solution of the inverse problem and the error of the initial data, as well as increasing of the error with removal of the observation point away from the surface. Besides, an error increase is observed with increasing (according to the number of approximation segments of boundary conditions) of the number of parallel operating units of the hybrid computer. This is related with an increase of the instrumental error.

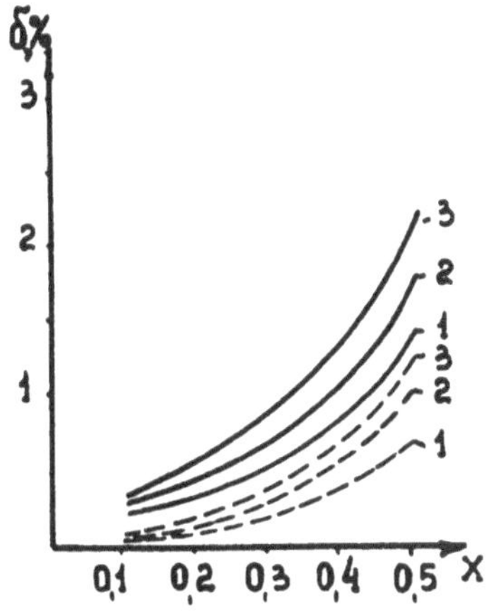

Fig. 1

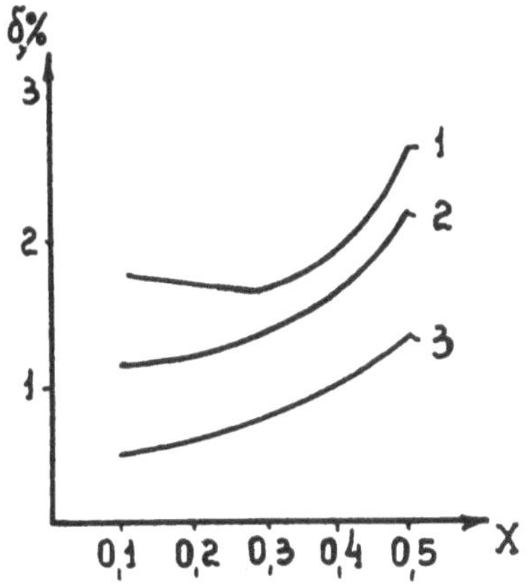

Fig. 2

Fig. 2 shows the dependences for the same cases of approximation of boundary conditions, but with account of errors of this approximation (the error of specifying initial data being 0.25%).

Evidently, the error of approximation of sought-for boundary conditions, together with the data errors plays a definite role when solving the inverse problem.

Thus, the hybrid solution of inverse problems by parametric optimization suggested in this paper may be regarded sufficiently efficient if accompanied by a systematic estimation of the above kinds of errors.

REFERENCES

1. Beky George A., Karplus Walter I.: Hybrid Computation. Moscow, Mir, 1970. 479 pp. (In Russian).
2. Matsevity Yu.M.,Tsakanyan O.S.: Hybrid Computing Systems for Investigating Physical Fields. Kiev, Naukova Dumka, 1983. 295 pp. (In Russian).

IDENTIFICATION OF THERMAL DIFFUSIVITY IN METAL HYDRIDE SYSTEMS

Matsevity Yu.M., Moultanovsky A.V., Rekada M.[1)]

The iterative modification of an optimal dynamic filter is used for determining the thermal diffusivity of metal hydrides by solving the inner inverse thermal conduction problem. Considered are the conditions under which experiments are carried out, the mathematical model of the phenomenon, and the algorithm of solving the inverse problem. The results of investigation are presented in the form of thermal diffusivity dependence on concentration of hydrogen in a metal hydride.

Keywords: metal hydrides, inverse thermal conduction problem, optimal filtration, iterative filter, state vector, prediction, thermal diffusivity, identified parameters.

When designing hydrogen storage and liquefaction systems, thermal sorption compressors and other units used for thermochemical compression of hydrogen, reliable data on the thermal and physical characteristics of metal hydrides are necessary.

Here we suggest to determine the thermal diffusivity of metal hydride $LaNi_5H_x$ by solving the inner inverse thermal conduction problem in which data on the thermal field of an object are used for determining its inner thermal transfer parameters. To obtain the thermal relationships used for solving this inverse problem, an experiment was carried out during which a flat layer of finely dispersed metal hydride powder with eight thermocouples arranged over its depth was heated by distributed heat sources, whereas the concentration of hydrogen in the metal hydride was estimated from the gas liberation of the installation.

The absolute error of temperature measurement during the experiment did not exceed ±9.45 K. The relative error of determining the concentration was 1%. The thermal conduction equation corresponding to the investigated process is written down as follows:

$$\frac{\partial}{\partial x}\left[\lambda(\nu_{H_2},T)\right]\frac{\partial T}{\partial x}+\rho q_s\frac{\partial \nu_{H_2}}{\partial \tau}=\rho c(\nu_{H_2},T)\frac{\partial T}{\partial \tau}, \qquad (1)$$

where λ, c - are the thermal diffusivity and specific heat of metal hydride, respectively; q_s - is the thermal effect of the phase transition of hydrogen from the bound state to the free molecular state; ν_{H_2} is the specific mass content of hydrogen in a metal hydride.

The problem was solved under the following assumptions:

a) the thermal effect of the phase transition of hydrogen is constant within the considered range of mass content, and is $q_s = 15.5 \times 10^6$ J/kg;

b) λ and c within the temperature range (298 to 332) K

1) Institute for Problems in Machinery, Academy of Sciences, Ukr. SSR, Dmitry Pozharsky Str. 2/10, 310046, Kharkov-46, USSR.

depend only on the concentration of hydrogen;

c) the temperature field is one-dimensional because the other dimensions of the layer are greater than the thickness by an order and more (i.e. the layer is considered to be an unlimited plate).

With account of these assumptions the thermal conduction equation within the considered range of temperatures reduces to

$$a(\nu_{H_2})\frac{\partial^2 T}{\partial x^2} + \frac{q_s}{c(\nu_{H_2})}\frac{\partial \nu_{H_2}}{\partial \tau} = \frac{\partial T}{\partial \tau}, \tag{2}$$

where $a(\nu_{H_2}) = \lambda(\nu_{H_2}/\rho c(\nu_{H_2}))$ - is the sought-for thermal diffusivity of a metal hydride. The relationship $c(\nu_{H_2})$ is a known one, and is determined by the relation

$$c(\nu_{H_2}) = 365 + 7\cdot 10^3\cdot \nu_{H_2} \quad \frac{J}{m^3 K} \tag{3}$$

The method of solving inverse thermal conduction problems (ITCP) adopted here is the method of optimal filtration, whose efficiency has been proved in numerous studies, e.g. in [2, 3].

The general algorithm of solving ITCP with the use of the iterative filter [2] is shown in Fig. 1. Being exceedingly noise immune with respect to errors of initial data, the method used here allows to exclude smoothening of experimental temperature dependences, and thereby to account for all actual details of the thermal process.

To search for a prediction of the temperature vector, a finite-difference approximation of equation (2) is used:

$$\left[\frac{a}{h^2}\right]T_{i-1}^{k+1} - \left[\left(\frac{1}{\Delta\tau} + \frac{2a}{h^2}\right)\right]T_i^{k+1} + \left[\frac{a}{h^2}\right]T_{i+1}^{k+1} =$$

$$= \left[-\frac{1}{\Delta\tau}\right]T_i^{k} - \frac{q_s}{365 + 7\cdot 10^3(\nu_{H_2}^{k+1} + \nu_{H_2}^{k})/2}\cdot\frac{\nu_{H_2}^{k+1} - \nu_{H_2}^{k}}{\Delta\tau}, \tag{4}$$

and to estimate the weight matrix

$$\left[-\frac{1}{\Delta\tau}\right]\cdot T_i^{k+1} + \left[\frac{T_{i-1}^{k+1} - 2T_i^{k+1} + T_{i+1}^{k+1}}{h^2}\right]\cdot a^{k+1} +$$

$$+ \frac{q_s\cdot \nu_{H_2}^{k+1}}{[\Delta\tau(365 + 7\cdot 10^3\cdot(\nu_{H_2}^{k+1} + \nu_{H_2}^{k})/2]} = \tag{5}$$

$$= \left[-\frac{1}{\Delta\tau}\right]\cdot T_i^{k} + \left[\frac{q_s}{(365 + 7\cdot 10^3(\nu_{H_2}^{k+1} + \nu_{H_2}^{k})/2)\Delta\tau}\right]\cdot \nu_{H_2}^{k}.$$

The unknown temperature values in the coefficient of a are the predictions of these temperatures taken from the preceding iteration $[T_i^{j/j-1}]_{K+1/K+1}$ or from the preceding time step $[T_i]^{K+1/K}$

The simulated interval (L = 0.011 um) was divided into 10 coordinate intervals; the duration of the considered part of the thermal process was divided into 36 time steps (h = 1 mm, $\Delta\tau$ = = 60 s). The initial approximations were $\hat{a}\%$ = 2×10^{-6} (m^2/s) and $\hat{T}\%$ = 293 K.

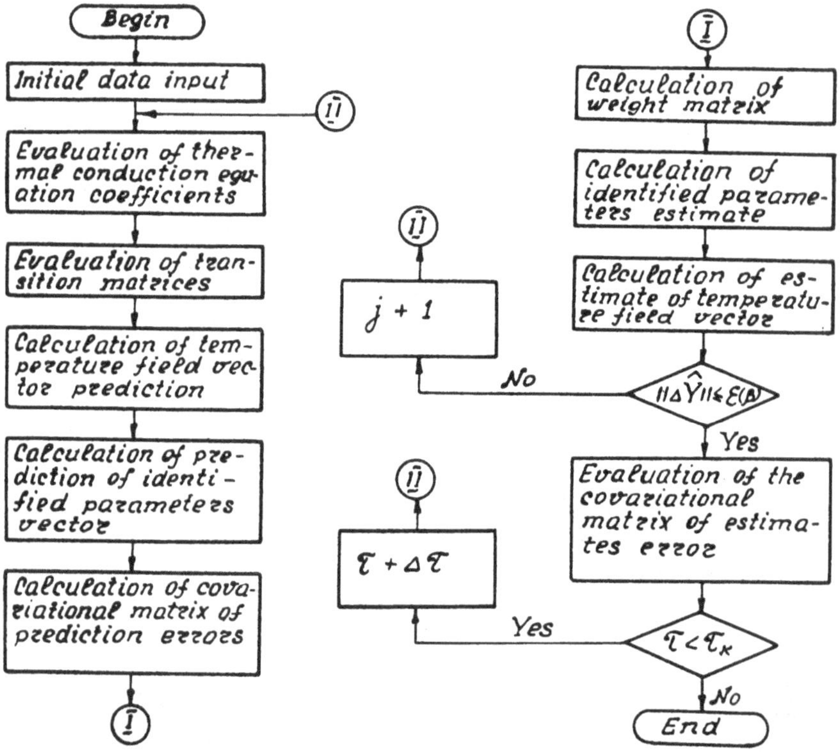

Fig. 1. $\|\Delta\hat{Y}\|$ - total error of measured and estimated temperatures; $\mathcal{E}(\beta)$ - threshold value; β - data error; j - number of iterations; τ_K - time.

The results of investigation are shown in Fig. 2, where curves 1, 2 and 3 were obtained with the help of an iterative filter with one, two and three measurements, respectively, the thermal measurement points coordinates being X_1 = 3 mm, X_2 = 5 mm, X_3 = 7 mm. The character of these dependences coincides with the form of the experimental curve obtained in paper [1] , and curve 5 obtained by gradient methods.

Studies directed towards determining the optimal number of measurements showed that an increase of the number of thermal measurement points in excess of two does not practically improve the estimates (curves 2 and 3 practically coincide as early as at the second time step). This presents us the possibility to use the remaining measured temperatures to evaluate the quality of the identified parameters. Such an estimate is carried out by comparing these temperatures with the results of the solution of the direct problem with the identified thermal diffusivity (the maximum relative error for one measurement was thereat 0.93%, for two - 0.5%, for three - 0.42%).

The results of identification of the thermal diffusivity of a metal hydride depending on the concentration of hydrogen have been used in designing automobile hydrogen accumulators.

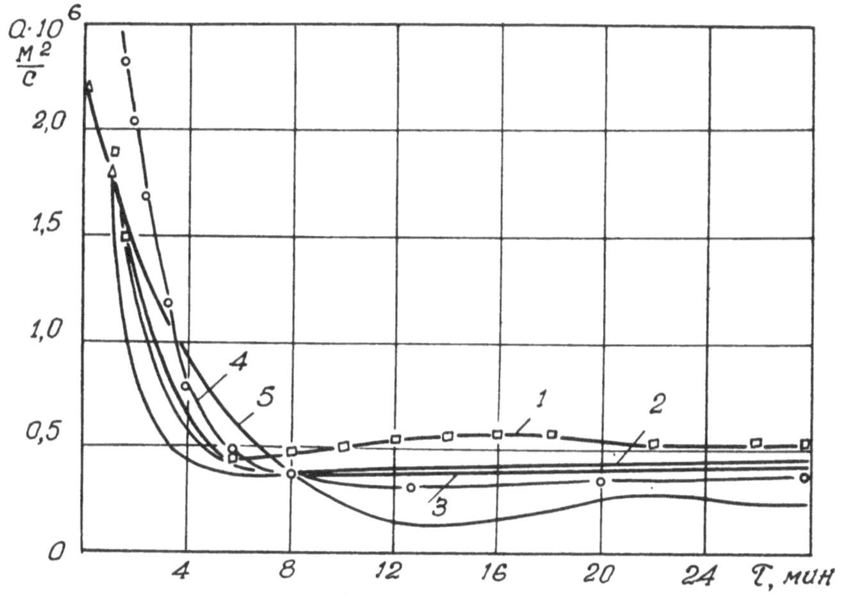

Fig. 2. Results of determining $a(\tau)$ for metal hydride $LaNi_5H_x$

REFERENCES

1. Artemenko A.N., Solovey V.V.: Heat-and-mass transfer during desorption of hydrogen from a finely dispersed metal hydride. Problems of Mechanical Engineering, 1982, 15, 91-95 pp. (In Russian).
2. Matsevity Yu.M., Moultanovsky A.V.: Identification in thermal conduction problems. Naukova Dumka, Kiev, 1982, 240 pp. (In Russian).
3. Matsevity Yu.M., Moultanovsky A.V.: On Identification Process Stability of Thermal Systems Parameters. Modelling, Simulation & Control, B, vol. 1, 1984, No. 1, p. 29-38 (In French).

CONTAINER CONTROL AND CARGO CLEARANCE USING COMPUTER PACKAGES

C S LALWANI
A K C BERESFORD*

The rate of growth in the number of containers shipped through the world's ports has, in recent years, necessitated the introduction of handling systems which are increasingly sophisticated. Computerisation has brought about a revolution in the efficiency of box handling by speeding communications, smoothing documentation and improving container selection and location. In this paper existing computer packages in UK ports based on simple database management are reviewed. These packages are non-transferrable to any other port. A generalized approach using Relational DBMS available commercially has been considered for possible applications.

Keywords: Computer database management systems, container tracking

1. Introduction

A revolution has taken place in recent years in industrial decision-making - a revolution that is largely due to the introduction of computers. The Port Industry has not been an exception. It was in the mid-seventies when a large number of ports introduced computers, but mainly in off-line applications such as payroll and some other accounting calculations. It was only at the beginning of this decade that the on-line application potential was realised by the port authorities in developed countries. To date a number of port authorities have introduced computers for on-line applications such as container control, cargo clearance, trade forecasting and information retrieval systems. This has not only improved the operational efficiency but also has acted as a very useful tool for management decision-making.

In a recent report published by ESCAP, United Nations[1], a list was given describing the areas where the installation of a computer could materially assist the workings of the management information system in ports. This list includes both off-line and on-line applications. An attempt is made here to review the areas where the installation of a computer has taken place to assist mainly the operational systems and also study the underlying methodologies and the basis required for such an installation. The areas for which computers have been used widely in the port industry are described below.

2. Container Control

The computer system implemented for container control in a number of ports worldwide is based on proven manual procedures. The largest container ports of the USA, Japan and Western Europe particularly, employ computerised systems. In the United Kingdom, Clydeport, London, Liverpool, Felixstowe and Southampton make extensive use of computers for this application.

*Department of Maritime Studies, University of Science & Technology, Cardiff CF1 3EU, UK

Computerised container control in the Port of Felixstowe was introduced in February 1982, but the work on the development of the package started in the late 1970's when the port was already handling 200,000 boxes per annum on the manual system. The advantages of the current system include providing all departments with instant access to the same information and any update is immediately available to all users via VDUs. The basic objective of container control computerisation is to assist in the efficient running of container terminals.

The eight departments that have access to the system at Felixstowe are: Terminal Manager, Terminal Information Office/Agent, Shipworking Foreman, Customs Office, Ingate, Outgate, Control, and Police

The departments are linked in the manner shown in Fig. 1. For users, the system is menu-driven and security coded. Each user has a set of funtions available to him which are tailored to his needs.

The Felixstowe container control system thus gives on-line facility for retrieval of information on vessels, exports, imports and vehicle monitoring. The system or operation started on Landguard Terminal with about 100,000 containers per year and was well-received by all workers and management. The system now processes about 450,000 containers per year and due to that the response to a retrieval has been very slow. Improvements are currently being made which should bring response times down to 20 seconds or less.

The hardware includes two Honeywell Level 64 mini computers with three disc drives and 28 VDU terminals. The computer package was developed at the port in consultation with the computing firm. The package in its present form is non-transferrable to any other port.

At all competitive container terminals the main objective is to maximise box throughput as a means to speed ship turnaround times. At Felixstowe, computerised container selection and location takes the form of a series of stacking policies. These are designed to suit shippers' needs and at the same time optimise space utilization on the container yard. Each policy attributes a different order of priority of box location by its length, height, weight and cargo characteristics. Where possible, separation is also by imports, exports and shipping line. The system is flexible, but its aim ultimately is to minimise container movement as a means to speed the loading/-unloading operation.

At Felixstowe, however, it is clear that there is room for improvement in the container control computer package since there is, for example, considerable mixing of import and export boxes. A computerised container selection system is most effective when used in conjuction with automated cranes. This fully automated operation of gantry cranes at the quayside for loading and discharging vessels is still at the experimental stage. In some ports, for example, Osaka and Richmond, the transtainers have computer control for movement of containers which is under the direction of the Terminal Manager. As containers are of standard lengths a means of indexing the crane position along its travel path can be of great help to the driver in accurate container placing. Further advances could be made in associated direct-link communication systems between crane drive, shipworking foreman and control. Short-wave radio-link is used in, for instance, Multi-Terminal, Rotterdam.

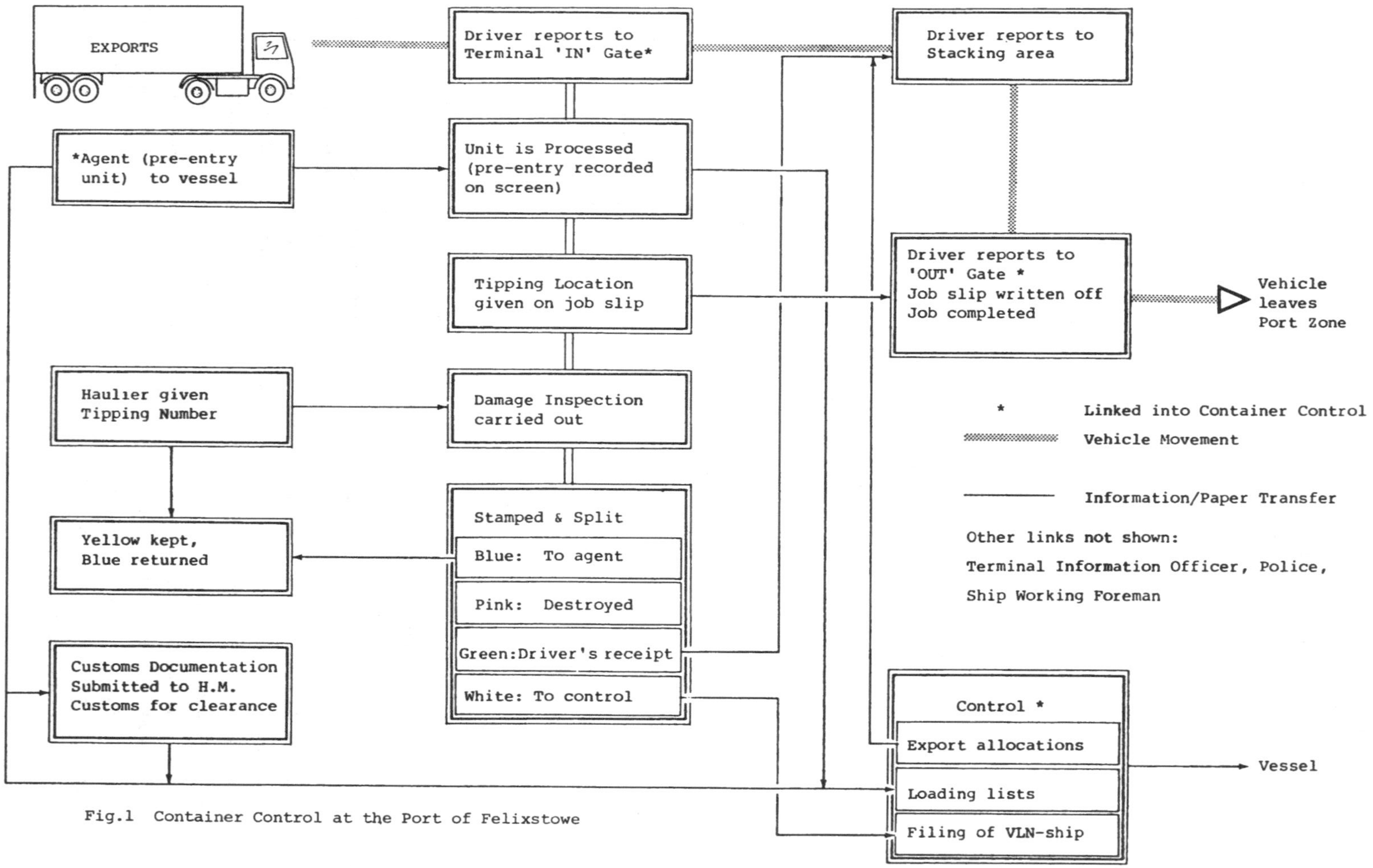

Fig.1 Container Control at the Port of Felixstowe

This generated the application for gantry crane guidance and some of the manufacturers (for example Ferranti's FD40 Mobile Gantry Crane) who have been studying this, have already installed such systems[2]; Eurokai at Hamburg is one. Automatic container identification is another area where the methods have been considered in relation to computer implementation, but such developments are still in their early stages.

In the Port of London Authority's multi-user terminal, a system to improve the clerical backup to the container handling operation has been running since September 1981. The computer program is given the name PLACER which stands for Port of London Authority's Container Event Recording[3]. The clerks interact with the PLACER system using VDU's. The system covers 3 major functions which are the status of an import container - from manifest to delivery of an export container - from initial receipt to confirmation of loading on a vessel, and the location of any particular container on the berth. The equipment used is a computer with 8046 processer with 2 disc drives and 22 VDU terminals. The computer system is used side by side with T-cards for stacking records. Some of the data processing functions, such as container rent invoices, dwell time statistics and other ad hoc reports are produced by means of providing the basic data by system links to the PLA mainframe computer.

In ports outside the UK, computerised container control systems have been implemented at the ECT Delta terminal in Rotterdam, the Matson terminal in Richmond, USA and at the 'K' line terminal in Kobe, Japan. At Antwerp, the Gylsen stevedoring company's terminal[4] has implemented a computer system for monitoring the container yard operations; it has a container masterfile, daily movement file, booking of order file, departure container file and unlocated export container file. The hardware includes a mini computer manufactured by Texas Instruments with a disc drive and eight VDU's (See Fig. 2).

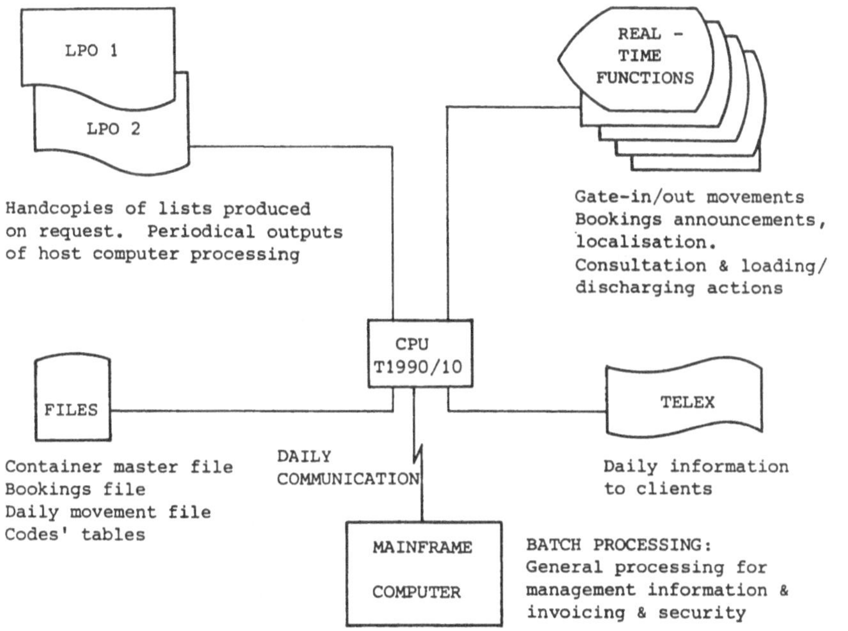

Fig. 2 Gylsen Stevedoring Company's Terminal. Container Control System. (Ref.4)

3. Cargo Clearance System

At the Port of Felixstowe a computerised cargo processing system is now in use. This system aims for cargo clearance within 48 hours of discharge at the port. The name given to this system is FCP80[5 6]. It is being implemented in two phases, the first phase introduces Direct Trader Input, whereby agents make their own customs entries direct to HM Customs computer. Now 90% of the port's total users are linked into the FCP8) System. (See Fig. 1). Both import and export entries are permitted. The agent requires the use of a VDU and a printer which may be in his office or in a location to which more than one agent has access. Customs procedures include the output on the agent's printer indicating that customs intended to check documents and/or to perform a physical examination of cargo and/or that customs will issue clearance automatically. The second phase comprises Inventory and Control of the following elements[5]:-

(i) Computer file creation for each vessel due to call at the port.

(ii) Manifest information file in respect of cargo from each port of loading.

(iii) A facility to pass equipment control information automatically to ships' agents.

(iv) A facility for customs entries to be compared automatically against the computer mainfest file ("writing off").

(v) A facility for customs and commercial release.

4. Data Base Management Systems (DBMS)

The container control package and FCP80 system at Felixstowe, the PLACER package at London and other existing packages are developed directly from the manual procedures folllowed at those ports. This restricts the use of the packages to those ports only and hence are non-transferrable to other ports having similar applications. Several of these packages are slow in retrieval time as the number of records in database grows in size due to the dependence on sequential search. In this section an attempt has been made to study the DBMS structure and to explore the possibility of using the most suitable one to container control and cargo clearance application.

Databases are classified into three categories according to the conceptual data structure used, namely, hierarchical network and relational. In the hierarchic approach, the data is represented by a simple tree structure in which each tree consists of a superior record together with subordinate records. Searching become progressively more time-consuming as more types of records are introduced and the hierarchy becomes more complex, so that the user may be forced to spend time and effort solving problems introduced by the increasing complications of the tree links.

By comparison, the network model is a more general structure, allowing the modelling of many-to-many correspondence using records, links and connectors. A connector represents the association between one record and another and contains data describing that association. The disadvantage is the complexity in the range of constructs (records, links) forming the network which makes the data manipulation language very complicated[7].

The relational approach is based on the assumption that tables or files that obey certain constraints may be considered as mathematical relations and hence the relation theory from mathematics can be used for various practical problems of dealing with data in such tables[8]. In a relational database any retrieval of information is a table, derived in some way from the database tables. For container control and cargo clearance data both network model and relational approach have applicability. Further research on application of commercially available relational packages such as RAPPORT (LOGICA Ltd) is in progress and results will be presented in the conference.

References

1. United Nations Economic and Social Commission for Asia and the Pacific "PORTMIS: Port Management Information System" Bangkok, 1981

2. Cant C J and Smith P W "The potential of microprocessors applied to container handling equipment" Proceedings of Seminar on the Application of the Use of Microprocessors in Engineering and Maintenance, 18 March 1982

3. Ford C M "The PLACER System" Proceedings of British Port Association Seminar on Application of Micro-computers to Smaller Ports, December 9th, 1982

4. Terminal Automation "The Mini-computer opens new terminal doors" Cargo Systems. September 1979

5. Port of Felixstowe FCP80 A System Outline, 1983

6. Port of Felixstowe FCP 80 Phase II Training Export Seminar 1984

7. Date C J An introduction to database systems, Addison-Wesley, 1979

8. Stewart M "Relational databases Part 1-10", Computer Weekly, Nov-Jan 1978/9

DESCRIPTOR VARIABLE ANALYSIS OF A NETWORK MODEL OF A FOOD PROCESSING AND DISTRIBUTION SECTOR[1]

Djordjija B. Petkovski[2] and Alexander H. Levis[3]

ABSTRACT

ALINET, a network modeling system, is used as a tool for analyzing energy use in wheat production and processing in a Yugoslav enterprise. The basic structure of the model is that of a material flow network where nodes represent the specific state of the material at that stage and links represent the processes that the material is undergoing. The descriptor variable approach has been used for determining the steady state flow of commodities through the system when decentralized decisions are taken into account. It is shown that supply and demand functions can be determined at the subsystem level, thus decomposing the problem of solving for the equilibrium.

Keywords: Networks, Simulation, Energy, Food Processing

1. INTRODUCTION

The rise in energy prices during the last decade and the limited availability of specific fuels have made it imperative that energy use in major sectors of the economy be examined to identify opportunities for conservation and for the introduction of energy conserving technologies. This has been the primary motivation for the development of ALINET, a network model of the food processing and distribution sector [1], [2]. Agricultural commodities flow through the network; the processing of these commodities is represented by a sequence of transformations, each of which changes the state of the material. A state is represented by a node, while transformations or processes are represented by links or branches. The use of node and links results in the representation of the food processing and distribution sector by a directed graph.

A new software package has been developed for determination of the steady state flow of commodities through the system. The descriptor variable representation [3], [4] that exploits system structure has been used. In this case, prices and quantities of commodities are considered as descriptor variables. This representation allows the consideration of chained structure, i.e., the system can be decomposed into a chain of subsystems. The motivation for identifying the chain structure and for using a descriptor variable formulation is that the overall equilibrium problem can be solved in a decentralized manner.

To demonstrate ALINET's use and the applicability of the related methodologies a pilot project was carried out with the cooperation of a Yugoslav enterprise.

2. THE FLOW MODEL

A new version of the computer software of ALINET has been developed so that it can be run in a limited computer environment, i.e., on a class of mini computers, such as Honeywell's Level 6 or Digital Equipment's PDP11 series. The step by step description of the new version of the computer software has been given elsewhere [5]. The first model implemented is that of wheat processing in a Yugoslav enterprise. More specifically, the movement of wheat from collection on the farm, milling, transportation of different types of flours, the manufacture of "multi-mixtures", different

1) This work was supported in part by the U.S.-Yugoslav Joint Board on Scientific and Technological Cooperation under Grant YOR-83/068
2) Faculty of Technical Sciences, Veljka Vlahovica 3, 21000 Novi Sad, Yugoslavia
3) Laboratory for Information and Decision Systems. MIT, Cambridge, MA 02139 USA

types of bread, pies and cakes, and final distribution of the plant products to their destinations are modeled. The structure of the graph is given in Figure 1.

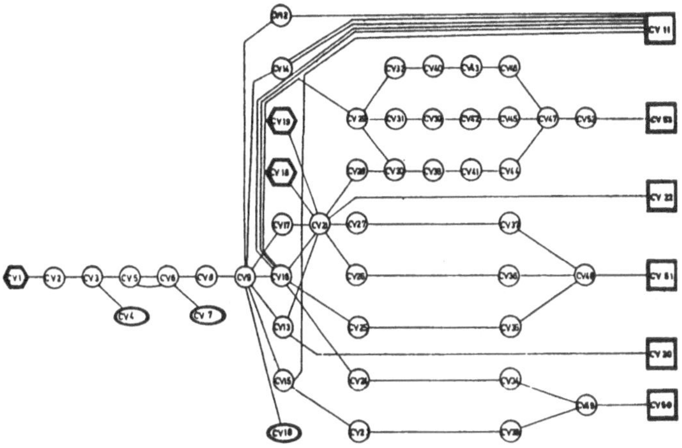

Figure 1. Structure of ALINET graph for Yugoslav Enterprise

Once the structure of the graph has been established, the supplies at the source nodes, the allocation coefficients for each link, and the attributes of the processes depicted by the links must be determined. The allocation coefficient denotes the fraction of the supply at the node allocated to each outgoing link. With the supplies specified at the source nodes and the prices and allocation coefficient estimated for each link, the material flows in each link and at each node were determined. Twelve such runs were made, one for each month in the baseline year June 1983 - May 1984. The simulation results were then compared to the data to verify that the model calculates material flows accurately.

The total amount of each energy form used in the system was obtained by summing over all links. The total direct energy use, by fuel type, and the cost from farm to consumer are given in Table 1. It includes energy used for transportation, drying and milling of grain, transportation of flours, multi-mixture manufacture, baking, and final product distribution. More detailed discussion of the material flows and energy use in the system has been given elsewhere [6].

TABLE 1. DIRECT ENERGY USE FROM FARM TO CONSUMER'

	ENERGY	COST
Electricity	8433 MWh	19.228×10^6 din
Natural Gas	869713 Nm^3	7.879×10^6 din
Diesel	54692 lit	2.6492×10^6 din

3. THE EQUILIBRIUM MODEL

In a typical national or regional model, there are many suppliers of commodities, many processing plants, and many consumers. One of the basic uses of a model such as ALINET is for the determination of the steady state flow of commodities through the system when the decentralized nature of the economic decisions is taken into account.

The standard modeling approach, using state-space systems, is not generally applicable for many economic systems which differ fundamentally from general physical systems. Therefore, there are cases where new concepts and methods need to be developed. An approach to the modeling, analysis and design of large-scale systems that exploits system structure is the descriptor variable one. It has been shown by Stengel [4], that the descriptor variable representation is particularly useful in determining the equilibrium of large-scale economic systems consisting of a supplier sector, producer sector, and a consumer sector, with no flow loops. Suppliers are characterized by supply functions, consumers by demand functions, while producers are described by demand function on their input side and supply functions on the output side. This allows the consideration of chained structures, i.e., sector can be decomposed into a chain of subsectors. The absence of loops leads to a chain structure corresponding to the goods market (see Figure 2).

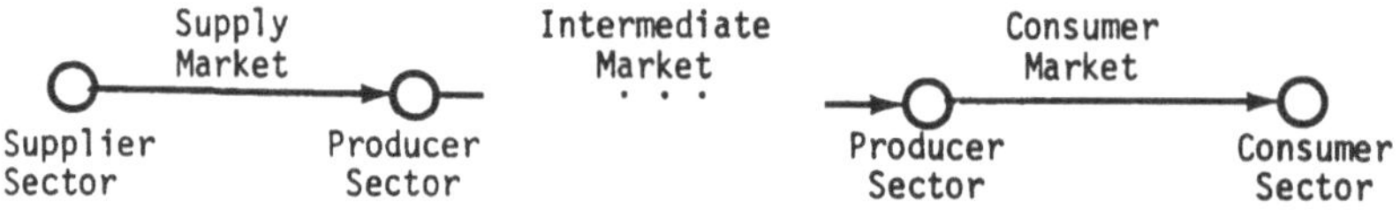

Figure 2. Economic Network Consisting of a Chain of Sectors

In the decentralized control approach [3], [4], supplier submodels respond to a demand for outputs by announcing the prices at which they will provide that flow of goods. Producers are given the flow rate of outputs they are to produce and the unit prices of their inputs, and they announce a demand for inputs and the prices of their outputs. Consumer sectors are price takers; they choose their consumption of goods.

Let q_i be a vector representing the flow of outputs in sector i and let p_i be the vector of corresponding prices. The model for suppliers in index set o is given by

$$p_o = f_o + S_o q_o \tag{1}$$

The producers in index set i, $1 \leq j \leq 4$ is given by

$$q_{i-1} = g_{i-1} + A_{i-1}p_{i-1} + B_i q_i \quad ; \quad p_i = f_i + C_{i-1}p_{i-1} + D_i q_i \tag{2}$$

Finally, the model for the consumer sector, index set 4, is characterized by

$$q_4 = g_4 + R_4 p_4 \tag{3}$$

The overall system can be written as $\underline{A}\,\underline{p} = \underline{f}$, where (4)

$$\underline{A} = \begin{bmatrix} I & -S_o & 0 & 0 & 0 & 0 & 0 & 0 & 0 & 0 \\ -A_o & I & 0 & -B_1 & 0 & 0 & 0 & 0 & 0 & 0 \\ -C_o & 0 & I & -D_1 & 0 & 0 & 0 & 0 & 0 & 0 \\ 0 & 0 & -A_1 & I & 0 & -B_2 & 0 & 0 & 0 & 0 \\ 0 & 0 & -C_1 & 0 & I & -D_2 & 0 & 0 & 0 & 0 \\ 0 & 0 & 0 & 0 & -A_2 & I & 0 & -B_3 & 0 & 0 \\ 0 & 0 & 0 & 0 & -C_2 & 0 & I & -D_3 & 0 & 0 \\ 0 & 0 & 0 & 0 & 0 & 0 & -A_3 & I & 0 & -B_4 \\ 0 & 0 & 0 & 0 & 0 & 0 & -C_3 & 0 & I & -D_4 \\ 0 & 0 & 0 & 0 & 0 & 0 & 0 & 0 & -R_4 & I \end{bmatrix} \quad \underline{p} = \begin{bmatrix} p_o \\ q_o \\ p_1 \\ q_1 \\ p_2 \\ q_2 \\ p_3 \\ q_3 \\ p_4 \\ q_4 \end{bmatrix} \quad \underline{f} = \begin{bmatrix} f_o \\ g_o \\ f_1 \\ g_1 \\ f_2 \\ g_2 \\ f_3 \\ g_3 \\ f_4 \\ g_4 \end{bmatrix} \tag{5}$$

The system (5) is solvable if and only if the matrix $\underline{A}$ is nonsingular. However, due to the high dimensionality of this matrix, it is useful to establish properties for the submatrices in (5) that insure stability and allow an efficient solution of the equilibrium flows and prices.

The motivation for identifying the chain structure and for using a descriptor variable formulation is that the overall equilibrium problem can be solved in a decentralized manner. Instead of a single large problem, a series of small problems, one for each subsector, can be solved. In each individual problem, only those relationships that affect its two adjacent subsectors are considered. In this case, sufficient conditions for solvability of the network are that S_k be positive definite and R_k be negative definite for all k, $0 \leq k \leq 4$, where each S_k is determined recursively from S_o by,

$$S_{k+1} = C_k(I-S_kA_k)^{-1}S_kB_{k+1}+D_k \tag{6}$$

and each R_k is determined recursively from R_4 by,

$$R_{k-1} = B_k(I-R_kD_k)^{-1}R_kC_{k-1}A_{k-1} \tag{7}$$

The corresponding values for the quantities and prices for sector 2, q_2 and p_2, are given in Table 2. The same results were obtained when the whole system of equations was solved directly.

TABLE 2. PRODUCER 2: QUANTITIES AND PRICES FOR EQUILIBRIUM

q_2 = (8506, 1711, 8930, 16830, 301.9, 356.6, 423.5, 41.72, 3463, 202.5, 252.8, 3835, 4195, 7.703)

p_2 = (8500, 28450, 24750, 26850, 168000, 188600, 20480, 27000, 36700, 27000, 27000, 92500, 46250, 30000, 5000)

5. REFERENCES

[1] A. H. Levis, I. S. Levis, E. R. Ducot, "Energy Conservation Analysis in the Food Processing Sector: A Simulation Approach," Int. Journal of Policy Analysis and Information Systems, Vol. IV, No. 1, 1980.

[2] A. H. Levis, E. R. Ducot, F. Webster and I. S. Levis, "ALINET: A Model of Assessing Energy Conservation Opportunities in Food Processing," Proc. ASAE National Energy Symposium, Kansas City, MO, September 1980.

[3] D. G. Luenberger, "Dynamic Equations in Descriptor Form," IEEE Trans. on Automatic Control, AC-22, No. 3, June 1977.

[4] D. N. Stengel, "Descriptor Analysis of Economic Systems," Ph.D. Thesis, Dept. of Engineering-Economic Systems, Stanford University, Stanford, CA, August 1978.

[5] D. Marganovic, Dj. Petkovski, D. Kukolj and T. Janos, "A Network Model for the Analysis of Energy Conservation Opportunities in the Food Processing Systems," XXVIII Yugoslav Conference ETAN, June 1984.

[6] Dj. Petkovski and A. H. Levis, "A Network Model for Assessing Energy Conservation Opportunities in Food Processing Annual Report for L984/85, U.S.-Yugoslav Joint Board on Scientific and Technological Cooperation, 1985.

PROCESS PREDICTION IN AUTOMATED INDUSTRIAL TRANSPORT SYSTEMS

Prof.Dr.sc.techn. H. Krampe [1]
Dr.-Ing. H.-G. Marquardt

1. Introduction

Automated industrial transport systems (GTS) serve for the linear and reticular development of production areas. Today, with microprocess computers efficient automation means are avaible for the control of these transport systems. The efficiency of the microprocess computers is in many cases not used completely because of the usual control tasks. Thus, it is possible, in parallel to the control tasks by means of microprocess computers, to solve such tasks that permit a higher quality of the process control.

2. Process control and identification in GTS

2.1. Problem posed

GTS consist of a variety of elements

- needed for constructing the transport network (connecting, ramifying, uniting and crossing elements),
- being able to store transport objects (storage elements),
- enabling the handling of tansport objects (operating elements).

Interaction between GTS and environment is described by vectors:

Input vector $\underline{x}$: Material flow to be received by the GTS
Output vector $\underline{y}$: Material flow leaving the GTS
Control vector $\underline{u}$: Technological conditions determining the transformation of the material flow from the input of the system to its output
Disturbance vector $\underline{\xi}$: Disturbances preventing or negatively influencing the transformation (failure of elements of the GTS).

It is characteristic for GTS that $\underline{x}$, $\underline{y}$ and $\underline{\xi}$ are stochastic quantities As furthermore the relations between the elements of a GTS are complicated already with simple systems, a closed representation of the sy-

[1]) Hochschule für Verkehrswesen "Friedrich List" Dresden
8010 Dresden Friedrich-List-Platz 1

stem behaviour in the form

$$\underline{y} = T(\underline{x}, \underline{u}, \underline{\xi})$$

is not possible. For this reason, state values of the elements are used for the description of the process in a GTS.

2.2. State values for the description of the process

The state values are quantities, the sizes of which completely characterize the process situation in a given moment. They provide all the necessary information for taking decisions on the process. That means that the number and kind of the state values are to be chosen in dependence on the specific behaviour of the process. The state values are elements of the quantity of permissible states Z_i with $z_i \in Z_i$. The quantity of all permissible states

$$Z = \{Z_1, Z_2, \ldots Z_n\}$$

form the discrete space of states

$$R_Z = Z_1 \times Z_2 \times \ldots \times Z_n .$$

At the moment $t = t_0$ (present time) the vector of the state values describes a point inthe discrete state space. The change of the state values in the field $t > t_0$ may be represented in the state space by trajectories.
The use of microprocess computers permits to observe the time behaviour of the state values, to evaluate it and to predict the future path of the trajectories in the state spaces. Thus, information is available that enables a considerable improvement of the process control.

3. Determination of the time behaviour of the state values

The determination of process parameters in complicated systems is a typical task of the process identification. Taking into account the a priori knowledge of the process, obtained for example from theoretical considerations, process quantities are measured during the process. The quantities measured may be already the state values of interest or the state values are calculated from them.
For the interpretation of the time behaviour of the state values, the measured quantities are evaluated by means of mathematical models providing the parameters of interest of the trajectory equations. Thus, the state of the process to be expected may be deter-

mined and used as a basis for further process control (Fig. 1).
The recursive regression and the exponential smoothing of second order are of special significance for a state estimation in GTS.

4. Examples of application

Given is a two-channel operations system with a highly varying heavy input flow. Only in peak hours the second channel is required. The control task is to put the second channel into operation in due time with increasing system load. Apparently, this is necessary before leaving the state space of the operating channel 1. With known quantities of the degree of occupancy b/b_{max} of channel 1 and with known average rate of arrivel and operation it may be determined when channel 2 is to be put into operation at the latest (Fig.2).

Ring structures are frequently used in commissioning systems and in flexible manufacturing systems of machine-building industry. The average queueing line at the input points of ring structures is determined by the

- average rate of arrival,
- average rate of operation,
- occupancy of the transport sections within the ring.

By means of digital simulation, these interconnections may be determined (Fig.3) and are memorized in a microprocess computer in the form of a state model |. Depending on the development of the process state it is possible to take prospective decisiond on the further process control, if, for example, disturbances should occur in the sinks of the system.

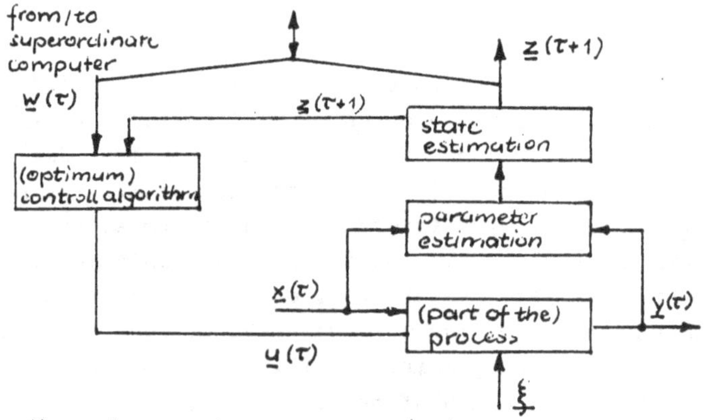

Fig. 1. State estimation and process control

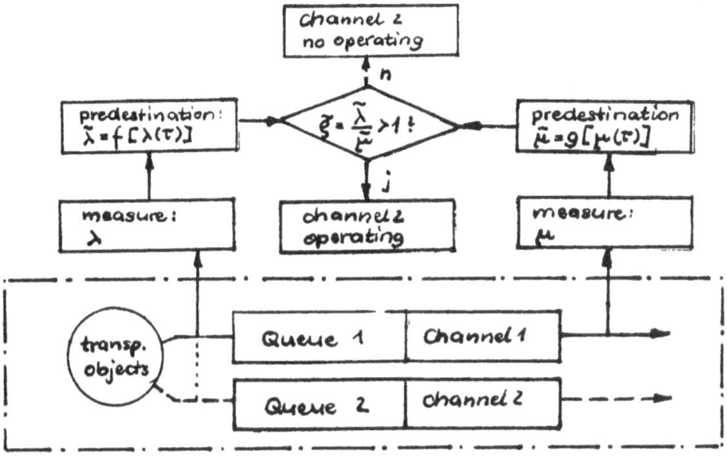

Fig. 2. Control of a two-channel operations system

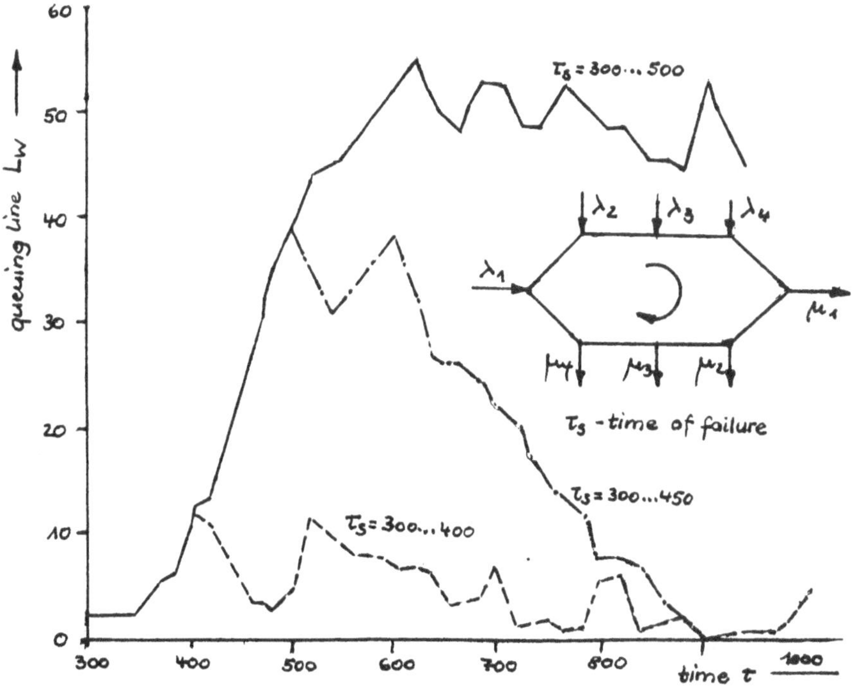

Fig. 3. Results of simulation for a ring structur

NEW APPLICATIONS OF DIGITAL SIMULATION IN CHEMICAL PLANT CONSTRUCTION

M. Wagner, Dr.; S. Arndt, Dr.; [1)]

Use of digital simulation for process design, for finding the best run of a plant, for designing the automatic control system. Formation of a model collection of chemical processes. Generation of the model structure, supply of the structure with valid parameters, control of the simulation run are implemented in a display-screen version. Examples: Design a multicontrol loop system for distillation column. Layout of a stirred tank process.

1. Application fields of digital simulation

The use of digital simulation has a growing importance in the phase of investment preparation of chemical plants. There are the following application fields:

. help for the process development by simulation experiments for the most important chemical engineering processes

. layout of the main equipment for chemical engineering under several conditions of load and disturbance of the plant.

- design of the automatic control system for the plant with regard to structure, algorithm and valid ranges of the control parameters
- determination of the optimal run variant of the plant with regard to energy consumption.

In the enterprise VEB CLG there is available the simulation system DIGSIM-ES for digital simulation of continuous systems, it is designed for ESER-computers.
The user can handle this system in a display screen version for the model - building process and for the control of the simulation runtime system. The program language is FORTRAN IV.

2. The structure of the models for the simulation system

The simulation system includes a collection of tested submodels in form of particular application programs for the dynamic behavior of equipment for chemical engineering, for chemical engineering processes, and automatic control devices. A user can apply these submodels for building his model. Generally, it is not necessary, that he knows mathematical details about modelling and solution methods. Simulation experiments include simulation runs for the dynamic behavior, for parameter optimization, and for sensibility analysis.
The model collection includes the following types:

1) VEB Chemieanlagenbaukombinat Leipzig-Grimma
Betrieb Anlagenbau, 7010 Leipzig, Georgiring 1-3

. elementary models, e. g. integrator, delay time, function generators

. process models, e. g. heat exchangers, stirred tank reactors, control valves

Process models are built by use of solution methods for ordinary differential equations and nonlinear or linear algebraic equations. The model structure for the technological lay-out is based on connecting the output variables of models with the input variables of models corresponding to the technological flow sheet.
Fig. 1 shows the outer model structure in form of particular application programs and the principle of coupling.

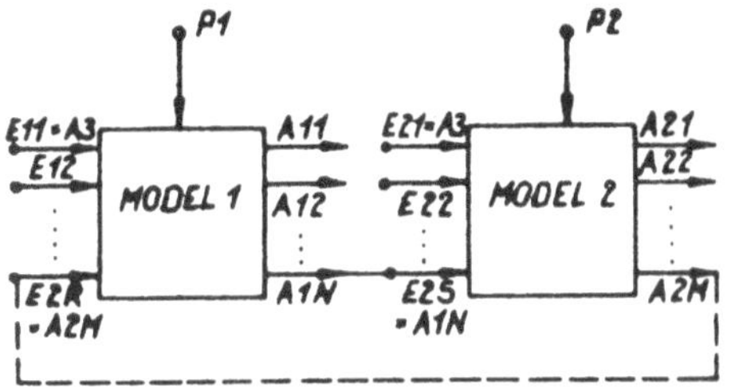

AIJ = designation of the output variables
EIJ = designation of the input variables
MODEL J = designation of the models
PI = parameter set for the models

Fig. 1 outer structure of the models and their coupling

3. Model-building and control of the simulation run-time system in a display screen version

In the model structure the output variables have different designations which can correspond to their technical significance. The model structure can include several models of the same type. The simulation system can be used in two steps. The first step is the interpreter step. Here the model structure is built in a display screen version. A connected program generator generates a complete FORTRAN-subroutine. This subroutine is linked with other programs to a complete simulation program. The second step includes the simulation run-time system. By selecting an identifying parameter during the simulation run-time, follow variants are possible:
- change of the output variables, which are to be observed
- change of the control and process parameters
- start up of a graphic representation of the results by line-printer
- run of special programs for calculating the dimensions of equipment
- continuation of the simulation run for the next time step
- cancel of the simulation variant

After cancel the simulation variant there are two possibilities for continuation:

- data input for the next variant
- change of the model structure.

4. Applications

Fig. 2 shows a part of the model structure for a multi control loop system of a distillation column. The temperature profil along the column is to be stabilised by the control system. The model structure includes a substandial part of the column only with 18 plates. The problem could be solved by the application of following types of process models:

- plate models for binary mixtures with enthalpy balances with and without side streams
- reboiler with control valve for heating steam
- PID-controller
- diverse elementary models

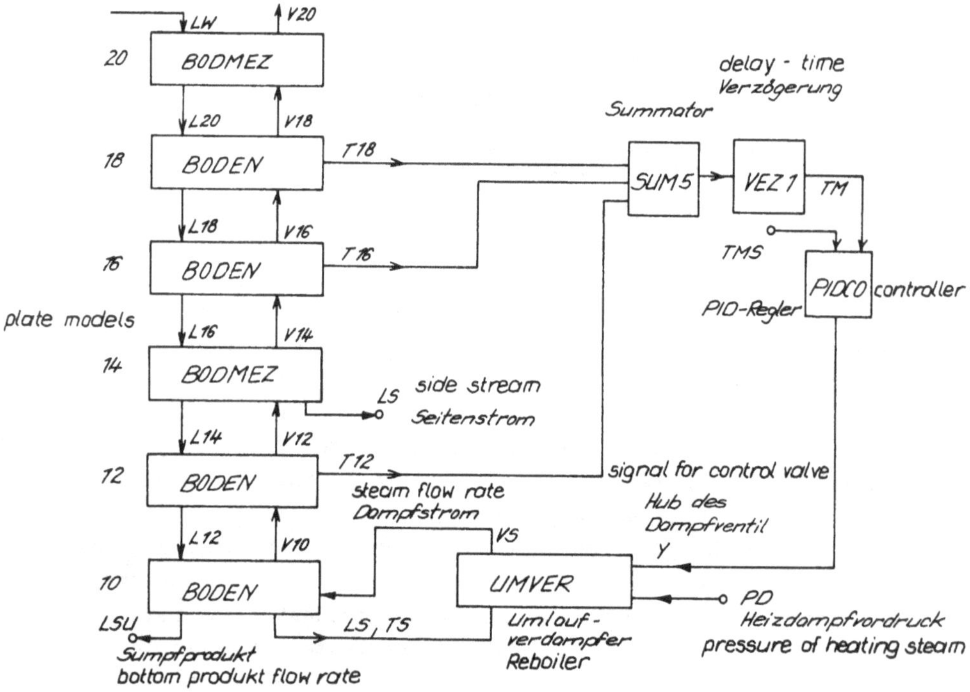

Fig. 2 part of the model structure for controlling distillation column

Simulation runs served for determining the transient responce and control behavior for disturbances in feed flow rate, composition, and supplying with heating steam. The best control variant was used for the column. This variant is characterised by the position of the temperature measurement points, the weighting factors for summation of the temperatures, and the controller parameters. The result was an essantial improvement of the control behavior compart to the previous variant. The simulation run-time for this nonlinear dynamic model is very long, about one minute for one minute in real time.

For the determining of the time for heating up a stirred tank process the following types of processmodels were used:
- steam heated heatexchanger
- ideal mixed stirred tank with heat loss to enviroment, and for continous product flow stream.

The dependence of the heating up time on the inaccuracy at determining of the heat tranfer coefficient has been investigated by sensibility analysis.

5. References

1. S. Arndt, O.-U. Langer

   Simulation und parametrische Optimierung verfahrenstechnischer Systeme mit dem Programmsystem DIGSIM in:

   Modellierung und Optimierung verfahrenstechnischer Systeme Akademie-Verlag Berlin 1978

2. M. Wagner

   Anwendung der Simulationstechnik im Chemieanlagenbau in:

   Modellbildung und Simulationstechniken Akademie-Verlag Berlin 1981

A MODELING WITHIN SYSTEMIC ASSESSMENT OF ALTERNATIVE TECHNOLOGIES

Vadim Nikolajew[1)]

Systemic assessment of alternative technologies based on scientific expertise elaborated on by the interdisciplinary research effort brings about a special kind of modeling ("loose modeling") that is adaqute to the features of the scientific and technological progress: high dynamics and uncertainty related to major technological breakthroughs, blend of heterogenous data flows generated by economic, technological and managerial structures. The distinguished feature of the modeling is strong emphasis on architecture of the complete solution. The modeling is given modular construction to meet the requirements of data flows, necessity to scale the technological events up and down, e.g. on the scope of national economy or detailed inquire into a technology, and professional interpretation of results.

1. Role of the modeling within systemic assessment of alternative technologies

Strategic thinking about new technological opportunities embraces a blend of social, political, economic and scientific issues to be grasped apart from ingenious inquiry into technologies for their own right. The resulting complexity feature of a growing number of problems challenging us requires a holistic systemic approach rather than isolated processing of parts possibly picked out at random.

The scientific expertise that is needed should be supported by interdisciplinary research aimed at bringing togher scientific knowledge and scientists separated hitherto by dash for specialization prevailing in development of the modern science.

A mix of interwoven qualitative and quantitative features rather heterogenous, sometimes even poor statistical backing, high dynamics and uncertainty embodied in the nature of technological breakthroughs shape the character of modeling that is necessary to raise the quality of assessment. Besides, the modeling is due to provide a common tool for performance of interdisciplinary research. [1]

Thus, the modeling within systemic assessment is aimed at adequate processing of quantitative data to be given the form of options or scenarios depicting implementation of technologies within societal and economic settings.

The logic of expertise implies:

- consulting of social, political, economic and scientific background,
- drawing up of options in a way

---

1) Dr. sc. oec., Zentralinstitut für Wirtschaftswissenschaften der Akademie der Wissenschaften der DDR, 1100 Berlin, Prenzlauer Promenade 149-152

IF we are going to implement a particular strategy of introduction of a new technology (technologies) e.g. dealing with the energy problem or developing microelectronics or biotechnology
THEN the ledger of consequences (direct, indirect and side-effects) would be...

represented in form of different options (scenarios) to enhance the "trial and error" way to reach a feasable solution in the framework of national needs and constraints. The options are subject of professional interpretation providing scientific and practical backing to trade-offs of options.

Fig. 1 places the procedure within the context of decision making process

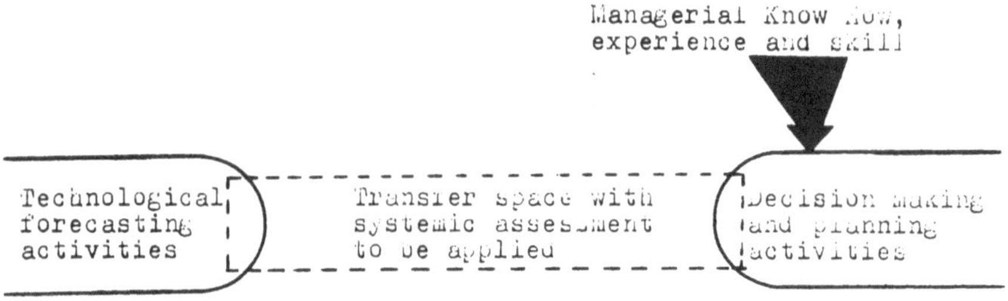

Fig. 1: Place of systemic assessment within decision making and planning activities

2. Some peculiarities of the modeling

The scientific expertise of technology assessment supported by a special kind of modeling called "loose modeling" because its tailoring to conditions of handling of "ill-structured problems" and is characterised by emphasis on architecture of complete solution rather than excellence of single composite methods per se.

The elements of the architecture are investigated and compiled on the basis of systems analysis. [2] Basically, in the modeling are used the following elements:

INPUTS via	ARCHITECTURE	OUTPUTS
Resources or production factors	of complete solution provided by a system of equations permitting chain calculations while performing simulations under different hypothesis or assumptions	Products
- investment		production value in monetary terms,
- R & D		physical volume,
- manpower		types of electronic components
- material		etc.
- energy ⎫ only		energy and chemical
- water ⎬ energy		useful products
- land ⎭ model		waste products

while the set of concrete inputs and "menus" of resulting outputs are to be tailored according to the architecture that provides the system solution.

Fig. 2 gives an overall view of principle scheme for systemic assessment of alternative technologies used in inquiry into the energy problem.

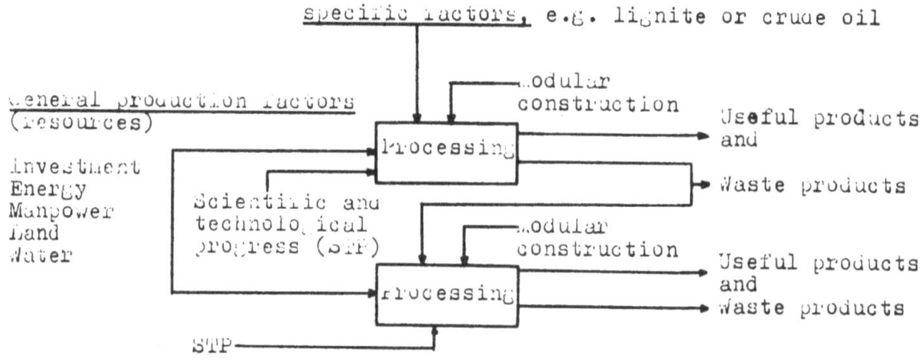

Fig. 2: Principle scheme for systemic assessment of alternative technologies (energy case) [3]

In the final analysis, three kinds of calculations were utilized:

- regression analysis or production functions,
- nonlinear trend calculations with different growth rates,
- part and whole relationship as weights of particular developments within higher level economic performance, e.g. investment in the energy sector in relation to investment of the economy as a whole.

The options can be proceeded in both directions:

- with given inputs (underlying resources constraints logic) calculation of possible outputs,

- with required outputs (needs) calculation of demanded inputs.

In the case of microelectronics after testing of about 100 regression equations and a great number of trend extrapolations about 50 regression equations along with growth rate and weight calculations were found appropriate to draw up low, high and "middle of the road" options in assessment of development of microelectronics.

Another point is that we are usually challenged by different nature of structures containing the factors to be dealt with. Three kinds of structures can be pointed out:

- technological structures, e.g. stages of crude oil refinary processes carried out on the basis of different technologies or basic technologies of electronic component production as they are familiar to engineers or natural scientists,

- product distribution structures. They are widely used in economic studies revealing interrelations among products as chains of intermidiate and final products, as proportions and distribution ratios among products within division of labor enabling smooth performance of an economy, e.g. products and production factors flows depicted in input-output tables or in production functions. These structures are conducive to economic assessment procedures but generally neglect differencies in efficiencies of various technologies. Frequently the product ratios are provided as average figures, "electric energy" etc. despite the fact that different efficiencies due to alternative technologies are at stake,

- management structures, e.g. organization of divisions or plants within greater production and organization units that establish management entities to be efficiently run. They are basic structures in decision making, because production factors and technologies are applied and products are planned according to these structures.

In systems oriented towards decision making the management structures should form the "basic floor" to accomodate other structures' elements.

[1] Parthey, H., Schreiber, K., edit. Interdisziplinarität in der Forschung, Akademie-Verlag Berlin 1983, p. 109-130.

[2] Nikolajew, V., Komplexes Prognostizieren des wissenschaftlich-technischen Fortschritts, Akademie-Verlag, Berlin 1977

[3] Similarity to WELMM, Analysis within global modeling, Energy in a Finate World, A Global Energy Systems Analysis, IIASA, Laxenburg, Austria p. 668 f.

A GLOBAL MODEL OF TECHNOLOGICAL ASYNCHRONOUS PROCESS SECTIONS IN CONNECTION WITH THE BINARY PROCESS ANALYSIS

Alder, Jochen[1)]

The paper deals with two models in connection with the binary process control and shows the aim basis and efficiency of each model.

Petri net, binary process analysis, binary process control

Some foundations of the binary process analysis

The method Binary Process Analysis (BPA) deals with processes that are subject to binary control and whose input and output variables are allowable for two values only. The term process includes sequences in technological plants, the necessary man-machine-communication and tasks of process safety (see/1/,/4/). The aim of the BPA is to get a mostly error-free task which does not depend on components of realizing the design of an automation equipment. The model of the process exactly fulfils this aim and consists of:

the process decomposition graph (PDG)	- parts of the process (PP) including control relations (communication)
the process control plan (PCP)	- a detailed control description belonging to each process part (PP) containing the control sequences
the technological scheme with points of automation	- specifications for any binary variable are to be listed

Exactly one control equipment is assigned to one PP and its behaviour results from the PCP on the basis of duality (see/1/,/3/). It is impossible, of course, to give a complete presentation of BPA, however, here are some foundations. The process will be divided or decomposed into a set M_T of PP's T^l:

$$M_T = \{T^l\} \mid T^l \neq T^m,\ l \neq m,\ l,m \in L = 01,02,\dots,|L| \qquad (1)$$

Necessary interactions of the PP's ensure the fulfilment of the demands of process control and are the so-called control relations $R_s = M_T \times M_T$. The system of (M_T,R_s) can be drawn as a process decomposition graph with M_T= set of nodes and R_s= set of directed arcs. As there is the empty control relation, too, the directed graph is partly connected.
All input control quantities of a PP cause a disigned change within this PP. That influence is called operation O_j and the binary-valued variables are the operation variables o_i:

$$O_j = \{o_i^{\,b_i}\} \mid i \in I_o,\ b_i \in \{\emptyset,1\} \qquad (2)$$

with a practicable interpretation:

1 := on!, open!, start!, .../ Ø := off!, shut!, stop!,.

The set O belongs to each PP: $O = \{O_j\} \mid j \in J = 01,02,03,\dots,|J| \qquad (3)$

1) Leipzig University of Technology, Section of Automation, DDR 7030 Leipzig, Box 66

Each sequence of operations is only demanded by the necessities of the binary process control and is, therefore, one result of using the BPA. The set O is structured by the set of process statuses $P \subsetneqq O \times O$, another result of the BPA. For all ordered pairs (O_j, O_k) each P_{jk} is called the transition process status (TPS):

$$P_{jk} = (O_j, O_k) \mid j,k \in J \qquad (4)$$

Especially, P_{jj} is the so-called holding process status. Each pair (O_j, P_{jj}) is called the PP-situation (PPS) S_j and each process status P_{jk} will be noted by the binary-valued variables p_i:

$$P_{jk} = (P_{P1}, P_{P2}, P_{P3}, \ldots, P_{Pt}, \ldots, P_{Pm}) \mid t \in I_{jk} \qquad (5)$$

$$\text{with} \quad P_{Pt} = \{p_i^{b_i}\} \mid i \in I_t,\ b_i \in \{\emptyset, 1\}$$

and the interpretation 1 := yes (true) / Ø := no (false).

Each P_{Pt} is called path with respect to the PCP. A set $P'_j \subsetneqq P$ belongs to each S_j and P'_j contains all the possible TPS's for all possible following operations of O_j. Therefore, a non-deterministic partial function d exists:

$$d(O_j, P_{jj}) = \{P_{jk}\} \mid P_{jj} \in \{P_{jk}\} = P'_j \qquad (6)$$

An example of blending and heating process

An interpretation of Petri nets will not be given here, but it is provided that Fig.1 is readable immediately. Fig.1 is taken from /2/ without any modification. The result of BPA is given in Fig.2 - 4. Fig.2 clearly shows the process decomposition taking place starting from the Petri net of Fig.1. Fig.3 shows additional demands described in /2/; they can be determined by comparing Fig.3 with Fig.1. The PCP shown in Fig.4 can be used for the two PP's, the charging of the components A and B, resp. In consequence, there are three interacting PP's represented in Fig.2. The control relations (2) between PPO1 and PPO3 are interactive and empty between PPO2 and PPO3. Now let us characterize the gain using BPA instead of the Petri net:

- Each PPS gives complete information on all possible process statuses designed and succeeding operations. Example: PPO3, $S_3 := (O_3, P_{33}) \longrightarrow \{P_{33}, P_{34}, P_{35}\}$ according to (6), and $P_{33} = p_5^0 p_6^0$, $P_{34} = p_5^0 p_6^1$, $P_{35} = p_5^1$ according to (5). The P_{jk} are always free of contradiction ($:= P_{33} \wedge P_{34} = 0$, $P_{33} \wedge P_{35} = 0$, $P_{34} \wedge P_{35} = 0$) and have the completeness ($:= P_{33} \vee P_{34} \vee P_{35} = 1$).
- Each PCP is an exact description of the function of controlling the PP. After the transformation of the complete technological values into abstract binary variables, there is the duty function of the PP control equipment for the design.

The Petri net as a global model of sequences of process sections

Definition: A process section (PS) is a sequence of an ordered subset of PP-situations given by the technological process connected with the man-machine-communication.

This definition means, considering Fig.2 - 4, that the PPS's belong to process sections as follows:

PS	initial stage	blending beginning	blending end	heating beginning	heating end	emptying beginning	emptying end
PPS[1)]	S0101	S0102	S0102	S0103	S0104	S0104	S0101
	S0201	S0202	S0203	S0201	S0201	S0201	S0201
	S0301	S0302	S0303	S0301	S0301	S0301	S0301

malfunction beginning	malfunction end
S0105	S0101
S0201	S0201
S0301	S0301

In Fig.2 each process section could be demonstrated by entering the sequence of effective situations (≙ stages) in Fig.2 (e.g. coloured). A Petri net will give more clarity because it can show concurrent sections and sequences of stages, resp., regardless of effective, complete PP-situations (Fig.5a). On the other hand, the result of the BPA is most suitable for the coordination of the process control. We give the following example: We need no call for the control of PP01 during the section blending and know the correct conditions from Fig.2 and 3. The real-time function of MC controllers, which you cannot get so easily and exactly from a Petri net, will be improved on this basis.

References

/1/ Alder, J.: Binäre Prozeßanalyse (in preparation)
VEB Verlag Technik, Berlin 1986

/2/ Henry, M., Baily, G., Sadiq, A.: Sequences in the control of processes The Chemical Engineer, 405(1984)9, pp. 9-12 (July)

/3/ Steusloff, H.: Rechnergestützter Entwurf von Automatisierungssystemen. Übersicht, Beispiele, Nutzbarkeit.
Fachberichte Messen, Steuern, Regeln: Fortschritte durch digitale Meß- und Automatisierungstechnik, Intercama-Kongress, Budapest 1983
Springer-Verlag Berlin, Heidelberg, New York, Tokio 1983, pp. 531-550

/4/ Weidlich, S.: Projektierungs- und Betriebserfahrungen mit Automatisierungssystemen.
Fachberichte Messen, Steuern, Regeln: Fortschritte durch digitale Meß- und Automatisierungstechnik, Intercama-Kongress, Budapest 1983
Springer-Verlag Berlin, Heidelberg, New York, Tokio 1983, pp. 682-696

---

1) The notation of the PPS: Slj with respect to (1) and (3)

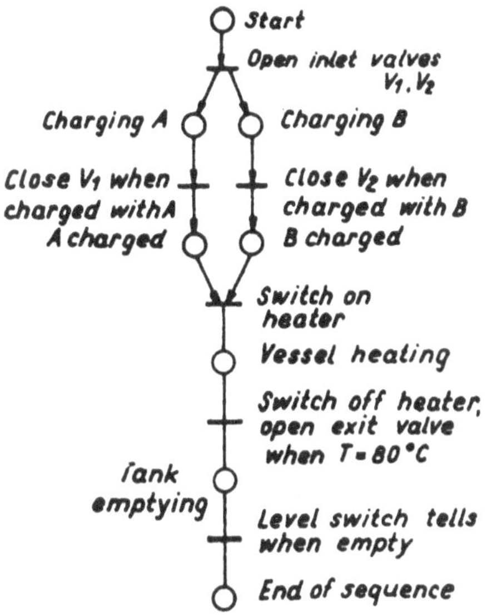

Fig. 1 Petri net of a blending and heating process

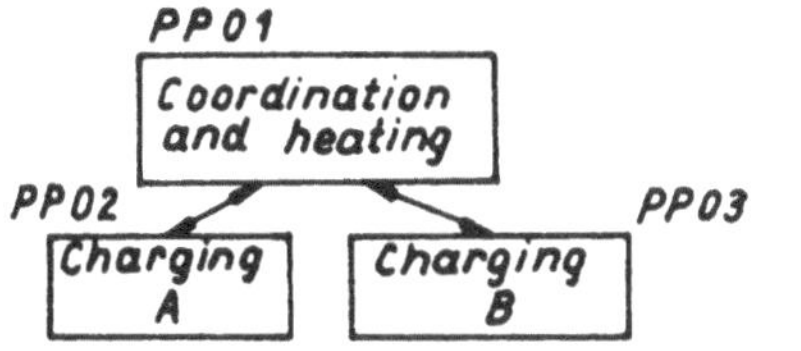

Fig. 2 Process decomposition graph according to process of Fig. 1

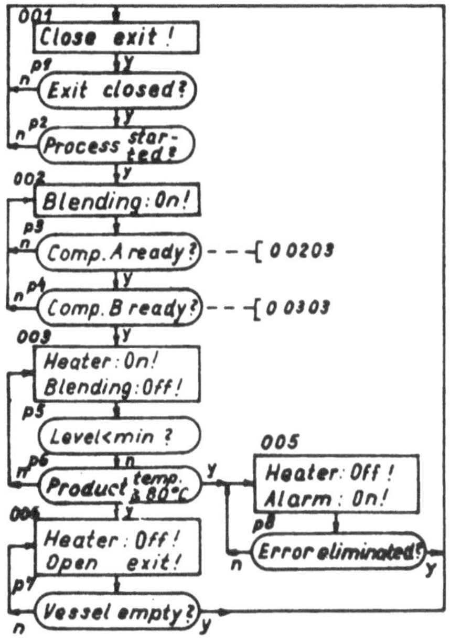

Fig. 3 Coordination and heating, PP 01

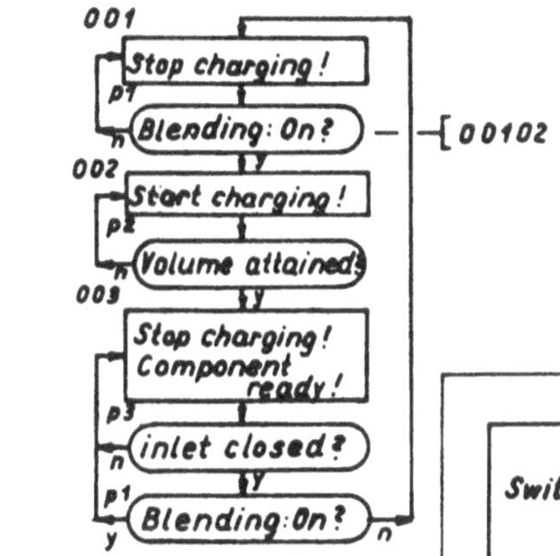

Fig. 4 Charging A, PP 02
Charging B, PP 03

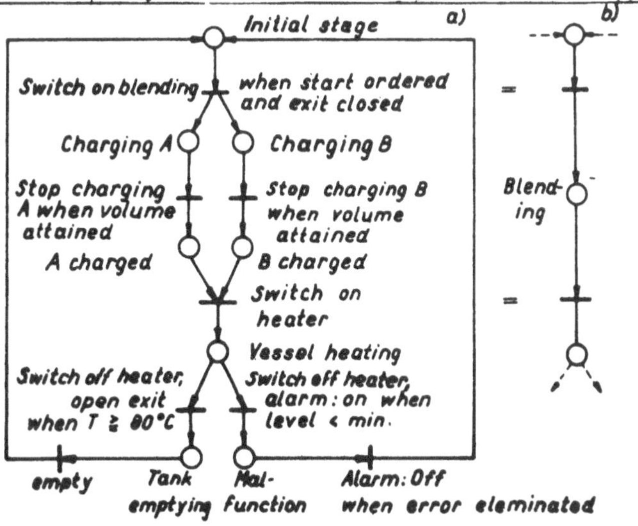

Fig. 5 a) Petri net, complete, as a result of BPA
b) Shortened

SIMULATION OF SLOW FLOW OF A FLUID PAST A SOLID SPHERE

dipl.eng.Mihai St.*, dr.eng.Mihai M.**, dipl.eng.Vasile Gh.*

Abstract. The problem of slow flow past a solid sphere is considered when the outer fluid is of infinite extent. Inertial effect is included in the analysis. A numerical method is employed to obtain solutions of the equation of motion. Solutions are obtained for the spherical coordinates components of velocity. The results obtained are in complete agreement with those obtained by other workers for the flow of the Newtonian fluid past a solid sphere.

The subject of the paper is very intrinsting for mass transfer units operations and chemical reactors. Inertial effects are included in the analysis of the problem of slow flow of a fluid past a solid sphere. Solutions are obtained for the spherical coordinater components of velocity. A numerical method is employed to obtain solutions of the equations of motion.

The numerical integration of the equations of motion is very difficult because the clasical methods generate numerical integration non-linear algebraic systems. The problem is very easy to solve if we integrate the motion and continuity equations in the following forms /1/ :

$$\int_S (\bar{\bar{p}} + \bar{\bar{\tau}})\overline{dS} - \int_V \rho \bar{g} dV = 0 \qquad (1.1)$$

and

$$\int_S \bar{v} d\bar{S} = 0 \qquad (1.2)$$

Let us define as dimensionless variables :

$$r = \frac{r}{R} \; ; \quad v = \frac{v}{v_o} \; ; \quad \tau = \frac{\tau R}{v_o} \; ; \quad p = \frac{pR}{v_o}$$

We define R to be the radius of the solide sphere and v_o to be the magnitude of the volocity of the undisturbed flow as $r \to \infty$.

The boundary conditions of this problem are :

$$\text{at} \quad r = R \quad v_r = v_\theta = 0 \qquad (2.1)$$

$$\text{at} \quad r \to \infty \quad v_r = -v_o \cos\theta \quad \text{and} \quad v_\theta = v_o \sin\theta \qquad (2.2)$$

After discretization we find for motion equations :

$$(p\bar{\bar{\delta}} + \bar{\bar{\tau}}_{rr} + \bar{\bar{\tau}}_{r\theta})_{i+1,j}\Delta S_{i+1,j} + \rho\bar{g}\cos\theta_j\Delta V_{i,j} - (p\bar{\bar{\delta}} + \bar{\bar{\tau}}_{\theta\theta} + \bar{\bar{\tau}}_{\theta r})\Delta S_{i,j-1} - \rho\bar{g}\sin\theta_{j-1}\Delta V_{i,j} - (p\bar{\bar{\delta}} + \bar{\bar{\tau}}_{rr} + \bar{\bar{\tau}}_{r\theta})_{i-1,j}\Delta S_{i-1,j} + \rho\bar{g}\cos\theta_j\Delta V_{i,j} + (p\bar{\bar{\delta}} + \bar{\bar{\tau}}_{\theta\theta} + \bar{\bar{\tau}}_{\theta r})_{i,j+1}\Delta S_{i,j+1} - \rho\bar{g}\sin\theta_{j+1}\Delta V_{i,j} = 0 \qquad (3.1)$$

end for continuity equation :

* Research Institute for Computer Technique, Calea Floreasca no 167, sect.2, 72321 Bucharest
** Politechnic Institute of Bucharest, Faculty of Chemical Engineering.

$$\frac{v^r_{i+1,j} + v^r_{i,j}}{2}\Delta S_{i+1,j} - \frac{v^r_{i-1,j} + v^r_{i,j}}{2}\Delta S_{i-1,j} - \frac{v^\theta_{i,j-1} + v^\theta_{i,j}}{2}\Delta S_{i,j-1} -$$

$$+ \frac{v^\theta_{i,j+1} + v^\theta_{i,j}}{2}\Delta S_{i,j+1} = 0 \qquad (3.2)$$

The components of stress tensor τ are defined by Newton's low of fluid. Utilizing the definitions of the components of stress tensor and set in order, equations (3.1) and (3.2) become :

$$A_{i+1,j}\, v^r_{i+1,j} + A_{i,j}\, v^r_{i,j} + A_{i-1,j}\, v^r_{i,j-1} + A_{i,j-1}\, v^r_{i,j-1} +$$
$$+ A_{i,j+1}\, v^r_{i,j+1} + B_{i+1,j}\, v^\theta_{i+1,j} + B_{i,j}\, v^\theta_{i,j} + B_{i-1,j}\, v^\theta_{i-1,j} +$$
$$+ B_{i,j-1}\, v^\theta_{i,j-1} + B_{i,j+1}\, v^\theta_{i,j+1} = AL_i \quad \text{where } i = 1, NI \text{ and } j = 1,NJ$$

and (4.1)

$$C_{i+1,j}\, v^r_{i+1,j} + C_{i-1,j}\, v^r_{i-1,j} + C_{i,j}\, v^r_{i,j} + D_{i,j-1}\, v^\theta_{i,j-1} +$$
$$+ D_{i,j+1}\, v^\theta_{i,j+1} + D_{i,j}\, v^\theta_{i,j} = 0 \quad \text{where } i = 1,NI \text{ and } j = 1,NJ \qquad (4.2)$$

After consideration the boundary conditions equations (4.1) and (4.2) become :

$$A_{i+1,j}\, v^r_{i+1,j} + A_{i,j}\, v^r_{i,j} + A_{i-1,j}\, v^r_{i-1,j} + A_{i,j+1}\, v^r_{i,j+1} +$$
$$+ A_{i,j-1}\, v^r_{i,j-1} + B_{i+1,j}\, v^\theta_{i+1,j} + B_{i,j}\, v^\theta_{i,j} + B_{i,j-1}\, v^\theta_{i,j-1} +$$
$$+ B_{i,j+1}\, v^\theta_{i,j+1} + B_{i-1,j}\, v^\theta_{i-1,j} = AL_i - A_{k+1,\ell}\, v_o \cos\theta_\ell -$$
$$- B_{k+1,\ell} v_o \sin\theta_\ell = 0 \qquad (5.1)$$

and $C_{i+1,j}\, v^r_{i+1,j} + C_{i,j}\, v^r_{i,j} + C_{i-1,j}\, v^r_{i-1,j} + D_{i,j-1}\, v^\theta_{i,j-1} +$

$$+ D_{i,j+1}\, v^\theta_{i,j+1} + D_{i,j}\, v^\theta_{i,j} = - C_{k+1,\ell}\, v_o \cos\theta_\ell - D_{k,\ell+1}\, v_o \sin\theta_\ell \qquad (5.2)$$

where $i = NI,NJ(NI-2)$, $j = NJ,NI(NJ-2)$, $k = NI-2$, $NI-1$ and $= NJ-1,NJ$.

The systems matrix is represented in figure 1. For solving we suggest the following techniques: on replace $v^\theta_{i,j}$ by $v^r_{i,j}$ and the obtained system is solved by Gauss method and than we use $v^r_{i,j}$ for finding $v^\theta_{i,j}$ by Choleski method /2/. For a discretization networck with 30/30 points the systems (5) are 1680 linear equations and the technique used finds the solution of the problem on the Romanian computer FELIX C-1024 in about two hours. The results obtained are in complete agreement with those obtained by other workers for the flow of the newtonian fluid past a solid sphere.

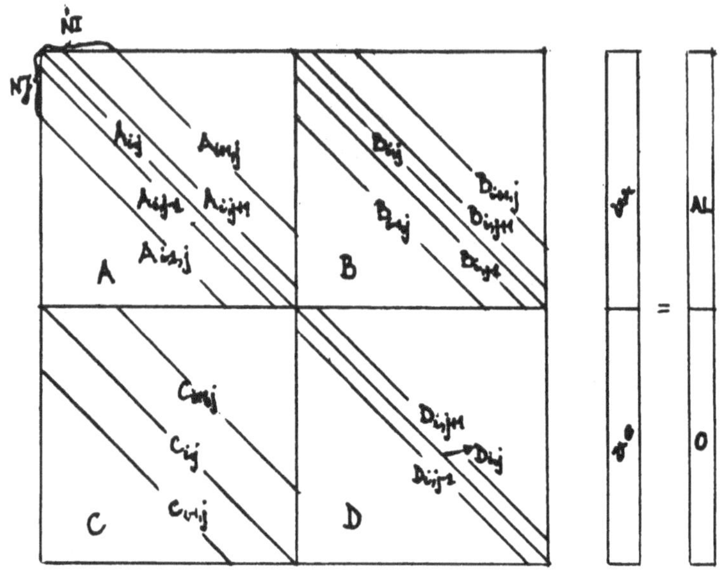

Fig.1. The systems matrix (4)

Notation

A,B,C,D - matrix

$A_{i,j}$, $B_{i,j}$, $C_{i,j}$, $D_{i,j}$ - element of matrix A,B,C,D

AL_i - element of vector AL

g - body force constant

I,i - counter for radial coordinate

J,j - counter for coordinate

p - pressure

r - spherical coordinate

S - surface

$v_r, v_\theta, v^r, v^\theta$ - components of velocity

V - volum

δ - Kroneker symbol

ρ - density

θ - spherical coordinate

τ_{rr}, $\tau_{r\theta}$, $\tau_{\theta\theta}$ - components of stress tensor τ

Literature

/1/ Landau L, E. Lifchitz: Mécanique des fluides, Éditions mir, Moscou 1971.

/2/ N. Andrei, C. Răsturnoiu - Sparce matrices and their applications, Editura Tehnică, București 1983.

APPLICATION OF A HYBRIDE COMPUTER FOR THE OPTIMIZATION OF MOTOR VEHICLES

Juergen Schulze [1)]

In order to support an optimum design for the oscillating technique of motor vehicles various computing programmes for the HRA 7000 (ADT 3000/R 4000) hybride computer have been worked out. Different computing programmes are available on interchangeable programming tables for the motor-cycle, motor-car, lorry and semitrailer. As regards the calculation of oscillations, non-linear simulation models with up to seven degrees of freedom for the simulation are used with road profiles measured at stochastic excitations. By way of example of a caravan the principal approach for the optimization of oscillations of motor vehicles is shown.

Keywords: Mathematical modelling, hybride simulation, real-time processing, calculation of oscillations, motor-vehicles, stochastic excitation, optimization

1. Problem and simulation model

The objective was to study the quadratic response for a caravan at different load, length of die rod damping. The non-linear shock-absorber curveshould be varied so as to obtain low wheel load variations at slight accelerations of cushioned mass.

In order to simulate vertical vehicle oscillations a two -axle tractor together with a single-axle caravan a plane simulation model having five masses and six degrees of freedom has been selected (Fig. 1). Plane simulation models provide sufficiently precise results bearing the comparison with the road test.

Moreover, the simulation model allows the calculation of oscillations of semitrailers and motor-cars. As regards the calculation of oscillations of the latter the trailer weight is set equal to zero.

The subsequent non-linearities are considered in the simulation model:

- rebounding the wheels on all three axles
- curved-linear characteristic curve of springs (2 straight sections)
- curved-linear characteristic curves of shock absorbers (3 straight sections)
- dry friction in the wheel suspension of the trailer axle.

1) Dipl.-Ing. Juergen Schulze, VEB Barkas-Werke, Main Factory of IFA-Kombinat PKW, Reseach Dept., Kauffahrtei 31, DDR-9040 Karl-Marx-Stadt

Tyre damping is neglected due to the low quantity as compared with the damping of the cushioned mass.

2. Equations of motion

The sequences of motions are described by a system of non-linear common differential equations of the second order. The equations of motion are formulated in a way of saving computing elements.

Vertical accelerations:

$$\ddot{Z}_1 = -A_1 F_1 + A_1 F_4 \qquad (1)$$

$$\ddot{Z}_2 = -B_1 F_2 + B_1 F_5 \qquad (2)$$

$$\ddot{Z}_3 = -D_1 \ddot{Z}_{53} - D_2 F_3 + D_2 F_6 \qquad (3)$$

$$\ddot{Z}_{44} = -E_1 \ddot{Z}_{45} + E_2 \ddot{Z}_{53} - E_3 F_4 \qquad (4)$$

$$\ddot{Z}_{45} = -G_1 \ddot{Z}_{44} - G_2 \ddot{Z}_{53} - G_3 F_5 \qquad (5)$$

$$\ddot{Z}_{53} = -H_1 \ddot{Z}_{45} + H_2 \ddot{Z}_{44} - H_3 \ddot{Z}_3 - H_4 F_6 \qquad (6)$$

Relative accelerations:

$$\ddot{s}_4 = \ddot{Z}_{44} - \ddot{Z}_1 \qquad (7)$$

$$\ddot{s}_5 = \ddot{Z}_{45} - \ddot{Z}_2 \qquad (8)$$

$$\ddot{s}_6 = \ddot{Z}_{53} - \ddot{Z}_3 \qquad (9)$$

Forces:

$$F_i = c_i s_i \qquad (i = 1 \ldots 3) \qquad (10)$$

$$F_i = c_i s_i + c_{iZS} s_{iZ} + k_{iD1} \dot{s}_{iD1} + k_{iZ1} \dot{s}_{iZ1} + k_{iZ2} \dot{s}_{iZ2} + F_{Ri} \operatorname{sign} \dot{s}_i \qquad (11)$$

$$(i = 4 \ldots 6)$$

$$F_K = -N_1 \ddot{Z}_{44} + N_2 \ddot{Z}_{45} + N_3 \ddot{Z}_{53} \qquad (12)$$

Relative paths:

$$s_i = Z_i - Z_{0i} \quad , \quad s_i \leq s_{i\,stat} \qquad (i = 1 \ldots 3) \qquad (13)$$

$$\left.\begin{aligned} s_i &\leq s_{i\,stat} \longrightarrow s_{iZ} = 0 \\ s_i &> s_{i\,stat} \longrightarrow s_{iZ} = s_i \end{aligned}\right\} \qquad (i = 4 \ldots 6) \qquad (14)$$

Relative velocities:

$$s_i \leq 0 \longrightarrow \dot{s}_{iD1} = \dot{s}_i \;,\; \dot{s}_{iZ1} = \dot{s}_{iZ2} = 0$$

$$0 \leq \dot{s}_i \leq \dot{s}_{izp} \rightarrow \dot{s}_{iD1} = 0 \;, \quad \dot{s}_{iz1} = \dot{s}_i \;, \quad \dot{s}_{iz2} = 0$$

$$\dot{s}_i > \dot{s}_{izp} \rightarrow \dot{s}_{iD1} = 0 \;, \quad \dot{s}_{iz1} = \dot{s}_{iz2} = \dot{s}_i \qquad (i=4..6) \qquad (15)$$

3. Excitation

In order to excite the simulation model single obstructions (positive and negative sinus half-wave) and road contours measured are used. They have been provided on a magnetic tape in a frequency-modulated memory. The road contours are digitalised and stored on a magnetic plate. An automatic sequence of the calculation for determining the optimum parameters is possible in a great variety of variants due to the quick access to the magnetic plate.
During this study of oscillations typical road contours of a good bituminous road and those of a poor one are selected for exciting the simulation model. The roads are used at a speed of between 40 and 80 km/h along a distance of approx. 600 m.

4. Computing programme

In order to carry out computations on the HRA 7000 hybride computer a DIWA programme /3/ meeting the subsequent functions is generated:

- calculation of the coefficients and potentiometer data
- setting the digital potentiometer and checking all potentiometer
- providing the excitation function, i. e., road contours or sinus half-wave as a single obstruction stored on a magnetic
- plate
- solving the equations of motion on the analog computer
- computation and recording the evaluation factors (r.m.s. values of wheel loads and of the acceleration of the cushioned mass)
- automatic variation of parameters given and automatic computation of several variants at stochastic excitation.

The excitation function for the front, rear and trailer axles is provided through the direct-start programmes SIER (sinus excitation) and STER (stochastic excitation), with the wheel-centre distances being considered by dead times. The road contour is provided through a scanning frequency depending on the driving speed at the outputs of the digital/analog converter as an electric tension for exciting the simulation model on the analog computer.

The non-linear characteristic spring-shock absorber curves can be changed by means of digital potentiometer so that an automatic variation of the parameters is possible. The operator provides the parameter number, the pitch and the limits of the parameter changes on the typewriter of the computer.

5. Evaluation and results

Generally, the objective for an optimization of the vehicle oscillations is to obtain a high oscillating comfort (low acceleration of the damped mass) and a high driving safety (low variations of wheel load).

Therefore, for the purpose of evaluation the r.m.s. values of the acceleration of the damped masses and those of the wheel load referred to the static load are plotted in a Cartesian coordinate system so that a parabel is obtained, e. g., with a variation in damping, the optimum for the quadratic response being in the vertex (Fig. 2). It is also possible to feed even additional weighting coefficients for the evaluation factors /1/. In order to find the r.m.s. values the mean effective values are determined on the analog computer, out of which the effective accelerations of the damped masses and the wheel loads are computed on the digital computer.

As the most essential result a deterioration of the quadratic response at the transition from a linear to a non-linear characteristic shock-abosrber curve at the trailer axle is found. In contrast to a damping force setting of tension/compression = 2.4 a reduction of the r.m.s. values of the acceleration of damped masses and wheel loads by 15 % on a poor bituminous road at a speed of 80 km/h has been achieved by linear damping.

The optimum damping force setting differs from the surfaces used here on a good and poor bituminous road only slightly. Generally, a lower damping is necessary at a lower speed and an improving state of the road.

6. Summary

By means of hybride simulation carried out in real time it is possible to compute a great number of variants within a relatively short time. As compared with the digital simulation on conventional digital computers a considerable reduction in computing time is attained by the hybride mode. As regards the size of the simulation model used in this paper the relation of the computing time from digital to hybride simulation can be assumed a value of 100.

Based on the results of the simulation the automotive designer can find an optimum design of automobiles in regard to oscillations. Thus, in case of new and further developments of vehicles target-oriented road tests can be carried out within an optimum range. As a result, a considerable saving in costs in the prototype building and during the road test is obtained due to a reduction of the amount of time and material required.

7. References

/1/ Helms, H.: Grenzen der Verbesserungsfähigkeit von Schwingungskomfort und Fahrsicherheit an Kraftfahrzeugen. Diss., TU Braunschweig 1974.

/2/ Voy, Ch.: Vergleichende mehrdimensionale Betrachtungsweise bei der Simulation vertikaler Fahrzeugschwingungen. Automobilindustrie (1977) 3, p. 51 - 66.

/3/ Anwendungsbeschreibung DIWA 4000 - HA/DIKO. VEB Robotron Dresden, Systemunterlagendokumentation 1979.

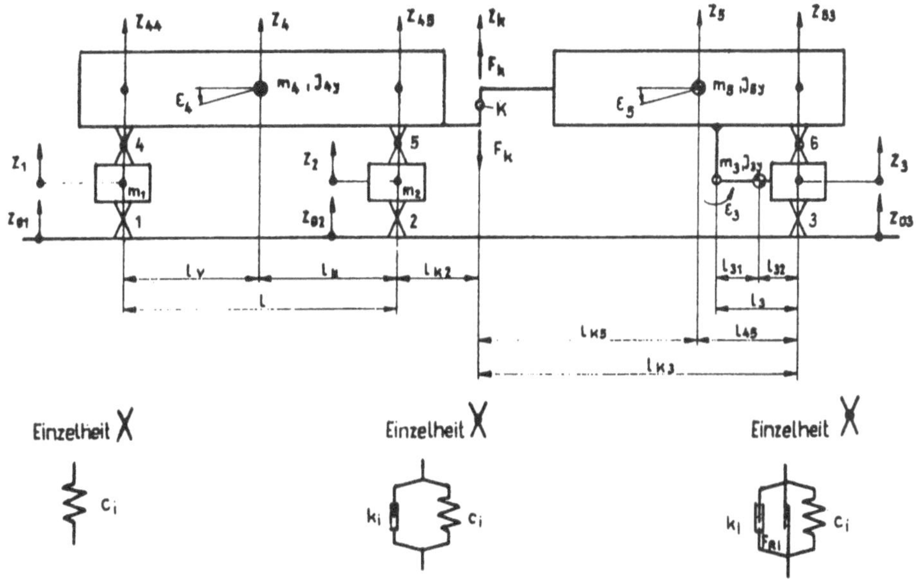

Fig. 1: Simulation model

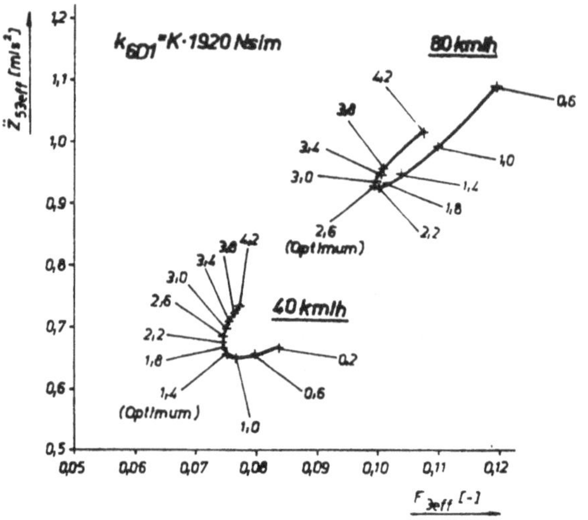

Fig. 2:
R.m.s. values of the acceleration of damped masses and of the wheel load at the trailer in case of variation of damping of damped masses in direction of compression

ANALYSIS OF DRIVER-VEHICLE SYSTEM PERFORMANCE BY MODELLING OF CAR-FOLLOWING MOTION

H. Kormanski , K. Rudzinska [1)]

Abstract : This paper is concerned with mathematical models of the control by human drivers while following another vehicle in single lane traffic. All the models are compared with real driver-vehicle behaviour recorded by an on-board data acquisition system. Non-linear programming procedures are applied to parameter identification of the models. Several criteria of the car-following motion performance are formulated and some effects of "the driver's parameters" on these indicies are investigated. The paper concludes with a brief discussion of some current areas of research and possible applications.

Keywords : car-following motion, human operator models, driver-vehicle system performance

1. Introduction

The development of various theories concerning traffic phenomena has received considerable attention during the last few years. One of the theories has come to be known as the follow-the leader theory of traffic. Such a theory pertains to single lane dense traffic with no passing, and is based on the assumption that each driver reacts in some specific fashion to a stimulus from the car /or cars/ ahead of him. Fig. 1 gives simplified representation of a human driver in a car-following task.

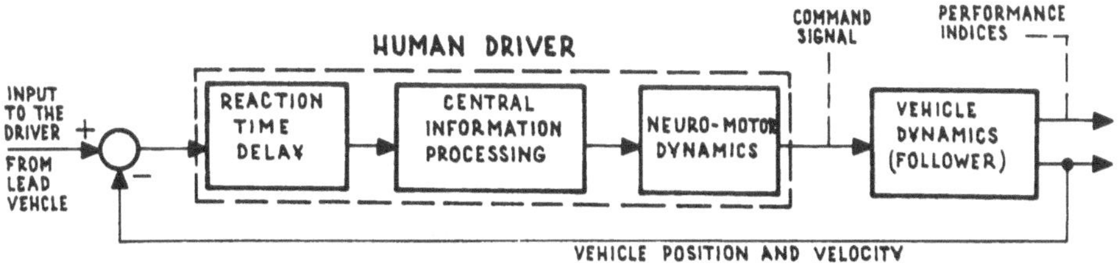

Fig. 1. CAR-FOLLOWING MODEL.

An understanding of the dynamics of the process is clearly important since improper car-following behaviour such as "tailgating" can lead to multiple accidents. It is well known that disturbances in the velocity of a lead car in a following platoon can be amplified as they propagate through the platoon primarily due to the dynamic characteristics of the drivers. Additionally, the unstable motion of vehicles causes increased fuel consumption and toxic emissions.

1) Institute of Computer Science
Technical University of Gdansk
80-952 Gdansk, Poland

2. Car-following models

We use the following notation to describe human control models :

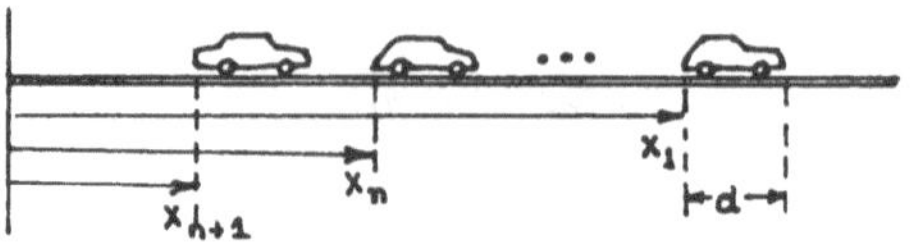

$\ddot{x}_n = a_n \qquad \ddot{x}_{n+1} = a_{n+1}$

$\dot{x}_n = v_n \qquad \dot{x}_{n+1} = v_{n+1}$

$\Delta x_n = x_{n+1} - x_n \qquad \Delta v_n = v_{n+1} - v_n$

x, v, a - position, velocity, acceleration (with subscript n - the leader, with subscript n+1 - the follower, without subscript - the model of the follower)

Δx_{nb} (Δx_b) - desired safe headway for real vehicle (for model)

τ - sum of reaction and perception time delay

C, K_x, K_v, α, β, γ, η, ξ, b - gain coefficients

D_x (D_v) - position (velocity) error deadbands.

Four approaches (the last one developed by the authors) to representing the drivers control strategies were taken under consideration :

1) Non-linear classic model [1]:

$$a(t+\tau) = C(\dot{x}_n(t) - \dot{x}(t))/(x_n(t) - x(t))$$

2) Linearized model of driver as optimal follow-up regulator [2]:

$$a(t+\tau) = K_v \Delta v + K_x(\Delta x - d - \Delta x_b), \quad \Delta x_b = \alpha v$$

3) Heurestic Bekey's model [3]:

$$a(t+\tau) = \begin{cases} 0 & \text{for } (\Delta x = \Delta x_b \text{ and } \Delta v = 0) \\ -0.5(\dot{x}_n(t) - \dot{x}(t))^2/(\Delta x - \Delta x_b) & \text{for } (\Delta x < \Delta x_b \text{ and } \Delta v > 0) \text{ or } (\Delta x > \Delta x_b \text{ and } \Delta v < 0) \\ -0.5(\dot{x}_n(t) - \dot{x}(t))^2/(\Delta x - \eta \Delta x_b) & \text{for } (\Delta x < \Delta x_b \text{ and } \Delta v < 0) \\ 0.5(\dot{x}_n(t) - \dot{x}(t))^2/((\xi - 1)\Delta x) & \text{for } (\Delta x > \Delta x_b \text{ and } \Delta v > 0) \end{cases}$$

$$\Delta x_b = \alpha v_n + d$$

4) Model with position and velocity error deadbands :

$$a(t+\tau) = K_v F_v(\Delta v(t), \Delta x(t)) + K_x F_x(\Delta x(t))$$

$$F_v(\Delta v, \Delta x) = \begin{cases} \Delta v - D_v^+ & \Delta v > D_v^+ \\ 0 & \text{for } D_v^- \leq \Delta v \leq D_v^+ \\ \Delta v - D_v^- & \Delta v < D_v^- \end{cases} \qquad F_x(\Delta x) = \begin{cases} \Delta x - \Delta x_b - D_x^+ & \Delta x - \Delta x_b > D_x^+ \\ 0 & \text{for } D_x^- \leq \Delta x - \Delta x_b \leq D_x^+ \\ \Delta x - \Delta x_b - D_x^- & \Delta x - \Delta x_b < D_x^- \end{cases}$$

$$D_v^+ = 1.5\beta(\Delta x - d) \quad D_v^- = -\beta(\Delta x - d) \qquad D_x^+ = 1.5\gamma(\Delta x - d) \quad D_x^- = -\gamma(\Delta x - d)$$

$$\Delta x_b = bv$$

3. Model verification

To verify the model some road experiments were carried out with a passenger car as the leader and an urban bus as the follower. The leader and the follower speed profiles were recorded simultaneously by on-board data acquisition electronic system. Next such parameter vectors for all the driver models were found which give the model outputs v(t) close to the follower speed profile $v_{n+1}(t)$. The recorded leader speed $v_n(t)$ was used as input signal to all the examined models. In each case the same longitudinal bus dynamics model was joint to simulated process. The best results were obtained for fourth human operator model. As we can see in fig. 2 it approximates the real driver remarkable well.

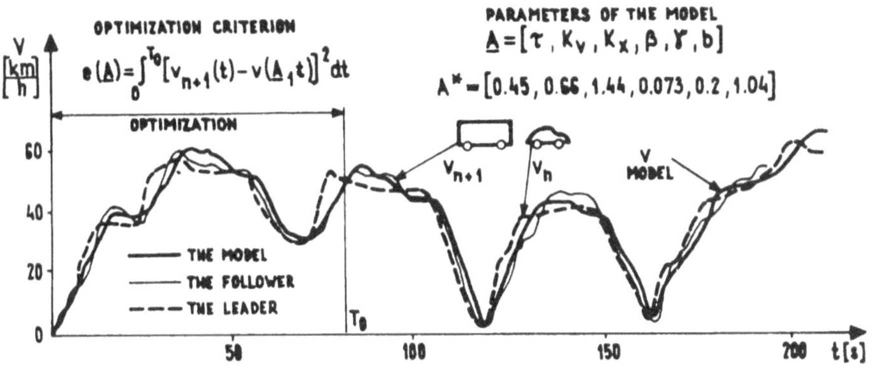

Fig. 2. THE MODEL (NR 4) AND THE ACTUAL FOLLOWER SPEED PROFILES.

4. Human behaviour effects on driver-vehicle performance.

Four quality indicies were taken under consideration. They are :

- mean fuel consumption : $Q = {}_0\int^T q(t)\,dt$
- mean NO emissions : $EN = {}_0\int^T \eta(t)\,dt$
- mean HC emissions : $EH = {}_0\int^T h(t)\,dt$
- mean headway (inversely proportional to traffic density) : $\overline{\Delta x} = {}_0\int^T \Delta x(t)\,dt$

where : $q(t)$, $\eta(t)$, $h(t)$ are fuel and toxic emissions flows

L is a distance covered in time T

The best human operator model fourth was used for the bus performance analysis. Its five parameters : τ, β, γ, K_x, K_v were examined. The ranges for the parameters were determined by initial simulation to eliminate the values which cause local or global instability of the following platoon.

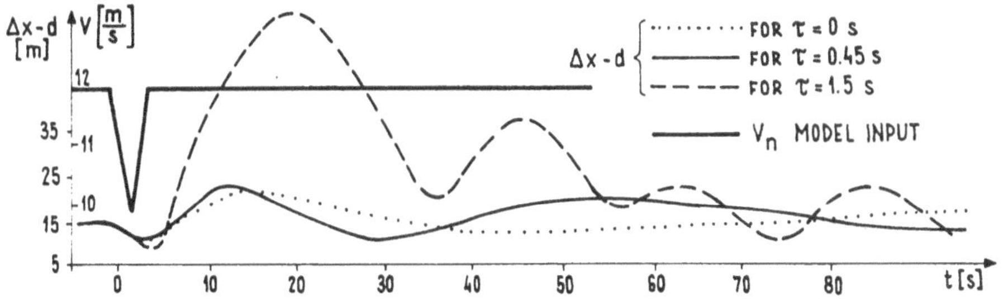

FIG. 3. EFFECTS OF THE LEADER VELOCITY DISTURBANCE ON THE FOLLOWER.

Typical traffic disturbance (the lead car slows down and then accelerates to its original velocity) was chosen as the input signal (see fig. 3). Three parameters : τ - total lag time, β - sensitivity of the velocity threshold and γ - sensitivity of the distance thresholds (both with respect to actual headway) show strong influence on the bus motion performance (see fig. 4). This effect mainly results from oscillation which appear while τ and β increase and γ decrease. Poor dynamics of bus makes impossible to investigate the effects of K_v and K_x gains on the criteria At present further research is continued with other vehicles to examine the problem.

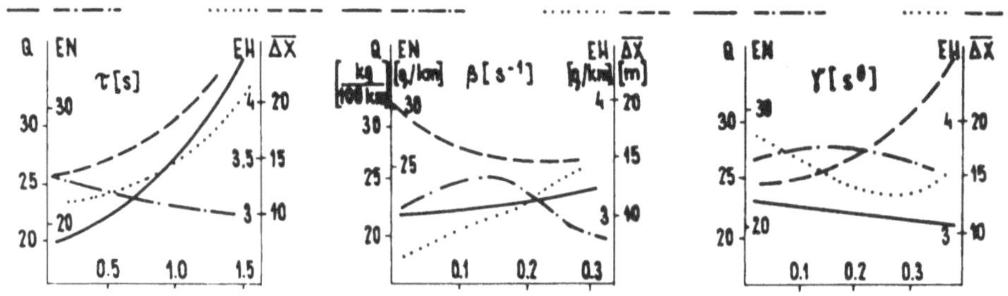

Fig. 4. EFFECT OF THE DRIVER PARAMETERS ON THE SYSTEM PERFORMANCE.

5. Conclusions

The implemented model although very simple can simulate a wide variety of driver types and car-following phenomena. The presented analysis may be very useful for designing automatic or semiautomatic vehicle guidance systems which not only reduce rear collisions but are energy-saving and environment protecting. Such systems should compensate effects of human time delay but also provide better information of an error in relative velocity and ignore small deviation of vehicle position from the save headway.

6. References

[1] Gazis D.C., Herman R., Rothery R.W. : "Nonlinear follow-the-leader model of traffic flow" Oper. Res., 9, 1961, pp. 545-567.

[2] Burnham G.O., Seo J., Bekey G.A. : "Identification of human driver models in car-following", IEEE Trans. on Aut. Contr.,1974,pp. 911-15.

[3] Burnham G.O., Bekey G.A. : "A heurestic finite-state model of the human driver in car-following situation", IEEE Trans. on SMC-6/8/,1976, pp. 545-567.

HYBRID TRAIN MOVEMENT SIMULATION FOR INVESTIGATION IN ENERGY SAVING DRIVING STRATEGIES

Mann, D. [1])

In the paper some details are given about train movement simulation with hybrid computer systems HRA 7200. The user with his practical experience is directly included by dialog in the simulation process. The aim is saving energy. Also at present excellent man-machine-relations are the basis for effective systematic experiments on the hybrid computer systems.

Keywords: Application of hybrid computer systems, train movement simulation, saving of energy

1. Introduction

Train movement simulation for obtaining

a) energy saving driving strategies for fixed schedule time,

b) energy favourable schedule running times

is a very important task for effective railway operation. With sufficient accuracy train movement and energy consumption can be described by ordinary differential equations. Many nonlinearities must be taken into account (tractive effort-speed-functions for each working step of the locomotive, the control of these steps, gradient ratio of the track, speed restrictions etc.). For the simulation of the problem mentioned we can use hybrid as well as digital computers.

The solution of the problem by application of optimization techniques (comp. e.g. /2/) will not be the subject of this contribution. In our case the user with his experience in dynamics of train movement is included in the simulation process directly by dialog. The essential dialog is performed during the simulated train run by the input of commands and the immediate graphical and numerical output of the results. From this point of view we have succesfully used hybrid computer systems HRA 7200. For these computer systems a complex of programmes, called ZUSI (train movement simulation) is elaborated (/3/). By automatic or manual variation of the driving commands driving strategies could be found yielding a 5 - 10 % saving of energy as compared to the average consumption (/1/). It is also suggested to use digital simulation techniques with a similar effective dialog (/4/).

[1]) Hochschule für Verkehrswesen "Friedrich List" Dresden
Sektion Mathematik, Rechentechnik und Naturwissenschaften
DDR-8010 Dresden, Friedrich-List-Platz 1

2. Model, equations

The movement from one station to another can be described by the equations

$$\frac{dv}{dt} = \begin{cases} f'_Z - f'_S - f'_W \\ - b \end{cases} \quad \text{for} \quad s \begin{matrix} < \\ \geq \end{matrix} s_j - \frac{v^2 - v_j^2}{2b} \quad , \tag{1}$$

$$\frac{ds}{dt} = v \tag{2}$$

with the initial conditions v(0)=0, s(0)=0 and

$$f'_Z = \frac{1}{\mu m_Z} F_Z(ST,v) \quad , \tag{3}$$

$$f'_S = \frac{g}{1000\mu} f_S(s) \quad , \tag{4}$$

$$f'_W = \frac{g}{1000\mu} (a_0 + a_1 v + a_2 v^2) \; . \tag{5}$$

For diesel traction the consumption of fuel mass m_K is described with good accuracy by

$$\frac{dm_K}{dt} = c(ST) \tag{6}$$

with constant specific consumption c(ST) for each working step ST of the locomotive. Further constants are the train mass m_Z, the resistance coefficients a_0, a_1, a_2, the braking deceleration b and the mass factor μ. Equation (1) involves braking corresponding to speed restrictions (speed v_j beginning at the track point s_j) and finally braking to a standstill at the end of the considered route. $f_S(s)$ is the track-dependent gradient ratio and $F_Z(ST,v)$ the tractive effort - speed - function for the working step ST of the locomotive. These functions are stored for the considered routes and the used diesel locomotives on the digital computer of the hybrid computer system. In our case for one locomotive up to 15 steps the tractive effort - speed - functions can be realized in the simulation process.

3. Simulation, dialog conditions

The simulation of train movement and energy consumption is realized on the analog part of the hybrid computer system. Preparation and control of analog simulation is organized by the digital computer of the system (see fig. 1). Simulation will be carried out with starting

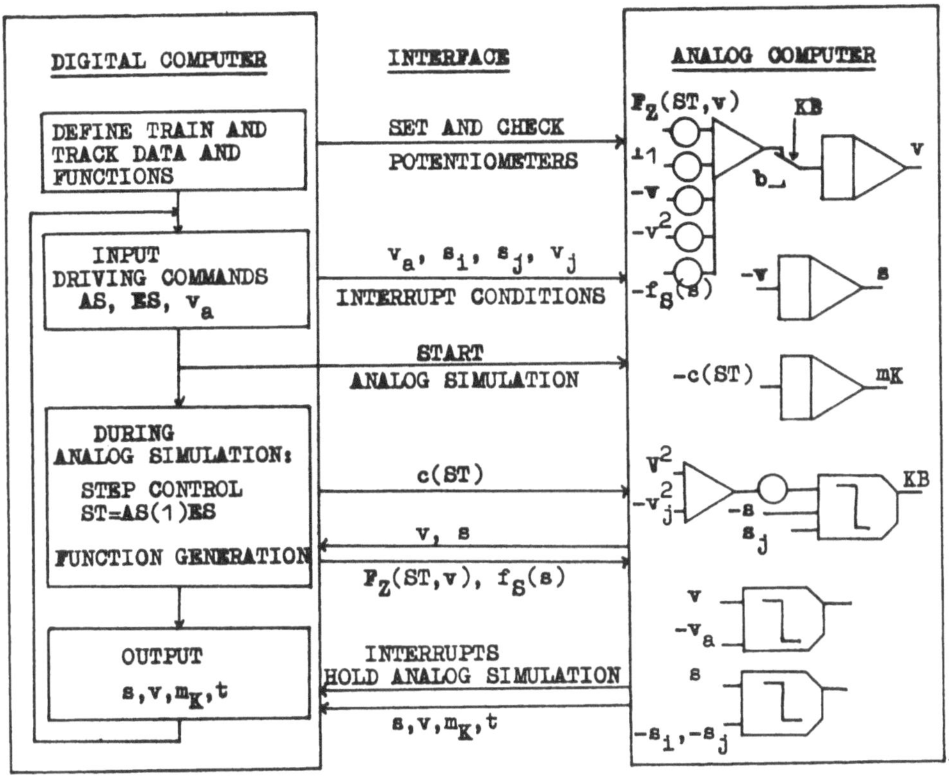

Fig. 1 Simplified scheme for hybrid train movement simulation

from station A to the stop in station B. During simulation there will be interrupts for the output of numerical results (trackpoint, speed, fuel consumption, time) and the input of new driving control commands. Such driving control commands generally consist of the three values AS, ES, v_a. This means tractive effort control from working step AS to working step ES and acceleration of the train till it reaches speed v_a. For AS=0 power will be shut off (begin of a coasting period). In the control procedure for maintaining each step ST = AS(1)ES-1 constant times Z(ST) are assumed.

Conditions for the dialog interrupts are

- reaching the speed v_a corresponding to the driving commands ,
- reaching of manually provided track points s_i,
- reaching of fixed track points s_j with the beginning of speed restrictions v_j.

The interrupts are generated by logic signals of comparators shown in fig. 1.

Essential parameters for variation are the final working steps ES and the speeds v_a. It is possible to repeat simulation at each point with corresponding initial conditions and to study the influence of the parameters (automatic variation). A great support for the user is also the immediate graphical output of resulting v-s-curves with the xy-plotter.

4. Conclusions

The hybrid train movement simulation described in this paper was succesfully applied for investigation in driving strategies for many routes and trains with different diesel locomotives. There are no difficulties to extend this simulation method to electric traction.

Literature:

/1/ Preysing, E.: Ermittlung energiesparender Fahrweisen bei der Dieselzugförderung mittels Zugfahrtsimulation. Schienenfahrzeuge (Berlin) 26(1982)6, 268-270

/2/ Horn, P.: Beitrag zur Lösung des Syntheseproblems der energieoptimalen Steuerung einer Zugfahrt. Dresden, Hochschule für Verkehrswesen "Friedrich List", Dissertation 1974

/3/ Mann, D.: Zugfahrtsimulation am hybriden Rechnersystem HRA 7200. Wissenschaftliche Zeitschrift der Hochschule für Verkehrswesen "Friedrich List" Dresden 32(1985)2

/4/ Bechtel, J.: Zugfahrtsimulation mit TSO am ES 1040. Wissenschaftliche Zeitschrift der Hochschule für Verkehrswesen "Friedrich List" Dresden 32(1985)2

CHOSEN TOPICS IN AN APPROXIMATE ANALYSIS OF A TELECOMMUNICATION CONTROLLED NETWORK WITH OVERFLOWS.

Eulalia J. Szmidt [1)]

Summary.

In this paper a new approach for an approximate analysis of a telephone network via symulation is presented. A compact method starting from measurements in a real network, leading to determining losses, overflows, distribution function of several overflows in a common line and enabling decomposition of total losses among separate components is proposed.

Key worlds : telephone network, primary line, secondary line, losses, overflows, composition.

Introduction.

In order to keep services on the best obtainable level a network with overflows is often proposed. The idea is a separate method of control on the one hand and it is an integral part of more complex methods [6] on the other. The idea boils down to collecting unserved parts of traffic from several lines /primary lines/ and putting them together to another line /secondary line/ which isn't so busy.

The commonest method used for examining the mentioned system is so called ERT [5]. The drawbacks of the method are connected with the assumptions [4] and impossibility of investigating individual losses in a total traffic. Also other methods aren't satisfactory enough. IPP [2] can't give results for all kinds of possible streams. Methods proposed by queueing theory are too complex and need parameters which aren't collected in the day's measurements. These are the main reasons which create needs for a new method of simúlation.

Notation.

R - number of links existing in a primary line,
m , v - mean and variance in a line with R links,
M , V - mean and variance in a line with ∞ links /without losses/,
$\delta(\cdot)$ - Dirac's function,
secondary line :
- without own calls - serves only unserved calls from primary lines,
- with own calls - the line has own characteristic (1) and serves unserved calls from other primary lines.

1) IBS PAN, ul. Newelska 6, 01-447 Warszawa POLAND

Assumptions :

- a primary line is fully available,
- calls can overflow only while a line is in its busy state /all links busy /,
- secondary calls aren't taken into account,
- the lines are mutually independent.

Model of a line state.

The proposed solution starts from an assumption of Normal distribution function of busy links in a line without losses. It is justified by the principle of maximum entropy - the most safe criterion in determining a distribution function under uncertain circumstances [1]. Also the measurements carried in a real working telecommunication network /in Warsaw node/ confirmed that result [4]. /It turned out that the knowledge of mean and variance of busy links is enough for practical purposes./

So a model i.e. distribution function of busy links in a line without losses /continous line in fig.1 / is [4] :

$$f(x) = (1/\sqrt{2\pi V})\exp\{-(x - M)^2/2V\} \tag{1}$$

But in practical situations we can only measure the parameters m , v of an overload line /interrupted line in fig.1 / and we want to know the parameters of an overflow /lined area in fig.1 /. So the parameters M and V must be described first.

It can be easy verified that knowing mean and variance of busy links in the overload line and the number of links in a line, under assumption of Normal distribution function of busy links in a line without losses we have :

$$M = m + VE_R$$
$$V = v/\left[1 - (R - m)E_R\right] \tag{2}$$

where :

$$E_R = \left\{(1/\sqrt{V})\exp\{-(R-M)^2/2V\}\right\} / \int_{-\infty}^{\beta} \exp\{-t^2/2\}\,dt$$
$$\beta = (R - M)/ V .$$

As a random variable Y describeing losses is : 0 for $X \leqslant R$, $X-R$ for $X > R$, the distribution function of losses is :

$$f(y) = a\delta(y) + (1/\sqrt{2\pi V})\exp\{-(y+R-M)^2/2V\} \tag{3}$$

where : a - probability that there are no losses in a line /fig.1: the area limited by the abscissa from 0 to R and the current line/

Formulas (1)-(3) make the standard composition of separate traffics possible. As formula (3) leads to complex results an approximation of an overflow traffic was proposed [4] :

$$f(y) = \gamma \exp\{-\lambda y\} \tag{4}$$

where:

$$\lambda = (\gamma/\sqrt{2\pi v})\int_{R}^{\infty}\exp\{-(x-m)^2/2v\}dx$$
$$\gamma = (1/\sqrt{2\pi v})\exp\{-(R-m)^2/2v\} \tag{5}$$

The presented approximation (4) states identity of losses in a primary line and the probability of useing a secondary line. In [4] other possibilities were also examined but they turned out to be less accurate than presented above.

Distribution function of busy links in a secondary line.

Having a distribution function of losses from a primary line (4) and a model of a primary line (1) we can /by standard composition/ obtain distribution function of busy links in a secondary line /with or without own calls/. For a secondary line collecting losses from k primary lines we have :

$$f(y) = [a_1 \sigma(y) + f_1(y)] * \ldots * [a_i \sigma(y) + f_i(y)] * \ldots * [a_k \sigma(y) + f_k(y)] \tag{6}$$

where : i - a line number i, $1 \leqslant i \leqslant k$,

a_i - see formula (3) ,

f_i - formulas (4)and(5) /γ_i and λ_i must be described for each primary line with losses/ or formula (1)for a line with own calls.

In order to obtain results from (6)all combinations of composition of l functions /$1 < l \leqslant k$/ must be calculated. /Full equations are presented in [4] . Equations (1) , (4) , (6) let us :

- examine general losses in a secondary line,
- fix a necessary number of links in a secondary line for chosen grade of service,
- describe a distribution function of losses in a secondary line,
- compare various strategies of control /e.g. different connection of streams/.

Some qualitative results are presented in fig.2 . From fig.2 and tab.1 we can see that a secondary line with own calls has less losses than a line without such calls which must serve the same amount of calls /a shape of a distribution function is important/ . That result proves well-known fact and speaks well for the method. Other quantity results are presented in [4].

Conclusions.

Presented models :

- give qualitative results comformablle to ones well-known from measurements in a real network and from other methods [4],
- are rather simple /for calculations/ and compact /lead directly from measurements in a real network to results needed for controling/,
- are "safe" /by useing principle of maximum entropy/,
- let quickly obtain quantity results,
- give grater possibilities than other methods :
  - secondary calls can be taken into account /it is proposed to treat these calls as an imaginary overflow in formula (6) /,
  - all kinds of overflows can be examined /all values of peakedness $z=v/m$ are possible - better than IPP method [2] ,
  - M and V /equations (2)/ are estimated what gives more exact results than method presented by Rahko [3] ,
  - decomposition of losses is possible [4].

The above features seem to point out the need for further developing of the model while a definite strategy of controlling is taken into account.

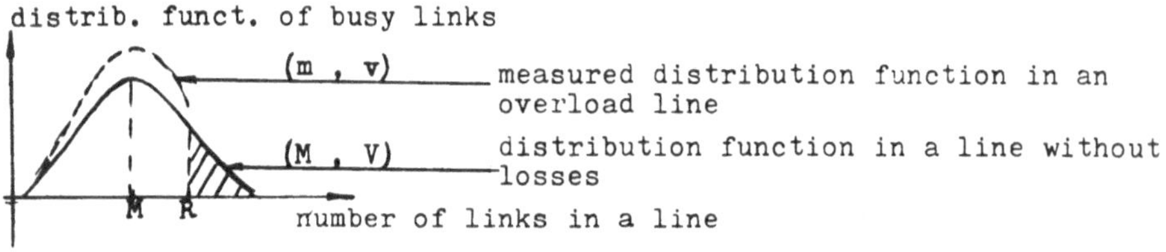

Fig.1. Measured and "real" distribution functions of busy links in a primary line.

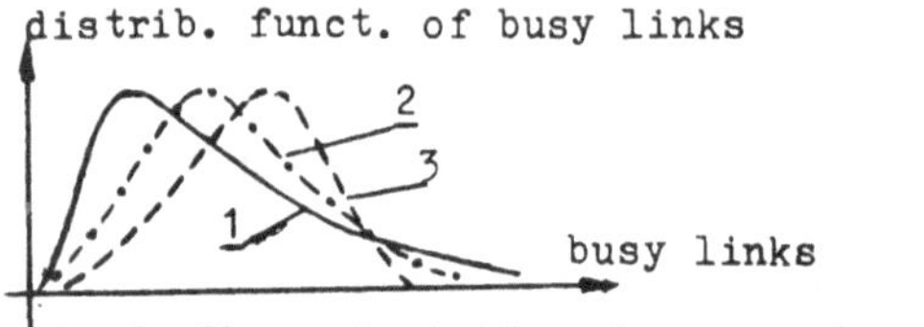

Fig.2. Characteristic of a secondary line with overflows. Parameters of "flows" are presented in tab.1.

Tab.1 Mean and variance of distribution functions of calls.

line	1		2		3	
	m	v	m	v	m	v
own calls	-	-	7	7.66	11	12.4
overflows	7 10 11	7.66 11.1 12.4	10 11 -	11.1 12.4 -	7 10 -	7.6 11.1 -
N	77		70		67	

N-number of needed links /1% loss probab

Literature.

[1] Jaynes E.T.:Where do we stand on maximum entropy? In: Levine R.D., Tribus M.: The maximum entropy formalism. MIT Press. 1979.

[2] Kuczura A.:The interrupted Poisson process as an overflow process. B.S.T.J. 52 /1973/ 3, 437-448.

[3] Rahko K.:The dimensioning of local teleph. traffic routes based on the distrib. of the traffic carried. Acta Pol. Scand. El 14. 1967.

[4] Szmidt E.J.:Modelling of chosen phenomena in a queueing system on the basis of a telecomm. network. Ph.D.Dissert.Warsaw Techn. Univ. 1984. /in Polish/.

[5] Wilkinson R.I.:Theories for toll traffic engineering in the USA. B.S.T.J. 35 /1956/ 2, 421-514.

[6] Wydro K. and others: A three-layer controlled telephone network. Warsaw Techn. Univ. 1980 /in Polish/.

MODELLING OF STRESS RANDOM HISTORIES BY SINUSOIDAL HISTORIES AT REAL TIME WHEN FATIGUE LIFE OF MACHINE OR CONSTRUCTION ELEMENTS IS DETERMINED

Ewald MACHA, Henryk STĘŻAŁA [1/]

1. INTRODUCTION

Replacement of random history of stress by a series of sinusoidal cycles or half-cycles is one of the most important operations when fatigue life of elements of machines or structures is determined. There are many methods of cycle counting /modelling of random history of stress by sinusoidal histories/. The following methods are applied most often:

- a/ range pair counting method
- b/ full cycles counting method
- c/ rain flow counting method
- d/ hysteresis loop counting method

These methods usually correspond with stress-strain characteristic of a material and identify the closed hysteresis loops as cycles and the open ones as half-cycles. In the most cases they give results conformable to the results obtained during the experimental investigations. According to the methods counting of cycles and half-cycles usually proceeds in a digital computer, when all the stress-time or strain-time recordings are ready.

Development of microprocessor techniques and application of microcomputers as counting elements give new possibilities of fatigue failure analysis concerning elements of machines and structures at real time. Thanks to that accumulation of a large number of data in the computer memory is avoided and the final results are obtained in a short time. It is very important especially when it is necessary to observe currently a degree of damage of the working machine or structure.

2. MODELLING OF STRESS RANDOM HISTORIES BY SINUSOIDAL HISTORIES

Only the extreme values of stress time history take place in the algorithms of cycle or half-cycle counting according to the methods mentioned above. A part of the random realization of stress and the corresponding stress-strain characteristic of the material are presented in Fig.1. The points of the extreme values of the stress are denoted by P (I-2), P (I-1), P (I), P (I+1). The series of the next extreme values, occuring in the random realization of the stress, is treated as the input data for all four algorithms of cycle counting. The range between the minimum point and the maximum one /they do not need to be the neighbouring points of the random realisation of the stress/ is treated as

1/ Technical University of Opole,
ul.Zrzeszenia Studentów Polskich 5, 45-233 Opole, Poland

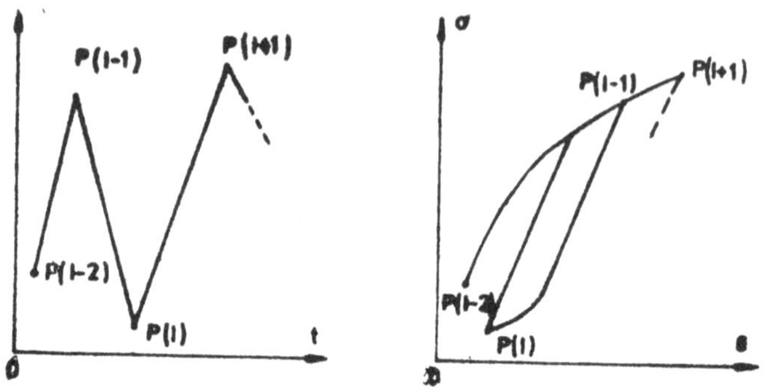

Fig. Part of random realization of stress and corresponding stress-strain characteristics of material

the sinusoidal full cycle or half cycle. The half of the absolute value of the range is the amplitude of the half-cycle or cycle and the mean value is the arithmetic mean of the stress values at those points.

RANGE PAIR COUNTING METHOD

According to this method, those stress ranges, for which a range having the same value but the opposite direction exists in the further history of the stress, are counted as cycles. Cycles are counted according to the following algorithm:

1. Read the value of the next extremal point /if out of data, stop/
2. Count the ranges:

   R1 = ABS (P (I-1) - P (I-2))

   R2 = ABS (P (I) - P (I-1))

   /if the vector contains less than 3 components, go to Step 1/
3. If R1 > R2, go to Step 1
4. Form cycle according to the range R1. Discard values P (I-1) and P (I-2) from the data vector. Go to Step 2.

The range pair method is very simple and it is its advantage. Digital computation can proceed as the numerical data arise. It is not necessary to know all the history of the stress before calculation. The disadvantage of the method is that not all the sections of the random realization of the stress are counted.

FULL CYCLE COUNTING METHOD

In case of the method it has been assumed that the random history of stress forms by overlaing of cycles having different amplitudes. Cycle counting proceeds at N steps. At the i-th step cycles with the range $R \leqslant i \cdot \Delta\sigma$, where $\Delta\sigma$ is the value of the minimum range, determined in advance, are counted.

After each step the random realization of the stress is simplified by elimination of those extremal points which have been the base of cycle forming. The full cycle counting method has two disadvantages. The first one lies in the fact that not all the segments of the random realization of the stress are counted. The disadvantage of numerical nature is that the values of the stress at all the extremal points of stress history should be known before counting. For the reason the method cannot be applied for calculations at real time.

RAIN FLOW COUNTING METHOD

The rain flow method is the most often applied method of cycle counting [1] .In paper [2] the algorithm of counting at real time has been proposed. It was worked out on the basis of the following remarks:

I If the pair of the extremal points is the basis of cycle counting according to the rain flow method,it will be such a basis also in case of the range pair counting method.

II The extremal points which are the basis of cycle counting according to the rain flow method are always between the pair of points forming a half-cycle.

The simplified algorithm of calculations for a case,when the minimum having the value equal to 0 is the first extremal point,is presented below. The variable PS takes the value of the point which is the beginning of the cycle. As for the variable L,L=1 when the maximum is that point, L=-1 when the minimum is the point of beginning.

1. Substitute: PS=0, L=-1
2. Read the value of the next extremal point /if out of data,stop/
3. If L=1,go to Step 6
4. If P (I) $\leq$ PS,substitute L=1 and go to Step 8
5. Form cycle according to the range pair counting method
6. If P (I) < PS,go to Step 5
7. Substitute L=-1
8. Form half-cycle according to range R=ABS (P (I-1) -PS). Substitute: PS=P (I-1) . Discard the value P (I-1) from the data vector. Go to Step 5.

HYSTERESIS LOOP COUNTING METHOD

The hysteresis loop counting method is a modification of the range pair counting method[3]. The cycle counting proceeds according to the same rules as in case of the range pair method with one exception: the first extremal point of random realization of the stress cannot be the beginning of any range. It eliminates the disadvantage of the range pair counting method. Unlike the last one,in case of the hysteresis loop counting method each segment of the random realization is oounted. Calcula -

tions according to the method can proceed at real time and they do not require accumulation of many data of random realization of the stress.

3. SUMMARY

Cycle counting can proceed at real time, except the full cycle counting method. In case of the hysteresis loop counting method and the rain flow counting method also half-cycles are counted. Their calculation is based on the physical dependence between stresses and strains of the examined material. All the ranges forming cycles correspond to the closed loops of stress-strain hysteresis and the ranges forming half-cycles correspond to the open loops of this hysteresis. The hysteresis loop counting method is more effective than the range pair counting method and it is more simple than the rain flow counting method in case of programming at real time.

REFERENCES

[1] DOWNING S.D., SOCIE B.F.: Simple rainflow counting algorithms, International Journal of Fatigue, Vol 4, N 1, 1982, pp.31-40

[2] MACHA E., BERGIER A.: Oprogramowanie SCOE-CAMAC-SM4. Wytrzymałość zmęczeniowa materiału, Raporty Instytutu Materiałoznawstwa i Mechaniki Technicznej Politechniki Wrocławskiej, 1981, nr7, pp.51

[3] OKAMURA H., SAKAI S., SUSUKI I.: Cumulative fatigue damage under random loads, Fatigue of Engineering Materials and Structures, Vol 1, 1979, pp.409-419

SYSTEM ANALYSIS AND SIMULATION - ELECTRIC ENERGY SYSTEM

A.S. Chaudhuri & J. Jhampati *

Abstract

This paper deals with the application of cybernetical tools to different aspects of electric power industry. The effectiveness of the recursive least square instrument variables algorithm has been shown for on-line flow simulation of a hydro-electric power plant. The growth model of electric energy consumption has been developed with multi-layer group method of data handling algorithm. A polynomial model of optimum complexity shows the trends of installed plant capacity and electric energy consumption in India up to 2000 A.D. A model of electric energy utilisation factor is also obtained.

Key Words: recursive least square, instrument variables, on-line flow simulation, parameter tracking adaptiveness, multi-layer group method of data handling algorithm, polynomial of optimum complexity.

I. Introduction

This paper describes a theoretical and experimental endeavour [1] to apply cybernetical tools to electric power industry. In section II the on-line simulation of hourly river flows for a run-of-the river hydroelectric plant has been developed. Section III describes the annual electric energyconsumptionmodel by a learning identification algorithm. The medium term and long term prediction models of annual installed plant capacity, annual consumption of electric energy and plant load factors have been developed in section IV by computer -aided self-organisation of mathematical models.

II. On-line Flow Simulation Model

The effectiveness of the hourly flow simulation technique developed by recursive least square instrument variables algorithm [2] with parameter tracking adaptiveness [3] in the estimation of the instrument variables has been demonstrated in obtaining an on-line simulation model of the hourly flows of the hilly river Teesta in North-Eastern Himalayan region at the Coronation Bridge Point on the basis of the observed data at different up-stream gauging stations. The strong correlation of the down-stream hourly flows, $Y(t_k)$ with those at the up-stream gauging stations, $U_1(t_{k-r1})$, $U_2(t_{k-r2})$,..., $U_2(t_{k-r2-1})$,... suggests that the hourly flow process can be represented as

* Jalpaiguri Government Engineering College, Jalpaiguri-735102, West Bengal, India.

$$Y(t_k) = \sum_{i=1}^{n} \beta_i Y(t_{k-i}) + \sum_{j=1}^{n} \sum_{i=0}^{n} \delta_{ji} U_j(t_{k-r_j-i}) + \eta(t_k) \quad (1)$$

where $\eta(t_k)$ is a current unknown realisation of the noise process.

After overcomming the problem of bias in the estimates of the parameters the recursive least square instrument variables algorithms are

$$\hat{\mathcal{L}}(t_k) = \hat{\mathcal{L}}(t_{k-1}) + \hat{P}^*(t_{k-1})\hat{z}(t_k)[1 + z^T(t_k)\hat{P}^*(t_{k-1})\hat{z}(t_k)]^{-1} \times \{Y(t_k) - z^T(t_k)\hat{\mathcal{L}}(t_{k-1})\} \quad (2a)$$

$$\hat{P}^*(t_k) = \hat{P}^*(t_{k-1}) - \hat{P}^*(t_{k-1})\hat{z}(t_k)[1 + z^T(t_k)\hat{P}^*(t_{k-1})\hat{z}(t_k)]^{-1} z(t_k)\hat{P}^*(t_{k-1}) \quad (2b)$$

$$\hat{z}^T(t_k) = [\hat{Y}(t_{k-1}), \ldots, Y(t_{k-n}), U_1(t_{k-r1}), \ldots, U_m(t_{k-rm-n})]$$

$$\hat{P}^*(t_k) \triangleq [\sum_{j=1}^{k} \hat{z}(t_j) z^T(t_j)]^{-1}$$

The instrument variables Y(.) in (.) has been obtained as,

$$\hat{Y}(t_k) = \sum_{i=1}^{n} \beta_i Y(t_{k-i}) + \sum_{j=1}^{m} \sum^{n} \delta_{ij} U_j(t_{k-rj-i}) + \sum_{q=1}^{Q} C_{tk-q}(Y(t_{k-q}) - \hat{Y}(t_{k-q})) \quad (3)$$

with $\hat{Y}(t_k) = a^T(t_{k-1}) z(t_{k-1})$ the algorithms are

$$a(t_{k+1}) = a(t_k) + S(t_{k+1}) z(t_k)[Y(t_{k+1}) - a^T(t_k) z(t_k)] \quad (4a)$$

$$S(t_{k+1}) = S(t_k) - S(t_k) z(t_k) z^T(t_k) S(t_k)[1 + z^T(t_k) S(t_k) z(t_k)]^{-1} \quad (4b)$$

$\hat{\mathcal{L}}(.)$ and $\hat{P}^*(.)$ have been initialised by least square method for a block of data and other parameters are initialised as $S(t_o) \triangleq I$ (unit matrix), $a(t_o) = 0$, $Y(t_j) = 0$ for $j = 0, -1, -2, \ldots$ When tested with field data for the river Teesta the model is observed to simulate adequately the hourly flow process with noise process $\eta(.)$ giving almost a white noise sequence.

III. Annual Electric Energy Consumption Model

This work presents a mathematical description of the annual energy consumption with population, gross national product, gross domestic saving and gross domestic capital formation as exogenous variables in the form of a polynomial of optimum complexity with the help of multi layer group method of data handling algorithm (GMDH) [4] . The computational features of multi-layer GMDH as applied in the present case is described below.

The input data from the Indian techno-economic scenario consist of the annual energy consumption (gross) in Mn. Kwh, population in millian, gross national product at factor cost (Rs. Crores) at 1970-71 prices, gross domestic saving and gross domestic capital formation as 1980-81. All the relevant data X(k) for the k-th year have been rationalised as

$$x(k) = (X(k) - X_{min}) / (X_{max} - X_{min}).$$

The annual electric energy consumption can be represented as

$$x(k)=f(y(k-1),y(k-2),\ldots,x_1(k),x_1(k-1),\ldots,x_3(k),x_3(k-1),\ldots) \qquad (5)$$

where k, k-1, k-2, ..., refer respectively to the current day, one day preceding the current day, two days preceding the current day and so on and $y(\cdot),x_1(\cdot),x_2(\cdot),x_3(\cdot)$ and $x_4(\cdot)$ are the rationalised data for annual electric energy consumption, population, gross national product, gross domestic saving and gross domestic capital formation respectively. After selection of arguments having correlation with the annual energy consumption the process equation becomes

$$y(k)=f(y(k-1),x_1(k),x_1(k-1),x_1(k-2),x_2(k),x_2(k-1),x_3(k)) \quad \ldots \quad (6)$$

or denoting the respective arguments as $y(k) = y$, $y(k-1) = x'_1$, $x_1(k) = x'_2, x_1(k-1) = x'_3, x_1(k-2) = x'_4, x_2(k) = x'_5, x_2(k-1) = x'_6, x_3(k)=x'_7$, the process equation becomes $y = f(x'_1,\ldots,x'_7)$.

There are $\binom{7}{2} = 21$ possible combinations of selecting two arguments at a time out of seven. For every such combination, the partial regression equation equation is written as

$$y_a=\alpha_{0a}+\alpha_{1a}x'_b+\alpha_{2a}x'_c+\alpha_{3a}x'_b x'_c+\alpha_{4a}x'^2_b+\alpha_{5a}x'^2_c \qquad (7)$$

where a = 1,2,...,21, while b and c are indices for all 21 combinations. The coefficients of α's are estimated by solving normal Gaussian equation systems constructed from the data set. From all the variables y_a seven more accurate ones are chosen which give low values of integral square error as

$$ISE = \left(\sum_{i=1}^{N} (y_{observed}(i) - y_{modelled}(i))^2 / \sum_{i=1}^{N} (y_{observed}(i))^2 \right) \qquad (8)$$

Seven intermediate variables of y_a layer chosen from the first layer give 21 combinations of y_a. Again in the second layer these become

$$z_a = \beta_{0a} + \beta_{1a} y'_b + \beta_{2a} y'_c + \beta_{3a} y'_b y'_c + \beta_{4a} y'^2_b + \beta_{5a} y'^2_c \qquad (9)$$

Calculation of the co-efficients of β and estimation of the accuracy of z_a are repeated as in the case of y_a. The seven z_a variables are then chosen for u_a layer. For all the layers the variables on the left hand side of the equations are kept equal to the value of the output variables.

It has been observed that as the layer increases ISE converge asymptotically to a very small value. To make the model suitable for practical application the ISE of 3.408346E - 04 at z_5 has been considered for point of termination of the process equation. Consequently, the annual energy consumption in India has been identified by the polynomial as

$$y = z_5 : z_5 = 7.490082E-03 + 1.439809 y'_1 - 0.496869 y'_6 - 35.759734 y'_1 y'_6 + 17.285701 y'^2_1 + 18.524885 y'^2_6$$

$$y'_1 = -0.031427 + 1.470818 x'_2 - 0.602856 x'_6 + 6.259670 x'_2 x'_6 - 3.220067 x'^2_2 - 2.814044 x'^2_6$$

$$y'_6 = 0.074196 + 0.934731 x'_4 - 0.131835 x'_7 + 0.063213 x'_4 x'_7 + 0.089502 x'^2_4 + 0.156888 x'^2_7 \qquad (10)$$

Modelling errors are found to have a variance of 1.000014 and mean error of 2.720174 E - 04 and are almost found to be uncorrelated for $i \neq j$.

IV. Medium Term and Long Term Prediction Model in Electrical Power Industry

An annual model of installed plant capacity of electrical energy for medium term (6 to 7 years) prediction has been obtained for India as a polynomial of optimum complexity [5] by self-organisation of polynomial models of increasing complexity on the basis of the selection critērion of minimum integral square error. The correlation co-efficients versus shift of instances of time show that the current year installed plant capacity is strongly correlated with the past three years installed plant capacity. Thus the process is assumed to have a finite difference form as

$$Y(t_{k+1}) = f(Y(t_k), Y(t_{k-1}), Y(t_{k-2}), t_k)$$

The function f(.) is sought in the class of quadratic polynomials of gradually increasing complexity. On the basis of the annual installed plant capacity data from 1961-1981 for India the model of optimum complexity has been found in the finite difference form as

$$Y(t_{k+1}) = 5584.157095 + 0.59965\,Y(t_k) - 0.330972\,Y(t_{k-1}) - 0.185849\,Y(t_{k-2})$$
$$+ 770.911284\,t_k + 1.05645165\,E-05\,Y(t_k)\,Y(t_{k-1})$$
$$- 8.57186948\,E-09\,Y(t_k)\,Y(t_{k-2})$$

On the basis of an annual growth rate of 8 per cent the long term (up to 2000 A.D.) prediction model of installed plant capacity has been obtained as

$$Y(t_{k+1}) = -238.764083 + 0.992665\,Y(t_k) + 0.617739\,Y(t_{k-1})$$
$$- 0.020582\,Y(t_{k-2}) + 0.086204\,Y(t_{k-3}) - 171.471615\,t_k$$
$$- 6.719727\,E-07\,Y(t_k)Y(t_{k-1}) + 0.3518649\,E-11\,Y(t_k)Y(t_{k-2})$$

The model of optimum complexity for long term annual energy consumption has been obtained as

$$Y(t_{k+1}) = -62.374577 + 1.115110\,Y(t_k) + 0.068099\,Y(t_{k-1})$$
$$- 0.114979\,Y(t_{k-2}) + 0.028869\,Y(t_{k-3}) - 128.655721\,t_k$$
$$- 3.701969\,E-08\,Y(t_k)Y(t_{k-1}) + 4.507208\,E-12\,Y(t_k)Y(t_{k-2})$$

The distinct periodicity of a 10 yearly cycle has been observed in the plant load factor. The annual plant load factor has been described by a finite difference equation incorporating sinusoidal terms as

$$P(t_k) = 0.215317 + 0.502090\,P(t_{k-1}) + 17.055098\sin(0.2\pi k)$$
$$- 16.103889\cos(0.2\pi k) - 23.161128\sin(0.2\pi(k-1))$$
$$+ 2.964121\cos(0.2\pi(k-1)) - 17.073073\,P(t_{k-1})\sin 0.2\pi k$$
$$+ 12.529761\,P(t_{k-1}) \times \cos(0.2)\pi k) + 20.968465\,P(t_{k-1})$$
$$\times \sin(0.2\pi(k-1)) - 4.75\,E-12\,P(t_{k-1})\cos(0.2\pi(k+1)).$$

V. Conclusion

The ral time experiment with the recursive least square instrument variable algorithm with associated parameter tracking adaptiveness allows one to assess the requirements and capabilities of microprocessor based estimation algorithms for real time monitoring of hourly river flows and on-line controlling of hydraulic structures of a large size run-of-the river hydroelectric plant.

The multilayer GMDH is found to simulate adequately the effects of interactions of different techno-economic parameters on annual electrical energy consumption.

The polynomial models of installed plant capacities and plant load factor will bring out many interesting features which have been kept outside the purview of our present research interest.

References

[1] J. Jhampati, "Application of Cybernetical Methods in power Industry, "Ph. D. Thesis submitted to the University of North Bengal, January, 1985

[2] P.C. Young, "Recursive Approach to Time Series Analysis," Bulletin of the Institute of Mathematics and its Application, Vol. 10, pp. 209 - 224, 1974.

[3] S.K. Sen Gupta and A.S. Chaudhuri, "Real Time Forecasting of Dissolved Oxygen Level of a Non-Tidal River," Journal of the Institution of Engineers (India), Vol. 62, pt. EN 3, June 1982, pp. 127 - 130.

[4] A.G. Ivakhnenko, "Polynomial Theory of Complex Systems, " IEEE Trans on Systems, Man and Cybernetics, Vol. SMC - 1, No. 4, October 1971.

[5] A.G. Ivakhnenko, "Present State of the Theory of Computer - Aided Self - organization of Mathematical Models," Soviet Automatic Control, Vol. 8, September - October, 1975.

MODELING OF A MEDIUM-TERM DEVELOPMENT OF COUNTRY'S ENERGY SYSTEM

Janusz COFALA[1)]

A set of models for analysing energy system development of Poland is presented. Models are used to support the scenario writing process and allow a check on the consistency of the different assumptions. Economic parameters are generated with the use of a model called STRUK. These parameters are used to compute the final energy demand, which is performed by another model called PROSK. Then the optimal structure of technologies of energy production, energy exports and imports, and in part utilization, is computed with the use of model DORSEK. The results obtained so far have been utilized in the process of preparation of the Polish Energy Programme up to the year 2000.

Keywords: Energy systems development, supply and demand models.

. Introduction

In the last decade energy has became one of the most important factors limiting the social and economic development. It has been caused mainly by the continuous process of drawing out of fossil fuels resources and the increasing costs of energy technologies. The processes of energy production and utilization have also more and more serious negative impact upon the environment. Therefore there is a need for establishing the proper relations between the development of energy system and the whole economy together with the natural and social environment.

Traditional methods of analysing such complex problems are not sufficient. New approaches have to be established in order to enable to analyse the development of energy system in a consistent and transparent mode, taking into account main technical, economic, social and environmental factors. A methodology which makes it possible is presented in this paper. It has been prepared by the Department of Energy Problems at the Institute of Fundamental Technological Research of the Polish Academy of Sciences. A set of models called SPSEK has been created. It is applied for analysing the country's energy demand and the possibilities of meeting if in the period of the next 20-25 years.

2. Description of the set of models

A scheme of the model set SPSEK is presented in Fig 1. This set consists of several mathematical models, including some elements of heuristic analysis. The analysis of the system's development begins with the definition of the scenario. The term "scenario" stands for a set of mutually interrelated assumptions depiciting the changes in such categories as:

- demographic development,
- rate of economic growth,
- structural changes in economy,
- preferences in the structure of consumption,
- world market prices of fuels,
- energy conservation,
-

1) Polish Academy of Sciences, Institute of Fundamental Technological Research, Swietokrzyska 11/21, 00 049 Warsaw, Poland

rate of implementation of new technologies in the sphere of energy production and in the sphere of energy consumption,

limitations in the growth of energy industries regarding, for example, the permissible rate of exploitation of the energy resources, levels of affecting the environment, capital expenditures etc.

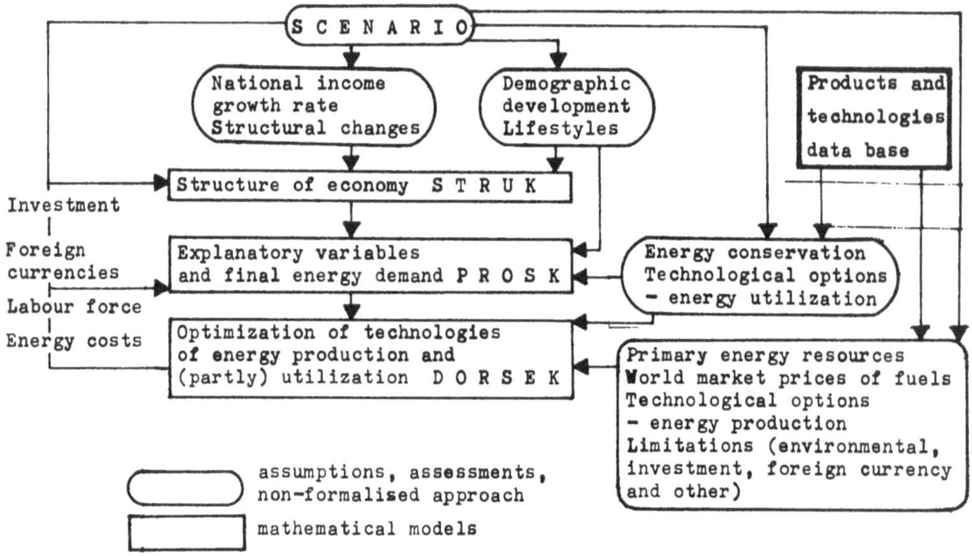

Figure 1 : Set of models for medium-term programming of a country's energy system development **SPSEK**

The elements of the scenario may be divided into 5 groups: population, economy, foreign trade, natural environment, and technology. While defining the scenario, the changes in each of these categories are depicted in a verbal-descriptive form e.g. the rate of economic growth: quick - slow; the variant of the growth of world market prices of fuels: high - low; the rate of implementing new technologies: quick - slow, etc. They are attributed concrete numerical values only during the formation of the scenario. Thus, its construction should be interpreted as a process of describing the development of the system in quantitive terms, where the main attention of the scenario-builder must be focused on achieving consistency, or internal compatibility, among the individual assumptions. In part, this consistency is assured by formalized mathematical models used in the process of scenario construction. Yet, with reference to some quantities, the consistency must be controlled by use of heuristic methods.

From the above categories it is possible to isolate a group of macroeconomic quantities. The relations among these quantities are analyzed by the structural model of economy STRUK. STRUK is an aggregated dynamic input-output model. It provides a representation of the formation and expenditure of national income in an aggregated form. Modelling is also used for the fixed capital formation, so as bring about the realization of the projected economic development. The model specifies 12 sectors, including 3 branches of the energy industry, in keeping with the aggregation presented in Table 1. The task of the model STRUK is to check if the initial assumptions concerning the growth rate of national income, foreign trade, and desired structure of consumption, are even realizable and, if so, under what conditions. The model's objective is to find solutions maximizing consumption in the period under study. Model generates the values of net output of each sector, which are used for the next step of computing the final energy demand in the economy. The values of some variables, or parameters of the model STRUK (for instance investments in energy industries or energy input coefficients in the input - output matrix) are calculated on the basis of information generated by another cooperating models, that is PROSK and DORSEK. In this way the models are coordinated in a series of iterative cycles.

Information obtained from model STRUK serves to determine the values of variables which explain the energy consumption. The explanatory variables denote the activity levels of the most energy-consuming economic processes in the production of goods and services. They can be measured in natural or monetary units. The natural units are used for measuring the production of especially important and energy-intensive products in the branches under study (for instance steel or cement), or the levels required to meet the energy-intensive needs of the population - space heating, transportation needs etc. The remaining values are expressed in the so-called real money values. In practice, dynamic indices referenced to the basic year, are used for this purpose.

The determination of the explanatory variables, measured in physical units, calls for analyses at the level of the specified economic sectors. Here the scenario-builder uses the results of extensive regression analysis of the relationships between the macroeconomic indicies and the values of the explanatory variables, as well as the information derived from other sources (planners' elaborations, experts, etc.). Next, an analysis of energy consumption is carried out for each sector and each variable. Due account is taken here of the relations among the major products in the analysed industrial branches, changes in technology, changes in unit consumption achieved trough conservation, etc. In this way the final energy demand is calculated. The computations are carried out with the help of model PROSK. In its present version the model takes into account more than forty activities (explanatory variables), and 15 energy sources. Both the variables and the energy sources can be aggregated, according to the model user's requirements.

Table 1: Aggregation of economy's sectors and branches

No	Sector or branch	No	Sector or branch
	Energy industry, of which	6	Chemical industry
1	- Coal industry	7	Building materials industry
2	- Fuels (liquid and gaseous) industry	8	Equipment goods industries
3	- Electrical power industry	9	Other industrial branches
	Rest of economy, of which	10	Construction
	Metallurgical industry, of which	11	Agriculture
4	- Ferrous metallurgy	12	Economy's other sectors
5	- Non-ferrous metallurgy		

The final energy demand serves as input information for the dynamic linear optimisation model DORSEK. DORSEK chooses the most advantageous technological structure in energy production, as well as in exports and imports of fuels. The model also takes into account the vital technological options in the sphere of energy utilization. The formulation of the model DORSEK is based on similar principles as utilized by other optimisation models (e.g. MESSAGE [3] or MARKAL [2]). It analyses more than seventy technologies and carries out independent balances for 20 kinds of fuels and energy. Great attention is paid, in the model, to the possibilities of modernisation and improvements of existing stock of energy facilities. The solution of the model DORSEK makes it possible to provide a fuel-energy balance sheet for the country at the level of primary energy, and to calculate the investment outlays, foreign currencies, and the like which are essential for a proper development and operation of the system. All the above-mentioned quantities are calculated for succesive 5-year periods of the analysed segment of time.

Some resulting quantities are next compared with the values assumed at the initialization of the scenario. In this way a check can be made to ensure that no glaring inconsistencies exist between the results and the assumptions. If considerable discrepancies do appear, the calculations are repeated, with the necessary corrections in the input data. The approach allows to obtain scenarios which are not inconsistent or incompatible. The models comprising the SPSEK set have been implemented upon a UNIVAC 1100 computer at the Planning Commision of the Council of Ministers in Warsaw. The technical-economic data, related to the goods, services and technologies studied by the SPSEK set, are supplied by the computer data base prepared at the Department of Energy Problems. A more detailed discription of the individual models of the SPSEK set is presented in [1].

3. Final remarks

The SPSEK model set has been used for a study of the development possibilities of the Polish energy system up to the year 2005. Results of computations for selected scenarios are given in [1]. These results confirm the correctness of the concept of the whole set of models. The application of a computer facilitates a speedy analysis of a number of scenarios, which, in effect, determines the area of permissible solutions and the choice of the most advantageous variant of development.

The models are utilized in cooperation with the specialists from national planning organs. The results obtained so far have been applied in the process of development of the Polish energy programme for the next 15 years. The models are being continually improved and adapted to the needs of the teams working directly for the decisionmakers. Also the data base is being consantly expanded.

4. Literature

[1] Cofala, J.: Methodology and Models for Analysis of Medium-Term Development of a Country's Energy System. IFTR Reports, Warsaw 1984.

[2] Fishbone, L.G., Abilock, H.: Markal, a Linear - Programming Model for Energy Systems Analysis: Technical Description of the BNL Version. Energy Research 5 (1981), 353-373.

[3] Schrattenholzer, L., The Energy Supply Model MESSAGE. RR-81-31, IIASA, Laxenburg 198 .

SIMULATION INVESTIGATIONS OF INTERCONNECTED POWER SYSTEMS. REDUCTION OF MATHEMATICAL MODEL

J.Halawa[1/], T.Halawa[2/]

The paper provides a mathematical model of a power system composed of three interconnected subsystems. Variations of loading power are the input functions for the subsystems while frequency swings are their outputs. Possible reduction of mathematical model order is discussed. Some results of simulation investigations attained whith help of a simulation language are given, as well.

MATHEMATICAL MODEL OF THE SYSTEM

The mathematical model of the subsystems and the coefficients of the transfer functions have been discussed in the following work [3] The model includes exclusively these elements which substantially influence the frequency swing. Characteristics of real power system have been adopted as coefficients for transfer functions of condensing and reheat turbines, valves, controllers and the rest subsystems with loads. No influence of water turbines in each subsystem has been included.

A block diagram of individual subsystem is encircled with a broken line in Fig. 1.

The block diagram of the subsystem comprises the following transfer functions:

G_{Rk} transfer function of primary controller

$G_{Ck} = \frac{1}{T_k s+1}$ transfer functions of control valves

$G_{TKk} = \frac{1}{T_c s+1}$ transfer function of condensing turbine

$G_{TRk} = \frac{1}{T_t s+1} \cdot \frac{c_1 s+c_0}{T_p s+1}$ transfer functions of reheat turbines

1/ Instytut Cybernetyki Technicznej Politechniki Wrocławskiej, Wrocław, Poland

2/ Instytut Automatyki Systemów Energetycznych, Wrocław, Poland

$$G_{Sk} = \frac{b_2 s^2 + b_1 s + b_0}{s^3 + a_2 s^2 + a_1 s + a_0}$$ transfer function of power system with loads

$\varepsilon_{1k}, \varepsilon_{2k}$ share coefficients for condensing and reheat turbine units.

The subsystems are interconnected by tie-lines denoted with b_{ij} in Fig. 1. Coefficients determining power ratio of the subsystems are marked with α_{ij} in this figure.

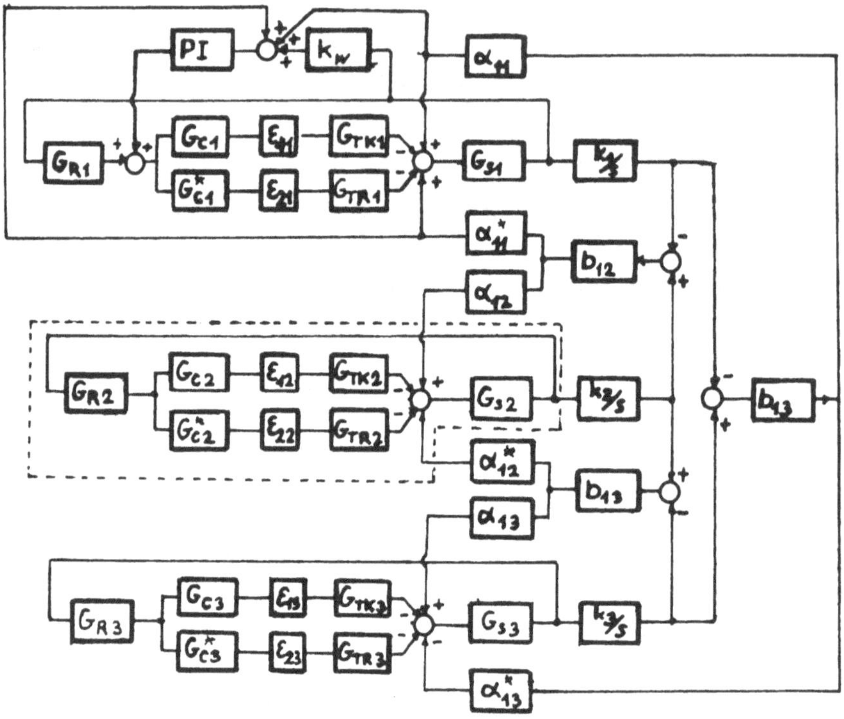

Fig. 1. The block diagram of the power system

REDUCTION OF MATHEMATICAL MODEL

The block diagram can be simplified. This would enable less number of functional blocks to be used for simulation process which is of importance, especially when analog computers are used for simulations. Likewise, simulation languages feature limitations as concerns the number of mathematical blocks. Simplifications of the mathematical model enables to shorten computation time of digital computers. No significant differencies in responses to step function nor in frequency characteristic are found for the original and simplified models.

In order to simplify the block diagram, the transfer functions of subsystems which do not subject to investigations on parameter influence are replaced by simplified ones which order is lower than original one. A continued fraction method is used in present paper to determine the simplified transfer functions. The transfer function has been expanded into a continued fraction of the form [1]:

$$G(s) = \frac{v_1|}{|v_0} + \sum_{i=1}^{k} \frac{v_{i+1}s|}{|v_i},$$

$$G(s) = \frac{1|}{|t_0} + \sum_{i=1}^{k} \frac{s|}{|t_i},$$

or

$$G(s) = \frac{c_0|}{|1} + \sum_{i=1}^{k} \frac{c_i s|}{|1}.$$

These expansions are equivalent, however they feature different numerical properties [2] . It should be noted that such expansions are not always possible [2]. Let us consider an example of subsystem simplification. When substituting specific parameters, the transfer function is as follows:

$$G_1(s) = \frac{L_1(s)}{M_1(s)}, \quad \text{where}$$

$$L_1(s) = 0.04375s^5 + 0.365s^4 + 0.82438s^3 + 0.2756s^2 + 0.075s + 0.01$$

and

$$M_1(s) = 0.3125s^6 + 2.65s^5 + 6.2675s^4 + 2.97081s^3 + 1.21893s^2 + 0.27184s + \\ + 0.0204.$$

This transfer function can be replaced by a convergent:

$$G_{23}^1(s) = \frac{0.14921s^2 + 0.022975s + 0.01033}{s^3 + 0.33501s^2 + 0.16964s + 0.02107}.$$

The table 1 provides responses of systems described with the transfer functions $G_1(s)$ and $G_{23}^1(s)$ to step function.

Table 1

t	y(t)	$y_{23}(t)$	t	y(t)	$y_{23}(t)$	t	y(t)	$y_{23}(t)$
0	0	0	16	0,42	0,43	32	0,47	0,48
2	0,23	0,24	18	0,47	0,47	34	0,49	0,49
4	0,36	0,35	20	0,50	0,49	36	0,50	0,50
6	0,38	0,37	22	0,51	0,50	38	0,50	0,50
8	0,36	0,35	24	0,49	0,49	40	0,50	0,49
10	0,33	0,33	26	0,47	0,48	42	0,49	0,49
12	0,33	0,34	28	0,46	0,47	44	0,49	0,49
14	0,36	0,38	30	0,46	0,47	46	0,48	0,49

As it results from the table, the approximation is satisfactory.

A time i frequency responses of a systems with the transfer functions $G_1(s)$ and $G^1_{23}(s)$ are plotted in Fig. 2 and 3.

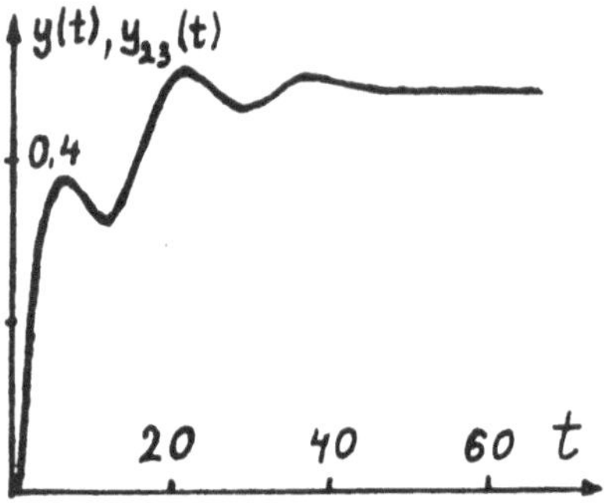

Fig. 2 . The time responses

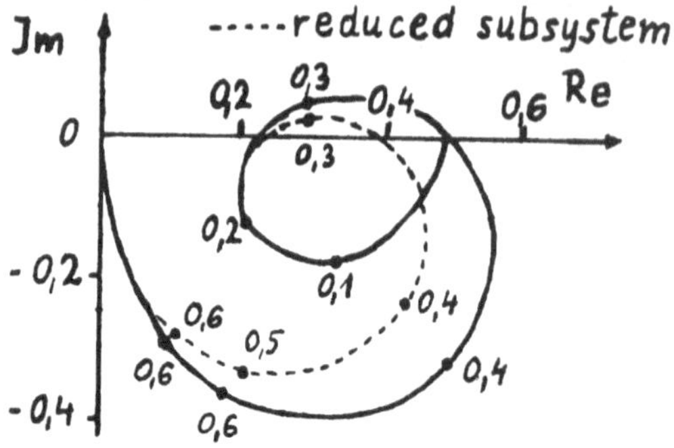

Fig.3. The frequency responses

If simulation investigates an influence of parameter variations in one subsystem while the other subsystems remain unchanged, the system under investigation is simulated with its full block diagram and the remaining subsystem models may be replaced with simplified models.

CONCLUSION

The reduction of order of subsystems down to the third order systems has enabled the time of simulation computations to be shortened. Excellent results of subsystem approximation has been achieved both in time and frequency responses within the interval of 0 to 0.2 rad/s. The essential frequency range in power system is narrower. Simulation investigations aim at determination of how time constants of turbines and the settings of controllers, both the central one and the local ones, influence the frequency variations.

REFERENCES

[1] Burtheel, A.: Division algorithms for continued fractions and the Pade table, Report TW41, Applied Mathematics and Programming Division, Katholieke Universiteit Leuven - Belgium /1978/

[2] Halawa, J., Trzmielak, A., Properties of simplified transfer functions as acquired with a method of continues fractions, Politechnika Wrocławska, I-6, preprint /1984/

[3] Halawa, T.: Wpływ parametrów systemu elektroenergetycznego na wahania częstotliwości, Prace IASE, Zeszyt 22 /1971/, 164-177.

SIMULATION METHODS OF LARGE ENERGETICS SYSTEMS BY MULTIPROCESSOR COMPUTERS

M.N.Kulik[1)]

The simulation methods of nonstationary and steady-state conditions in large gas-transfer (GTS) and electric power (EPS) systems (large energetics systems (LES) by multiprocessor computers (MPC) are considered. The algorithms of calculation paralleling are studied on the base of diakoptics and equivalentation. The structure and main technical characteristics of MPC designed to solve the problems of operational control and planning for LES development are described. Gas main, gas-transfer system, electric power system, implicit integration method, sparse matrix, diakoptics, equivalentation, multiprocessor computer, problem-oriented processor.

Modern GTS represent the union of power gas mains with complex coupled structure. The gas main consists of a few tens of linear sections (LS) and compressor stations (CS). Nonisothermic gas motion in LS pipe-line is described by a system of three nonlinear two-dimensional partial differential equations and by algebraic gas state equation which relates the unknown gas pressure, flow rate, temperature and density [1]. When analyzing the nonstationary conditions CS are described by the non-linear algebraic equations which allow one to define the CS operational parameters by means of characteristics and parameters of gas pumping sets. For the points of mutual pipe-line coupling (network nodes) the equations of pressure equality, mass and energy conservation are additionally used. The set of above equations describing LS, CS and the units of GTS form its mathematical model. The models of large GTS contain hundreds and thousands of partial equations and non-linear algebraic equations which have to be studied as a single complex system. The application of the numerical methods to the partial differential equation systems results in the equivalent systems of algebraic equations with a few of thousands of derivatives.

The transient processes in large GTS are very prolonged, the time of a special process running is often about some days. Therefore the integration of differential equation systems should be made with a relatively great time step that calls for the application of implicit methods. The application of implicit methods lowers the accuracy at large

1) Institute of Modelling Problems in Energetics Academy Science UkSSR, Brest-Litovski pr., 102, 252680 Kiev-57, USSR

steps. New simulation methods for large GTS were developed that allow to study the nonstationar conditions in large complexly-coupled systems of gas mains. They provide high accuracy by working with large (a few tens of minutes) time steps.

The combination of implicit and explicit methods of T-transform serves as a methodological basis for the algorithms constructed [2]. The essence of the explicit and implicit methods of T-transform is the organization of function-image discrete value calculation on the base of the solution obtained at a previous iteration (time layer). The peculiarity of this process is the definition of subsequent discrete values in terms of the previous ones by recurrent dependences. Here, for the quasi-linear systems of partial differential equations that describe the gas-dynamics processes in pipe-lines, the analysis is reduced to the successive solution of the third order algebraic equation sets. Besides the obvious simplicity the explicit methods of differential transforms can control the accuracy order, the method order being self-defined according to a given absolute error. The studies performed by the analysis of real GTS conditions, showed that when using explicit methods of T-transforms it is reasonable to work with time step of minute order. The attemps to increase the time step result in an essential increase of the discrete value and of arithmetical operation number.

The problem of time step increase is solved by means of implicit methods of T-transforms. The computational schemes are proposed which enables one to work with very large time steps. On the base of the implicit methods of T-transforms the computational algorithms were constructed that provide practically unlimited accuracy. For gas-dynamic equations however, the dimensionality of the problems solved is essentially increased. Therefore the second-fourth order implicit methods of integration were developed with acceptable accuracy and labour-consuming characteristics. These methods possess an absolute stability and allow to operate with very large steps of the order of a few tens of minutes, hours, days. It is possible to use an infinite step that is often used to calculate the stationar GTS condition by the setting method.

The algorithms developed are of a practical importance where the second-fourth order implicit methods of T-transforms are used as basic and the solutions are formed on the base of their iterative application. The high solution accuracy is provided by its formation as a linear solution combination obtained with divisible steps on the base of the implicit methods with a mean accuracy [3].

On the base of the above explicit and implicit methods of T-transforms and their combinations the programm-computational systems were developed designed for the analysis of the nonstationar conditions of large GTS's with a high accuracy provided (standard calculations), for proximate calculations made under the conditions of operational control and for calculations of a mean accuracy by solving the problems of GTS condition design and planning. These programm-computational systems are designed for implementation on computers of EC ЭВМ type, on other universal computers and multiprocessor computers.

The steady-state EPS conditions are described by non-linear algebraic equation sets in polar coordinates. The application of Newton-Raphson method with block-diagonal representations of Jacobi matrix is effective for their solution. The analysis of EPS dynamic stability by implicit methods and the evaluation of a current power system state by the measurement results give also the sets of non-linear algebraic equations, their solution being performed by the above method.

When the GTS and EPS conditions are calculated the sets of linear algebraic equations are solved iteratively with their matrices having many zero elements. Therefore, in LBS simulation programms the algorithms using the matrix sparsity are widely used [4].

The main problem of calculation organization on multiprocessor system is the algorithm paralleling for homogeneous loading of all processors [5]. By LBS simulation it is proposed to decompose the initial system into subsystems by means of diakoptics methods. Another important way of calculation paralleling which was implemented in LES simulation programms is the subscheme equivalentation on individual processors with subsequent analysis of the transformed scheme on the multiprocessor computer by means of diakoptics methods.

The simulation of large GTS under the conditions of operational control is associated with solving the equation set of large dimensionality in small time periods. Therefore it is necessary to use the most effective computing facilities for the analysis. The development of problem-oriented multiprocessor computers is an effective way of constructing large-powered computing facilities.

In the developed MPC, the elder EC ЭВМ model (EC-1045, EC-1055) is used as a central processor (nucleus) with which the programm-oriented processors are coupled in a system through standard channels. The problem-oriented processors have parallel structure, they perform the operations of addition, multiplication, division, rooting, the calculation of trigonometric functions by means of hardware. The performance of the problem-oriented processor is about 2 millions operation per second by Gibson, its element basis consists of microprocessor sets. The MPC configuration includes up to 10 problem-oriented processors. The MPC of above structure satisfies the demands of LBS' operational control and development planning.

Literatura

[1].Васильев О.Ф., Бондарев Э.А., Воеводин А.Ф., Каниболотский М.А. Неизотермическое течение газа в трубах, СО "Наука", Новосибирск, 1973.

[2].Пухов Г.Е. Дифференциальные преобразования функций и уравнений, "Наукова думка", К., 1980.

[3].Марчук Г.И. Методы вычислительной математики. "Наука", М., 1981.

[4].Tewarson R.P. Sparsematrixes. Academic press, New York and London, 1973.

[5].Пухов Г.Е., Кулик М.Н. Гибридное моделирование в энергетике. "Наукова думка", К., 1977.

SYSTEMS MODELLING OF THE ELECTRIC POWER OBJECTS

V.E.Tonkal[1], Yu.G.Blavdzevitch[2], V.G.Derzski[2]

ABSTRACT

In order to carry out the complex researches of the influence of the different factors on the object-system operation efficiency in the electric power system (EPS), the multi-level systems models which represent the connections and relations between the elements of an object modelled are required.

As an example of the systems approach to the modelling of the EPS one may consider the experience of developing the set of the model and program modules united into the system which describes the electromagnetic and heat processes in the turbogenerators; the characteristics of the assemblies connected to the electric power station, the static & dynamic operation conditions of the EPS and the distribution of the active & reactive power in the EPS units. The program modules are united by the common data base. The control programs provide for the systems analysis of a problem in question in the dialogue man-machine mode of operation.

Key words: modelling, power station, engineering solution, optimization, probability, evaluation

The complexity of planning, developing & optimum control of the power system is due to the large scale of the power industry and its intensive development conditions, the differences in engineering, economic & operation condition characteristics of the separate elements of the power system. For solutions of such problems, decomposition method is used, the equivalenting procedures are applied to the power station assemblies, the power networks etc. Along with the progress in solving a certain kind of problems by the powerful software, there is deficiency in provision for the complex researches necessity of which progressively increases at present time.

Such situation has given rise to the investigations being carried out in the Institute of Modelling Problems in Energetics of the Ukrainian Academy of Sciences on developing the methods & software for information-modelling system (IMS) for analysis of the mutual influence in the the EPS elements of the different hierarchy levels & control of the optimum engineering solutions.

The first phase of the IMS is oriented to solve the following research problems of planning, designing & control:

1. Load distribution optimization of the power generated by the equipment installed.

2. Optimization of the operation conditions of the EPS in changing the composition & parameters of the equipment installed.

3. Optimization of the electric network configuration.

1. Institute of Electrodynamics of the Ukrainian Academy of Sciences, 252680, Kiev-57, Brest-Litovski str.102, USSR.

2. Institute of Modelling Problems in Energetics of the Ukrainian Academy of Sciences, 252680, Kiev-57, Brest-Litovski str. 102, USSR

4. Specification of the parameters & characteristics of the objects & the systems for optimzation of the problems in designing & control.

In this case, the variable parameters in the engineering solutions may be: 1. The operation condition parameters; 2. The equipment parameters; 3. The electric network configuration; 4. The fuel characteristics.

The factors of the engineering solution efficiency in general case are as follows: 1. The fuel consumption; 2. The technological consumption of the electric power for its transportation & distribution; 3. The voltage levels in the EPS units; 4. The expenditure for implementation of the engineering solutions.

In the course of developing the methodical support of the IMS, the following points have been developed:

- The principles of construction of a systems mathematical model for the complicated electric devices & systems;
- The identification algorithm of the electric engineering objects & systems;
- The probabilistic-statistic methods of the data estimation;
- The conseption of the estimated-experimental analysis of the electromagnetic & heat processes in the electric devices for construction of the mathematical models differing in algorithm-united calculation of the field, the losses & the heating with correction in the course of an analysis of the preliminary assumptions on the base of the experimental data;
- The effective algorithm and the programs for the computer-based implementation of the finite-element method, application of which in the field calculation problems provides the reduction of the machine time for solving the linear algebraic equation sets at the cost of application of the double dynamic factorization, the automation of processing & representing the calculation results in the visual & convenient form for analysis.

The analysis of the random load & voltage variations in the electric power system, the characterisic inaccuracy of the fuel consumption relative increase in the power stations (due to imperfection in measuring methods), the indeterminancy of the parameters in the operation conditions & networks (due to the errors of the measuring devices and instruments) has shown that when optimizing the power station loading in active power, the probability of the EPS operation with the fuel overexpenditure exceeds the probability of the operation with saving the fuel. But in case of deterministic specification of the initial data in accordance with the calculation, the saving of the fuel consumption is expected.

Allowing for the EPS probability properties makes it possible: to select the set which considerably affects the initial data final results, to optimize the reliability level of the power consumption by the consumers, to estimate the expediency etc. of any changes in the parameters

& the structure of the EPS elements.

Necessity of solving the whole set of the interconnected problems in the EPS inevitably leads to the transition of a great number of prolems from the technical category into the technical-economy category. In this case, the efficiency function obtains the complicated implicit form But the traditional program set which models the physic-technical laws of the EPS operation demands the powerful technical basis, the large memory size & computing time, very much labour designing work and development of the program interface of the components with their union into the program set, the collection and computing of a huge number of initial data, the high level of personnel qualification. But if one takes into consideration the fact that the greater part of the initial data possesses some indeterminancy which considerably minimizes the calculation results obtained by the traditional method, application of the simplified static models is found to be more effective.

In constructing the static models, one uses the methods of planning the experiment, the group account for the arguments, the parabolic interpolation by Chebishev's polinomial, the regression analysis. The characteristic features of the static models developed are simplicity, compactness, explicit form of the efficiency function, possibility of taking into account the initial data indeterminacy etc.

Complex statement of a problem for analysis of the electric power systems determines the system model structure of an object in question in which the different level subsystems are separated from the electric power system, namely: the power stations (PS), the electric power consumers, the electric networks, the turbogenerators (TG), the converters etc.

As for the TG, the PSs and respectively the EPS are the external system. In the models of the upper levels, the generators are represented by the parameters which determine the transient process nature (electromagnetic & electromechanic ones) in the EPS nodes as well as by the permissable load constraints of the assemblies in accordance with the operation reliability specifications with allowance for the parameters changes on the PS bus bars which are prescribed by the EPS performance operation conditions. Accordingly, the variation of the phisical processes (electromagnetic & temperature fields) and the TG parameters respectively in changing their operation condition or element & unit structure exert a direct influence on the conditions of the PS assembly reliable operation and EPS operation condition efficiency.

The program set "The turbogenerator operation condition parameters' as a software subsystem of the IMS allows to check the permissible performance loads on the condition of its element heating in the course of optimizing the EPS operation condition. The temperature estimated values are differed from the check test data by not more than 5 %.

SIMULATION AND DECISION MAKING ALGORITHMS
FOR ENERGY SYSTEMS OF INDUSTRIAL PLANTS

Andrzej Ziębik [1)]

Summary

The paper deals with the application of methods of systems analysis for the planning, control and design of energy economy in industrial plants. Simulation and decision making algorithms for the energy system of an industrial plant are based on a mathematical model of energy economy. In the paper the following models and methods are considered: mathematical models of energy economy for planning and production control, a matrix method of calculating the unit costs of energy carriers and a system approach for the determination of the structure of energy economy in the preliminary design.

Keywords: energy systems, energy economy, mathematical modelling.

1. Introduction

The production system of an industrial plant can be divided into the technological subsystem and energy subsystem (energy economy). The production plan of the energy subsystem results from the demand of energy carriers in the technological subsystem. Because of the complexity of connections between the energy branches, control and planning of the production of energy carriers a computer system must be used. In this system simulation and decision making algorithms for the energy system of an industrial plant are used. These algorithms are based on mathematical models of the energy economy of an industrial plant.

2. Mathematical model of a long - term balance plan of energy economy

A long-term balance plan of energy economy for a month, a quarter or a year is prepared. Average values of technical coefficients may be used in this case. The mathematical model results from the balances of energy and fuels. It is a linear model [2]. This model can be written in matrix notation as follows:

$$G + P + FG + QE_1 + \bar{F}\bar{G} + \bar{Q}E_2 + D = AG + XE_1 + \bar{A}\bar{G} + \bar{X}E_2 + H, \quad (1)$$

where: G, P, D - column vectors of peak production, basic production and external supply of the energy carriers; $\bar{G}$ - column vector of the production of the technological subsystem; F, $\bar{F}$ - matrices of the coefficients

1) Institute of Thermal Engineering, Silesian Technical University, ul. Konarskiego 22, 44-101 Gliwice, Poland.

of the by-production of energy carriers; Q, $\bar{Q}$ - matrices of the by-production of energy carriers independently of the production; A, $\bar{A}$ - matrices of the coefficients of the consumption of energy carriers; X, $\bar{X}$ - matrices of the consumption of energy carriers independently of the production; H - column vector of the final production of energy carriers; E_1, E_2 - column vectors with unity elements; the dashes above the symbol denote the kind of subsystem (without any dash - energy subsystem, with a dash - technological subsystem).

After the solution of equation (1) elements of the vectors G and D are obtained. This is a descriptive model. If the vectors G, P and D are calculated simultaneously, this problem is solved by means of linear programming. The linear aim function is formulated as a minimum of the energy feeding costs. Equation (1) is a total limitation. The production capacity and limits of external supplies as limitations are used too.

3. The matrix method of calculating the unit costs in the energy subsystem of an industrial plant

This method is based on the mathematical model of the energy balance of an industrial plant [3]. The balance of the production costs of energy carriers leads to the following matrix equation:

$$(AG^D + X)^T B^{-1} (G^D k_G + P^D k_P + L^D k_U + D^D k_D + K_T) + K = G^D k_G + P^D k_P + (FG^D + Q)^T k_U , \qquad (2)$$

$$B = G^D + P^D + L^D + D^D , \qquad (3)$$

$$L = FG + QE_1 + \bar{F}\bar{G} + \bar{Q}E_2 , \qquad (4)$$

where: k_G, k_P, k_U, k_D - column vectors of unit costs of the peak production, basic production, by-production and external supplies; K_T - column vector of the distribution costs of energy carriers; K - column vector of constant costs; the index "D" denotes a diagonal matrix.

The elements of the vectors k_P, k_U are connected with the respective elements of the vector k_G by linear dependences. These dependences result mainly from the principles of the division of costs in combined processes. The calculating value resulting from (2) is the vector k_G.

4. System approach to the determination of the structure of energy economy

The algorithm of the system approach to the determination of the structure of energy economy in the preliminary design contains solutions of the following problems:

- determination of the structure of the binary "input-output" matrix,

- structure analysis of the binary "input - output" matrix,
- determination of the time-distribution function of the demand of energy carriers,
- determination of the elements of the "input-output" matrix,
- determination of the optimal capacity of the arrangement of energy.

It is the aim of this model to determine the optimal variant of the structure of energy economy in an industrial plant. The aim function is expressed by the following equation:

$$K_R = (\varrho_P + \beta_P)\, I_P + (\varrho_G + \beta_G)\, I_G + (\varrho_D + \beta_D)\, I_D + \alpha_P\, \dot{P}_N + \alpha_G\, \dot{G}_N + \alpha_D\, \dot{D}_N + \varepsilon_P\, P + \varepsilon_G\, G + \varepsilon_D\, D + S_R \longrightarrow \min\,, \quad (5)$$

where: $\varrho, \beta, \alpha, \varepsilon$ - row vectors of the discount rates, rates of the constant costs of repairs and maintenance, labour and energy and of the operating costs; I - column vectors of capital costs; $\dot{P}_N$, $\dot{G}_N$, $\dot{D}_N$ - column vectors of the nominal capacity of the basic and peak energy arrangement and of the nominal external supply; S_R - loss of technological production due to a lack of energy carriers; the indices "P", "G" and "D" denote respectively basic production, peak production and external supply.

Equation (1) is a total limitation. The Lagrange method of decomposition [1], as well as the matrix method of calculating the unit costs of the energy carriers as a coordination procedure, have been applied. On the level of particular energy carriers, the aim function may be expressed in the form of the following equation:

$$L_j = (\varrho_{Pj} + \beta_{Pj})\, I_{Pj} + (\varrho_{Gj} + \beta_{Gj})\, I_{Gj} + (\varrho_{Dj} + \beta_{Dj})\, I_{Dj} + \alpha_{Pj}\, \dot{P}_{Nj} + \alpha_{Gj}\, \dot{G}_{Nj} + \alpha_{Dj}\, \dot{D}_{Nj} + \varepsilon_{Pj}\, P_j + \varepsilon_{Gj}\, G_j + \varepsilon_{Dj}\, D_j + S_{Rj} + P_j \Big(\sum_{i=1}^{n} a_{ij}^{P}\, k_i - \sum_{i=1}^{n} f_{ij}^{P}\, k_{Ui}\Big) + G_j \Big(\sum_{i=1}^{n} a_{ij}^{G}\, k_i - \sum_{i=1}^{n} f_{ij}^{G}\, k_{Ui}\Big) \longrightarrow \min\,, \quad (6)$$

where: a_{ij}^P, f_{ij}^P, a_{ij}^G, f_{ij}^G - elements of the matrix A^P, F^P, A^G, F^G; k_i - weighted average unit cost of the energy carriers.

The decomposition algorithm is solved by means of an iterative procedure. The unit costs are fixed on the level of the particular energy carriers. The limits of external supplies of energy are local limitations in the optimization problem.

5. Mathematical model of energy economy for production control

The mathematical model of energy economy for production control is composed of a set of nonlinear algebraic equations resulting from the balances of the energy carriers [3]. This model may be considered as a descriptive one, but it can be also used for optimization purposes. The energy balances for a shift and for twenty-four hours are assembled by means of one-hour balances. The balance equation for time-intervals of one hour and for the energy carrier "i" looks as follows:

$$\dot{\Gamma}_i + \sum_{j=1}^{n} \dot{U}_{ij}(\dot{\Gamma}_j) + \sum_{k=1}^{m} \dot{\bar{U}}_{ik} + \dot{D}_i =$$

$$= \sum_{j=1}^{n} \dot{Z}_{ij}(\dot{\Gamma}_j) + \sum_{k=1}^{m} \dot{\bar{Z}}_{ik} + \dot{Y}_i + \Delta\dot{C}_i + \dot{H}_i + \dot{V}_i \ , \qquad (7)$$

$$\dot{\Gamma}_i = \dot{G}_i + \dot{P}_i \ . \qquad (8)$$

where: $\dot{G}_i$, $\dot{P}_i$ - peak and basic production of energy carriers; $\dot{U}_{ij}$, $\dot{\bar{U}}_{ik}$ - by-production of energy carriers in the energy and technological subsystems; $\dot{D}_i$ - external supply of energy carriers; $\dot{Z}_{ij}$, $\dot{\bar{Z}}_{ik}$ - consumption of energy carriers in the energy and technological subsystem; $\dot{Y}_i$ - consumption of energy carriers for the use of the whole plant; $\Delta\dot{C}_i$ - increment in the energy storage of energy carriers; $\dot{H}_i$ - final production of energy carriers for external consumers; $\dot{V}_i$ - loss of energy carriers.

In the case of the descriptive model the following quantities can be calculated: either the own production of energy carriers or an external supplementary supply of energy carriers or an amount of energy carriers from energy storage. Therefore the set of balance equations is divided into three groups of equations.

The equations of the model used for optimization are solved by means of the decomposition of the total optimization problem. The method of Lagrange s multipliers and the matrix method of calculating the unit costs of energy carriers are used.

Literature

[1] Popyrin, L.: Matematicheskoje modelirowanie i optimizacja teploenergeticheskich ustanowok. Energia, Moskwa 1978.

[2] Szargut J., A.Ziębik: Linear mathematical model of the material and energy balance in iron-works (in Polish). Archiwum Energetyki (1972) 2, 103 - 131.

[3] Ziębik A., J.Gwóźdź: Mathematical model of the energy economy of ironworks for control and scheduling. 7 th International Seminar "Algorithms for Production Control and Scheduling", Karlovy Vary 1984.

DYNAMIC STATE ESTIMATION OF THE ELECTRIC POWER SYSTEM

Zoran Morvaj [1)]

SUMMARY

The main topics of this paper is a developement of the procedure for dynamic state estimation of the electric power system. The power system under normal operating condition is considered. A dynamic estimator, based on Kalman filter, is developed in two version.The procedure is tested by digital simulation and some tests results are given.

1. INTRODUCTION

Controling a power system means keeping it in a operational state which enables a reliable and economical supply of consumers by quality electrical energy. When the power system under normal operating conditions is treated , the complex bus voltage is suficient to expres its state. These state variables cannot be measured directly. Instead, noisy measurements that are functions of system state variables are available The problem of determining state variables from noisy measurements is known as state estimation. One of the possible methods for solving it is Kalman filter. Succesful use of the Kalman filter requires knowledge of a methematical model of a power system state time behaviour, and sufficient statistics of all stochastical quantities. Since the system is described by the state variables approach, the mathematical model of system behaviour can be expressed in terms of state transiton. That means that one must specify how one state transformed into another as time passes.

The discussion in this paper is presented in three main sections. First, the Kalman filter technique is shortly described. Secondly, the determination of the power system model and some other relevant quantities is described. Finally, some testing results are given and shortly comented.

2. KALMAN FILTER

Consider a discrete dynamic system whose state $x(i)$ is described by

$$x(i+1)=F(i+1,i)x(i)+w(i) \tag{1}$$

$F(i+1,i)$ is a state transition matrix, and $w(i)$ is a zero mean vector random sequence, which represents system disturbances, with known covariance matrix Q. Measurements data contain random error v, and are related to the system state by

$$z(i)=H(i)x(i)+v(i) \tag{2}$$

The statistics of $v(i)$ is known, and $H(i)$ is a known matrix.

The optimal estimation of the state variables is given by

$$\hat{x}(i+1)=F(i+1,i)\hat{x}(i)+K(i+1)(z(i+1)-H(i+1)F(i+1,i)\hat{x}(i)) \tag{3}$$

where

$$K(i+1)=M(i+1)H(i+1)^T(H(i+1)M(i+1)H(i+1)^T+R(i+1))^{-1} \tag{4}$$

$$M(i+1)=F(i+1,i)P(i)F(i+1,i)^T+Q(i) \tag{5}$$

$$P(i)=M(i)-K(i)H(i)M(i) \tag{6}$$

K, M and P are optimal filter gain matrix, covariance matrix for state prediction errors and covariance matrix for estimation errors, respectively. The equations (3), (4),

1) Faculty of Electrical Engineering, Unska 3, 41000 Zagreb, Yugoslavia

(5) and (6) are known as the Kalman filter.

For a succesful implementation of the Kalman filter to power system dynamic state estimation, two mathematical models are required, (1) and (2), and also three covariance matrices, R, P(0) and Q, have to be known. Determination both of the model of measurements (2) and covariance matrix for measurement errors R is well known, [2], so it will not be described here. It can also be easily shown that for the value of the covariance matrix for initial state: $P(0) \gg R$, already the first measurement z(1) will lead to a large reduction from $P(0) \gg R$ to $P(1) < R \ll P(0)$.

2.1. Suboptimal Kalman filter with sequential measurement processing

An important consideration in developing an on-line dynamic state estimator is the need to reduce both storage size and computation time. As long as one works with full matrices the storage requirements will increase rapidly with the power system size Therefore, if the algorithm is to be applied to a large system , the following changes must be done:

- sequential instead of simultaneous measurement processing (that means that each set of data are treated separatly so we never need the whole filter gain matrix K, but only one of its columns, in the storage)
- use of the suboptimal Kalman filter (the covariance matrices P, Q and M are assumed to be diagonal and are treated like column matrices)
- exploiting the sparsity of the network admittance matrix Y.

Wit these changes storage requirements are highly reduced and the amount of computations also decreases. The following calculation must be done for each set of data which are entered into the computer in on-line fashion:

$$\hat{x}(i+1)_{j+1} = x'(i)_j + K(i+1)_{j+1}(z_{j+1} - h_{j+1}(x'(i)_j))$$

$$K(i+1)_{j+1} = \frac{M(i+1)_{j+1} H^T(i+1)_{j+1}}{H(i+1)_{j+1} M(i+1)_{j+1} H^T(i+1)_{j+1} + \sigma^2_{j+1}}$$

$$M(i+1)_{j+1} = F(i+1,i)P(i)_j F^T(i+1,i) + Q(i)$$

$$P(i+1)_{j+1} = M(i+1)_{j+1} - K(i+1)_{j+1} H(i+1)_{j+1} M(i+1)_{j+1} \qquad (7)$$

where j=0,...,m (m=number of measurements), F, M, P and Q are diagonal matrices treated like colimn matrices, H_j is the jth row of H, h_j is the jth component of H_j, z_j is the jth component of the vector z, $K(i)_j$ is the jth column of the optimal filter gain matrix, x′ and $\hat{x}$ are the predicted and estimated state vectors, respectively. The computational process goes from x′(i) to $\hat{x}(i+1)$, and consits of a sequence of intermediate processes which start with j=0 and end up with j=m.

3. MODEL OF POWER SYSTEM STATE TIME BEHAVIOUR

The model of system state time behaviour is given by the equation (1). To determine this model it is necessary to determine a state transition matrix F. To enable prediction of any desirable state at every moment inside one day, the state transition matrix has to be determined a day in advance. Therefore, it is a reasonable to connect the state transition matrix determination with daily load forecasting.

From the forecasted hourly loads, the complex voltage of every node for every hour can be calculated. Then one can calculate the requested value of the state transition matrix at full hours. The question is, however, what the values inside one hour section are. One solution is to calculate them by the means of some interpolation methods. But classical methods of interpolation (Lagrange, Chebyshev, Fourier) are not sufficiently accurate for this purpose [1]. The COONS-AHUJA method, from the class of

computational geometry, is utilised, instead (for details see [5]).
As a result of this method, continous functions ($f(t)$), which express the nominal time behaviour of the state variables, are obtained. These functions form a basis for state transiton matrix determination. If the sampling rate of measurements Δi is known, it is possible to calculate the state transition matrices for the successive states in advance

$$\begin{matrix} t_1=i & x(i)=f(t_1) \\ t_2=i+\Delta i & x(i+\Delta i)=f(t_2) \end{matrix} \Longrightarrow F(i+\Delta i,i)=\left[\frac{x(i+\Delta i)}{x(i)}\right] \tag{8}$$

If it is desirable to predict a system state for d steps in advance, it is clear that the matrix $F(i+d,i)$ has to be known. On the basis of the property (9) of the state transition matrix

$$F(i,j)F(j,k)=F(i,k)$$

the $F(i+d,i)$ can be calculated as

$$F(i+d,i)=\prod_{k=1}^{d} F_k(i+k,i-(k-1)) \tag{10}$$

When the state transition matrix is known, the system model (1) is determined. There still remains the determination of the covariance matrix of system disturbances.

3.1. Determination of Q matrix

The determination of the Q matrix is vary important, especially in the case of an uncertain dynamic system, beacause it can strongly degrade the filte's performances. It can be seen from (5) that $M(i+1)\geq Q$ for $P(i)\geq 0$. This means that the Q matrix sets a performances limit on the prediction accuracy. Since the state of the system is predicted on the basis of the system model (1), then the Q matrix is also a measure of confidence in the system model. Besides, there are many various unpredictable random disturbances as well as the numerical factors that arise due to errors in modelling, which increase the uncertainity of power system states. Their influence has to be included by the means of the Q matrix, too. Under these circumstances, the best way to determine the Q matrix is by the filter optimization, after careful experimental examinations.

4. EXPERIMENTAL RESULTS

The developed model and dynamic estimator has been tested by digital simulation on a part of a real network with 12 nodes and 17 branches. The forecasted hourly loads have to be known. Measuremnet data are simulated on the basis of previous power-flow studies. The intention was also to examine the suitability of mini computers for such problems, as well as the applicability of the proposed algorithm for mini computers. So the dynamic estimator has been developed in two versions: in the first one the optimal Kalman filter is used and measurements are simultanously processed, in the second one the suboptimal Kalman filter is used and measurements are sequrntially processed. All tests were performed on the Hewlett Packard 9845B.

First, the continous functions of the nominal system state changes are determined see fig. 1). That enables a calculation of the state transition matrix. Secondly, the optimal filter gain matrix is determined. Here the influence of the Q matrix on the filter properties has been noticed. Namely, if the chosen Q matrix is too small, the optimal filter gain matrix diverges - the values of its elements becomes smaller and smaller. That means that the influence of new measuremnets become smaller and the influence of predicted states stronger, (from (3)). Therefore the relation Q/R is very important. In this paper the optimal filter gain matrix is dtermined with Q/R=4.

Only the estimation of the actual state in every interval and prediction of the state for the next interval have to be done on-line. This way the on-line calculation is largely reduced as well as the memory requirements. The time and space complexity of the algorithm based on suboptimal Kalman filter and sequential measurements processing is shown on figure 2. This version has considerably smaller time and memory requirements.

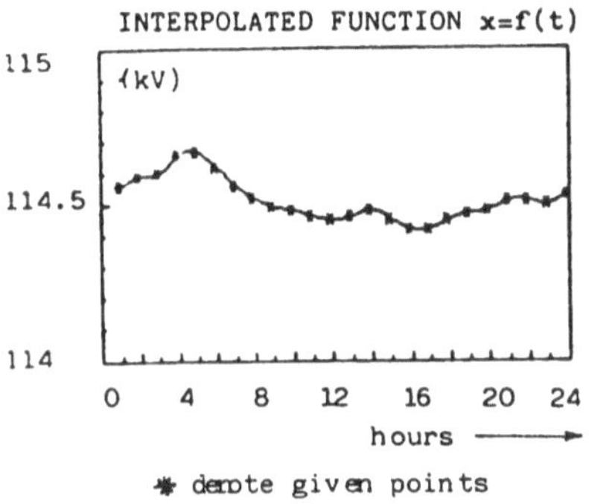

Fig.1. **Interpolated function of the nominal system state time behaviour**

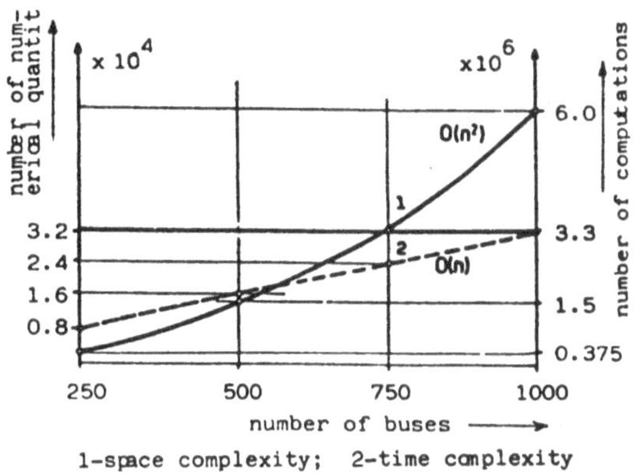

Fig.2. Time and space complexity of the suboptimal Kalman filter

5. CONCLUSION

This paper has presented the derivation of the mathematical model of the power system state time behaviour, which enables prediction of the future system states,and also power system dynamic state estimation. All tests were performed on a mini computer because the intention was also to examine their convience for such problems, and the applicability of the proposed algorithm for mini computers. Therefore, the dynamic estimator has been developed in two version. The performed experiments and the time and space complexity analysis have demonstrated satisfactory performances of the developed model and dynamic estimator.

6. REFERENCES

[1] Degoricija, D.: Reliability and Efficiency of the Methods for the Presentation of Nonmathematical Curves, Disertation, Zagreb, 1980. (in Croatian)

[2] Handschin, E.: Real Time Data Processing Using State Estimation in Electric Power System. Procc. of symp. of real time control of electric power system, Elsevier publ. comp. Amsterdam, 1972

[3] Morvaj, Z.: A Mathematical Model of the Electric Power System for Dynamic State Estimation. to be published in Electric Power System Research Journal, vol.3, 1985

[4] Sorensen, H.W.: Kalman Filtering Techniques. Academic Press, New York, 1966

[5] Turk, S.: Computer Graphics - Basis of Theory and Application. Školska knjiga, Zagreb, 1980. (in Croatian)

A NEW DIGITAL CONTROL STRATEGY FOR A NUCLEAR BOILING WATER REACTOR

M. Gopal* and P. Pratapachandran Nair**

A method is proposed for the design of constant gain optimal feedback control laws for discrete-time systems which reduce trajectory sensitivity to small system parameter variations and ensure the closed-loop eigenvalues at the prescribed locations. The selected performance index is of standard linear regulator type, modified to include a quadratic sensitivity term. An efficient computational procedure based on direct cost oprimization using gradient type algorithm is also reported. A case study of a Boiling Water Reactor power plant has been presented to illustrate the application of the design procedure.

Key words: Control Systems, Sensitivity, Optimal Control, Multivariable Control, Digital Control, Nuclear Reactor Control.

The Digital Control Algorithm

The linear model of the plant can be formulated in the standard form as

$$\dot{x}(t) = A(\mu)x(t) + Bu(t) \tag{1}$$

where $x(t) \in R^n$ is the state vector, $u(t) \in R^m$ is the control vector; B is a constant matrix and A is a matrix function of a scalar parameter μ with nominal value μ_o; $A(\mu) = A$; $A(\mu_o) = A_o$. $A(\mu)$ is assumed to be differentiable. It is assumed that

$$\text{rank } B = m; \quad \text{rank } [B, A_oB, \ldots, A_o^{n-1}B] = n \tag{2}$$

The quadratic performance index which includes a sensitivity term penalizing excessive variations in x(t) due to changes in parameter μ can be expressed as

$$J = \int_0^\infty (x^T(t)Qx(t)+u^T(t)Ru(t)+\sigma^T(t)S\sigma(t))\, dt \tag{3}$$

where $\sigma(t) = \dfrac{\partial x(t)}{\partial \mu}$ is the trajectory sensitivity vector. Q and S are n x n positive semi-definite matrices and R is an m x m positive definite matirx. The discrete-time behaviour of the continuous-time system is modelled by the set of first-order linear difference equations

$$x(k+1) = Fx(k)+Gu(k) \tag{4}$$

$$\text{where} \quad F = e^{AT}; \quad G = \int_0^T e^{At}dt.B \tag{5}$$

The discrete-time equivalent of performance measure (eqn. 3) is obtained as[3]

$$J = \frac{1}{2} \sum_{k=o}^{\infty} [x^T(k)\hat{Q}x(k)+2x^T(k)Mu(k)+u^T(k)\hat{R}u(k)+\sigma^T(k)S\sigma(k)] \tag{6}$$

For a given set of real eigenvalues $\{\lambda_i\}$; i = 1,2,...,n, there exists a control law

$$u(k) = Kx(k) \tag{7}$$

such that the set $\{\lambda_i\}$ coincides with the spectrum of (F_o+G_oK).

For multi-input systems, this K will not be unique and one can utilize this design freedom to minimize a given performance index J. There exists a K that simultaneously assigns the spectrum $\{\lambda_i\}$ to (F_o+G_oK) and minimizes the given J. Using this control law, the nominal parameter closed-loop system becomes

$$x(k+1) = (F_o+G_oK)x(k) = D_ox(k); \; x(o) = x_o \tag{8}$$

The trajectory sensitivity vector $\sigma(k)$ is governed by the equation

$$\sigma(k+1) = (F_o+G_oK)\sigma(k)+(F_\mu+G_\mu K)x(k) = D_o\sigma(k)+D_\mu x(k); \; \sigma(o) = o \tag{9}$$

*Dept of Electrical Engineering, Indian Institute of Technology,New Delhi 110016, India.
**Dept of Electrical Engineering, Engineering College, Trichur 680009, Kerala, India.

where $F_\mu = \frac{\partial F}{\partial \mu}\Big|_{\mu_o}$; $G_\mu = \frac{\partial G}{\partial \mu}\Big|_{\mu_o}$

It can be shown that for the closed-loop system (eqn. 8) when augmented with the trajectory sensitivity equation (eqn. 9), the expected value of the optimum J is given by

$$\hat{J} = \frac{1}{2} \operatorname{tr} P_{11} \tag{10}$$

where $P = \begin{bmatrix} P_{11} & P_{12} \\ P_{12}^T & P_{22} \end{bmatrix}$ is the solution of discrete Lyapunov-type equation

$$\begin{bmatrix} D_o & 0 \\ D_\mu & D_o \end{bmatrix}^T P \begin{bmatrix} D_o & 0 \\ D_\mu & D_o \end{bmatrix} - P = - \begin{bmatrix} Q+MK+K^TM^T+K^TRK & 0 \\ 0 & S \end{bmatrix} \tag{11}$$

Simultaneous assignment of eigenvalues and eigenvectors has been attempted by many authors. Using the approach of Fahmy and O'Reilly[1], identify the class of feedback matrices $\mathcal{K} = \{ K : (F_o+G_oK)$ has the assigned eigenvalues$\}$ through parameterizing the allowable class of closed-loop eigenvectors. An optimum $K \in \mathcal{K}$ is then determined that minimizes the performance index $\hat{J}$. By this approach the feedback matrix K is obtained as

$$K = [w_{11} \; w_{21} \; \cdots \; w_{n1}] \, [S(\lambda_1)w_{11} \; S(\lambda_2)w_{21} \; \cdots \; S(\lambda_n)w_{n1}]^{-1}$$
$$= WV^{-1} \tag{12}$$

where w_{ij} are m-dimensional free parameter vectors and $S(\lambda_i) = (\lambda_i I_n - F_o)^{-1}G$.
For convenience, we define the parameter vector as

$$= [w_{11} \; \cdots \; w_{1m} \; \cdots \; w_{n1} \; \cdots \; w_{nm}]^T \tag{13}$$

Then for each value of the parameter vector α - for which det $V \neq 0$, there corresponds a feedback matrix $K = WV^{-1}$ which assigns the required spectrum to (F_o+G_oK). To obtain the optimal control, $\hat{J}$ is to be minimized with respect to the parameter vector α. In the minimization procedure, the necessary conditions of optimality are obtained by setting $\frac{\partial \hat{J}}{\partial \alpha}$ equal to zero. Using perturbation analysis analytical expressions are developed for $\hat{J}$ and $\frac{\partial \hat{J}}{\partial \alpha}$ so that a gradient technique such as Davidon-Fletcher-Powell algorithm[4] is used to determine K by direct minimization of $\hat{J}$. The gradient vector $\frac{\partial \hat{J}}{\partial \alpha}$ has been obtained as

$$\frac{\partial \hat{J}}{\partial \alpha} = [\theta_{11} \; \cdots \; \theta_{1m} \; \cdots \; \theta_{n1} \; \cdots \; \theta_{nm}]^T \tag{14}$$

where

$$\theta_{ij} = \operatorname{tr}[\{(G_o^TP_{11}D_oL_{11}+G_o^TP_{12}D_\mu L_{11}+\hat{R}KL_{11}+M^TL_{11})^T+L_{12}D_o^TP_{12}^TG_o$$
$$+(G_o^TP_{12}^TD_oL_{12}+G_\mu^TP_{11}D_oL_{12}+G_o^TP_{22}D_\mu L_{12}+G_\mu^TP_{12}D_\mu L_{12})^T$$
$$+(G_o^TP_{22}D_oL_{22}+G_\mu^TP_{12}D_oL_{22})^T\} R_{ij}]$$

where L_{11}, L_{12}, L_{12}^T and L_{22} are appropriate partitioning of matrix L and can be obtained by the solution of Lyapunov-type equation

$$\begin{bmatrix} D_o & 0 \\ D_\mu & D_o \end{bmatrix} \begin{bmatrix} L_{11} & L_{12} \\ L_{12}^T & L_{22} \end{bmatrix} \begin{bmatrix} D_o & 0 \\ D_\mu & D_o \end{bmatrix}^T - \begin{bmatrix} L_{11} & L_{12} \\ L_{12}^T & L_{22} \end{bmatrix} + \begin{bmatrix} x_o x_o^T & 0 \\ 0 & 0 \end{bmatrix} = 0$$

$$R_{ij} = \frac{\partial K}{\partial w_{ij}} = \frac{\partial W}{\partial w_{ij}} V^{-1} - K \frac{\partial V}{\partial w_{ij}} V^{-1}$$

Case Study: Design of a Sensitivity-reduced Controller for a BWR Plant

An experimentally validated linear mathematical model[2] of Big Rock Point BWR is considered here. The design procedure described in the earlier section is applied to this system to obtain a set of feedback gains for the controller, to give a response with reduced trajectory dispersion owing to changes in the parameter, fuel time constant from its nominal value.

The linearized system equation is given by

$$\dot{x}(t) = Ax(t)+Bu(t)$$

where A and B are the system and control matrices of appropriate order and x and u are deviation from steady-state of the system and control vectors respectively.

$$x^T = (\Delta N, \Delta C, \Delta T_F, \Delta Q, \Delta V_{in}, \Delta T_I, \Delta P, \Delta w_{FW}, \Delta T_D) \quad \text{and}$$
$$u^T = (\Delta POWP, \Delta A)$$

where N = neutron power (MW), C = delayed neutrons (MW), T_F = temperature of fuel (^oF), Q = heat transfer flow rate in the reactor core (MW), V_{in} = velocity of water at inlet to core (ft/sec), T_I = temperature of water at inlet to core (^oF), P = reactor pressure (psi), w_{FW} = feed water flow rate (lb/sec), T_D = temperature of water at inlet to plenum chamber (^oF), A = cross-sectional area of steam flow at the throttle valve, a control variable (in^2) and POWP = power applied to recirculating pump (ft-lb/sec).

The optimal feedback gains with and without sensitivity measure have been obtained. The closed-loop system with feedback gain matrix K*, designed without sensitivity measure is simulated for an impulse disturbance of 5^oF in inlet plenum chamber temperature. Also the closed-loop system with feedback gain matrix K_s^*, incorporating sensitivity considerations, is simulated for the same disturbance. The nominal and off-nominal responses for 20% change in the fuel time constant have been compared in the two cases. The deviation from the nominal response of the desensitized system is found to be smaller than that for the earlier case.

Conclusions

The design technique which combines the optimal control and pole placement is applied to multivariable discrete-time systems. The design freedom which remains after closed-loop pole allocation is used to minimize a given quadratic performance index which is of standard linear regular type, modified to include a quadratic sensitivity term. The class of all controllers is parameterized and expressions are derived for the gradients of the performance index with respect to the control parameter vectors. A minimizing algorithm is then developed to compute the optimal controller. The proposed design procedure has been illustrated by a case study of a BWR power plant.

References

/_1_7 Fahmy, M., and J. O'Reilly: "On eigenstructure assignment in linear multivariable systems", IEEE Trans. Automat. Contr., Vol. AC-27, 1982, pp. 690-693.

/_2_7 Girijashankar, P.V., G. Srikantiah, and M.A. Pai: "Optimal control of a boiling water nuclear reactor", Annals. Nucl. Energy, Vol. 3, 1976, pp. 531-538.

/_3_7 Gopal, M.: Modern Control System Theory, New Delhi : Wiley Eastern, 1984.

/_4_7 Horisberger, H.P., and P.R. Belanger: "Solution of the optimal constant output feedback problem by conjugate gradients", IEEE Trans. Automat. Contr., Vol. AC-19, 1974, pp. 434-435.

LIGHT WATER REACTOR PLANT STATUS MONITORING

Duřt, Z., Kracík, V.

INTRODUCTION

The efficient and safe operation of a nuclear power plant is the primary demand on the operating staff. To perform this task, the staff must track vast quantity of information about the status of various processes and sbusystems. The complexity of this job is continually increasing because of growing number of power plant performance and safety requirements.

Because it is known, that the operator is the key factor in achieving a high level of plant availability and safety, there are sponsored research and development projects to improve the operator-process interface. An exemple is the hierarchical analysis system, which uses computers to assist the operator in the prompt and precise recognition of abnormal phenomena. In this application the great number of plant status and alarm information is processed by a system to identify deviations which are significant for the mode of plant operation. By integrating the information from many signals, the system may provide timely and precise messages about the plant status.

There are two techniques used to a disturbance identification. In the first system each parameter is analysed every T seconds using current plant operating data. When sufficient conditions are satisfied the system gives message to the operator. This approach ensures a complete analysis in each step, but it takes progressively longer time in the case that the number of plant variables increases.

The hierarchical analysis system searches for a disturbance by means of testing on several levels. The search is on each level limited to only those parameters which are preliminarly chosen for that level. This approach is not so much sensitive to the size of the data basis or the number of variables. (Fig. 1.)

On the first level there is "operation without failures" possible. In this case the searching ends. On the third level the system can give a message "failure of an unknown part". Then the operation staff must complete the data base with an additional information.

MATHEMATICAL METHODS FOR PROCESSING OF A DATA BASE AND EVALUATION

For the hierarchical analysis system we have to process data base. Needed data set of monitored parameters we can obtain from operating plants or from models.

Let us consider only one level of hierarchical system, which has R possible states (classes). Each of those classes is described with M-dimensional vector $\mathbf{x}$ and we suppose that these vectors have M-dimensional Gaussian distribution $N(\mu_r, \sigma_r)$ with average value vector μ_r and dispersion matrix σ_r, $r = 1,2,\ldots,R$. Note that dimension M of a vector $\mathbf{x}$ must be generally greater or equal to (R-1). This requirement makes no diffeculties for large systems (for example a power plant). But there arises another problem with reduction of vast quantity of monitored parameters.
One of possible approaches is the transformation of M-dimensional vector $\mathbf{x}$ to N-dimensional space ($N < M$), where N is chosen in such a way to minimize the number of false messages. In the literature there are described several methods for this transformation. /1/.
Let us consider K chosen values of vector $\mathbf{x}$, signified as $\mathbf{x}_k$, k = 1, 2,...,K. Chien and Fu method uses transformation matrix built from eigenvectors of autocorrelation matrix

$$\mathcal{K}(\mathbf{x}) = \frac{1}{k} \sum_{k=1}^{K} \mathbf{x}_k \mathbf{x}_k^T \tag{1}$$

Transformation matrix is described by expression

$$\mathbf{B}(\mathbf{x}) = \begin{vmatrix} \mathbf{v}_1^T \\ \mathbf{v}_2^T \\ \vdots \\ \mathbf{v}_N^T \end{vmatrix} \tag{2}$$

where $\mathbf{v}_1^T \cdots\cdots \mathbf{v}_N^T$ are eigenvectors of matrix $\mathbf{\kappa}(\mathbf{x})$ corresponding to the first N greatest eigenvalues λ_i, $i = 1,2,\ldots,N$.
Method of variance ratio uses transformation matrix which minimizes probability of false decision, that is indirect proportionated to the ratio

$$\frac{\text{variance between classes}}{\text{variance within classes}}$$

The variance between classes is described with the aid of the expression

$$\mathbb{P}(\mathbf{x}) = \sum_{r=1}^{R} P(\omega_r)(\hat{\mu}_r - \hat{\mu})(\hat{\mu}_r - \hat{\mu})^T \tag{3}$$

where $P(\omega_r)$ is a priori probability of r-th class,

$$\hat{\mu}_r = \frac{1}{K_r} \sum_{k=1}^{K_r} \mathbf{x}_k \quad \text{and} \quad \hat{\mu} = \frac{1}{K} \sum_{k=1}^{K} \mathbf{x}_k$$

are average value vectors of the r-th class and of the whole data set respectively.
The variance within classes is defined with expression

$$\mathbf{\kappa}^1(\mathbf{x}) = \frac{1}{K} \sum_{r=1}^{R} \sum_{k=1}^{K_r} (\mathbf{x}_k - \hat{\mu}_r)(\mathbf{x}_k - \hat{\mu}_r)^T \tag{4}$$

The variance ratio is then determined by means of the trace of the matrix

$$\rho(\mathbf{x}) = tr\,((\mathbf{\kappa}^1(\mathbf{x}))^{-1}\,\mathbb{P}(\mathbf{x})) \tag{5}$$

Transformation matrix is described with expression (2), where $\mathbf{v}_1, \cdots \mathbf{v}_N$ are $\mathbf{\kappa}^1$ orthonormal eigenvectors of matrix $[(\mathbf{\kappa}^1(\mathbf{x}))^{-1}\mathbb{P}(\mathbf{x})]$ corresponding to the first N greatest eigenvalues λ_i, $i = 1,2,\ldots,N$.
As a result of both methods we obtain transformation matrix $\mathbb{B}$. Real plant signals are then transformated by means of that matrix and then there it is possible to compute the probabilities of existence for each class.

CONCLUSION

Both methods were processed in the form of programme for the computer RPP-16S. Needed data sets were simulated on the analog computer MEDA with a model of nuclear reactor of power plant VVER-440.
Results we have obtained, have shown that both methods are sensitive enough to discriminate the rated on parameters operation with noise and the disturbances.
Making comparison of these two methods, Chien and Fu method achieved more false messages, but it never gave the message "failure of an un-

known part". Method of variance ratio were more precise, but in some states it couldn't determinate the kind of failure. Therefore Chien and Fu method was recommended for higher levels of the hierarchical analysis system because of existing only limited number of possible states. Method of variance ratio is advisable for lower levels of the system because of preciseness and possibility to find out new failures not catalogised in the data set yet.

LITERATURE:

/1/ Kotek, Z., Brůha, L., Chalupa, V., Jelínek, J.: Adaptivní a učící se systémy. Praha, SNTL 1980.

PROGRAM SET FOR SYSTEMS ANALYSIS OF PERMISSIBLE LOAD CONSTRAINTS IN LARGE TURBOGENERATORS

Dr. Yu.G.Blavdzevitch, V.Y.German, Dr. V.V.Timokhin[1]

Abstract

On the base of a systems model which describes the electromagnetic & temperature fields in the elements & units of the large turbogenerators, the program set which makes it possible to carry out the analysis of the influence of the turbogenerator construction & operation condition on the characteristics of the physical processes determining the turbogenerator operation reliability & efficiency has been developed.

Key words: systems analysis, turbogenerator, permissible loads, program set, electomagnetic processes, heat processes

When operating the large turbogenerators(TG) at the power station, the operation conditions which are different from the rated ones are used. There is a problem of estimating the TG operation reliability under the specified operation conditions, i.e. the problem of determining the permissible loads. This load value is dependent of constraints on the current of the stator & rotor windings, the static stability for power, as well as the TG active zone heating temperature, especially that of the stator core end laminations under operation condition with power factor more than rated one & under the underexcited operation conditions.

In the Institute of Modelling Problems in Energetics of the Ukrainian Academy of Sciences, the permissible load calculation methods in terms of the program set for large turbogenerators with allowance for the above mentioned constraints have been developed & implemented. These methods include the electromagnetic & temperature calculations, take into account the magnetic circuit saturation along the main flux path, the magnetic flux variation depending on the operation condition through the air gap, the real geometry of the TG elements & units, the certain cooling system (radial, axial or combined one).

The base of these methods are the systems models of the electromagnetic & heat processes in large turbogenerators in the development of which by decomposition, we have obtained the mathematical models of physical processes in the different elements & units of the large turbogenerators which were called as the model modules. In a number of cases, to adjust these modules, the imitation modelling method has been used. In addition, the original method for calculating the stator core

1. Institute of Modelling Problems in Energetics, Ukrainian Academy of Sciences. 252057. Kiev, Brest-Litovski str., 102, USSR

end lamination which consists in the algorithmical combination of both heat & electromagnetic calculations with use of the results of TG heat tests called a method of calculation-experimental analysis of the physical processes in the large TGs has been developed. The structure flowchart of the systems model is shown in Fig.1.

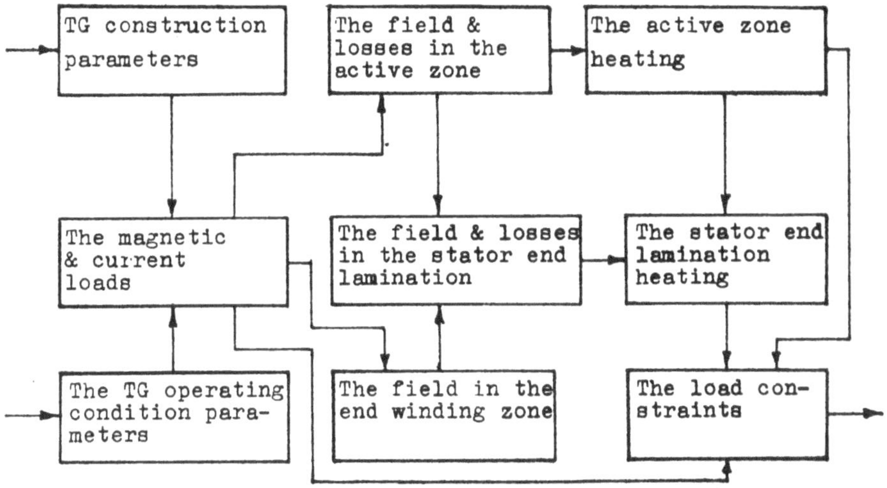

Fig.1 Structure flowchart of a turbogenerator systems model.

When implementing the systems model in terms of a program set, each model module generated the adequate program modules. In the course of operation, the job algorithm for numerical modelling is reduced by the control program which consists of the different program modules, service program and their interfacing.

The characteristic features of the program set with respect to the program modules are:

1. The independent development of the program (as well as model) modules for different subobjects (entity) & modelling methods.

2. The algorithmic interfacing of the program modules by the control program on the principle of "input-output".

3. The presence of the buffer storage which is used by different program modules & contains the data arrays .

4. An arrangement in which a certain program modules use the other program module results as well as variant calculation results or experimental data contained in the data array in the different forms.

The important peculiarity of the turbogenerator system model is that the subsystems models taking intermediate positions in the hierarchy of an model structure between the system model & model modules possess a certain degree of independence of one another. Therefore, their program implementation is performed in the form of a number of the application packages corresponding to the blocks in Fig.1. The control program of the whole system interacts with the package control programs through the program interfaces. Such an arrangement provides the important property of openness for the program set as a system and improves its adaptability to the manufacture & survivability.

The software system of the set & information paths in it are shown in Fig.2. The data flow, the command flow, the checking information paths & the subroutine transmission in composing the packages are indicated by the straight lines with ordinary arrows, the same lines with the double arrows, the wavy lines & the wide unblackened arrows respectively.

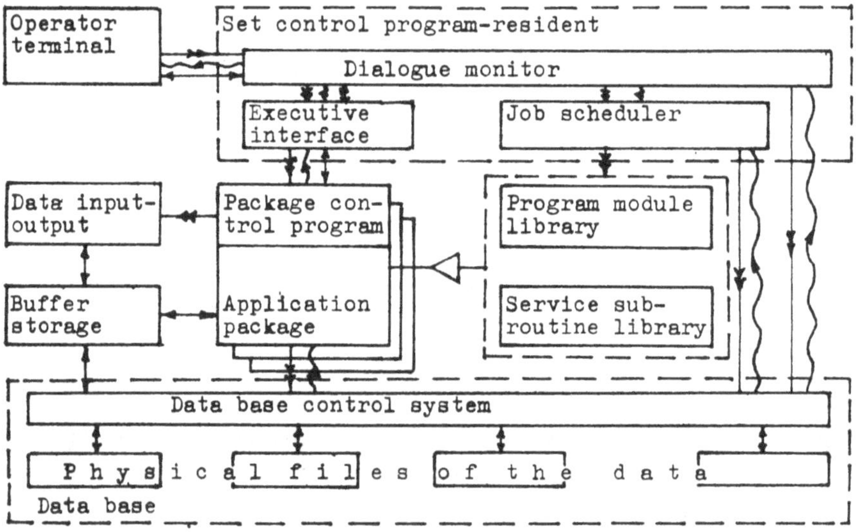

Fig.2 Set software stucture flowchart

The kernel of the software of the set is the control program-resident. The control program-resident calls: the dialogue monitor which carries out the operator interfacing through the terminal, the job scheduler which composes the call chain for the program modules or packages, the executive interface which directly calls either set components.

The buffer storage is designated for storage of the data array which is used in the specified job & provides for the logical independence of the data base. Its structure is two-level one: the common blocks in the working memory & the high-speed magnetic carrier files.

The data base possesses a certain degree of independence of the other set components & interacts only with the control program of the set & its packages. In this case, the control system plays a part of interface of these interactions. Due to the system structural construction, the data base may be readily expanded with minimum change number.

The mathematical model of the physical processes in TG includes the numerical methods of solving the boundary-value problems, the complicated equivalent circuits, the empirical dependencies. However, there are problems which demand the short-term analysis of the TG loads under operation providing the engineering calculation accuracy, maximum convenience for the user, the minimum machine time, the orientation to the mini-& microcomputers etc. These are regression models obtained by the above set that satisfied to these requirements. As an example, the relation between the temperature of the stator core end lamination & power factor ( ) has been found. This relation is shown in Fig.3 in comparison with the experemental data.

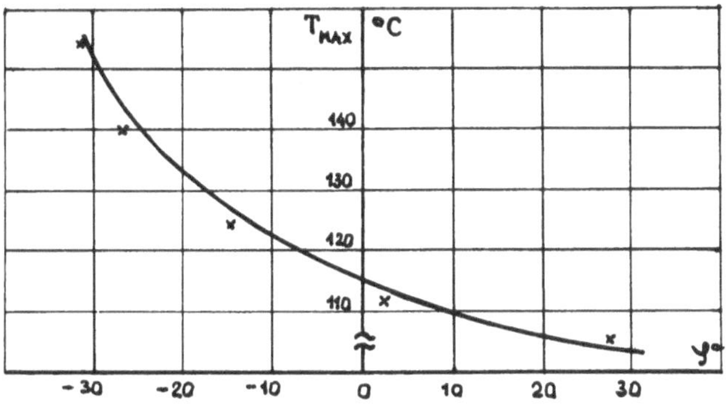

Fig.3 Relation between the stator core end lamination heating temperature & power factor

$$T_{max} = 15,06\varphi^{*} + 50,82\varphi^{*-1} + 48,1, \quad \varphi^{*} = 1 + 0,0156\varphi .$$

The positive values of φ correspond to the operation condition of generating the reactive power to the supply line, while the negative values correspond to the consumption of the reactive power from the supply line. The experiment & calculation data are different by not more than 7 %.

The program set developed has been effectively used for the study of the physical process parameters & the operation condition characteristics of large electrical machines in the designing & operation stage. This set is also oriented so as to be used as a subsystem in the information-modelling systems for the optimal engineering solution control in energetics.

STRATEGIE FOR ADJUSTMENT OF THE COMPLETE BOILER CONTROL SYSTEM IN POWER UNITS

H.W. Röder[1) G.S. Rösel[2)

Abstract. A method for obtaining optimal controller parameters for complex designed control systems is described and applied to the parameter setting-up in a power control system of a fossil fuel fired power unit. The plant model consists of nonlinear ordinary differential equations and as performance index to be minimized a generalized quadratic index is used. The nonlinear model is linearized for each combination of search parameters for which the performance index is determined by solving a Lyapunov equation. The search is carried out in discrete parameter space axis parallel. The computed optimal values were tested by simulation runs and were verified by application to the power unit.

Keywords. Optimisation, power station control, nonlinear model, Lyapunov-equation.

1. Introduction

The initiation of the designed automatic control system in the power station is an important practical problem. Besides the correct operation of the whole system the quality of controlled warrented process parameters is to be verified in the distinctive modes of plant operation. The warrented parameters concern several control loops, which are coupled by the technological plant. If each controller is adjusted separatly, the couplings will effect the parameter adjusting negatively. Improvement of dynamical qualities of one loop involves deterioration of dynamical qualities in the other loops. In this place the adjusting engineer starts an iterative process of alternating parameter changing and will finish his work if he gets a satisfying compromise using his practical experiance. For shortening the initiation plant simulation and optimisation calculations may be helpful.

2. Parameter adjustment using the performance index (1)

Advantageous for parameter setting-up in control loops is the formulation of dynamical quality as quadratic performance index (1).

$$I = \int_0^\infty (\underline{y}-\underline{y}_{ss})^T Q \, (\underline{y}-\underline{y}_{ss}) dt \qquad (1)$$

(1) is often applied to optimize the parameters as well as the structure of the control system [1],[2],[5],[11]. $\underline{y}$ includes all the process variables, whose dynamical qualities have to be warrented. The choise of all elements of Q depends on the individual practical problem and can not be stated generally. The fundmental relation, that by increasing q_i the output y_i will be taken more into consideration during the optimizing calculation, loses its significance with increasing dimension and multivariability of the process. Nevertheless practicable

1) VEB Bergmann Borsig (KKAB), DDR 1140 Berlin, Allee d. Kosmonauten 32
2) IIR der AdW der DDR, DDR 1199 Berlin, Rudower Chaussee 5

methods to determine Q may be discovered [6]. For the nonlinear and multivariable process (1) may be calculated directly, if $\underline{y}(t)$ is known. But the numerical integration of the nonlinear process model is necessary for it. Because this calculation is to be executed for each permissible parameter combination, this is too expensive in practice. Yet (1) may be determined in the analytical way proceeding from the nonlinear model (2)

$$\dot{\underline{x}} = \underline{f}(\underline{x},\underline{u},\underline{p},t) \quad ; \quad \underline{y} = \underline{g}(\underline{x},t) \tag{2}$$

with state vector $\underline{x}$, output vector $\underline{y}$, disturbance vector $\underline{u}$, adjusting parameter vector $\underline{p}$, to the linearised model (3), which is linearised for the topical combination of adjusting parameters $\underline{p}$.

$$\dot{\underline{x}} = A\underline{x} + B\underline{u} \quad ; \quad \underline{y} = C\underline{x} \tag{3}$$

The performance index (1) now may be written as (4)

$$I = \int_0^\infty (\underline{x}-\underline{x}_{ss})^T Q_x (\underline{x}-\underline{x}_{ss})dt \tag{4}$$

wherein

$$Q_x = C^T Q\, C \tag{5}$$

If the model is stable and the initial conditions vanish, then eq.(6) is valid 9 .

$$I = \underline{r}^T(0)\; P\; \underline{r}(0) \tag{6}$$

with

$$\underline{r}(t) = \underline{x}(t) - \underline{x}_{ss} \tag{7}$$

P is the solution of the Lyapunov equation (8)

$$P: \qquad A^T P + P\, A = C^T Q\, C \tag{8}$$

The solution P does not depend on either the input matrix B or the disturbances $\underline{u}$, but the value of I in (6) depends on $\underline{r}(0)$ as to which combination of disturbances the optimisation will be done. If optimizing takes place for each u_i separately, then it is not possible to conclude from eq.(9) to (6) in general, because of the nonlinear relation between u and I.

$$I_i = (A^{-1}b_i u_i)^T P\; (A^{-1}b_i u_i) \tag{9}$$

In contrast to the separate parameter optimisation of isolated control loops the setting-up by (9) possesses advantages nevertheless, since for valuation of adjustment use will be done not only of one output variable but also of the complete state or output vector respectively. The conclution from eq.(9) to eq.(6) may be done in two different ways. Either all the single disturbances are collected in $\underline{u}$ and eq.(9) is determined with this disturbance combination or the engineertechnical compromise is expressed by creation of a new performance index as a function of the I_i using weighting factors w_i (10)

$$I = \sum_i w_i I_i \tag{10}$$

A priori none of the both solution ways can be preferred.

3. Plant model

Models will already be used during design stage to calculate technological plant parameters, to curry out simulation studies and to design control systems. These models are to be modified yet, because the extend especially to describe dynamical processes in conection with control studies increases exeedingly. Suited models were develloped and described in [3],[4],[5],[7] and [8]. Notice, that with a view to parameter adjustment of the master control system the different states of controls (e.g. manual/automatic or setpoint extern/intern) have to take into consideration in modelling especially.

4. Setting-up strategy for control plant parameters

A feasible strategy for setting-up control parameters has to ensure, that in accordance with the proceedings of plant mounting some control loops are already ready to operate, while others do not work yet and this state is changed until finally the automatic plant is installed completely. This is to regard in the model and in the optimizing.
The numerical device for calculation of setting-up parameters consists of the following major parts:

- calculation of the nonlinear model (2)
- determination of linear models (3)
- search algorithm and performance index calculation (9),(10)
- simulation of nonlinear model using optimized parameters $\underline{p}$.

The linearized model matrices A, B and C are generated by numerical linearizing. The Lyapunov equation (8) is solved by the method of Smith [10]. As pretended components of $\underline{u}$ all contractual arranged transitions of disturbances or setpoints are used. The weighting matrix Q is a diagonally one. It is determined by a sensitivity study. Main idea is to get with each element of Q a significant influence on I. For this reason a fixed adjustment is used and each element of Q is changed about some decades. For each variation of q_{ii} the index I_i is calculated for all single u_i. In analysing the results the value of q_{ii} is fixed.
The evaluation of I and the determination of improved parameters $\underline{p}$ is realised by a search algorithm developed for this purpose. The search is carried out in the discrete parameter space axis parallel. The discrete approach offers the benefit, that the search space may be subdivided eligible and the parameter may be calculated just as precise as they are applied later in the plant. The investigated optimal values are examined by nonlinear model simulation before they were adjusted to the plant.

5. Application of the setting-up strategy

The strategy described above was applied to a fossil fuel fired 500 MW power unit. The model (3) consists of 34 state-, 8 output- and 15 input variables. As inputs, at first, only setpionts were regarded, thus opti misation was carried out as to setpiont changes. The investigations were done with 6 controllers and 5 disturbance variables. As start combination of search all controller gains were set to 1.0. Q was determined at first by feeling as (11)

$$Q = \text{diag} \left[2. \quad 200. \quad 1. \quad 8. \quad 0.1 \quad 7.5 \quad 2.5 \right] \tag{11}$$

and as the result of the snnsitivity study Q was defined anew as (12)

$$Q = \text{diag} \left[0.2 \quad 20. \quad 1. \quad 50. \quad 20. \quad 0.1 \quad 80. \quad 30. \right] \tag{12}$$

The weighting factors w_i were determined to equalize the I_i for the start combination of sarch parameters (13)

$$\underline{w} = \left[10.2 \quad 1.6 \quad 3.6 \quad 1.3 \quad 2.3 \right] \tag{13}$$

The setting-up procedure was carried out in two steps. At first, the parameter range from 0.1 to 20. was covered by a net of 10 values. In the second step the range between near-optimal points was used. The calculated optimal parameter set $\underline{p}$ was (14)

$$\underline{p} = \left[3.35 \quad 0.4 \quad 3.74 \quad 5.04 \quad 5.4 \quad 3.78 \right] \tag{14}$$

Simulation runs proved, that the control loops, which were optimized to setpoint changes, also produce favourable system reaction in overcoming disturbances. The calculated set-up parameters were verified by acceptance into the power plant and proved successful to date.

6. References

[1] Dettinger, R., E. Welfonder:"Ermöglichung viel steilerer Leistungsgradienten durch strukturoptimale...". BWK 29 (1977) 2, 33-39.
[2] McDonald, J.P., H.G. Kwatny, J.H. Spare:"A nonlinear model for reheat-boiler-turbine-generator-systems". JACC 23 (1971), 219-236.
[3] McDonald, J.P., H.G. Kwatny:"Design and analysis of B-T-G controls using optmal theory". IEEE Tr. Aut. Contr. AC-18 (1973), 202-208.
[4] Röder, H.W., G.S. Rösel:"Theoretische Prozessanalyse eines Dampferzeugers". 3. Tagung Algor. Prozessanalyse, Dresden 1978.
[5] Röder, H.W., G.S. Rösel:"Modellbildung und Entwurf von Mehrgrössenregelungen für Dampferzeuger". 24. IWK, TH Ilmenau (1979), 115-121.
[6] Röder, H.W., G.S. Rösel:"Entwurf opt. Regelungen mittels Strategie zur Festlegung der Bewertungsmatrizen". MSR 22 (1979) 9, 207-212.
[7] Röder, H.W., G.S. Rösel:"Modellbildung und Entwurf von Mehrgrößenregelungen für Dampferzeuger". Dissertation A, 1980, (nur über die Autoren erhältlich).
[8] Röder, H.W., G.S. Rösel:"Anwendung von 6 Entwurfsverfahren für Mehrgrößenregler auf ein DE-Modell". MSR 25 (1982) 11, 494-499.
[9] Schwarz, H.:"Mehrfachregelungen". Bd.2, Springer Verlag, 1971.
[10] Smith, R.A.:"Matrix Equation XA+BX=C". SIAM J. 16 (;968) 5, 198-201.
[11] Tysso, A., J.Ch. Brembo, K. Lind:"The design of a multivariable control system for a ship boiler". Automatica 12 (1976) 4, 211-224.

MODELING OF PHYSIOLOGICAL PROCESSES BY THE 'CONTROLLED COMPARTMENT' - APPROACH

F. Breitenecker, Technical University Vienna*

For modeling and simulating physiological processes dynamically in the time domain three methods are available: curve fitting (fitting measured time series), concept fitting (using concepts readily amenable to mathematical analysis, for instance feedback control mechanism) and compartmental modeling. The last method gives maximum insight into the behaviour of the process, but very often the model parameters cannot be determined because of only few measured data. In this paper an approach as compromise between concept fitting based on control theory and compartmental modeling using bottom- up techniques for modeling is proposed to build up models with few parameters and sufficient accuracy, the so- called 'controlled compartment'- approach. The advantages of this approach are shown in modeling heart rate regulation during exercise, in modeling pathological blood pressure behaviour in patients with a stenosis, in modeling glucose production rate in rat livers and in modeling blood glucose concentration in diabetic men during stress.

1. Methods for modeling physiological processes

Mathematical models for physiological processes are formulated from a priori knowledge of physiological processes, from current experimental data from a combination of both. Naturally, the greater the level of a priori information inclused in the model, the greater is the explanatory and predictive power of the model.
Following /4/ three levels of mathematical representations have been successfully used for modeling physiologicall processes: curve fitting, concept fitting and compartmental modeling.
Curve fitting is an overhauled method, where relatively simple mathematical formulas (polynominals, (sum of) exponential functions,..) but also nonlinear differential equations are used for fitting the measured data (time series). This process corresponds to 'black- box'- modeling, that means it is unknown whether the model structure (the form of the mathematical formulas and equations) will have any simple relation to the physiological structure of the process and whether the model parameters have any physiological importance (relation to physiological parameters).
Because of these disadvantages this method should be used with care and has, in general, to be seen an overhauled method. Nevertheless, useful informations about physiological parameters of special processes can be estimated by the model parameters using special (interpolation) formulas (for instance, for plasma disappearance of labeled unconjugated bilirubin, /4/).
Another approach for modeling physiological processes is concept fitting which involves describing systems in terms of concepts which are readily amenable to mathematical analysis. For static and quasi- static behaviour there statistical analysis (parametric and nonparametric) between measured data, physiological parameters, etc. is a very often used successfull mathematical concept and tool.
For dynamic behaviour for instance control theory can be used. There corresponding to known inputs (excitations,...) and known outputs and states (measured time series) a feedback control mechanism is build up (realisation problem of control theory).

* Technical University Vienna, Karlsplatz 13, A- 1040 Vienna

For instance, metabolic processes can be described in such a way, even though the regulatory processes involved are known to be much more complex than this simplifying concept would suggest (/4/). An advantage of this approach is that many of the assumptions made lead usually to simple models, in general linear control systems. In this case a lot of mathematical knowledge can be used now: minimal realisations, observers for uncertain or missing time series, stability considerations, etc. But on the other hand the 'linearity' of the physiological process is associated with only small fluctuations in the variables of the system.

The third approach, compartmental modeling (intrinsic biomathematics) gives maximum insight into organisation and behaviour of complex processes. Wherever possible, this approach involves developing mathematical representations of unit- process dynamics- that is, representations of each of the subprocesses giving submodels, the so- called 'compartments' which are usually mathematically linked by flow or bilance equations. Thus, the model reflects the structure of the process and the model parameters especially those characterising the linking of the compartments have a physiological meaning. Obviously, there are practical limitations: on the one hand it may not be possible to develop a very large system model entirely in terms of unit processes (compartments) because of the inordinate complexity of the system, on the other hand a very large number of model parameters has to be identified- a problem which often cannot be solved because of only few measured different time series.

2. The 'controlled compartment'- approach

Considering the modeling approaches 'concept fitting with nonlinear control theory', and 'compartmental modeling' then both have 'natural' advantages. Compartmental modeling reflects the structure of the process to be modelled, consequently each state (store or auxiliary store in a compartment) and most of the model parameters have a physiological meaning. Control systems (linear and nonlinar one) work with the perhaps most natural principle, the feedback principle. A combination of both approaches, that means starting with compartmental modeling and using nonlinear control systems for characterising the dynamics within a compartment, would be a very accurate and the most natural approach. The disadvantages of both methods are the amount of states and model parameters and some uncertainities in large scale nonlinear control systems. That yields to the fact, that the model parameters cannot be identified, especially for specific and individual time series. Consequently these models are sufficient for modeling the qualitative behaviour of a very large number of natural states (compartments), but they are insufficient to model the quantitative behaviour of only few states under specified excitations. To overcome the fore- mentioned disadvantages of the usually advantageous (combination-) approach it is now proposed to use instead of the top- down modeling technique a bottom-up technique: instead of building up a large scale model with a combination of compartmental modeling and (nonlinear) control theory and reducing the number of states and parameters (by cancelling terms which seem to be neglectable) in order to get an identifiable number of parameters the modeling procedure starts with only few compartments with a dynamical description in terms of control theory. If the resulting model is not accurate enough, the number of states (compartments) has to be increased successively. This modeling technique, the so- calld 'controlled- compartment' approach, consist of the following steps:

(1) decomposing the process into as few compartments as possible- corresponding to the number of measured time series (outputs) available and to the number of excitations (inputs)
(2) modeling the dynamics within the compartment with (nonlinear) control theory
(3) modeling the interactions between the compartments by input- output relations
(4) identification of the parameters due to individual time series (test series)
(5) validation of the results and of the model itself: is the model accurate enough for the time series? -have the states and model parameters 'enough' physiological meaning? -is the model able to describe the qualitative behaviour under modified inputs?
(6) if validation fails,restart with (1) or correct dynamics and states in the compartments (restart with (2)) or modify the input- output relations (restart with (3))

It is to be noted that the resulting compartments are 'generalised' one, they are more than the state of an idealised store (/4/, /6/) sometimes they may characterise fictive states too.
The advantages of the outlined approach are obvious. The 'controlled compartment'-approach uses a well choosen compromise between compartmental modeling and control theory based on bottom- up modeling technique due to the aim (depending on excitations (inputs) and time series available) of the considerations.

3. Applications

In the following the advantages of the outlined approach for modeling physiological processes are shown. Simulation and identification of the models were performed using the hybrid simulation language HYBSYS (/5/). The CSSL- standard- type language HYBSYS provides model declaration and problem investigation in an efficient and hardware- independent high- level language. If analog (parallel) integration is choosen as integration algorithm HYBSYS performs automatically the problem set- up using an autopatch system and automatic scaling. Two reasons make analog integration interesting for the models under consideration: points of discontinuity and non- smoothness (bounds, switchings) cause no problems for integration, the time for a simulation run is independent from the model- an ideal fact for identification by optimisation procedures.
The first process to be modelled is the relation physical load- heart rate with physical load (bicycle test) and heart rate as measured time series (/2/). The process is decomposed into the feedforward- compartment (reaction of the fast nerval component on the quantitative load indicator $u_1(t)$) modelled by a first order lag (giving $x_1(t)$ as output fig. 1a) and into a feedback- compartment (reaction of the slow parasymphatic component on the quantitative load $u_2(t)$ and the nerval component $x_1(t)$) modelled by a first or second order lag (giving $x_2(t)$ or $x_3(t)$ as output, fig. 1b, 1c). The overall output p(t) is given by $p=x_1+x_{2,3}+p_R$, where p_R is the heart rate at rest. The model parameters (gain factors and time constants of the control loops) can be interpreted physiologically, interesting relations between these model parameters and the physiological parameters age, weight, height, expectation for maximal heart rate and for maximal performance can be shown. Due to the aim of the model (modelling the relation physical load- heart rate) the model can be seen validated. It is to be noted, that the compartments used characterise fictive pathological states.
The second process is a pathological blood pressure behaviour in patients with a stenosis.

A special kind of congenital stenosis localised below the left subclavian artery, the so- called coarctation of aorta, causes a pathological blood pressure difference ('gradient') between upper and lower extremities at rest, during and after load. The aim was to build up a model for this pathological behaviour after a treadmill test, where this gradient oscillates. Available were measured blood pressure in brachial artery (upper extremities) and in posterior tibial artery (lower extremities) starting at the end of the load until achieving the resting pressure (/1/,/3/). This process is decomposed into a 'brachial compartment' (characterising the nonpathological blood pressure behaviour in the brachial artery, fig. 2-1) modelled by a first order lag and a 'tibial compartment' (characterising the pathological oscillation in the posterior tibial artery, fig.2-2) modelled by a first and a second order lag. The output of the bracchial compartment is the blood pressure in bracchial artery m(t) (time series available); the output of the tibial compartment is the blood pressure in the posterior tibial artery p(t) (time series available) with the internal states q(t) (fictive nonpathological blood pressure behaviour) and v(t) (pathological oscillation). A third compartment, the gradient compartment (fig. 2-3) calculates the gradient. The time constants and gain factors characterise the nonpathological transient of blood pressure, while amplitude, frequency and damping factor of the oscillation characterise the pathological blood pressure behaviour (/1/). These model parameters not only have a specific physiological meaning, they allow to correct assumptions on the physiological parameters used for characterising the desease (gradient at rest, gradient immediately after load, time for achieving resting gradient). It turns out, that the gradient after load has no signifigance and has to be replaced by the amplitude of the pathological oscillation which can be measured, too (/3/). Nonparametric statistical analysis (cluster analysis) shows that the gradient at rest is the most important physiological parameter: it allows to predict the parameters of the pathological oscillation (/3/). Consequently the degree of obstruction caused by the stenosis can be classified (roughly) without performing the treadmill test. It is obvious, that this model is more than validated. Furthermore it is to be noted that nonlinearities are involved by bounds on the blood pressure.

The third process is the glucagon- stimulated hepatic glucose production of an isolated rat liver. Available measured time series are glucose production and a so- called 'messenger' signal (cAMP- release) characterising the changed glucagon stimulus before 'entering' the liver. The process is decomposed into a 'cAMP- compartment' (characterising the reaction of the messenger on the glucagon stimulus, fig. 3-1) modelled by a nonlinear relation between infused and utilised glucagon dose (fig. 3-1a), by a first order lag with differentiating element (fig. 3-1b) and by a limiter (fig. 3-1c) and into a 'glucose compartment' (characterising the reaction of glucose production on the messenger, fig. 3-2) modelled by a first order lag (fig. 3-2b) extended by nonlinear feedback and gain (fig. 3-2a, 3-2c).

Again it can be shown that the model parameters, time constants and gain factors of the control loops, have physiological meaning (/6/), the model can be seen validated, too (/6/).

An interesting fact has to be noted: in the first stage of modeling this process only time series for glucose production were used; one of the artificial internal states used in the modeling procedure turned out to be of physiological importance- it was nothing more than the 'messenger' for the glucagon where time series were available, too (comp. /7/).

The fourth process is the blood glucose concentration in diabetic men. The modeling procedure starts with the fore- mentioned 'rat- liver- model'; compartments characterising the neoglucogenesis and the glucose utilisation are to be added. The model is tested up to now with efficient results (comp. /7/).

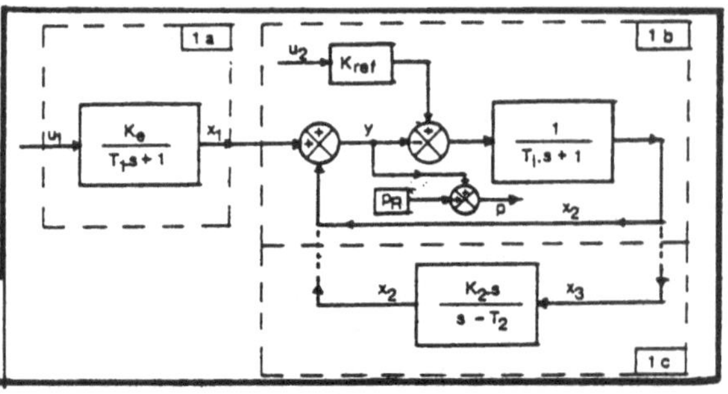

Fig.1: Model for relation load - heart rate

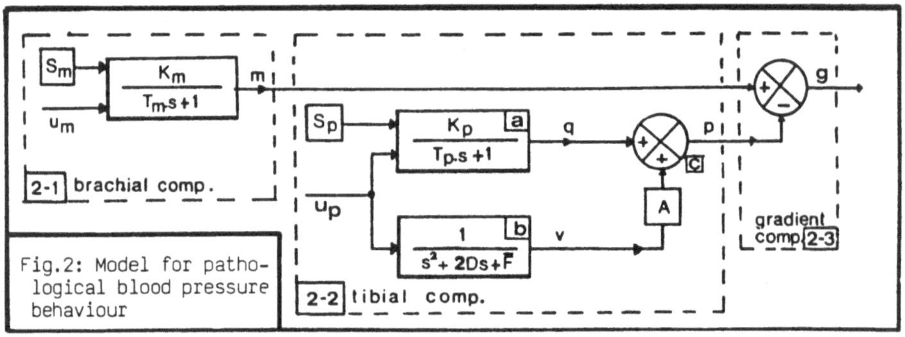

Fig.2: Model for pathological blood pressure behaviour

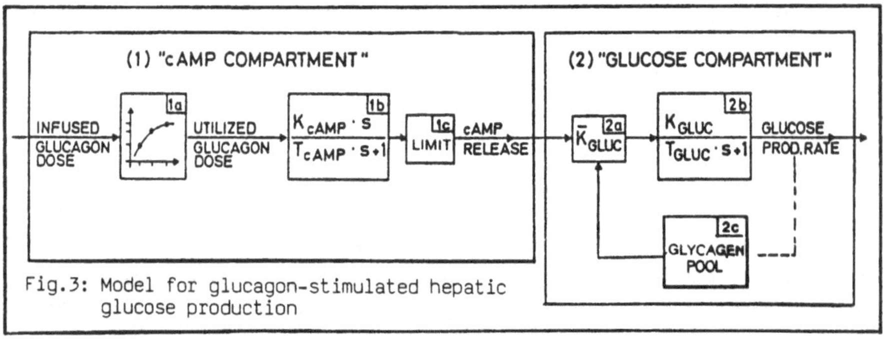

Fig.3: Model for glucagon-stimulated hepatic glucose production

/1/ Breitenecker F., Kaliman J.: Simulation and analysis of pathological blood pressure behaviour after treadmill test in patients with coarctation of the aorta. Cybernetics and System Research 2, North Holland, Amsterdam (1984), pp 423.

/2/ Breitenecker F., Kaliman J., Reisner G.: Modellbildung und Simulation der Herzfrequenz unter Belastung. Informatik- Fachbericht 85, Springer, Berlin (1984).

/3/ Breitenecker F., Kaliman J., Ettl W.: Results of modeling and simulating pathological blood pressure behaviour in patients with coarctation of the aorta. Proc. Int. IMACS Conf. 'European Simulation Meeting on Simulation in Research and Development', August 27-30, Eger, Hungary; North Holland, to appear 1985.

/4/ Carson E.R., Jones E.A.: Use of Kinetic analysis and mathematical modeling in the study of metabolic pathway in vivo. The New England Journal of Medicine (1979), pp 1016.

/5/ Solar D., Berger F., Blauensteiner A.: HYBSYS- interactive simulation software for a hybrid multiple- user system. Informatik- Fachbericht 56, Springer, pp. 234.

/6/ Troch I., Breitenecker F., Gampe J. et al: Modeling by simulation of hepatic glucose production in vitro. Proc., see /3/.

/7/ Troch I.: Modeling and /for/ by simulation. These Proceedings.

SIMULATION OF THREE-DIMENSIONAL CELL GROWTH WITH APPLICATION TO OPTIMAL TUMOR TREATMENT

W. Düchting* and Th. Vogelsaenger

SUMMARY: In this paper a new approach is made to use methods of systems analysis, control theory, automata theory and computer science for solving problems in the field of biology and medicine. The main goal of this contribution is to simulate different kinds of cancer treatment (surgery, radiation therapy, chemotherapy) and to forecast the optimum method and time of treatment prior to clinical therapy. An illustrative example is given showing the simulation of tumor treatment by surgery.

Keywords: Simulation, modelling, biology, medicine, cancer, cell growth

1. GENERAL BACKGROUND

From the background of mathematics cancer models have been developed describing the complex process of cell growth (4). In addition the authors (2) apply methods of control theory, automata theory and computer science to the cancer problem. In earlier papers (references are cited in (1)) it was possible to develop a closed-loop control circuit for normal and abnormal (malignant) cell growth and to interpret tumor diseases as structurally unstable control circuits. The simulation results led to characteristic curves representing the number of cells as a function of time.

2. TASK

Cancer therapy requires additional instructions on the exact spatial position of a tumor in a tissue. Therefore the task in this paper is to construct a model describing the three-dimensional growth of a tumor cell in a vascularized tissue segment and to simulated different kinds of tumor treatment.

3. LIMITATIONS

Starting with the difficult task of modelling cell growth some oversimplified assumptions must be made: (i) There is only a limited 40x40x40 cell space available which is equivalent to a 1 mm^3 tissue volume or to about 60 000 cells. (ii) The construction of the blood vessels is assumed to be static. (iii) The influence of metastases, side effects, immunological reactions and drug resistance is neglected.

*Department of Electrical Engineering, University of Siegen, Hölderlinstr. 3, D-5900 Siegen, West-Germany

4. DESIGN PHASE

It is impossible at present to find a complete overall solution describing the complex biological phenomenon of cell growth. Therefore, the problem is subdivided into several parts each of which can be solved separately

4.1 Cell-cycle models

The cell cycle is defined to be the time in the life of a cell from one mitotic division until the end of the next mitotic division. It consists of distinct phases (G1, S, G2, M). The noncycling cells (GØ, N) are called quiescent cells. Figure 1 represents a block diagram describing the cell-cycle process of a tumor cell. It should be stressed that a normal cell has a limited ability of multiplication. In this paper it is transferred to a noncycling end-state after 4 divisions. This condition does not hold true for tumor growth because a tumor cell is theoretically able to divide indefinitely. The relevant data used in this model refer to cell kinetic literature (3). They are compiled in TABLE 1. The phase durations are assumed to be normally distributed with an appropriate standard deviation.

4.2 Cell production rules

The cell-cell interactions are described by a list of cell production rules which govern the regeneration of normal cells and the multiplication of tumor cells.

Normal Cells: Before entering into mitosis each active (proliferating) normal cell checks if there is a free position in the corresponding row or column. If so, the division takes place and the newly created daughter cell shifts all cells one position further to the nearest empty position.

Tumor Cells: A tumor cell is able to divide even if there is no empty space available for the positioning of a daughter cell. But the division is possible only if the distance between a dividing tumor cell and the blood vessels (capillaries) is less than 3 cell layers. All tumor cell residing in a distance of more than 3 cell layers enter the resting phase "GØ" because of lacking oxygen and nutrient supply.

5. ILLUSTRATIVE SIMULATION EXAMPLE

At T=0 time units 11 tumor cells have been positioned into the center of a tissue segment near capillaries. They gradually grow to the tumor shown in Figure 2. In a subsequent simulation run a surgical removal of the right half of the tumor shall be performed at T=201 time units (Figure 3(a)). In a very short period the stem cell pool delivers normal cells into the right region (s. cross section of Figure 3(b)). The capillaries removed as well cannot be replaced in this short period. Therefore the tumor cells shifted into this space are transferred to the stage of necrosis very quickly. Further treatment is necessary to kill the viable tumor cells which are dominantly concentrated in the left half of the tissue segment (Figure 3). It can be concluded that it seems possible to predict the optimal method and time of tumor treatment by computer simulation prior to clinical therapy. At present we are trying to stimulate tumor biologists to start sophisticated <u>in-vitro</u> experiments for validating our <u>in-vivo</u> simulation results

6. REFERENCES

(1) Düchting, W.: Modeling of cancer problems and its progress. In: "Computer Modeling of Complex Biological Systems", S.S. Iyengar (ed.), CRC Press, Boca Raton, Florida (1984a), 55-75.

(2) Düchting, W. and Th. Vogelsaenger: The present status of simulating three-dimensional tumor growth and treatment. In: "Proceedings of the 1984 UKSC Conference on Computer Simulation", D.J. Murray-Smith (ed.), Butterworths, London (1984b), 219-227.

(3) Lloyd, D., Poole, R.K. and St.W. Edwards: The Cell Cycle Division. Academic Press, London 1982.

(4) Segel, L.A.: Modelling dynamic phenomena in molecular and cellular biology. Cambridge University Press, Cambridge 1984.

TABLE 1. Cell-phase durations in hours ($T_C=T_{G1}+T_S+T_{G2}+T_M$).

PHASE DURATION IN HOURS	T_{G1}	σ_{G1}	T_S	σ_S	T_{G2}	σ_{G2}	T_M	σ_M	T_C	$T_{G\emptyset}$	$\sigma_{G\emptyset}$	T_N	σ_N
NORMAL CELL	10	2	8	1	4	1	1	0	23	24	8	16	2
TUMOR CELL	4	1	4	1	1	1	1	0	10	5	2	40	4

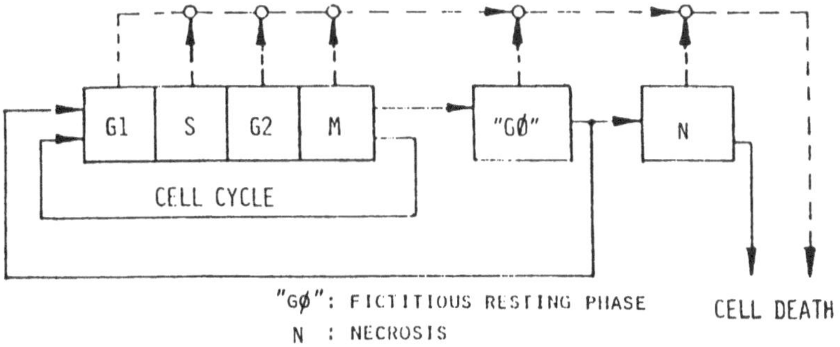

Fig. 1 Cell-cycle model of a tumor cell.

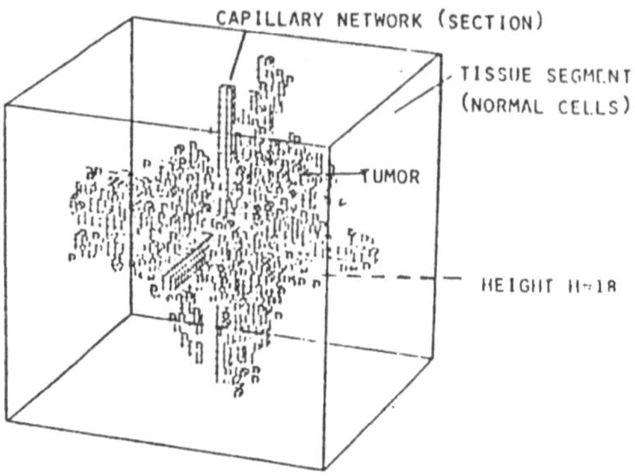

Fig. 2. Proliferation of 11 tumor cells in a vascularized tissue segment.

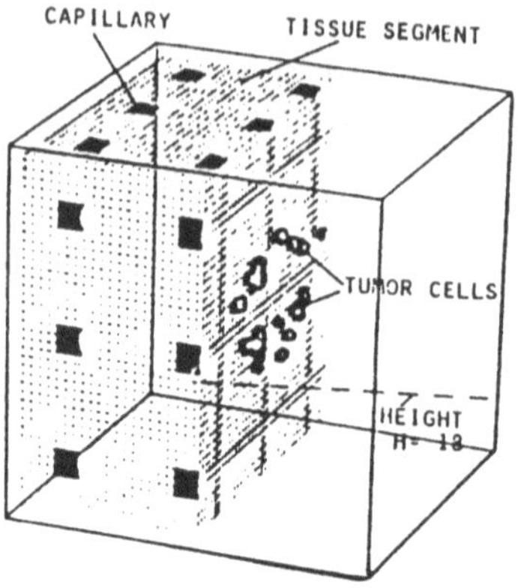

Fig. 3(a). Surgical removal of the right half of the tissue segment at T=201 time units.

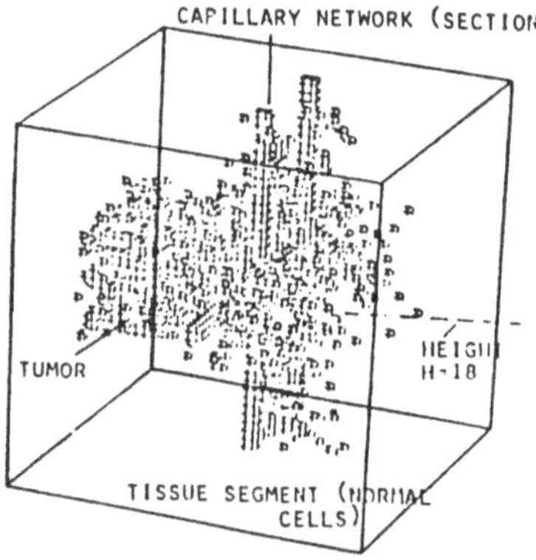

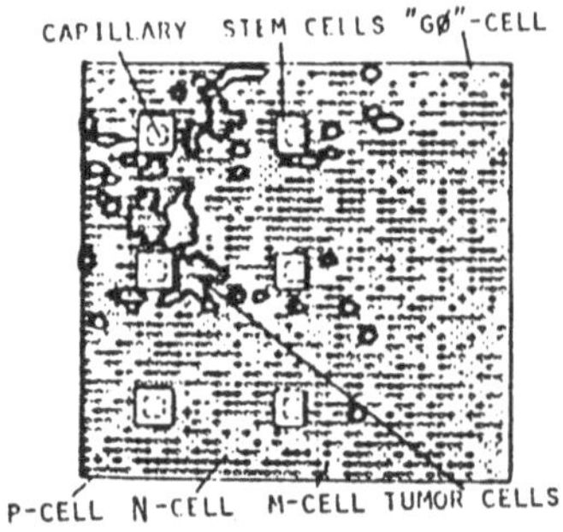

CROSS SECTION

Fig. 3(b). Tumor configuration at T= 500 time units.

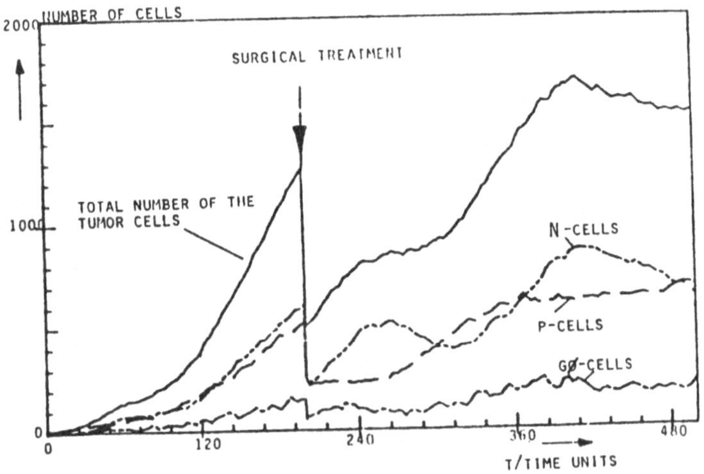

Fig. 3(c). Number of tumor cells as a function of time.

Fig. 3(a)-(c). Simulation of a surgical treatment.

COMPUTER MODEL OF RESPIRATION

Hieronymi, U. and H. Mrochen 1)

A computer model for simulation of human respiration has been implemented in a desktop computer (HP 9845B) using BASIC. Artificial ventilation as well as spontaneous breathing can be simulated under different conditions with varying input data. Respiration and circulation are characterized by commonly used respiratory, metabolic, and haemodynamic parameters. The model is intended to be used for the simulation of states of acute respiratory insufficiency and the respective consequences of variations of respiratory therapy.

Keywords: Computer model, human respiration, respiratory insufficiency

In human respiration many complex processes are involved. The oxygen uptake from the environment, the expiration of carbon dioxide, the ventilation of the lung and its control, the gas exchange with the blood, and the distribution of gases to the body are nonlinear processes described by numerous variables.

The result of simultaneous changes of several parameters (e.g. airway pressure, tidal volume, respiratory rate) is difficult to estimate or to predict, especially in the case of severely disturbed lung function. Such evaluation, however, turns out to be an essential problem in intensive care medicine in the treatment of acute respiratory insufficiency, an often met serious complication with high mortality. An important part of therapy of this life-threatening disturbance is artificial ventilation. To cause an optimum effect the artificial ventilation has to be flexibly adapted to the special needs of the respective patient and his varying state of illness. After assessing the actual physiologic state of the patient by a certain set of measured and calculated parameters, corresponding ventilator settings have to be found to maintain and restore the vital function respiration.

Though urgently needed, up to now only few aids of limited use are available for finding optimum ventilator settings and for facilitating this difficult task. One way to meet this challenge should be simulation. But computer simulation can only be used for assistance in the interpretation of changes in this complex physiologic system, if a suitable mathematical model of respiration is available. In addition to the reflection of normal physiologic reactions the model is required to allow for simulation of both states of acute respiratory insufficiency and artificial ventilation and to give results within an acceptable time when implemented in a relatively small and inexpensive digital computer. According to the last condition neither the type of model of Grodins [3] nor the modelling approach of Farrell

1) Research Department for Intensive Medicine, Friedrichshain Hospital, Leninallee 49, DDR-1017 Berlin, GDR

and Siegel [2] are practicable. Grodins' model consists of a system of differential-difference equations with variable time delays, Farrell and Siegel who were the first to "combine simulation studies and clinical data on patients to obtain a better interpretation of observed cardiorespiratory changes" [2] used a Continous System Modeling Program.

Dickinson [1] introduced a comprehensive computer model of human respiration which was originally designed as a self-instructional device for educational purposes. The model essentially consists of a compartmental model of lung function, which is connected by circulation with a tissue pool consuming oxygen and producing carbon dioxide. The pulmonary function is realized by the interaction of three idealized compartments representing dead space (ventilated but unperfused part of the lung), shunt (perfused part of the lung without ventilation), and ideal oxygenation (part of the lung with proper relation between ventilation and perfusion). Known algorithms from literature (e.g. [5],[6]) are used for the conversion of blood gas tensions into contents and reverse and for the calculation of the acid-base state in the blood pools. Empirically derived functions describe the influence of lung perfusion, rate of airflow, alveolar oxygen tension and (if used) positive end-expiratory pressure on effective dead space and venous admixture. Thus the model gives a very wide representation of the processes of respiration. The possibility of artificial ventilation is provided. Generally used respiratory, metabolic, and haemodynamic parameters form the input and output signals for the characterization of respiration and circulation. The potential clinical value of this model has already been shown [4]. The published FORTRAN version has been implemented in relatively small digital computers [1].

This encouraged us to select Dickinson s model as a starting point in the search for the possible use of computer simulation in connection with the described clinical problems. We translated the program into BASIC and successfully adapted it for the use on a desktop computer Hewlett-Packard 9845B. It requires about 160 KByte of memory. The dialog has been enhanced. The quasi graphic output shows the changes of respiratory rate (F), minute ventilation (V), arterial oxygen tension (O), and arterial carbon dioxide tension (C) during a selectable time interval displaying the choosen iteration intervals of maximum 10 s to minimum 2 s simulated time (see figure 1).

```
(kPa)(0)            (4)            (8)           (12)            (16)
      0        20        40        60        80        100       120
min.sec         .    .    .    .    .    .    .    .    .    .    .
  3.10.        VF       C     O
  3.20.        VF       C        O
  3.30.        VF       C           O
  3.40.        VF        C            O
  3.50.        VF        C             O
  4. 0.        VF       C               O
  4.10.        VF       C                O
  4.20.        VF       C                 O
  4.30.        VF        C                O
  4.40.        VF        C                O
  4.50.        VF        C                O
  5. 0.        VF        C                O
  5.10.        VF        C                O
  5.20.        VF        C                O
  5.30.        VF        C                 O
  5.40.        VF        C                 O
  5.50.        VF        C                 O
  6. 0.        VF        C                 O

 Final values for this run were :

 arterial PO2 =    9.90 kPa ( 74.2 mmHg); O2 content =  19.1 Vol%
 art. O2 Sat. =   95.0  %
 arterial PCO2=    5.22 kPa ( 39.1 mmHg); CO2 content=  47.1 Vol%
 arterial pH  =    7.40     ( 40 nM); arterial HCO3- =  23.6 mmol/l

 respiratory rate =    20.0 /min;  tidal vol.   =   900   ml
 total ventilation=    18.0 l/min; actual CO    =     6.2 l/min
 tot. dead space  =   603   ml; act. ven. admxt.=    20.0 %
```

Figure 1: The sample output shows the changes of arterial blood gases after transition from spontaneous air breathing to artificial ventilation with increased inspiratory oxygen fraction (FIO2 = 0.45) in a case of simulated ventilation-perfusion mismatch which initially caused an arterial oxygen tension of 6.4 kPa (48 mmHg).

It is under investigation in which ways the model can be brought into states corresponding to acute respiratory insufficiency. The next step will be to learn whether the model, once driven to pathologic conditions of such extreme extent, can reflect reactions which are compatible with clinical observations. Only then the intended benefit can be achieved, to provide an additional tool for the clinician in the decision on optimum artificial ventilation.

References:

[1] Dickinson, C.J.: A computer model of human respiration MTP Press Limited, Lancaster (England) 1977

[2] Farrell, E.J., J.H. Siegel: Investigation of cardiorespiratory abnormalities through computer simulation, Comp. Biomed. Res. 5 (1973) 161-186

[3] Grodins, F.S., J. Buell, A.J. Bart: Mathematical analysis and digital simulation of the respiratory control system, J. Appl. Physiol. 22 (1967) 260-276

[4] Hinds, C.J., D. Ingram, L. Adams, P.V. Cole, C.J. Dickinson, J. Kay, J.R. Krapez, J. Williams: An evaluation of the clinical potential of a comprehensive model of human respiration in artificially ventilated patients, Clin. Science 58 (1980) 83-91

[5] Kelman, G.R.: Digital subroutine for the conversion of oxygen tension into saturation, J. Appl. Physiol. 21 (1966) 1375-1376

[6] Kelman, G.R.: Digital computer procedure for the conversion of PCO2 into blood CO2 content, Respir. Physiol. 3 (1967) 111-115

A PROGRAM PACKAGE FOR THE ELECTROSTIMULATION OF THE ACOUSTIC NERVE

F. RATTAY *)

Simulation on hybrid computer is used to get the statistical response of a nerve fiber. Generation of the stimulus in the computer is possible as well as from extern sources. Various models for the excitation of nerve are available and with one signal several runs are made to get histograms describing the firing rates of the fibers of the acoustic nerve.

Keywords: Acoustic nerve - hybrid simulation - electrical stimulation

INTRODUCTION

Completly deaf patients get hearing sensations by electric signals which are produced in implanted prosthesis. These signals are similar to their acoustic sources but restrictions in the range of loudness and frequency of the electrical stimulated nerve require compression and filtering of the signals. There are also methods of sending square impulses or modulated currents instead of an analogy to the acoustic signals.

The classical way to simulate the answer of a single fiber to an electrical stimulus is by the well known Hodgkin Huxley equations, /1/, containing a partial diffusion equation. Simulation not concerning the propagation of signals but only the excitation of fiber is possible with the reduced set of 4 ordinary nonlinear differential equations. A modification by introducing a further ionic current for myelined nerves is known as Frankenhaeuser Huxley model. /2/ The nonlinearities of these models include several exponential functions and divisions of the form 0/0 and therefor a computer simulation is more effective by solving a reduced system. Starting with a modified Van der Pol equation Fitzhugh found an appropiate second order model. /3/ A modified Fitzhugh system includes most of the essential properties of the bigger models. /4/

One single nerve fiber is able to send an action potential (AP) if the excitation is big enough - and further sending of AP's is blocked for a time of about 1 msec. The statistical behaviour of the whole nerve, consisting of about 30.000 single fibers, is responsible for the high quality of understanding the acoustic perceptions in the non-handicaped ear.

Studing the information achieved through several parallel stimulated axons requires statistical response of the model. Hybrid simulation seems to be the only effective answer to the big number of necessary simulations to

---

*) Technical University Vienna, Karlsplatz 13, 1040 Vienna, Austria

get significant histograms describing the reactions of the fibers of the acoustic nerve when they are stimulated by the same source signal.

THE HYBRID COMPUTER OF THE TU-VIENNA

A program package simulating the statistical behaviour of a boundle of axons was especially concepted for the hybrid computer of the Technical University of Vienna.

The computer consists of an analog EAI-680 unit with 25 integrators available and a 16 bit digital PACER. The analog part has total digital control (autopatch) and is to be programmed by the user oriented language HYBSYS. /5/ Discretization of functions is possible with maximal 1024 breakpoints. Extern inputs and outputs are used for direct communications with analog signals for real time problems.

So - if not produced computationally- the stimulus signal comes directly from microphone, speech processor, recorder or in digitalized form with max. 4*1024 breakpoints.

User's FORTRAN written subroutines are activated by HYBSYS statements by overlay techniques. Within an overlay subroutine commands for making analog runs, setting of parameters, reading out of variables, etc. are useful tools to manage hybrid simulation sequencies.

HYBRID SIMULATION OF THE EXCITED NERVE

The advantage of hybrid simulation is parallel working of all analog components. As shown in fig. 1 the stimulus signal is produced or filtered (e. g. with a Butterworth filter of 5-th order to cut frequencies higher than 4 kHz). The differential equation concerning one of the models is solved and the times at which action potentials are generated are marked. All will be done in parallel if the capacity of the computer is not exhausted.

To study the statistical behaviour of fibers a noise term is added to the models. /4/ The reactions of parallel fibers which are stimulated by the same electrical signal are simulated by sequential computation of one axon. For this purpose the source signal e.g. coming from the speechprocessor and being filtered by the analog computer is stored digitally, if it is not produced synthetically on the analog machine. Digitally stored datas are filtered again before entering the model of nerve. If the excitation voltage reaches a certain level an action potential is produced and the corresponding time is marked. These times are stored digitally and after one set of runs a histogram shows either the times of reactions (posthistogram) or the times between AP's (intervalhistogram).

The program package can be used without statistic applications, too. A sinusoidal signal with growing frequency or amplitude is produced on the analog machine to study threshold phenomens. Superposition of complicated

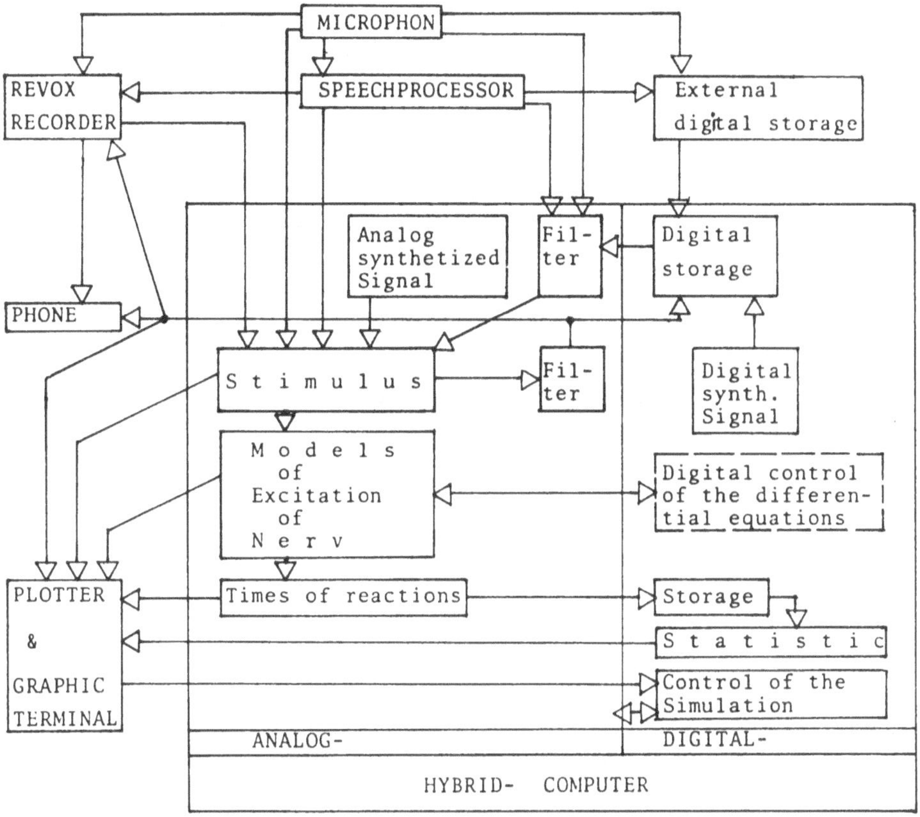

fig. 1

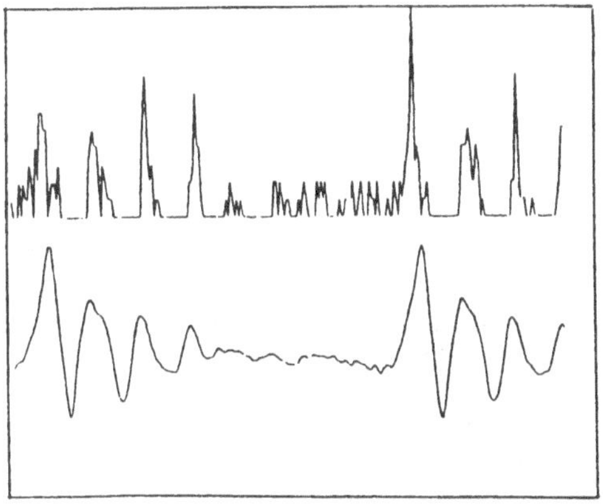

fig. 2

stimulus signals are digitally synthetized and they are stored on a 'PHOnetic' file to compare the results of several runs.

The control of the whole simulation is made by a graphic terminal.

After choosing a special simulation model and the setting of parameters the following commands allow to produce fig.2.

PHO,3	activates the subroutine PHO and sets the digital analog converter function (DACF) to the 3-rd set of datas, representing the vocal 'a'.
HIS,80	The subroutine HIS produces a statistic of times of AP's (posthistogram) and manages 80 runs of the analog computer. One analog run takes 10 msec.
PLOT SD	Plot of the histogram variable 'SD'
PLOT S	Plot of the stimulus 'S'

In case only few users are working this simulation is finished within 3 minutes.(Hybrid timesharing).

The upper curve of fig. 2 correspond to the added AP signal of 80 parallel fibers evoked by the same electrical signal, which is a current - lower curve - proportional to the soundpressure of 'a'. The figure shows activieties of fibers only in the positive phase of the source signal. This phenomen exists also in the normal ear.

DISCUSSION

The speed of producing histograms like the one in fig. 2 is fascinating but there were several difficulties to overcome. The following techniques were necessary

- o use of function-generators to avoid division 0/0 and unaccuracy by to many nonlinear elements
- o feedback at multipliers for better scaling
- o cutting of the full range of some variables
- o use of a correction term for the stationary state
- o control of the simulation by digital computation

Another restriction is the limitation of breakpoints for digitalized input. Sharing the information on 4 input variables allows to send 4*1024 breakpoints into the analog computer which is enough for studing one-syllable signals which are in a band width of 4 kHz.

REFERENCES

/1/ Hodgkin A.L., Huxley A.F. (1952) A quantitative description of membrane current and its application to conduction and excitation in nerve. J. Physiol. 117, 500-544

/2/ Frankenhaeuser B., Huxley A.F. (1964) The action potential in the myelined nerve fibre of Xenopus laevis as computed on the basis of voltage clamp data. J. Physiol. 169, 431-437

/3/ Fitzhugh R. (1961) Impulses and physiological states in theoretical models of nerve membrane. Biophys. J. 1, 445-466

/4/ Hochmair-Desoyer I.J., Hochmair E.S., Motz H.,Rattay F. (1984) A model for the electrostimulation of the nervus acusticus. Neuroscience 13, 553-562

/5/ Solar D. (1983) HYBSYS hybrid simulation system manual. EDV-Zentrum, Abt. Hybridrechenanlage,Techn. University of Vienna

BIFURCATIONS AND CHAOS IN DIFFERENTIAL SYSTEMS REPRESENTING THE ELECTRICAL ACTIVITY OF NERVE CELL MEMBRANES.

Arun V. Holden[1]

Abstract. Excitation equations are systems of nonlinear o.d.es that represent the biophysical properties of nerve and muscle membrane. They can have equilibrium, periodic, quasi-periodic and chaotic solutions. Numerical evaluation of eigenvalues show that periodic solutions develop from equilibria by Hopf bifurcation. Further bifurcations can lead to quasi-periodic or chaotic attractors. Methods of identifying complicated attractors are illustrated.

The excitable membranes of nerve cells may be represented by relatively high order (>3) nonlinear differential systems that have been obtained from an empirical, biophysical description of the displacement (capacitative) and ionic currents that flow across the nerve membrane. The electrical activity of nerve membrane may be quiescent, or consist of small amplitude oscillations, a repetitive, regular discharge of large amplitude action potentials, regular or irregular patterned bursting discharges, or irregular amplitude-modulated oscillations.

These behaviours may be identified with numerical solutions of the differential system that represents the nerve cell membrane. The Hodgkin-Huxley membrane equations [1] are a four-dimensional nonlinear differential system that represents the excitable membrane of the squid giant axon: they are an intensively studied system, not a typical system.

A resting membrane potential represents a stable equilibrium solution of the membrane excitation equation, or a stable electrochemical steady state. It is unstable if any of its eigenvalues have positive real parts. At a Hopf bifurcation a **single complex conjugate pair** of eigenvalues cross the imaginary axis, and so the equilibrium become unstable and small amplitude periodic solutions emerge. These periodic solutions themselves may be stable, when they appear as small (less than 10 mV) amplitude oscillations, or they may be unstable, when they provide a transient path to large amplitude periodic solutions that correspond to periodic action potentials [3].

The eigenvalues of the Jacobian matrix may be evaluated at a stationary point of the excitation equation as a parameter, say the applied membrane current density **I**, is changed. The Hopf bifurcation points occur at values of the parameter where a single complex conjugate pair of eigenvalues crosses the imaginary axis.

The Jacobian matrix for the Hodgkin-Huxley system, evaluated at a stationary point where $dV/dt = dm/dt = dh/dt = dn/dt = 0$, is:

1. Reader in General Physiology, Centre for Nonlinear Studies, The University, Leeds LS2 9NQ, U.K.

$$\begin{bmatrix} -\partial F/\partial V & -\partial F/\partial m & -\partial F/\partial n & -\partial F/\partial h \\ \gamma_m m'_\infty & -\gamma_m & 0 & 0 \\ \gamma_n n'_\infty & 0 & -\gamma_n & 0 \\ \gamma_h h'_\infty & 0 & 0 & -\gamma_h \end{bmatrix}$$

with $\gamma_x = \alpha_x + \beta_x$, for x=m,n,h, and F(V,m,n,h) given by the Hodgkin-Huxley membrane equations.

The eigenvalues of this Jacobian were calculated using algebraic methods, and the characteristics were evaluated according to the methods of Marsden and McCracken [5]. Calculations were performed in BASIC on a microcomputer (Zilog MCZ 1/20), and in FORTRAN 77 on the University of Leeds mainframe Amdahl V/470.

If two parameters are varied, say I and the Nernst potential for potasssium, V_K , the bifurcations will fall on a curve in the plane I - V_K.

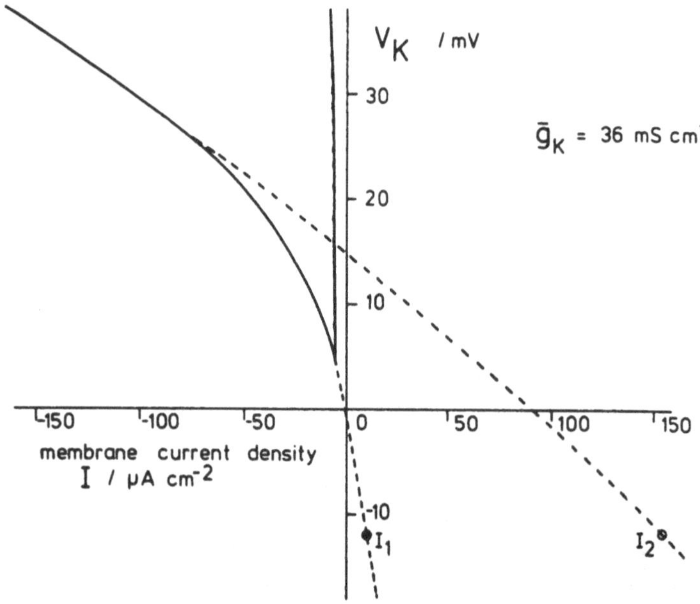

The solid lines enclose a region where there are three equilibrium solutions, and the dashed lines are the Hopf bifurcation curves. I_1 is a bifurcation into small amplitude periodic solutions that are **unstable** - they provide a path into the large amplitude periodic solutions that correspond to a repetitive discharge of action potentials. For parameter values just below such a subcritical bifurcation there are two stable solutions: a stable equilibrium (resting potential) and stable, large amplitude periodic solutions (repetitive discharge of action potentials). Close to, and below, a subcritical Hopf bifurcation repetitive activity may be annihilated by an appropriate perturbation.

Hopf bifurcations also occur when the maximal potassium conductance (proportional to the membrane channel density) [4], or the extracellular Calcium activity, are reduced [3]

The Hodgkin-Huxley system responds to a sinusoidal forcing by periodic, quasi-periodic or chaotic solutions: similar behaviour is seen in the biological preparation. A phase plane display of the chaotic response shows the banding characteristic of a strange attractor:

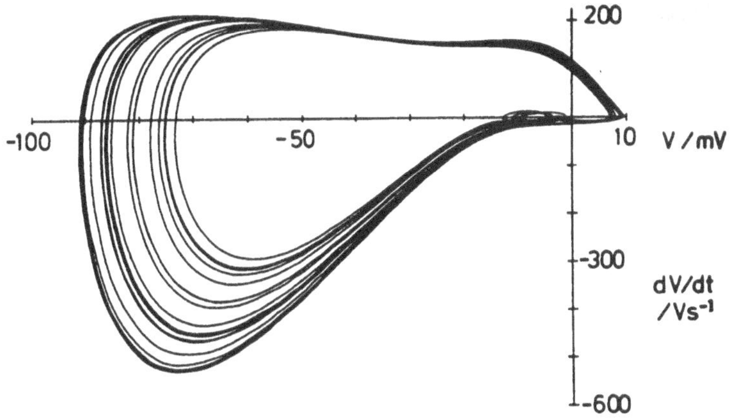

and the strange attractor may be reconstructed from measurements of V(t) [2]. However, the state space of the Hodgkin-Huxley system is four dimensional, so the shape of the strange attractor is difficult to visualize, and so now I consider the excitable membrane of mammalian myelinated nerve fibres that may be represented by a third order system.

This system responds to sinusoidal forcing by periodic, quasi-periodic and chaotic solutions: at a driving frequency of 290 Hz the solution looks chaotic, and the attractor in the three-dimensional (V,m,h) state space appears strange. However, these solutions are quasi-periodic - this is clearly seen by a plot of the amplitude of a local maximum against the amplitude of the preceeding local maximum (a Lorenz plot). At a driving frequency of 260 Hz the solutions are chaotic, and the Lorenz plot consists of separate line segements.

The chaotic and quasiperiodic attractors may be distinguished by their Lyapunov exponents, which have a (+ 0 -) spectrum for the chaotic attractor, and a (0 0 -) spectrum for the toroidal attractor. Although numerical evaluation of the Lyapunov exponents is feasible [6], it is computationally expensive, and so graphical methods of identifying strange attractors (by their self-similarity, Lorenz plots and Poincaire sections) are almost as useful.

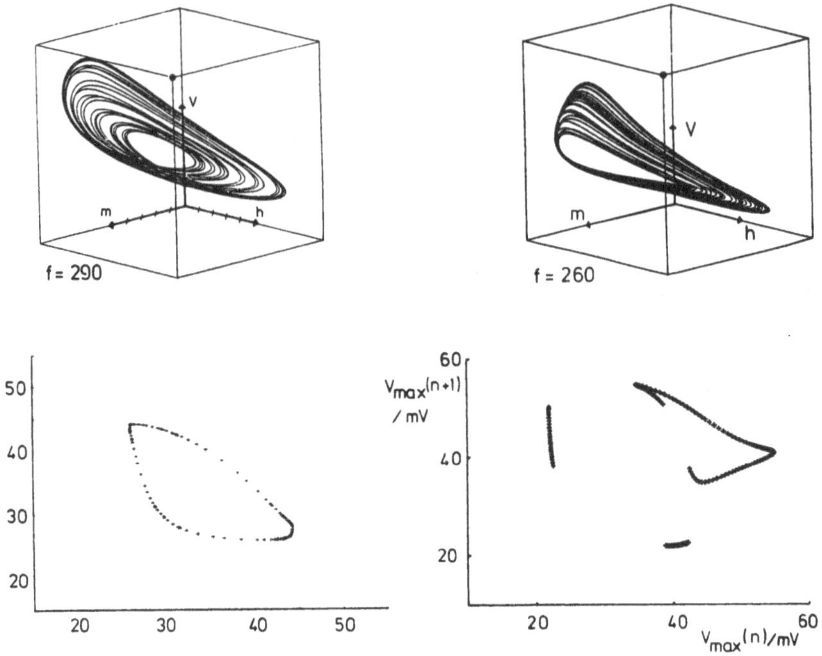

[1]Hodgkin, A.L. and Huxley,A.F.: A quantitative description of membrane current and its application to conduction and excitation in nerve. J. Physiol. **117** 500-544 (1952).

[2]Holden,A.V. and Muhamad,M.A.:The identification of deterministic chaos in the activity of single neurones. J. Electrophysiol. Techniques **11** 135-147 (1984)

[3]Holden,A.V. and Winlow,W.: Neuronal activity as the behaviour of a differential system. IEEE Trans **SMC-13** 711-719 (1983)

[4]Holden,A.V. and Yoda,M.:Ionic channel density of an excitable membrane can act as a bifurcation parameter. Biological Cybernetics **42** 29-38 (1981)

[5]Marsden,J.E. and McCracken,M.:The Hopf bifurcation and its applications. Berlin:Springer-Verlag (1976)

[6]Wolf,A.:Quantifying chaos with Lyapunov exponents. in:Chaos, Ed. A.V.Holden. Manchester Univ. Press (1985).

CABLE MODELS AND THEIR USE FOR ANALYSIS OF NEURONAL INTEGRATION

Andreas Schierwagen[1)]

Cable theory provides the mathematical methods that allow to evaluate the role of passive electrical membrane properties and dendritic geometry in neuronal information processing. Some of the cable models frequently employed in neurophysiology are reviewed, with special stress laid on the segmental cable model. It is concluded that a dendritic neuron can comprise a multitude of computational units, in contrast to the traditional view on the neuron as the functional elementary unit of the central nervous system.

neuron - electrical signals - mathematical descritption - cable models - parameter identification - neuronal information processing

0. Introduction

Nerve cells use two types of electrical signals for communication: passively conducted graded potentials, which arise at interfaces such as synapses (i.e. junctions between cells) and actively conducted action potentials of constant amplitude generated at the neuron's spike trigger zone.

In contrast to the traditional view, recent research assigns an essential signal processing role to the local, graded potentials which interact by specific synaptic arrangements, the local circuits. Dendritic regions as the main locations of synapses, however, are rather difficult to investigate directly by physiological methods.

Yet, during the last 20 years, a quantitative approach to neuronal integration of passive potentials has been developed [1].

Different types of cable models derived from this theory are reviewed and their use as tools in evaluating some important aspects of a neuron's signal processing capability is demonstrated. Exemplified by a class of sensori-motor neurons of the cat [2], special stress is laid on a detailed segmental cable model.

1. Basic equation

The usual mathematical description of the integrative neuron function proceeds from the fact that neuronal membranes possess electrical properties allowing one-dimensional cable theory to be employed.

The equation for membrane potential $V = V(x,t)$ in a cylindric segment of length l and diameter d is (see [1] for a derivation)

$$\lambda^2 \frac{\partial^2 V}{\partial x^2} = \tau \frac{\partial V}{\partial t} + V \qquad (1)$$

where x represents distance in axis direction of the segment ($0 \leq x \leq l$), t is time, $\lambda = [(d/4)(R_m/R_i)]^{1/2}$ is the length constant and $\tau = R_m \cdot C_m$ denotes the membrane time constant. R_m, C_m and R_i are characteristic 'material' constants (membrane resistance, membrane capacitance and intracellular resistivity). Equation (1) can be solved with little difficulty for most relevant boundary conditions. Its applicability,

1) Karl Marx University, Carl Ludwig Institute of Physiology
DDR-7010 Leipzig Liebigstr.27

however, is restricted to the assumed simple geometry of a cylindric process.

2. Cable models

In the case of branching structures such as dendrites, the situation becomes much more complex. One has to include boundary conditions at each branch point and each terminal, resulting in very complicated expressions for the corresponding solutions.
Up to date, models of different complexity and extent of employment has been developed to cope with the problems.
A frequently applied model in neurophysiology is the Rall model of the nerve cell [1]. In short, it can be characterized as follows: For R_m, C_m and R_i assumed to be uniform over the neuron surface, the cell soma is represented by a single RC-element, and each dendritic tree is treated as an 'equivalent' cylinder, with identical boundary conditions at all terminations. The mapping of a tree onto a cylinder requires a particular branching structure: at each branch point the 3/2 power of the parent cylinder diameter must equal the sum of the 3/2 power of the daughter cylinder diameters.
Provided each of the equivalent cylinders terminates at the same electrotonic distance $X = \int_0^x \lambda^{-1}(\xi)d\xi$ from the soma, a further simplification can be made: they can be treated collectively as a single equivalent cylinder.

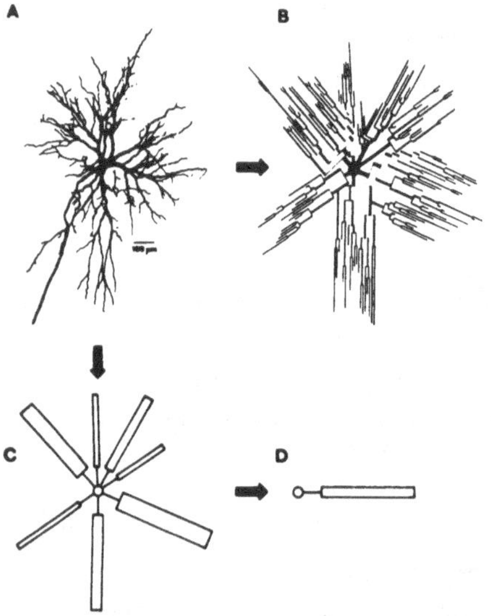

Figure 1. Approximation levels for neuronal cable models. Sequence A→C→D illustrates the transformation steps made in the case of idealized branching, whereas A→B represents the derivation of a detailed segmental cable model as used in the general case, see text. The neuron displayed in A belongs to a class of cat neurons involved in sensorimotor control [2], [3], [4].

Unfortunately, these assumptions do not seem valid for most real neurons. Another way of analyzing a dendritic neuron is to approximate its parts by regular geometrical bodies (Figure 1.B). In this case, equation (1) has to be applied iteratively to each cylindrical segment of the dendritic trees.

3. General solution of the cable equation

The Laplace transform of equation (1) with respect to time is

$$\frac{d^2\tilde{V}(x,s)}{dx^2} = \gamma(s)\,\tilde{V}(x,s) \ . \qquad (2)$$

In terms of hyperbolic functions, its general solution is

$$\tilde{V}(x,s) = A(s)\cosh(\gamma(s)x) + B(s)\sinh(\gamma(s)x), \qquad (3)$$

where $s = j\omega\tau$ is normalized frequency, $\omega = 2\pi f$ and $\gamma(s) = (1+s)/\lambda$. Combining (3) with the corresponding transform $\tilde{I}(x,s)$ for axial current, the input admittance at $x=0$, $\tilde{Y}_{in}(s) = \tilde{I}(0,s)/\tilde{V}(0,s)$ can be calculated as

$$\tilde{Y}_{in}(s) = \frac{\gamma(s)}{r_i}\ \frac{\sinh(\gamma(s)l) + Y_t^*(s)\ \cosh(\gamma(s)l)}{\cosh(\gamma(s)l + Y_t^*(s)\ \sinh(\gamma(s)l)} \qquad (4)$$

$Y_t^*(s)$ is the normalized terminating admittance, which has to be assumed for the most distal segments, and calculated otherwise.

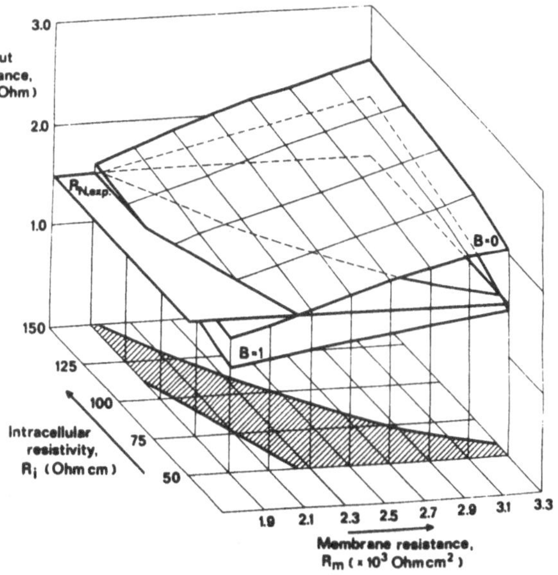

Figure 2. Three-dimensional plot of calculated R_N vs. assumed R_m and R_i for sealed (B=0) and killed (B=1) dendrite terminations, see text.

4. Applications

The list of problems investigated in neurophysiology by means of cable models includes identification of electrical membrane parameters [3] prediction of size and relative efficacy and analysis of interaction of synaptic potentials [4], [5], evaluation of the function den-

dritic spines may have in neuronal plasticity [6], and others more. As an example, the parameter identification problem is discussed:
By sequential application of equation (4) for $\omega = 0$, the input conductance of each of the major dendrites, G_D, can be determined. Summation of all conductances including that of soma gives the total neuron conductance

$$G_N = G_{soma} + \sum G_D \quad . \tag{5}$$

The inverse of G_N is the neuron input resistance R_N, which is a function of the three variables R_m, R_1 and B, with the latter being the assumed value $Y_t^*(s)$ for the dendrite terminals. In Figure 2, B=0 ('sealed end' condition, $d\tilde{V}/dx|_{x=x_{ter}} = 0$) and B=1 ('killed end' condition, $\tilde{V}|_{x=x_{ter}} = 0$) is taken. The intersections of the two corresponding R_N surfaces with the plane $R_N = R_{N,exp.}$ = 1.5 MOhm denote the parameter region (hatched) which is consistent with the experimental R_N measurements.
On the whole, the results derived both by cable models and experiments have suggested a new view on a neuron's information processing capabilities. In contrast to the traditional view on the neuron as the functional elementary unit of the nervous system, it can be concluded that a dendritic neuron may comprise a multitude of computational units which act independently and simultaneously.

References

/1/ Rall, W. (1977): Core conductor theory and cable properties of neurons. In: Kandel, E.R. (ed.), Handbook of physiology. The nervous system, vol.1, part 1. Amer.Physiol.Soc., Bethesda, 39-97.
/2/ Grantyn, R., A.Grantyn and A.Schierwagen (1983): Passive membrane properties, afterpotentials and repetitive firing of superior colliculus neurons studied in the anesthetized cat. Exp.Brain Res. 5o, 377-391.
/3/ Schierwagen, A., A.Reichenbach und R.Grantyn (1984): Elektrophysiologische Aspekte der integrativen Funktion von Neuronen des Vertebraten-ZNS. Wiss.Z.KMU Leipzig, Math.-Naturwiss.R.33, 485-497.
/4/ Schierwagen, A. (1985): Model investigations on factors controlling synaptic integration in cat sensori-motor neurons. Technical Report, Hungarian Academy of Sciences, Central Research Institute of Physics, Budapest, in the press.
/5/ Koch, C., T.Poggio and V.Torre (1983): Nonlinear interaction in a dendritic tree: localization, timing and role in information processing. Proc.Natl.Acad.Sci.USA 8o, 2799-28o2.
/6/ Turner, D.A. (1984): Conductance transients onto dendritic spines in a segmental cable model of hippocampal neurons. Biophys.J. 46, 85-96.

SIMULATION OF SOME FUNCTIONS OF HUMAN THINKING

G.V. Avaliani[1)]

A heuristic experiment with experienced specialists is suggested for establishing methods with the help of which a man recognises objects. The experimental results are used in construction of models.
Key words:pattern, recognition, heuristics.

The paper deals with possibility to simulate some functions of human thinking. Namely it concerns pattern recognition function. The available machine procedures do not utilize as usual the experience stored by a man while dealing with similar problems. However a qualified specialist uses his experience and intuition for high precision decision of pattern recognition problems. Such a situation arises for instance in diagnosis of image procession.

Heuristic approach has been proposed in the Institute of Cybernetics of Georgian Acadamy of Sciences (USSR) to analyse specialists' experience for machine procedures. An experiment with experienced specialists of recognition problems makes the basis of the approach. The purpose of the experiment is to study an object, reveal its features essential for the object and establish the means, ways and regularities used by an experienced specialists in recognition.

For such an experiment the situation usual for specialist has to be secured. The technical means include: a)directions of experiment, b)realizations of the recognition object, c)recording devices. The directions depend on a given object. The common is that, a person under study is explained the purpose of the experiment; the method is given (in terms usual for the tested person) to describe the recognition object; he is requested after data has been studied to say aloud all his reasonings for problem decision, present the sign description of the object, to trace the relationship between the signs, evaluate according the given scale the degree of his belief towards each sign (or its meaning) and so on. The tested is given a set of typical realizations of the object together with the directions before experiment starts.

Set of realizations of the object must fully include all the cases of recognition met in practice. Recording devices represent firstly a tape-recorder for reasonings to fix, various time counters to research a number of psychological parametres of a man. If it is possible better the tested man not be aware he is being recorded.

According to the methods the experimentator may propose some questions. The experiment is possible simultaneously to 2-3 persons. The difference in opinions of the tested group may bring some interesting results. The experiment allows:

1.The experimentator perceives the object deeper;

2. Get sign description of the object;

[1)]Institute of Cybernetics of Academy of Sciences GSSR, S.Euli Str.5, 380086, Tbilissi, USSR.

3.Make the known signs more exact;
4. Study the relationships between signs;
5. Get classification or signs and sign meanings;
6. Get degrees of trustworthiness of sign meanings;
7. Trace and fix the ways and methods used by an experienced specialist in recognition process.

By means of experiment a number of regularites are revealed which are formalized at certain meanings of signs into logical functions determining the recognition object. Heuristic method was applied in various problems (medical diagnosis image processing). The analysis of data elicited some general ways of decision making used by a man in recognition. These are conventionally diveded into groups. The first approach - when a tested person introduced the notion "Generalized class A". The reasoning goes as follows: let us say we deal with either A or $\bar{A}$; if the available set of signs (sign values) allows us distinguish A (or $\bar{A}$) then a new notion of either subclass A' is introduced and the procedure is repeated. This dichotomical process leads to some concrete image.

The second approach involves a method called "illimination method". The tested person suggests a hypothesis -„it is not the object a_1" and then presents his argumentations. If the answer is negative the procedure is ended. If it is positive then a new hypothesis is declared - "it is not the object a_2" and so on untill we reach some concrete image. Some of experts uses an approach called "reverse illimination method". The first hypothesis here is "it is a_1" and the procedure ends if the answer is positive and a new hypothesis (the answer being negative) is declared - "it is a_2" and so on. Here the tested applies a full look over. The second and the third approaches are used in the case the number of objects is little. Sometimes all three approaches are used on different stages of recognition. The approaches were used for machine procedures. The models obtained take into account: a)signs used by an experienced specialist, b) degrees of trustworthiness of the sign (sign meaning) form the specialist's point of view; c)principles used in the correct decision making.

The models are simple for realization including dialogue systems. The models have been utilized in a number of practical tasks.

ON THE MODELBUILDING OF ELEMENTARY AND COMPOSED HEALTH-CARE-SYSTEMS

Ch. Dahme, Th. Hager, K. Fuchs-Kittowski [1)]

It is shown as models of elementary and composed health-care-systems (HCS) can be build up in consideration of the principles of the unit of subjective and of objective as soon as of formalized and of non-formalized. The efficiency of an elementary HCS is characterized by the duration of the medical activity of this HCS for one patient and by its capacity especially by so-called necessary health-care-factors and health-care-factors which promote efficiency. Proceeding on a health-care-situation characterized by a demand-situation and the efficiency of the HCS it is shown as a HCS can react on such a health-care-situation (especially if the efficiency is not enough). For this, strategies of the adaptation and of the co-operation are developed.

Keywords : modelbuilding, elementary and composed health-care-systems, strategies of the adaptation and of the co-operation, necessary health-care-factors, HC-factors which promote efficiency

A health-care-system (HCS) is such a social system (see Dahme 1985) which is able to carry out prophylactic, diagnostic, therapeutic and/or metaphylactic activities for the maintenance, the promotion and the recovery of the state of health of a population.
An elementary HCS is such a HCS that is orientated on a certain patient typ and consists of at least one physician, who is able on grounds of his qualification to carry out the medical activities for patients of this typ, and of such health-care-factors that are necessary for the carrying-out of these activities (necessary HC-factors).
A HC-situation for an (elementary) HCS is characterized by

a) the efficiency of the (elementary) HCS
b) the demands on this HCS (i.e., the demand-situation)

The efficiency P of an elementary HCS is a function of its capacity C and of the average time of treatment T (the duration of the medical activity of this HCS) for one patient of this typ :

$$P = C / T \qquad \text{and} \qquad C = X_B \cdot \min_i \left\{ 1,\ X_i / (X_B \cdot \overline{X_i}) \right\}$$

at which X_B is the available number of the physicians [2)]
X_i is the available part of the ith necessary HC-factor
$\overline{X_i}$ is the part of X_i which is necessary for the treatment of <u>one</u> patient [2)]

If the demand-situation ϕ is opposed to the efficiency P then two situations can be distinguished :

---

1 Humboldt-Universität zu Berlin, Sektion WTO, DDR-1o86 Berlin

2 In this case we go out from the situation that one pysician can carry out medical activities only in regard to <u>one</u> patients in the same time !

$\phi \leq P$ (i.e., the efficiency is enough for this demand-situation)
$\phi > P$ (i.e., the efficiency is not enough for ...)

In this connection the following questions are of interest :

1.st How long does the stage $\phi > P$ ($\phi/P = Z^1(t) > 1$) last ?

2nd How often do a situation with $\phi > P$ occur with which duration and intensity and with which temporal distance between such situations ?

3rd Which possibilities of the behaviour (for reaction) does an elementary HCS have in regard to the situation $\phi > P$?

a) unadaptable
b) quasi-adaptable (see 4th and 5th)
c) adaptable (see 6th)

4th Does a bottle-neck exist relating to C ?

(i.e., is $X_B > C$?) In this case there are necessary HC-factors with $X_i < \overline{X_i} \cdot X_B$. Then $X_i^3 = \overline{X_i} \cdot X_B - X_i$ is the additional demand for the i th necessary HC-factor to achieve $X_B = C$.

5th Which additional demand on HC-factors is necessary to achieve

a) $\phi = P$ or b) $\phi = P$ and $N^w(t) \leq \overline{N}^w$?

In the case of a) $C^1(t) = C + \phi(t) \cdot \Delta t - N^e(t)$ is the required capacity to achieve $\phi = P$ ($N^e(t)$ is the number of discharged patients). In order to achieve $C^1(t)$ the following parts on necessary HC-factors are necessary :

$X_B(t) = C^1(t)$ and $X_i^1(t) = X_B^1(t) \cdot \overline{X_i}$. Then
$X_B^2(t) = X_B^1(t) - X_B$ and $X_i^2(t) = X_i^2(t) - X_i$ is the deficit or

the additional demand on necessary HC-factors to achieve $\phi = P$.
If $X_B^2(t)$ or $X_i^2(t)$ is negative, then $X_B^2(t)$ or $X_i^2(t)$ indicates the part that is not used.
In the case that in addition is $N^w(t) \leq \overline{N}^w$ ($N^w(t)$ is the number of waiting patients and $\overline{N}^w$ is an reasonable length of a queue) then the following unequation must be vatid for $C^1(t)$:

$$C + \phi(t) \cdot \Delta t - N^e(t) \leq C^1(t) \leq C + N^w(t) - \overline{N}^w + \phi(t) \cdot \Delta t - N^e(t)$$

6th Which possibilities of adaptation exist for an elementary HCS to react on $\phi > P$?

a) relating to the pysician

- the extension of the working time from T_A^n (contract hours) to T_A : Let $Z^4 = T_A / T_A^n$ then it holds for the efficiency

$$P^1(t) = Z^4 \cdot P(t) = T_A / T_A^n \cdot C / T$$

- the shortening of the time of treatment from T to T^1 through the reduction of typical medical activities for this patient typ that is in general attended with a reduction of the quality (grade) of the treatment and with a risk.

$Z^5 = T^1 / T$ indicates the degree of the reduction

b) relating to necessary HC-factors
the opening of additional reserves $X_i^4(t)$ (as emergency beds)

c) relating to
through a re-distribution of patients if it is possible
Let $\phi^1 = P$ the stream of the patients that can be treated and
$\phi - \phi^1$ the stream of the patients re-distributed (In this case a "potential" queue is built up !)
then $Z^6 = 1 - \phi^1/\phi$ indicates the degree of re-distribution.

For an elementary HCS strong restrictions can exist especially in the dependence of the patient typ, as :

$Z^1(t) \leq 1$ all patients arriving must be treated

$Z^3(t) \leq 1$ the leangth of the queue must not get greater than a given number

$Z^5(t) \geq 1$ the time of treatment must not be shortened

$Z^6(t) = 0$ no patient must be temporally re-distributed (defered) in regard to his treatment

Such strong restrictions influence the possibilities of adaptation !

If an elementary HCS consists of necessary HC-factors only and $C = X_B$ as soon as $\phi > P$ then the following question arises :

7th How can the efficiency of this HCS be increased without X_B to change (contrary to 5th) and without the quality of the medical care to reduce (contrary to 6a) ?

The answer of this question leads to so-called HC-factors which promote efficiency. Such HC-factors makes possible through their introduction :

a) the increasing of the capacity of an elementary HCS without X_B to have to change

b) the reduction of the time of treatment (in certain bounds) without the quality of medical care to influence negativly but also

c) the quality (the level) of the medical care to increase

d) the conditions of the activity (of work) of an elementary HCS to improve

In regard to a) and b) the HC-factors which promote efficiency can be characterized by the following :

$S_j^B(1,\overline{Y_1})$ is the part of the activity of the pysician which can be undertaken by a person and/or by a means that represents this j-th HC-factor, which promotes efficiency, with

$$\overline{Y_j} = \min_j \left\{ Y_j \;\middle|\; S_j^B(1,Y_j) = S_j^B(1,Y_j+1),\ Y_j > 0 \right\}$$

Y_j is a lot of the j-th HC-factor which promotes efficiency (in general the available part)

$\overline{Y_j}$ is the necessary part of Y_j that is able to substitute $S_j^B(1,\overline{Y_j})$ or to reduce the time of the medical activity of the pysician

$$\text{from } T \text{ to } T \cdot (1 - S_J^B(1, Y_J))$$

The efficiency of such a substitution can often be given only comparative and in some cases metric.

Suppose that several elementary HCS's exist relating to the same patient typ. Then there are several possibilities of behaviour for these HCS's (see Dahme 1982) :

a) independent (the HCS's do not react on each other)

b) competitive (the HCS's indirectly interact, i.e., only over the patients : every HCS tries to get so much patients that it achieves $\phi = P$)

c) co-operative

The porpose of a co-operation may be :

1st the common mastering of the demand-situations through a re-allocation of patients (in certain bounds) between the HCS's co-operating

2nd the common using of HC-factors
Suppose that a HCS does not utilize fully some of his HC-factors. Then such a HCS can try to gain (or to co-operate with) other HCS's in order to utilize such HC-factors together.
On the other side the introduction of a HC-factor, which promotes efficiency, can be useful only then if several HCS's utilize this HC-factor together.
Through such common utilization of HC-factors it emerges composed HCS's.

Basing on such investigations we are going to build up models for a network (for a system) of elementary and/or composed HCS's. For such a modelbuilding it is necessary to investigate :

a) typical demand-situations for elementary and composed HCS's in regard to different patient typs

b) territorial aspects

c) different possibilities of the allocation (principles of the allocation) of the resources in consideration of the possibilities of the behaviour of an elementary HCS and of the efficiency as soon as of the demand-situations.

In order to corroborate our investigation we have carried out empirical investigations in regard to special medical settings.

References

/1/ Ch. Dahme, On the particularity of the modelbuilding of elementary social systems, Berlin 1985, in this volume

/2/ Ch. Dahme, Zur Modellbildung sozialer Systeme und Prozesse dargest. an Problemen des Gesundheitsschutzes, Diss. A, Berlin 1982

/3/ Ch. Dahme, Th. Elm, I. Goos, Th. Hager, Some questions of systems-analysis of obstetrical care, international symposium "Simulation of Systems in Biology and Medicine", Praha 1984

SIMULATION IN EPIDEMIOLOGICAL STUDIES OF CHRONIC DISEASES

František Hauser[+]

Chronic diseases represent the most pressing health problem in the industrialized countries. Epidemiological studies of chronic diseases are complicated by long observation periods and multitude of risk factors. Simulation can be helpful in such studies but due to a low level of current knowledge in the area, simulation results have limited use. Development of epidemiological models of chronic diseases involves classification of individuals, determination of risk factors and quantification of risk factor effects. The models then describe population projections with respect to the given health problem. Simulation results can be used to test hypotheses concerning risk factors and genetic links, to evaluate screening and prevention programs, and to promote interdisciplinary collaboration.

Keywords: simulation of health care systems, epidemiological models, medico-demographic models

CHRONIC DISEASES

Eradication and causal treatment of a large number of infectious diseases have brought about increasing prominence of chronic diseases. This trend is further enhanced by a growing proportion of elderly people in population since the elderly suffer most from all kinds of chronic and degenerative diseases. In the developed countries such diseases have already become the main medical problem.

The most important differences between infectious and chronic diseases are in etiology, therapy and prevention. While the former usually have one main causal agent and other factors play only minor secondary roles, the latter are truely multifactorial. Several environmental, behavioral, genetical, etc., factors have typically been shown to be associated with the development of chronic diseases. The same factors can often lead to the development of different diseases and on the other hand elderly people often suffer from polymorbidity, i.e. from several chronic diseases. Therapy is prolonged and can require long time nursing or hospitalization. Early diagnosis, while desirable, is usually difficult so that the onset of the disease remains unrecognized. Earlier diagnosis as a rule results in a better prognosis.

Uncertainty about risk factors and multiplicity of risk factors cause serious difficulties in prevention of chronic diseases. In addition, most factors are influenced not by the health care system but by other branches of national economy and by individual's behaviour.

The increased life span thus leads to increased prevalence of chronic

+Hybrid Computation Lab., Inst. for Social Medicine and Organization of Health Services, Vítězného února 54, 121 39 Praha 2, Czechoslovakia diseases. In this way new problems for health care systems arise. They consist mainly in growing requirements for hospital and nursing services and concentrate especially in internal, neurological and psychiatric departments. To evaluate public health policies and to estimate demand for health care resources necessary for treatment of chronic diseases, it is useful to gain some more cognition of the inherent dynamics of these diseases from the viewpoint of their epidemiology. New insight and better understanding of these problems can be provided by simulation of epidemiological processes.

DEVELOPMENT OF MODELS

The first step in the development of an epidemiological model of a chronic disease is specification of the studied health problem and setting of criteria for classification of individuals with respect to disease phases usually connected with important health status changes. In addition, it can be useful to classify healthy individuals with respect to disposition for contracting the disease, unless population non-homogeneity is accounted for in some other way, including e.g. regional, social or genetic classification criteria. Coefficients (probabilities) of transitions between classes are generally age, sex and time dependent. However, transition rates depend also on a complex of external (environmental, social, etc.) conditions. Among the external conditions it is possible, though not easy, to identify certain factors which affect transition rates especially strongly. Hence, the second step in model development is identification of risk factors together with an attempt to quantify their effect.

Classification of individuals is based on the current knowledge of pathophysiology, determination of risk factors on the current knowledge of etiology. Transition rates can be provided by routine statistics or estimated from indirect indices such as sick leave data, data on invalidization, etc. Quantification of the effect of external factors should be based on the analysis of statistical and epidemiological data but if such data are not satisfactory due to incompleteness or incoherence or are entirely lacking, they may be substituted by expert opinion.

With such a model it is then possible to calculate population projections. The projections derive from basic demographic characteristics of the population: original age - sex structure, mortality and fertility rates. During the calculation transition coefficients cause restructuring of population from the viewpoint of the studied health problem in consistence with hypotheses about the evolution of external factors.

Two types of mathematical description can be used to represent evolution of population structure in time /1/. The first type assumes discrete changes in health status and as the fundamental equation uses a translational one-dimensional first-order partial differential equation for each class of individuals. The second type assumes continuous changes in health status and the fundamental equation for each class of individuals is a special case of a two-dimensional translation-diffusion second-order partial differential equation.

USE OF MODELS

Current knowledge available in the field of epidemiology of chronic diseases is not sufficient to allow creation of verified simulation models. In the lines of general principles for the use of simulation models in the investigation of medical systems it is thus not possible to use these models for scientific prediction and the applications are limited by the possible simulation results which are only as follows/3/:

- survey of missing knowledge
- refutation of a hypothesis on which the model was constructed
- quantified model as a more accurate hypothesis about the system
- estimation of future development obtained by means of the quantified model.

With this in mind we come to the conclusion that at present the main purpose of making epidemiological models of chronic diseases should be to gain knowledge. It means above all testing plausibility of hypotheses, evaluating compatibility of assumptions and using the results as inspiration for further studies. After these words of caution it is now possible to review specific uses of epidemiological simulation models.

Epidemiological investigations can result in speculations about causes of changes in mortality from and/or incidence of a given disease. Evaluation of various hypotheses about relative influence of risk factors on population projections helps to distinguish between various possibilities. Simulation of assumed changes of the external factors then implies what might be observable in connection with the particular change.

The influence of any risk factor is different in every individual. Special methods were developed to measure the distribution of certain risks in population. It is possible to ask what will happen if that risk distribution would be changed, e.g. by adopting special measures to lower the risk in the most endangered population groups.

Genetic links are often the subject of special epidemiological studies but the results for acquired chronic diseases are frequently contradictory. Population projections for more than one generation may cast some light on the problem. In addition it should be noted that the spread of genetic disposition for certain diseases might even be supported by advances in treatment as fertility of afflicted individuals had been severely decreased before the treatment was available.

Human behaviour is an important factor in disease prevention or promotion. Healthy life style with regular exercise and right diet gives other prospects than stress, wrong habits and addictions. Such influences can be transformed to risk factors and connected in a broader context with evaluation of prevention programs. Experimentation with real prevention programs is extremely difficult to conduct due to long observation periods and problems generally connected with long follow-up studies. The same arguments hold also for the consequences of implementing therapy innovations.

Another area is evaluation of screening programs aiming at identification of early disease stages and decreasing proportion of untreated individuals. In most instances early detection means a better prognosis, slowing or even halting the degenerative process. A suitable timing is important in all screening programs and it can be proposed on the basis of simulation. Simulation is even more important for estimating transient phenomena, i.e. the impact of screening program initiation.

In more elaborate programs for complex disease control, it is of high interest to investigate conditions insuring rubustness of the program with respect to changes in uncontrollable external factors. It can be done by performing a kind of sensitivity analysis. Sensitivity analysis is useful also for determining data requirements for improved epidemiological analysis. Simulation results can suggest what information to measure or collect and when.

Perhaps the most important use of simulation in epidemiological studies of chronic diseases is mediation of interdisciplinary links. Interest in this subject is shared by physicians, health care managers, information engineers and others. As many factors influencing the onset and course of chronic diseases lie outside the health care system, experts from other fields (food industry, ecology, education, etc.) must participate in programs of chronic disease control and simulation seems to be an appropriate tool for developing interdisciplinary collaboration /2/.

REFERENCES

/1/ Hauser,F., Mader,R., Kotva,M.: Modules for the design of epidemiological models with age structure. In: System Science in Health Care (Van Eimeren,W., Engelbrecht,R., Flagle,C.D., eds.), Springer, Berlin, 1984.

/2/ Koonce,J.F., Yashin,A.I., Walters,C.J., Rusnák,M.: Modelling of Public Health: Call for Interdisciplinary Actions. CP-84-1, International Institute for Applied Systems Analysis, Laxenburg, 1984.

/3/ Kotva,M.: Systems simulation in the investigation of biological systems. In: Modelling and Data Analysis in Biotechnology and Medical Engineering (Vansteenkiste,G.C., Young.P.C., eds.), North--Holland, Amsterdam, 1983.

ANALYSIS OF MANPOWER TURNOVER IN AN INDUSTRIAL PLANT

Urszula Hajdrowska-Mikolajczak [1)]

Abstract: a relationship between a length of service and a leave of workman has been presented with the help of cohort analysis with respect to the following phenomena: education, sex and kind of employment. The results of investigations have been used to illustrate an intensity of manpower turnover: this last factor has an influence on employement planning.

1. Purpose of the paper

The aim of this paper is a solution of the problem of leaving from industrial plant with respect to the length of service of the employees. The process of leaving is also called as a process of manpower turnover: this term is associated with the fact that employees after leaving from industrial plants also take part in the labour market, and usually once more take work in the other industrial plant. The cohort analysis has been chosen as a method of investigation of the process of leaving from industrial plant. This method has been chosen from computational complexity point of view smaller comparing with other methods, for instance census analysis.

2. The algorithm of cohort analysis

The background of application of cohort analysis /1/ is a computation of the number of persons L_i leaving from the industrial plant as a function of the length of service x_i $(i=0,1,\ldots,k)$, expressed in years. Number of leaving allows also to compute the number of survivings Z_i in the industrial plant, which can be described as $Z_i = \sum_{j=1}^{k} L_j$.
The following step of the algorithm is to assign the probability density function f_i which allows to evaluate a probability that employee will leave in the time interval (x_i, x_{i+1})

$$f_i = L_i / c_i Z_o \text{ for } i=0,1,\ldots,k \text{ and se } (f_i) = (L_i(1-L_i/Z_o))^{1/2} \qquad (1)$$

where se (f_i) is the standard deviation, and $c_i=(x_{i+1}-x_i)$ for $i=0,1,\ldots,k$.
The survivor function G_i is the next stage of our investigation: it serves to describe the probability of survivor up to x_i, this magnitude can be computed as

1) Technical University of Poznań, Institute of Organization and Management, 60-965 Poznań, Poland.

$G_i = Z_i/Z_o$ for $i=1,2,\ldots,k$ and se $(G_i) = (G_i(1-G_i)/Z_o)^{1/2}$ (2)

for $l=1,2,\ldots,k$

where se (G_i) is the standard deviation.

Next we compute the conditional probability of leaving at length of service x_i which defines a probability that employee with the length of service x_i leaving from the factory before x_{i+1}. This can be assigned as follows

$$q_i = L_i/c_i Z_i \quad \text{for } i=0,1,\ldots,k \tag{3}$$

together with standard deviation se (q_i) by the following formula

$$\text{se } (q_i) = (c_i q_i (1-c_i q_i)/Z_i)^{1/2}/c_i \quad \text{for } i=0,1,\ldots,k \tag{4}$$

Additionally we are interested in how tendency to leaving m_i from the industrial plant varies in consequent intervals of length of service: this factor is said to be a force of separation: it can be oomputed as follows

$$m_i = L_i/(c_i Z_{i+1} + 1/2 c_i L_i) \quad \text{for } i=0,1,\cdots,k-1 \tag{5}$$

together with standard deviation se (m_i) by the following formula

$$\text{se } (m_i) = m_i \quad L_i \quad \text{for } i=0,1,\ldots,k-1 \tag{6}$$

The next step of the algorithm is to apply χ^2 test with the aim to verify an equality hypothesis of types of distributions under consideration /2/.

3. Conclusions

The upper described algorithm has been practically verified using empirical data from one of the industrial plant (1181 questionaires). the analysis of these data has been conducted with respect to the following phenomena: (i) education, (ii)sex, (iii) kind of employment of the employee (technical or administrative). The results of computations concerning process of leavings and survivings from industrial plant for different levels of education have bee illustrated in Tab. 1.

Tab. 1 Estimates of probability functions for workman of elementary (E) and university (U) educations

x_i	f_i E	f_i U	se(f_i) E	se(f_i) U	G_i E	G_i U	se(G_i) E	se(G_i) U	q_i E	q_i U	se(q_i) E	se(q_i) U	m_i E	m_i U	se(m_i) E	se(m_i) U
0	.33	.11	.02	.04	1.00	1.00	.00	.00	.33	.11	.02	.04	.28	.11	.02	.04
1	.28	.27	.02	.06	.66	.88	.02	.04	.42	.31	.02	.06	.34	.26	.02	.07
2	.13	.15	.01	.05	.38	.60	.02	.06	.36	.25	.03	.07	.30	.22	.03	.08
3	.05	.19	.01	.05	.26	.45	.01	.06	.22	.43	.03	.10	.20	.35	.03	.11
4	.05	.09	.01	.04	.18	.25	.01	.06	.29	.38	.04	.13	.25	.32	.04	.11
5	.05	.05	.01	.03	.13	.15	.01	.05	.39	.37	.06	.17	.32	.31	.06	.18
6	.03	.03	.01	.02	.08	.09	.01	.04	.38	.40	.07	.21	.32	.33	.08	.23
7	-	-	-	-	.04	.05	.01	.03	-	-	-	-	-	-	-	-

Fig. 1 presents a surviving function of employees represented with the help of lognormal probability scale. The deeper analysis of Tab. 1 and Fig. 1 indicates the great difference of leaving process from industrial plants with respect to the factor of education. Fig. 2 presents a force of separation for education. Tab. 2 and Fig. 3 represents the results of kohort analysis for Tab. 2 Estimates of probability functions for women and men

x_i	f_i W	f_i M	$se(f_i)$ W	$se(f_i)$ M	G_i W	G_i M	$se(G_i)$ W	$se(G_i)$ M	q_i W	q_i M	$se(q_i)$ W	$se(q_i)$ M	m_i W	m_i M	$se(m_i)$ W	$se(m_i)$ M
0	.11	.24	.02	.01	1.00	1.00	.00	.00	.11	.24	.02	.01	.11	.21	.02	.01
1	.24	.24	.02	.01	.88	.75	.02	.01	.28	.31	.03	.01	.24	.27	.03	.01
2	.14	.18	.02	.01	.63	.51	.03	.01	.23	.35	.03	.02	.20	.30	.03	.02
3	.09	.09	.01	.01	.48	.33	.03	.01	.18	.28	.03	.02	.17	.24	.03	.02
4	.06	.07	.01	.01	.39	.23	.03	.01	.17	.30	.03	.02	.16	.26	.04	.03
5	.06	.04	.01	.01	.32	.16	.03	.01	.04	.29	.02	.03	.03	.25	.02	.03
6	.08	.04	.01	.01	.25	.11	.02	.01	.06	.39	.03	.04	.06	.32	.03	.04
7	.05	.04	.01	.01	.16	.07	.02	.01	.33	.57	.07	.05	.28	.44	.07	.06
8	.05	.11	.01	.01	.11	.03	.02	.01	.57	.61	.09	.08	.44	.46	.11	.11
9	-	-	-	-	.04	.01	.01	.01	-	-	-	-	-	-	-	-

women and men: from these considerations follows that women have greater tendency to professional stability comparing with men.

The same regularity has been observed for administrative employees (8%) comparing with technical employees (2%)

Fig. 4. All cohorts have been verified with the help of χ^2 test which every time has shown on different types of analysed distributions.

The upper-mentioned regularities allow to formulate, from different structural point of views, model of leaving of manpower taking into account a length of service in the factory: this result is a helpful tool in a manpower planning in the industrial plant.

-Fig. 1 Graphical comparison for the level of education

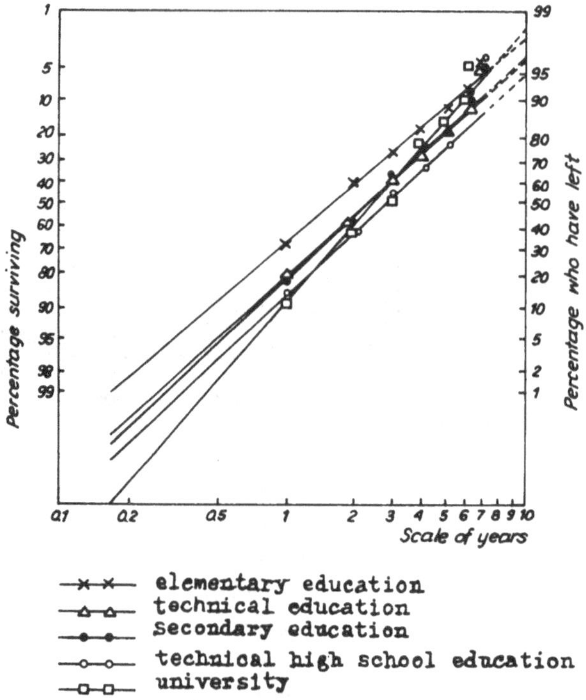

Fig. 2 Force of separation for the level of education

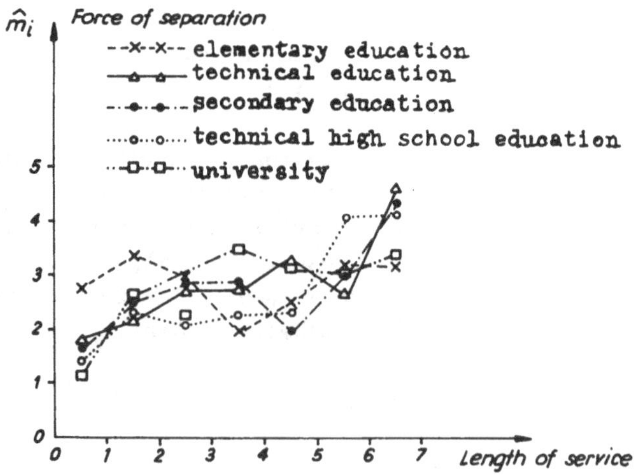

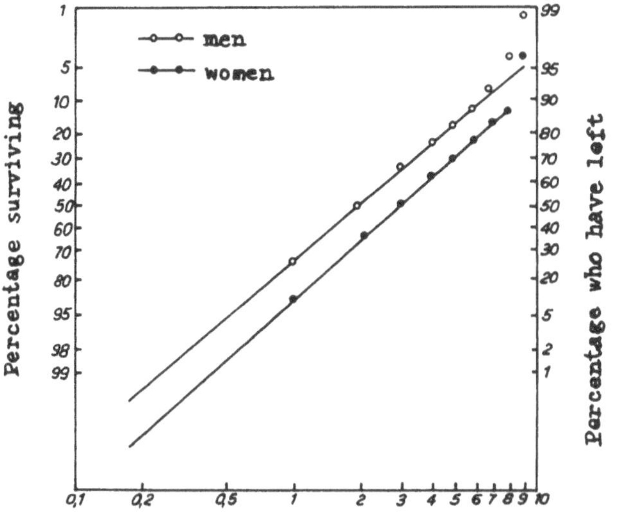

Fig.3 Comparison of leaving processes from industrial plant for women and men

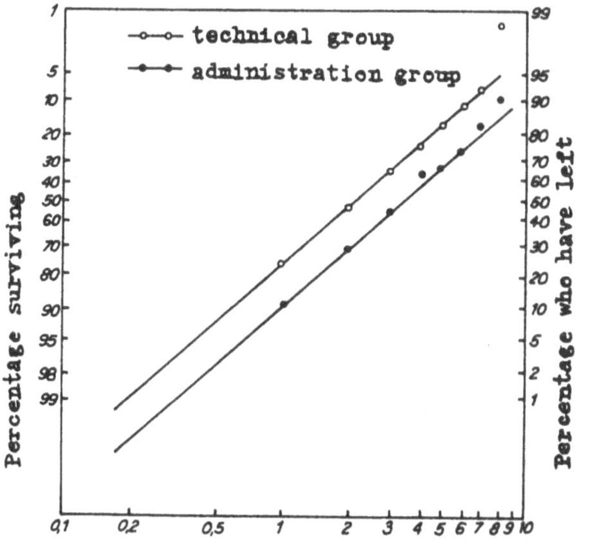

Fig.4 Graphical comparison of leaving processes from industrial plant for technical and administration groups

REFERENCES

/1/Bartholomew,D.,J.,and A.F.Forbes:Statistical Techniques for Manpower Planning.1979.
/2/Greń,J:Mathematical Statistics.Models and Exercises.Polish Scientific Publishers,Warsaw 1975 (in Polish).

QUANTITATIVE DESCRIPTION OF CHAOS IN A BIOCHEMICAL SYSTEM

H. Herzel, W. Ebeling[1)], Th. Schulmeister[2)]

ABSTRACT: We investigate the chaotic dynamics of a set of nonlinear differential equations. The orbital instability is characterized by Lyapunov exponents. We discuss several generalizations: the local expansion rate and the sensitivity on different scales. The dynamics can be visualized as next-amplitude maps. Further we present a correlation function and estimate the predictability.
We found that the sensitivity of chaotic trajectories survives under the influence of small fluctuations and that periodic "windows" disappear. Therefore generalizations of Lyapunov exponents in the presence of noise are used.

KEY WORDS: chaos, local Lyapunov exponents, next-amplitude map, correlation function, predictability, fluctuations

We consider a three-variable model that describes regulating mechanisms of a living cell [11] and discuss the influence of short-correlated Gaussian noise $G_i(t)$.

$$\begin{aligned} \frac{dx}{dt} &= 1 - Bx - xy^2 - Exy + z + G_1(t) \\ \frac{dy}{dt} &= A(xy^2 - y + D) + G_2(t) \\ \frac{dz}{dt} &= F(Exy - z) \end{aligned} \tag{1}$$

Computations show that eqs. (1) exhibit self-sustained oscillations, coexistence of different attractors, period-doubling and deterministic chaos [11] . The influence of noise on oscillations was in detail discussed elsewhere [2,6] . Here we focus on the quantitative description of chaos at certain parameter values (A=4; B=0.35; D=0.1; E=1.5 or 1.5095; F=0.2) and the role of fluctuations.
Chaos can be characterized by the Lyapunov spectrum and the attractor dimension, the latter can be estimated by a formula proposed by Kaplan and Yorke [5] . Using the standard algorithm [13] we find

$$\begin{aligned} &\lambda_1 = 0.0283 \pm 0.0005\,;\ \lambda_2 = 0\,;\ \lambda_3 = -0.248 \pm 0.001 \\ &d_{KY} = 2 + \frac{\lambda_1}{|\lambda_3|} = 2.114 \pm 0.002 \end{aligned} \tag{2}$$

1) Sektion Physik, Humboldt-Univ., 1040 Berlin, Invalidenstr. 42, DDR

2) Inst. für Molekularbiologie, 1115 Berlin, Lindenberger Weg 70, DDR

A positive Lyapunov exponent indicates the instability of orbits. More information than the averaged quantity λ_1 the local Lyapunov exponent $\lambda_1(\underline{x}(t))$ contains [4,6,10]. We compute it by solving the linearized dynamical equations. They describe the behaviour of infinitesimal deviations $\underline{q}(t)$ from the trajectory $\underline{x}(t) = (x,y,z)$.

$$\frac{dq_i}{dt} = \sum_j \frac{\partial f_i}{\partial x_j}(\underline{x})\, q_j = \sum_j J_{ij}(\underline{x})\, q_j \quad ; \quad \frac{dx_i}{dt} = f_i \tag{3}$$

$$\frac{d}{dt}\|\underline{q}\| = \frac{d}{dt}\left(\sum_{i=1}^{3} q_i^2\right)^{\frac{1}{2}} = \frac{\sum_{i,j=1}^{3} q_i J_{ij} q_j}{\|\underline{q}\|^2}\|\underline{q}\| = \lambda_1(t)\,\|\underline{q}\| \tag{4}$$

The quantity $\lambda_1(\underline{x}, \underline{q})$ measures how a perturbation $\underline{q}$ exponentially shrinks or grows in the course of time. It depends on the trajectory $\underline{x}$ and the direction of $\underline{q}$. Averaging leads to the usual Lyapunov exponents. Fig. 1 shows that contraction and expansion change rapidly. The inhomogenity can quantitatively expressed by the standard deviation of $\lambda_1(t)$ which was called "Non-Uniformity Factor" (NUF) [10].

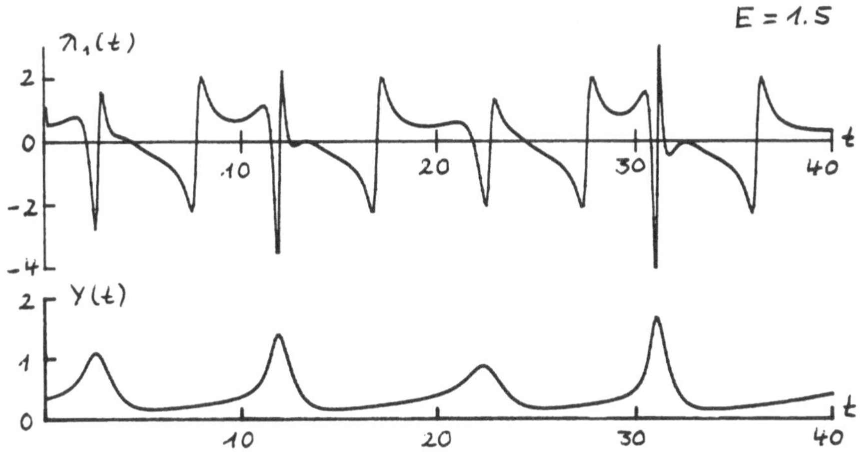

Fig. 1 The local expansion rate $\lambda_1(t)$ and the corresponding y(t)

The chaotic dynamics has been studied by next-amplitude maps (Fig. 2,4) [6]. We calculated the Lyapunov exponent of the maps from the slope of the curves. Fig. 2a shows a two-band structure. In Fig. 3a the decay of correlations in one of the bands is presented using eq. (5).

$$C(n) = \frac{1}{m}\sum_{i=1}^{m} Ymax_{2i}\, Ymax_{2i+2n} - \left(\frac{1}{m}\sum_{i=1}^{m} Ymax_{2i}\right)^2 \tag{5}$$

As a result of the instability of chaotic orbits the finite precision of any measurement leads to a limited predictability. The quantity I(n) (see [1, 3] for the definition) gives the amount of information contained in a prediction made n iterations into the future. For a given partition (20 boxes) the information is nearly lost after 10 iterations (Fig. 3b full line). One bit is preserved due to the two-band structure of the map [7] . The decay can be estimated by the Lyapunov exponent (broken line) [12] .

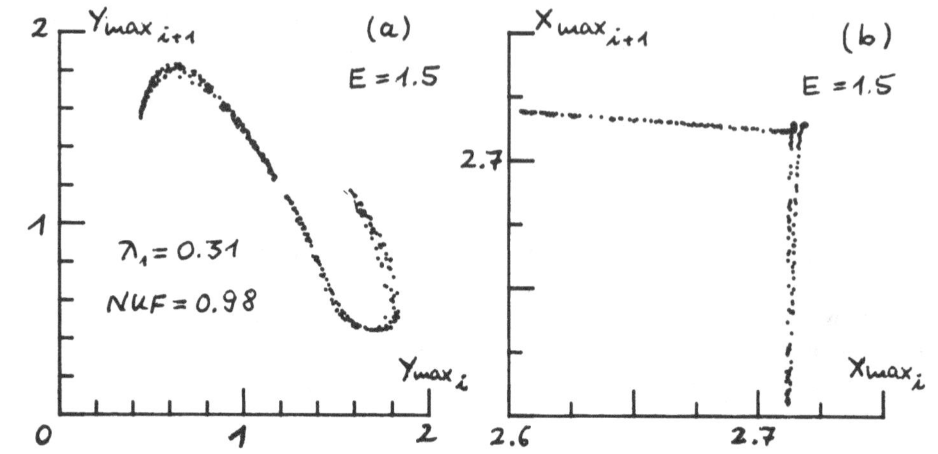

Fig. 2. Plot of the maxima against the maximum before

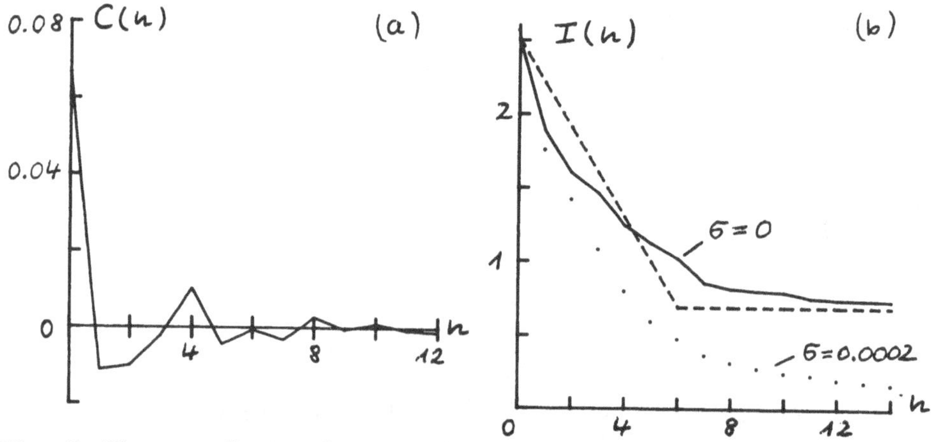

Fig. 3 The correlation function (m=1000) and the decay of information using the map of Fig. 2a

Real systems are always in contact with a fluctuations source (e.g. a fluctuating supply in our model). There are several contributions which show that noise acts as an disordering field for chaotic dynamics [1,9]. Recently the opposite effect namely "noise-induced order" was found in the B-Z model [8] . As a first estimation of the influence of noise we calculated a quantity $\lambda_1(\varepsilon)$ which measures the sensitivity of the orbit on different scales ε [13] .

$$\lambda_1(\varepsilon) = \lim_{n\to\infty} \lim_{\tau\to 0} \frac{1}{n} \sum_{i=1}^{n} \frac{1}{\tau} \log \frac{\| \underline{x}_2(t_i+\tau) - \underline{x}_1(t_i+\tau) \|}{\varepsilon} \qquad (6)$$

with $\| \underline{x}_2(t_i) - \underline{x}_1(t_i) \| = \varepsilon$

In the B-Z model $\lambda_1(\varepsilon)$ decreases with ε. In our model the expansion remains nearly constant up to $\varepsilon = 0.01$ ($\lambda_1(0.01) = 0.032 \pm 0.001$). We simulate the perturbations by adding every 0.1 time units Gaussian pseudo-random numbers with a standard deviation σ. The influence of noise can be visualized in the corresponding next-amplitude maps (Fig.4). Due to the band-merging (Fig. 4a) the information of one bit vanishes (dotted line in Fig. 3b). At E = 1.5095 the maximal Lyapunov exponent (without noise) is zero indicating a periodic window. Noise leads to a behaviour that is quite similar to the chaos at E = 1.5 (compare Fig.2a and 4b).

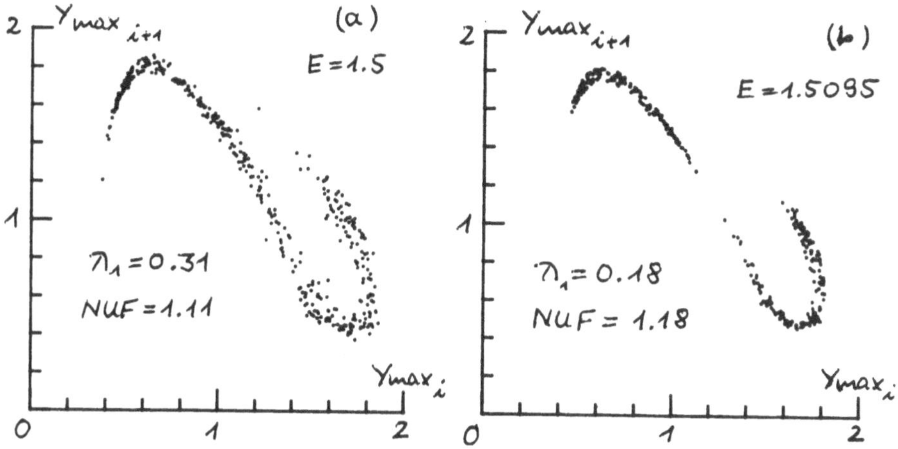

Fig. 4. Next-amplitude maps in the presence of noise

Generalizations of Lyapunov exponents in the presence of noise are complicate even in one-dimensional maps [1]. Here we calculate λ_1 from the slope of the maps (Fig. 4) and by averaging over the local Lyapunov exponents along the noisy trajectory (E= 1.5; $\sigma = 0.001$; $\lambda_1 = 0.028 \pm 0.002$; E = 1.5095; $\sigma = 0.0004$: $\lambda_1 = 0.018 \pm 0.002$). Keeping in mind that the period from one maximum to the next is about 10 it is seen that both methods lead to similar results. Thus we found that chaos survives at small noise levels and that windows disappear.

References:

1 Crutchfield, J.P., Packard N.H., Physica 7D (1983) 201-223
2 Ebeling, W. et al., studia biophysica, 98 (1983) 147-154
3 Farmer, J.D., Z. Naturforsch. 37a (1982) 1304-1325
4 Fujisaka, H., Progr. Theor. Phys. 70 (1983) 1264-1275
5 Grassberger, P., Physica 13D (1984) 34-54
6 Herzel, H. et al. in Proc. UNESCO-Conf. on System Theory, Eisenach 1984, Akademie-Verlag, Berlin 1985
7 Leven, R., B. Pompe to appear in Ann. der Phys. 1985
8 Matsumoto, K., I. Tsuda, J. Stat. Phys. 31 (1983) 87-106
9 Mayer-Kress. G., Haken, H., J. Stat. Phys. 26 (1981) 149-171
10 Nicolis, J.S., et al., Z. Naturforsch. 38a (1983) 1157-1169
11 Schulmeister, Th., studia biophysica, 72 (1978) 205-206
12 Shaw, R., Z. Naturforsch. 36a (1981) 80-112
13 Shimada, I., T. Nagashima, Prog. Theor. Phys. 61 (1979) 1605-1616

MODELLING OF THE TRANSIENT RESPONSE OF THE NEUROMUSCULAR SYNAPSE

Bernd Fritzsche [1)]

The neuromuscular synapse provides a unidirectional flux of signals (action potentials) from the nerve to the muscle. In the sense of cybernetics this junctional structure is an important nonlinear biological transfer element. A mathematical description of its operation would be of interest for a better understanding of experimentally observed transmission characteristics and of possible pharmacological influences. The present paper deals with results of simulations in the range from miniature end-plate potentials to end-plate potentials in vivo.
Keywords: modelling and simulation in biology - neuromuscular transmission - miniature end-plate potential - end-plate potential

1. Introduction

Motor units are special functional systems which convert innervation patterns into muscle contractions. A motor unit consists of a motoneuron, its motor axon and the muscle fibres excited by the motoneuron. The impulse discharge pattern of the motoneuron (one or more action potentials) is transmitted to the muscle fibres via the neuromuscular synapses or motor end-plates. This occurs in a chemical way by release of the transmitter acetylcholine from the presynaptic area. After release, acetylcholine diffuses through the synaptic cleft into the subsynaptic membrane of the postsynaptic site, where it forms an unstable chemical complex with receptor molecules. The acetylcholine-receptor complex causes a configuration change in the subsynaptic membrane resulting in an increase of the end-plate conductance, an end-plate current and a subsynaptic depolarization (end-plate potential). The acetylcholine released by the decay of the unstable chemical complex is hydrolyzed by acetylcholinesterase to choline and acetic acid which are taken up into the presynaptic terminal and used for resynthetizing the transmitter. The enzymatic destruction of acetylcholine terminates the configuration change of the subsynaptic membrane and leads to a time limitation of its increased conductance and, consequently, of the synaptic current and of the depolarization. Postsynaptic action potentials are evoked if the subsynaptic depolarization exceeds the threshold of the sarcolemma. In vivo spikes generated by a motoneuron always induce action potentials, and hence contraction, in the postsynaptic muscle fibres of a given motor unit.

A miniature end-plate potential is caused by random release of an elementary quantum acetylcholine from the non-depolarized presynaptic area. This minimum subsynaptic depolarization is unable to generate an action potential at the postsynaptic site.

For details see reviews by Gage [3], Kuffler et.al. [4], Takeuchi [6], Bernard [1] and Neumann [5] .

---

1) Carl-Ludwig-Institut für Physiologie der Karl-Marx-Universität DDR-7010 Leipzig Liebigstraße 27

2. The Model

The interrelated chemical and electrical processes may be described by using the following equations with one accessory assumption (see [2]):

$$\dot{c}_A^e = - k_A c_A^e + k_A c_A \tag{1.1}$$

$$\dot{c}_A = k_A c_A^e - k_A c_A + k_{-1} c_X + k_{-2} c_Y - k_{+1} c_A c_R - k_{+2} c_A c_E \tag{1.2}$$

$$\dot{c}_X = - k_{-1} c_X + k_{+1} c_A c_R \tag{1.3}$$

$$\dot{c}_Y = - (k_{-2} + k_{+3}) c_Y + k_{+2} c_A c_E \tag{1.4}$$

$$c_R = c_R^o - c_X \tag{1.5}$$

$$c_E = c_E^o - c_Y \tag{1.6}$$

$$\dot{E}_M = - \frac{G_{SC} + G_K N_L V_{SM} c_X(t)}{C_S} E_M + \frac{G_{SC}}{C_S} E_R \tag{1.7}$$

accessory assumption: If the solution $E_M(t;t_o,E_{Mo})$ exceeds the threshold E_C of the sarcolemma a postsynaptic action potential is generated.

In this system of algebraic and differential equations means

c_A^e concentration of free acetylcholine in the synaptic cleft,

c_A concentration of free acetylcholine in the subsynaptic membrane,

c_R concentration of free receptors,

c_X concentration of formed acetylcholine-receptor complexes,

c_E concentration of free acetylcholinesterase,

c_Y concentration of formed acetylcholine-acetylcholinesterase complexes,

c_R^o total concentration of receptors (integral of motion),

c_E^o total concentration of acetylcholinesterase (integral of motion),

k_A velocity constant for the diffusion of acetylcholine out of the synaptic cleft into the subsynaptic membrane,

k_{+r} kinetic constant of the forward reaction r $(k_{+r} > 0 \ \forall r)$,

k_{-r} kinetic constant of the backward reaction r $(k_{-r} \geq 0 \ \forall r)$,

E_M subsynaptic depolarization caused by presynaptic release of acetylcholine,

E_R resting membrane potential of the sarcolemma,

C_S capacity of the subsynaptic membrane,

G_{SC} resulting conductance of all extrasynaptic resistances (parts of the extrasynaptic sarcolemma, of the intracellular and extracellular spaces) which influence the formation of the subsynaptic depolarization,

G_K conductance step which is caused by formation of one acetylcholine-receptor complex,

N_L Loschmidt's constant,

V_{SM} volume of the subsynaptic membrane.
G_{SV} as the end-plate conductance may be described by the expression (see [2])

$$G_{SV}(t) = G_K N_L V_{SM} c_X(t) \quad . \tag{2}$$

The physically possible initial values of the differential equations of (1.1 - 1.7) must satisfy the relations

$$0 \leq c^e_{Ao} < v_A^{-1} \qquad 0 \leq c_{Ao} < v_A^{-1}$$

$$0 \leq c_{Xo} \leq c^o_R \qquad 0 \leq c_{Yo} \leq c^o_E$$

$$E_R \leq E_{Mo} < E_C$$

where v_A is the mole volume of acetylcholine.

3. The Simulation Results

With

$$c^e_{Ao} = c^o_A > 0 \quad c_{Ao} = 0 \quad c_{Xo} = 0 \quad c_{Yo} = 0 \quad E_{Mo} = E_R$$

as initial values the equations (1.1 - 1.7) were simulated by computer for biologically relevant parameters which are derived from neuromuscular synapses of the rat diaphragm or of the frog sartorius.

3.1. Miniature end-plate potential (mepp)

a) maximum end-plate conductance
   experimental: 1×10^{-7} - 2×10^{-7} S

   simulated : 8.5×10^{-8} - 1.5×10^{-7} S
b) duration of the mepp
   experimental: longer as 10 ms
   simulated : 15 ms
c) peak time of the mepp
   experimental: 1 ms
   simulated : 0.79 ms
d) maximum subsynaptic depolarization
   experimental: 0.5 - 1 mV
   simulated : 0.76 mV

3.2. End-plate potential in vivo (epp)

a) maximum end-plate conductance
   experimental: 1×10^{-5} - 5×10^{-5} S

   simulated : 2.6×10^{-5} - 4.6×10^{-5} S
b) duration in which the subsynaptic depolarization is reaching the threshold
   experimental: 0.25 - 1 ms
   simulated : 0.17 ms

Taking into consideration the time for release (0.2 ms), the peak time of the mepp and the duration in which the subsynaptic depolarization is reaching the threshold are also in a very good agreement with the experimentally observed time course of adequate subsynaptic depolarizations.

3.3. Results of simulations in the range from miniature end-plate potentials to end-plate potentials in vivo

It is remarkable that the model reproduces the experimental observation after which a maximum subsynaptic depolarization of about 15 mV is recorded if the end-plate conductance has a value of 2.6×10^{-6} S.

4. Summary

A model of neuromuscular transmission had been proposed which considers morphological structure and interrelated chemical and electrical mechanisms. The adequate kinetic scheme had been described by means of a system of nonlinear first order differential equations. The network of corresponding electrical processes had been treated as a linear first order differential equation with one variable coefficient. By the analytical investigation of derived differential equations detailed informations about the qualitative behaviour of the solutions were obtained. The results of modelling characterize the neuromuscular synapse as a "perfect" transfer element, since all the possible dynamic solutions tend to exactly one singular point - the "resting state", independent of initial values and all other parameters. The special time-dependent solutions simulated by computer with biologically relevant parameters are in good agreement with experimentally determined transient response of the neuromuscular synapse. The investigations were finished with an outlook of the possible pharmacological application of the developed model (see [2]).

5. References

[1] Barnard, E. A.: Neuromuscular transmission - enzymatic destruction of acetylcholine. In: The peripheral nervous system, pp. 201-224. Hubbard, J. I., ed. New York: Plenum Press 1974

[2] Fritzsche, B.: Qualitative und quantitative Analyse des nichtlinearen Obertragungsgliedes "Neuromuskuläre Synapse". Promotion A, Sektion Physik der Karl-Marx-Universität, Leipzig 1983

[3] Gage, P. W.: Generation of end-plate potentials. Physiol. Rev. 56 (1976) 177-247

[4] Kuffler, S. W., Nicholls, J. G.: From neuron to brain. Sunderland, MA: Sinauer Associates 1977

[5] Neumann, E.: Chemical representation of ion flux gating in excitable biomembranes. In: Molecular aspects of bioelectricity, pp. 27-43. Schoffeniels, E., Neumann, E., eds. Oxford: Pergamon Press 1980

[6] Takeuchi, A.: Junctional transmission. I. Postsynaptic mechanisms. In: Handbook of physiology and clinical physiology series. Vol. I/I, pp. 295-327. Baltimore: The Williams & Wilkins Company 1977

A COMBINATORIAL APPROACH TO THE PROBLEM OF SELF-ASSEMBLY OF SUBMOLECULAR STRUCTURES

Ja.S.Smetanich [1], V.V.Kornilov [2] and Ja.B.Kazanovich [1]

A formal scheme is proposed in an attempt to reflect the real features of submolecular structure self-assembly. The main results concern the theorem yielding a polynomial-time algorithm for identifying a structure generated by self-assembly, and the symmetry properties of such a structure.
Keywords: submolecular structure, self-assembly, symmetry.

Our approach to the problem of self-assembly may briefly be described as follows [3,4]. Geometric figures constructed from separate building blocks (subunits) are considered in 3-D Euclidean space. Subunits may be of arbitrary shape, and the entire figure is assumed to be a solid. By the assembly of a figure a process of consecutive bonding of subunits is meant that is determined by the rules of their pair interaction. When the assembly is over, i.e. a figure cannot be further enlarged according to the given bonding rules, the figure is called a structure. A structure is unique if the process of assembly cannot give any other structure. The unique structure may be interpreted as a submolecular structure generated by self-assembly.

Further in the paper the spheres of various radii will be used as subunits with the only objective to simplify the narrative. The figure assembled from a finite set of spheres will be called a framework. The spheres in the framework are either contacting or not crossing, and their bonds are assumed to be rigid (the spheres cannot rotate in respect to one another). Two frameworks are regarded as identical if they can be superimposed by a proper motion (the first kind of motion). Let us index each point of contact in a sphere by a letter of some alphabet. The framework indexed in such a manner is called a structure on the framework or simply a structure. We call the indexed sphere a subunit and its points of contact faces. Two structures are identical if they have identical frameworks and the indexes of corresponding faces are the same.

Now we are going to describe the rules of self-assembly. Let a and b be two contacting subunits of a structure. Being extracted together from the structure, they form a figure called an elementary complex and denoted by $a \oplus b$. It should be emphasized that there is a difference between a free and a contacting face. For example, in the elementary complex shown in Fig. 1 the faces α and η are contacting, while β and φ are free. The structure S is said to be genera-

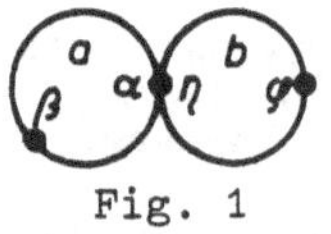

Fig. 1

1) Institute of Biophysics, USSR Academy of Sciences
2) Research Computing Centre, USSR Academy of Sciences
Both: 142292 Pushchino, Moscow Region, USSR

ted by original data $\mathcal{D} = \langle A, Q \rangle$ where A is the set of subunits in S, and Q is the set of elementary complexes of S.

Consider an arbitrary figure assembled from the subunits of A so that all its elementary complexes belong to Q. This figure is called a complex corresponding to $\mathcal{D}$. A complex that cannot be further enlarged is called a structure corresponding to $\mathcal{D}$. Note that the complex may not be a part of S, as well as the structure corresponding to $\mathcal{D}$ may not be identical to S (in particular, it may have free faces). If the structure corresponding to $\mathcal{D}$ has no free faces it is certainly a structure on the framework. The following sentence has been proved in [3].

Theorem 1. Let $\mathcal{D}$ be the original data of a structure S on the framework. Then an arbitrary complex corresponding to $\mathcal{D}$ can be enlarged up to a structure corresponding to $\mathcal{D}$.

We say that a structure is unique if there is no other structure corresponding to $\mathcal{D}$. The set of unique structures will be denoted by $\mathcal{U}$. To formulate the criterion for identifying unique structures, we need some additional notions.

Consider original data $\mathcal{D} = \langle A, Q \rangle$ and a subunit $a \in A$. We say that the subunit a has single-valued neighbourhood at its face α (that is fixed for the moment) if there is only one elementary complex $(a \oplus x) \in Q$ such that x contacts with a at the face α. Otherwise, multi-valued neighbourhood is involved.

Let $a \oplus x$ and $a \oplus y$ be two elementary complexes of a structure S on the framework where x and y contact with the same face of a, so, a has multi-valued neighbourhood (Fig. 2a). We introduce the operation of substitution of y for x in S. This operation implies that x is removed from $a \oplus x$ in S and y is inserted instead. If there are subunits of S that prevent y from being inserted, they should be removed too (but no other subunits). Denote by L the complex resulting from the substitution and contacting with the new-formed elementary complex $a \oplus y$ (Fig. 2b). We say that substitution of y for x can be realized if the complex $L \oplus (a \oplus y)$ is a part of S.

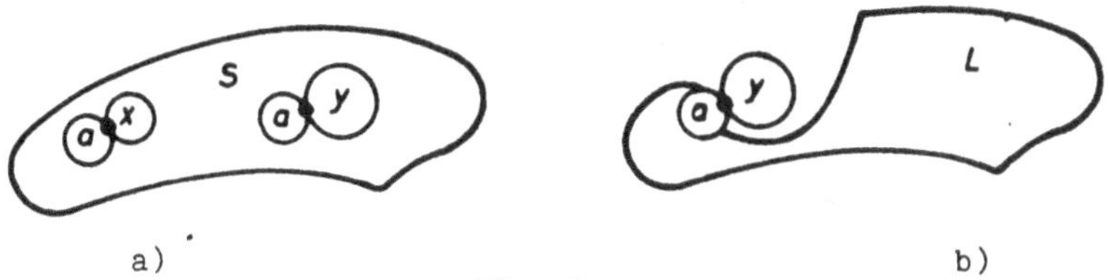

Fig. 2

Using the theorem 1 and the operation of substitution, we have proved the following theorem [4].

Theorem 2. A structure S on the framework is unique iff all the substitutions in S can be realized.

Applying this theorem, we have designed a polynomial-time algorithm for deciding whether S is unique or not. It has been shown in [4] that the number of elementary computing operations needed to make this decision is less than $O(m^2 n^2)$ where n is the number of subunits in S and m is the total number of their faces.

Now we turn to the symmetry properties of structures.The symmetry group of a structure S is a finite group of structure "self-coincidences" point group of the first kind or group of proper rotations). We write $\Gamma(S)$ for the maximal order symmetry group of S . Note that each group of proper rotations can be the symmetry group of a structure.

The group $\Gamma(S)$ acting on the subunits of S subdivides them into equivalent classes called orbits [5] . Each subunit of S can be of one of the following three types: (1) a subunit with no symmetry axes of S going through it; (2) a subunit lying along one of the symmetry axes of S and (3) a single subunit locating in the symmetry centre of S .Using known facts on the number of elements on an orbit [5] , we derive the following formula for the total number M of subunits in S :

$$M = \kappa N + \sum_{i=1}^{m} \frac{N}{p_i} + \delta,$$

where N is the order of $\Gamma(S)$, κ is the number of orbits with subunits of the first type, m is the number of orbits with subunits of the second type, p_i is the order of a symmetry axis for the subunits of the second type locating on the i -th orbit and δ is the number of the subunits of the third type (δ is 0 or 1).

For a special class of unique structures a complete information on the position of subunits has become available. Let a structure S be assembled from subunits with asymmetric arrangement of points of contact. We suppose besides that the points of contact of a subunit in S are different and that each subunit in S has single-valued neighbourhood at all its faces. The set of such structures will be denoted by O . It is easy to show that $O \subset U$. The following theorem has been proved in [4] .

Theorem 3. Let S be a structure on the framework with the symmetry group $\Gamma(S)$. Then $S \in O$ iff there are no identical subunits on different orbits of $\Gamma(S)$.

As consequence for $S \in O$, we obtain that the number of different subunits in S is equal to the number of orbits K generated by $\Gamma(S)$,and that the total number of subunits in S is κN where N is the order of $\Gamma(S)$.

In conclusion we will note that the proposed formal scheme may be used as a starting point for modelling self-assembly of submolecular structures such as capsids of spherical viruses, multienzyme complexes and microtubes [1,2] .

Literature.

[1] Caspar, D.L.D.: Movement and self-control in protein assemblies. Quasi-equivalence revisited. Biophysical J. 32 (1980) 1, 103-138.

[2] Oliver, R.M. and Reed, L.J.: Multienzyme complexes. Electron microscopy of proteins.V. 2., ed. by J.R.Harris, Academic Press 1982.

[3] Smetanich, Ja.S. and Kornilov, V.V.: A combinatorial approach to the problem of self-assembly of submolecular structures. Local interaction (in Russian). Preprint, Sci.Centre Biol.Res., USSR Acad.Sci., Pushchino 1983.
[4] Smetanich, Ja.S., Kornilov, V.V. and Kazanovich, Ja.B.: A combinatorial approach to the problem of self-assembly of submolecular structures. Uniqueness and symmetry (in Russian). Preprint, Sci.Centre Biol.Res., USSR Acad.Sci., Pushchino 1984.
[5] Weyl, H.: Symmetry. Princeton University Press, New Jersy 1952.

A MATHEMATICAL MODEL SYSTEM USED TO STUDY BIOMASS RESOURCES IN HUNGARY

Zsolt Harnos [1)]

Summary: The paper is intended to give an account of the overall structure of the model system used in the recently accomplished "Biomass Study". In 1981, a research program coordinated by the Hungarian Academy of Sciences under the title "Long-term Perspectives in Utilization of Materials of Biological Origin", or Biomass Study in short, was started. In the Biomass Study, the question has been examined that reckoning with the natural limits to production, how production and utilization of biomass should be shaped in order to reach different strategic targets. In other terms, the dynamics of the biomass production-transformation-utilization system was put into the focus of investigation with special attention to the interactions between human activities and natural conditions.

Keywords: Model system of biomass utilization, control problem, hierarchical system, multiobjective optimization.

For the sake of analyzing the expected development of biomass production and utilization a model-system was worked out describing

- plant production
- animal husbandry
- the movement of biomass in the area of plant production, animal husbandry and food industry .

The system were described by a hierarchical system of control problems, with

- the dynamic interrelations between the state transition of production and habitat described by the plant production model.
- the most important relationships playing a role in the breeding process of the different species described by the model of animal husbandry.
- the block of biomass utilization having a controlling character. Plant production, animal husbandry and other possibilities for utilization are linked, and balance within the system, the full circle of biomass is ensured by this part.

The mathematical description of the plant production model:

Control conditions

$$B\,\underline{u}_N(t) \leq \underline{u}_N^o(t)$$
$$\underline{u}_N(t) \geq \underline{0}$$

State equation

$$\underline{x}_N(t+1) = D\,\underline{x}_N(t) + E\,\underline{z}_N(t) + C\,\underline{u}_N(t) - \underline{d}(t)$$
$$\underline{x}_N(t_o) = \underline{x}_N^o$$

1) Bureau for Systems Analysis
State Office for Technical Development
Budapest, P.O.B. 565. H-1374 Hungary

Equations describing the functioning of the system

$$\underline{x}_N(t) = F \cdot \underline{z}_N(t)$$
$$A\ \underline{z}_N(t) = \underline{b}(t)$$
$$H_N\left(z_N(t),\ y_N(t),\ u_N(t)\right) \leq 0$$

The output of the system /the product mix/ is described by

$$\underline{y}_N(t) = G(t)\ \underline{z}_N(t)$$

together with the constraints

$$\underline{y}_N^o(t) \leq \underline{y}_N(t) \leq \underline{y}_N^1(t)$$

The meaning of the different groups of conditions is as follows: The characteristics of the production site in plant production - i.e. area of land available and its composition according to fertility classes - are represented by the state variables of the system.
The individual groups of variables have the following meaning:

$\underline{x}(t)$	quality composition of lands
$\underline{z}(t)$	actual sowing structure, also containing nutrient levels and agrotechnology in an implicite form
E	effects on the land of the agrotechnology applied
C	amelioration and other qualitative changes of the land
d(t)	the vector describing the decrease of the land

The equality

$$\underline{x}_N(t) = F\ \underline{z}_N(t)$$

in the equations describing the functioning of the system refers to the utilization of available area, while the condition

$$A\ \underline{z}_N(t) \leq \underline{C}(t)$$

to the relationships of sowing structure, and the condition

$$H\left(\underline{z}_N(t),\ \underline{y}_N(t),\ \underline{u}_N(t),\ S(t)\right) \leq 0$$

to the connections between nutrient supply and product mix. Representing the product mix. Representing the product mix, $\underline{y}_N(t)$ is the output of the system, to be computed from sowing structure and average yields on the basis of the equality

$$\underline{y}_N(t) = G(t)\underline{z}_N(t)$$

The matrix $\underline{G}(t)$ is for describing the development of yields, with the elements determined by using the prognoses and development curves elaborated within the framework of the Survey of Agrotechnological Potential project, according to production site types and nutrient supply levels.

The product mix, y_N is controlled by

- domestic consumption,
- fodder needs by animal husbandry,
- processing capacity of the food industry,
- raw material needs by different industrial sectors, and
- export-import possibilities.

The structure of the model of animal husbandry is similar to that of the plant production model, and therefore I do not go into the details. Relations connecting the sectors are given in the form of inequalities

$$H_o\left(y_N(t),\ y_A(t),\ f_N(t),\ f_A(t)\right) \leq 0$$

The models presented here have been elaborated on two levels of aggregation. Relationships are described on the macro level by the more aggregated version, while the more detailed versions are used to describe the production within regions or sectors. This solution was necessitated by two causes:

- the detailed model describing the production and utilization of biomass in the whole country is practically unsolvable because of its large size,
- real decision processes are better represented by a hierarchical structure.

Central planners make their decisions on the basis of macro level relationships concerning the allocation of central resources, and formulate at the same time production targets. Production structure is then set up by the sectors according to their preference system - not being necessarily identical with central goals. We tried to represent these two sides of the coin at developing the model system.
The structure of the model system is as follows

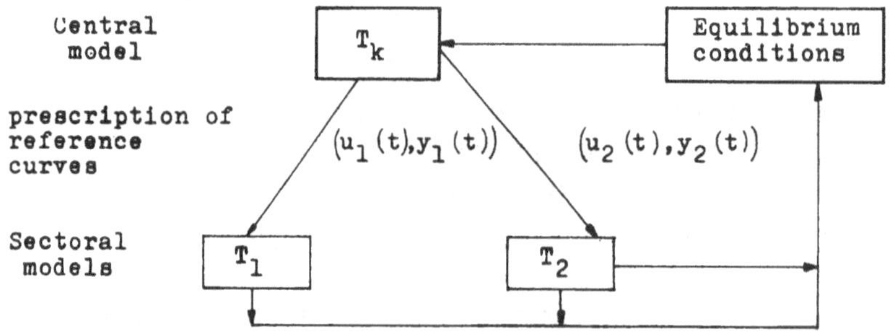

The model can be represented as one single system of inequalities. In this case the matrix of coefficients has the following structure:

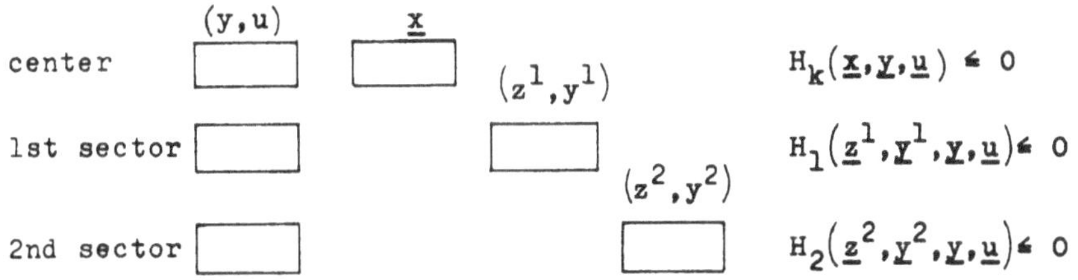

Contradicting interests can be best represented in the system, if a vector valued goal function is given, and in such a case a multiobjective optimization is to be solved.

It can be proved that if certain rules are observed while aggregating, then the set of the solutions of the sectoral problems belonging to a solution of the central problem is not void, i.e. the hierarchical problem has a solution.

REFERENCES

[1] Csáki, Cs., Harnos Zs., Láng I.: Agricultural Development and Ecological Potential : The Case of Hungary. Kieler Wissenschaftsverlag Vauk,/1984/

[2] Haith, D.A.: The modelling of environmental impacts of crop production. Proc. Modelling Agricultural - Environmental Processes in Crop Production /CP-82-S5/

[3] Harnos, Zs.: The Survey of the Agroecological Potential of Hungary - A brief summary. Collaborative Paper, IIASA /CP-82-21/ /1982/

[4] Harnos, Zs., Pór, A., Sivák, J.: A Hierarchical Multiobjective Planning Model. XII. Hungarian Conference on Operations Research, Balatonfüred /1983/

[5] Nijkamp, P.: Theory and application of environmental economics. Vol.1., North-Holland Publ. Comp. /1977/

[6] The Biopotential of Hungary: Present and Future. Club of Rome Conference, Budapest /1983/

DEVELOPMENT OF AN OPTIMAL LINEAR CONTROLER FOR ANTIBIOTIC FERMENTATION

V. Schulz[1)]

SUMMARY

The optimization method based on the Pontryagin's continuous maximum principle and was used to predict the optimal specific growth rate profile for a typical antibiotic fermentation model with a given nominal set of start values and parameters. If this set is unknown or known only to be in some domains for the process, then it is necessary to use an optimal linear controler (OLC). The behaviour is demonstrated by simulation runs.

Good results of model aided optimization of antibiotic fermentation presume that the model for the process must accurate reflect the main effects of growth and product formation. If the nominal set of start values and parameters in the process model is unknown or known to be in some domains or disturbed randomly, then the open-loop-control gives nonoptimal results. In this case the development and application of an optimal controler is necessary. In this paper the global solution for an OLC published by Burmeister and Winkler /1/ was used for improvement of process productivity taking into account low perturbations of nominal set of start values and parameters.

The following phenomenological model for the nominal optimization problem is proposed:

$$dx_1/dt = (\mu(t)-\beta)x_1 \; ; \quad x_1(0) = x_{10}^0$$

$$dx/dt = \mu(t)x_1 \; ; \quad x(0) = x_0^0$$

$$dp/dt = k_p(\mu,x)x_1 \; ; \quad p(0) = p_0^0 = 0$$

$$Q = p(t_f)/t_f \xrightarrow[\mu]{} \max$$

where the characteristic function is:

1) Department of Biophysics, Central Institute of Microbiology and Experimental Therapy, Academy of Sciences of the German Democratic Republic, GDR 6900 Jena

$$k_p(\mu,x) = k_{pmax}\,\mu / \mu_p(2-\mu/\mu_p)(1/2 + 1/\pi \, \mathrm{arctg}(a(x_{max}-x)))$$

In this model (x_1,x,p) means the state vector (concentration of active for growth and product formation biomass, concentration of total biomass, concentration of antibiotic product); Q as performance criterion is the productivity of antibiotic formation at the final time t_f; μ as the control variable for optimization is the specific growth rate; the characteristic function k_p is the specific product formation rate as a function of μ /2, 3/ and x /4/; $(ß,k_{pmax},\mu_p,a,x_{max})$ is the nominal parameter vector and (x_{10}^0,x_0^0,p_0^0) the nominal start vector. The optimization is a problem with free t_f and free end point of state vector. The optimal fermentation profile (Fig. 1) was calculate by Pontryagin's continuous maximum principle with the program system MOPSI /5/.

If the state vector or the control variable was disturbed, then the performance criterion is always lower than in nominal solution. Where as the perturbation of parameter vector gives greater or smaller performance criterion (Fig. 2, 3 and 4). The OLC is the first order of optimal controler in the neighbourhood of the nominal optimal control profile for low differences between the predicted and actual, nonoptimal state vector (X_1,X,P). For measurable state vector (X_1,X,P) the OLC was calculated by:

$$\mu(X_1,X,P,t) = \mu_{opt}(t) + R_1(t)(x_1-X_1) + R_2(t)(x-X) + R_3(t)(p-P)$$

where (R_1,R_2,R_3) is the optimal linear control vector of the process and must be calculated before starting the fermentation. The quality of OLC for a certain process depends on the kinds of perturbations (Fig. 2, 3 and 4). The improvement of performance criterion varies from 20% to 100% if the perturbation act in more than one state equation. In opposite to that improvement is very low if the perturbations act only in one state equation. As shown in Fig. 2, 3 and 4 the performance criterion using the OLC is not decreased for all perturbations of the process.

The application of the OLC for antibiotic fermentation is sufficient. The OLC improves the performance criterion significantly, if the model represent the main effects and if perturbations act in more than one state equation. The developed OLC could be implemented in a microcomputer.

REFERENCES

/1/ Burmeister, H.-L., J. Winkler: Optimale Regelung zustandsgestörter Prozesse. msr, 21 (1978) 10,550-555.

/2/ Biryukov, V, V.: Equation of relationship between biomass growth and biosynthesis of metabolites under inhibition effect. Antibiotici (russ.), (1980) 10, 738-743.
/3/ Ohno, H., E. Nakanishi, T. Takamatsu: Optimal Control of a Semibatch Fermentation. Biotechnol. Bioeng., 18 (1976), 847-864.
/4/ Ryu, D.Y., A.E. Humphrey: Examples of Computer-aided Fermentation Systems, J. appl. Chem. Biotechnol., 23 (1973), 283-295.
/5/ Schorcht, R.: MOPSI - Programmbeschreibung. Programmbibliothek ZIMET Jena, 1983.

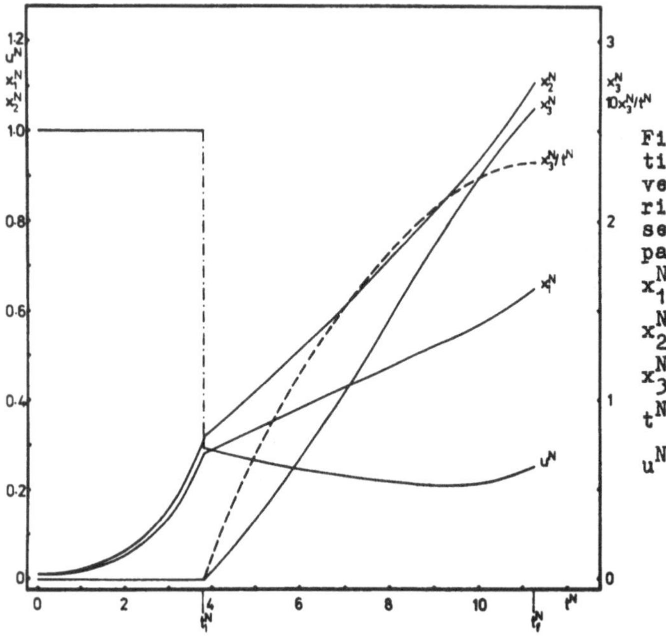

Fig. 1: Normalised optimal profile for state vector and control variable for a nominal set of start values and parameters. Here is

$x_1^N = x_1/x_{max}$,

$x_2^N = x/x_{max}$,

$x_3^N = p/u_{max}/k_{pmax}/x_{max}$,

$t^N = t/u_{max}$,

$u^N = /u/ / u_{max}$.

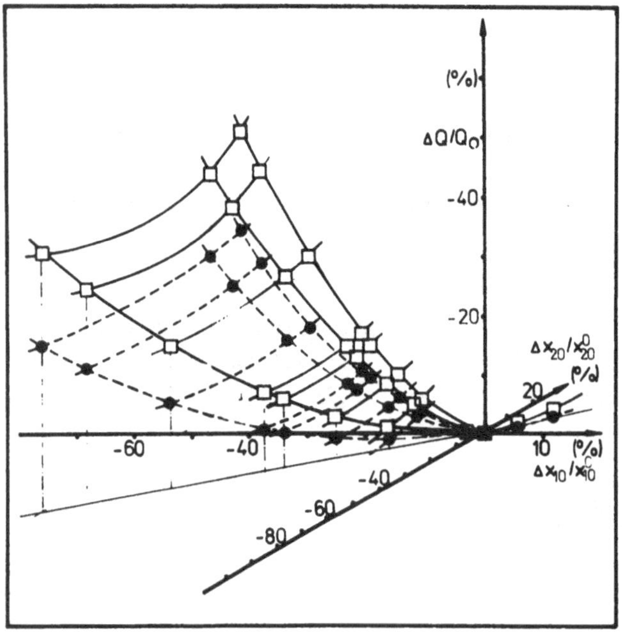

Fig. 2: The variation of performance criterion independence on the start value disturbance (□ - without OLC, ● - with OLC).

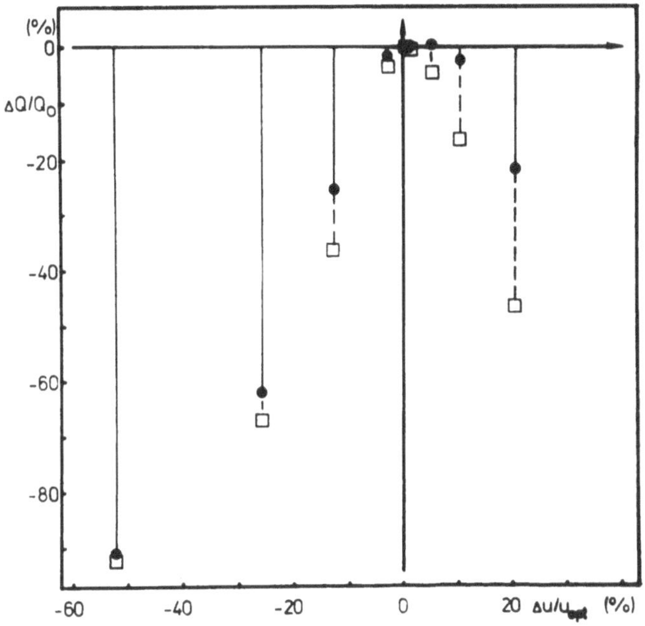

Fig. 3: The variation of performance criterion in independence on the control variable disturbance (□ - without OLC, ● - with OLC).

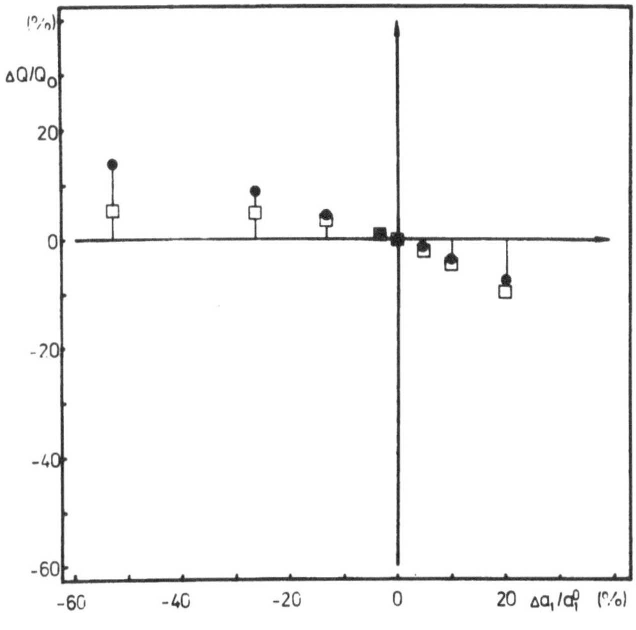

Fig. 4: The variation of performance criterion in independence on the perturbation of the parameter ß. Here is $a_1 = ß / u_{max}$ (□ - without OLC, ● - with OLC).

List of Authors

Band 1

Inverse and Improperly Posed Problems in Differential Equations

Proceedings of the Conference on Mathematical and Numerical Methods, held in Halle/Saale (GDR) from May 29 to June 2, 1979

Edited by Gottfried Anger
296 Seiten, 1979

Band 2

Fundamentals of Computation Theory FCT '79

Proceedings of the Conference on Algebraic, Arithmetic and Categorial Methods in Computation Theory, held in Berlin/Wendisch-Rietz (GDR) September 17–21, 1979

Edited by Lothar Budach
576 Seiten, 1979

Band 3

Decidability and Generalized Quantifiers

By Andreas Baudisch, Detlef Seese, Hans-Peter Tuschik, and Martin Weese
XII, 235 Seiten, 1980

Band 4

Zur Elastizitäts- und Thermoelastizitätstheorie

Potentialtheoretische Methoden in Anwendung auf verschiedene Modelle

Von Lothar Jentsch und Johannes Maul
157 Seiten, 1980

Band 5

Systems Analysis and Simulation 1980

Proceedings of the International Symposium held at Berlin (GDR)
September 1–5, 1980

Edited by Achim Sydow
496 Seiten, 1980

Band 6

Nichtlineare diskrete Optimierung

Von Eberhard Girlich und M. M. Kowaljow
218 Seiten, 1981

Band 7

Completeness Problems in the Structural Theory of Automata

By Jürgen Dassow
148 Seiten, 1981

Band 8

Darstellung von Lösungen linearer elliptischer Differentialgleichungen

Von Günther Wildenhain
92 Seiten, 1981

Band 9

Introduction to the Spectral Theory of Operators in Spaces with an Indefinite Metric

By I. S. Iohvidov, M. G. Krein, and H. Langer
120 Seiten, 1982

Band 10

Nonlinear Time-discrete Systems

A General Approach by Nonlinear Superposition

By Michael Gössel
115 Seiten, 1982

Band 11

Preprimal Algebras

By Klaus Denecke
162 Seiten, 1982

Band 12

Ergodic Theory and Related Topics

Proceedings of the Conference held in Vitte/Hiddensee (GDR)
October 19–23, 1981

Edited by Horst Michel
232 Seiten, 1982

Band 13

Singularly Perturbed Differential Equations

By H. Goering, A. Felgenhauer, G. Lube, H.-G. Roos, and L. Tobiska
176 Seiten, 1983

Band 14

Leben wir in einer Volterra-Welt?

Ein ökologischer Zugang zur angewandten Systemanalyse

Von Manfed Peschel und Werner Mende
233 Seiten, 1983

Band 15

Markov Decision Problems with Countable State Spaces Optimality Criteria — Algorithms — Clustering

By Hans Michael Dietz and Volker Nollau
174 Seiten, 1983

Band 16

Stochastische Verfahren zur Suche nach dem Minimum einer Funktion

Von Ryszard Zieliński und Peter Neumann
133 Seiten, 1983

Band 17

Invertibility and Asymptotics of Toeplitz Matrices

By Albrecht Böttcher and Bernd Silbermann
200 Seiten, 1983

Band 18

Structural Induction on Partial Algebras

Introduction to Theory and Application of Partial Algebras — Part II

By Horst Reichel
205 pages, 1984

Band 19

Algebraic Methods for Toeplitz-like Matrices and Operators

By Georg Heinig and Karla Rost
212 pages, 1984

Band 20

Frege Conference 1984

Proceedings of the International Conference held at Schwerin (GDR), September 10—14, 1984

Edited by Gerd Wechsung
408 pages, 1984

Band 21

Numerical Analysis of Selected Semilinear Differential Equations

Edited by Christian Großmann, Thomas Riedrich, and Hartmut Schönheinz
203 pages, 1984

Band 22

Singular Ordinary Differential Operators and Pseudodifferential Equations

By Johannes Elschner
200 pages, 1985

Band 23

Lotka-Volterra-Approach to Cooperation and Competition in Dynamic Systems

Proceedings of the 5th Meeting of UNESCO's Working Group on Systems Theory held on the Wartburg, Eisenach (GDR) at March 5—9, 1984

Edited by Werner Ebeling and Manfred Peschel
271 pages, 1985

Band 24

Convergence Structures 1984

Proceedings of the International Conference on Convergence Structures and Applications held in Bechyné (Czechoslovakia), September 17—21, 1984

Edited by Josef Novák, Werner Gähler, Horst Herrlich, and Jan Mikusiński
250 pages, 1985

Band 25

Parcella '84

Proceedings of the International Workshop on Parallel Processing by Cellular Automata and Arrays held in Berlin (GDR), September 25—27, 1984

Edited by Wolfgang Händler, Tamás Legendi, and Gottfried Wolf
239 pages, 1985

Band 26

Multiobjective and Stochastic Optimization Based on Parametric Optimization

By Jürgen Guddat, Francisco Guerra Vasquez, Klaus Tammer, and Klaus Wendler
171 pages, 1985

Band 27

Systems Analysis and Simulation 1985 I: Theory and Foundations

Proceedings of the 2nd International Symposium on Systems Analysis and Simulation held in Berlin (GDR), August 26—31, 1985

Edited by Achim Sydow, Manfred Thoma, and Robert Vichnevetsky
496 pages, 1985

Band 28

Systems Analysis and Simulation 1985 II: Applications

Proceedings of the 2nd International Symposium on Systems Analysis and Simulation held in Berlin (GDR), August 26—31, 1985

Edited by Achim Sydow, Manfred Thoma, and Robert Vichnevetsky
496 pages, 1985

Band 29

Pseudo-differential Operators and Mellin Operators in Spaces of Functions with Conormal Singularity

By Stephan Rempel and Bert-Wolfgang Schulze
about 480 pages, 1986

Systems analysis and simulation become evident in an increasing degree as an effective method and tool by mastering large complex systems. Among the scientific foundations methods for computer-aided multiple-objective decision making, modelling and control of large-scale decentralized or hierarchical systems, respectively, and simulation by means of parallel processes are mostly developed. Object-oriented simulation languages and -systems as well as interactive man-machine communication are useful conditions for an effective problem solving of model users and designers.
This volume presents newest applications of modelling, simulation, and decision making concerning the following systems:

— Economic systems
— Environmental systems
— Ecological systems
— Water systems
— Global and regional systems
— Industrial and technical systems
— Agricultural systems
— Energy systems
— Medical and biotechnological systems

Volume I contains mathematical methods and methodological foundations.

Systemanalyse and Simulation erweisen sich in zunehmendem Maße als wirkungsvolle Methode und Instrument zur Beherrschung großer komplexer Systeme. Unter den wissenschaftlichen Grundlagen werden die Methoden zur computergestützten mehrkriteriellen Entscheidungsfindung, zur Steuerung und Modellierung großer dezentraler bzw. hierarchischer Systeme sowie zur Simulation mittels paralleler Prozesse und neuer Computerarchitekturen intensiv entwickelt. Sowohl objektorientierte Simulationssprachen und -systeme als auch interaktive Kommunikationsmöglichkeiten bieten dem Modellentwickler und dem Modellnutzer wesentliche Voraussetzungen für eine effektive Problemlösung.
Im vorliegenden Band werden neueste Anwendungen der Modellierung, Simulation und Entscheidungsfindung für die nachfolgenden Systeme vorgestellt:

— Ökonomische Systeme
— Umweltsysteme
— Ökologische Systeme
— Wassersysteme
— Globale und regionale Systeme
— Industrielle und technische Systeme
— Landwirtschaftliche Systeme
— Energiesysteme
— Medizinische und biotechnologische Systeme

Der erste Band enthält mathematische Methoden und methodologische Grundlagen.

ISSN 0138-3019

Kolibri 360 GmbH
Ferdinand-Jühlke-Straße 7
99095 Erfurt, Deutschland
produktsicherheit@kolibri360.de